Advanced Information Processing

Series Editor

Lakhmi C. Jain

Springer
Berlin
Heidelberg
New York
Barcelona
Hong Kong
London
Milan
Paris
Tokyo

Shu-Heng Chen • Paul P. Wang (Eds.)

Computational Intelligence in Economics and Finance

With 130 Figures and 80 Tables

Springer

Shu-Heng Chen
Department of Economics
National Chengchi University
Taipei, Taiwan 11623
R.O.C

Paul P. Wang
Department of Electrical
and Computer Science
P.O. 90291, Durham, NC 27708-0291
USA

Library of Congress Cataloging-in-Publication Data applied for

Die Deutsche Bibliothek - CIP-Einheitsaufnahme

Chen, Shu-Heng, Wang, Paul P.: Computational Intelligence in Economics and Finance
- Berlin; Heidelberg; New York; Barcelona; Hong Kong; London; Milan; Paris; Tokyo: Springer, 2002 (Advanced information processing)
ISBN 3-540-44098-48

ACM Subject Classification (1998): I.2.1, G.3, J.1

ISBN 3-540-44098-4 Springer-Verlag Berlin Heidelberg New York

Springer-Verlag Berlin Heidelberg New York
a member of BertelsmannSpringer Science+Business Media GmbH

Cover Design: KünkelLopka, Heidelberg
Typesetting: Computer to film by author's data
Printed on acid-free paper SPIN 10853536 45/3142PS 5 4 3 2 1 0

Foreword

Computational intelligence (CI) emerged about two decades ago as an alternative to the traditional artificial intelligence (AI) paradigm. Early on, its focus was from a physical engineering perspective where the goal was to integrate the flexibility of judgment and response into conventional software programs and to create a computing environment that had immediate access to extensive databases and the capacities to learn and adapt. Given the successful development of CI in physical engineering, particularly in areas like pattern recognition and nonlinear forecasting, it was not long before economists like Halbert White were experimenting with economic predictions using neural networks and other CI technologies, and the era of economic and financial engineering (E&FE) began.

Of course, economics and finance presented a unique set of problems from a CI perspective so it generally was not possible to simply port solutions over from physical engineering without first modifying them. In this context, important E&FE modeling considerations include the heuristic nature of the approach, the emphasis on nonlinear relationships, data issues, domain knowledge, and behavioral changes. The remainder of the foreword reflects on these issues and then comments on the current volume.

Heuristic Approach. Many of the problems in E&FE involve nonlinearities and since there is no canonically optimal approach to the development of models for nonlinear problems, solutions can vary considerably from problem to problem. In this sense, the technology is highly heuristic and often ad hoc and, since the field is still in its embryonic stage, there is considerable room for improvement. Thus, the formulation of a science of CI for E&FE modeling should be considered as a work in progress. Moreover, while the focus is on CI technologies, this cannot always be to the exclusion of more traditional approaches. Regardless of the sophistication, if the signal-to-noise ratios are so poor that reasonable relationships cannot be derived, it may be necessary to resort to more conventional technologies to attain adequate performance.

Emphasis on Nonlinear Relationships. A related issue is the assumption in many E&FE problems that there are important nonlinearities both between the observables (independent variables) and the dependent variable, as well as nonlinearities among the observables. Here, the emphasis is on not making unjustified assumptions about the nature of those nonlinearities, and technologies are used that have the capacity, in theory at least, to extract the appropriate interaction terms adaptively.

Data Issues. There are many issues with E&FE data if there is no theoretical framework to constrain the solution, since the resolution of the problem depends on and is highly sensitive to the nature of the sample data. As a consequence, considerable resources are devoted to processing the data, with an emphasis on missing and corrupted data and the removal of bias from the sample. Additionally, since multiple sources of data often are involved, the consistency of the differential semantics across these sources has to be verified.

Domain Knowledge. In E&FE, the technologies do not always achieve their ends because of the signal-to-noise ratios in the sample data. Of necessity, in these instances, the approach is to constrain the solution by introducing expert knowledge into the process. For example, domain segmentation might be employed, which involves the identification of segments within the decision space where the implicit relationship between variables is constant. Such segmentation has been demonstrated to provide enormous amounts of value-added performance. So, again, it is not quite a theoretical framework but it is more a heuristic framework to help constrain the solution space.

Behavioral Changes. Behavioral changes exemplify the class of characteristics that are unique to E&FE. Empirical evidence suggests that a very important aspect of predicting behavior is not simply the current status of an individual, but how that status changes over time. So, a number of aggregates have been derived to help capture that characteristic and the model must accommodate the predictive variables that are sampled over time to monitor the trend.

The CI Research in Economics and Finance. A good bit of the CI research in economics and finance has focused on innovative ways to deal with the foregoing types of issues, and the articles in this volume reflect the current trend of this research. Here, not only is the basic triad of fuzzy logic, neural networks, and genetic algorithms represented, but also there is an exploration of technologies that have only recently been brought to bear in this application area, like ant algorithms, grey models, lazy learning, rough sets, support vector machines, and wavelets. All in all, these articles provide a diverse cross-section of the state-of-the-art of CI technologies and current economic and finance application areas.

Of course, the state-of-the-art is a moving target, and, whether it is the development of new technologies, or new hybrids thereof, or novel application areas, it is only through conferences like CIEF'2002 that we can stay abreast of the changes. Ideally, we attend the conference, so that we can communicate, challenge each other's ideas, and conceive new tasks. But when this is not possible, we depend on the proceedings of the conference to keep us up to date. The editors and authors of this volume have produced an excellent product in this regard.

Penn State University, October 2002 *Arnold F. Shapiro*

Table of Contents

Part I. Introduction

Part II. Fuzzy Logic and Rough Sets

Part III. Artificial Neural Networks and Support Vector Machines

Part V. Sequence Matching and Feature-Based Time Series Models

Part VIII. Agent-Based Models

List of Contributors

Fernando Álvarez
Banco Central de Venezuela
Oficina de Consultoría Económica
Venezuela
faapnew@hotmail.com

Julián Andrada-Félix
Universidad de Las Palmas de Gran Canaria
Islas Canarias, Spain

Jane M. Binner
Department of Finance and Business Information
Nottingham Trent University
Nottingham, NG14BU, UK
jane.binner@ntu.ac.uk

Cafer E. Bozdağ
Istanbul Technical University
Industrial Engineering Department
80680, Maçka, Istanbul, Turkey
bozdagc@itu.edu.tr

Néstor Carrasquero
Universidad Central de Venezuela
Facultad de Ingeniería
Caracas 1041-A, Venezuela
nestor@neurona.ciens.ucv.ve

Lijuan Cao
Institute of High Performance Computing
89C Science Park Drive
118261 Singapore
caolj@ihpc.nus.edu.sg

Nen-Jing Chen
Department of Economics
Fu-Jen Catholic University
Hsin-Chuang, Taiwan
ecos1005@mails.fju.edu.tw

Shu-Heng Chen
AI-ECON Research Center
Department of Economics
National Chengchi University
Taipei, Taiwan 11623
chchen@nccu.edu.tw

Hilary Cheng
Department of Business Administration, Financial Data Mining Laboratory
Yuan Ze University
Chung-Li 320, Taiwan
hilary@saturn.yzu.edu.tw

Chih-Chou Chiu
Institute of Commerce Automation and Management
National Taipei University of Technology
Taipei, Taiwan
chih3c@sun.cc.ntut.edu.tw

Kai Chun Chiu
Department of Computer Science and Engineering
Chinese University of Hong Kong
Shatin, N.T., Hong Kong, P.R. China
kcchiu@cse.cuhk.edu.hk

Michael Doumpos
Financial Engineering Laboratory
Dept. of Production Engineering and Management
Technical University of Crete
University Campus
73100 Chania, Greece
dmichael@ergasya.tuc.gr

Fernando Fernández-Rodríguez
Universidad de Las Palmas de Gran Canaria
Islas Canarias, Spain
ffernandez@dmc.ulpgc.es

Alicia M. Gazely
Department of Finance and Business Information
Nottingham Trent University
Nottingham, NG1 4BU, UK
alicia.gazely@ntu.ac.uk

Mounir Ben Ghalia
Engineering Department
University of Texas-Pan American
Edinburg, TX 78539, USA
benghalia@panam.edu

Morikazu Hakamata
Graduate University for Advanced Studies
4-6-7 Minami-Azabu, Minato-ku, Tokyo 106-8569, Japan
hakamata@ism.ac.jp

Andrew Hughes Hallett
Department of Economics
Vanderbilt University
Nashville, TN, USA
and CEPR, London, UK
A.HughesHallett@Vanderbilt.edu

Hongxing He
CSIRO Mathematical and Information Sciences
Commonwealth Scientific & Industrial Research Organisation
GPO Box 664, Canberra ACT 2601
Australia
Hongxing.He@csiro.au

Yasuo Kadono
Management Science Institute Inc.
5-31-6 Taishido, Setagaya-ku, Tokyo 154-0004, Japan
kadono@msi21.co.jp

Cengiz Kahraman
Istanbul Technical University
Industrial Engineering Department
80680, Maçka, Istanbul, Turkey
kahramanc@itu.edu.tr

Li Kang
Institute of Systems Engineering
Tianjin University
Tianjin 300072, P.R. China
kangli@baf.msmail.cuhk.edu.hk

Graham Kendall
Department of Computer Science
University of Nottingham
Nottingham, NG8 1BB, UK
gxk@cs.nott.ac.ukemail

Genshirou Kitagawa
Institute of Statistical Mathematics
4-6-7 Minami-Azabu, Minato-ku,
Tokyo 106-8569, Japan

Tzu-Wen Kuo
AI-ECON Research Center
Department of Economics
National Chengchi University
Taipei, Taiwan 116
kuo@aiecon.org

Tian-Shyug Lee
Department of Business
Administration
Fu-Jen Catholic University
Hsin-Chuang, Taiwan
badm1004@mails.fju.edu.tw

Weiqiang Lin
Department of Computing
Macquarie University
Sydney, NSW 2109, Australia
wlin@comp.mq.edu.au

Raquel F. López
Leon Faculty of
Economics and Business
Administration
Campus de Vegazana
University of Leon s/n
24071 Leon, Spain
dderfl@unileon.es

Yi-Chuan Lu
Department of Information
Management, Financial Data Mining
Laboratory
Yuan Ze University
Chung-Li 320, Taiwan
imylu@saturn.yzu.edu.tw

José A. Moreno
Universidad Central de Venezuela
Facultad de Ingeniería
Caracas 1041-A, Venezuela
jose@neurona.ciens.ucv.ve

Dimitrios Moshou
Katholieke Universiteit Leuven
Leuven B-3001, Belgium
dimitrios.moshou@agr.kuleuven.ac.be

Mehmet A. Orgun
Department of Computing
Macquarie University
Sydney, NSW 2109, Australia
mehmet@comp.mq.edu.au

Arnold Polanski
Facultad de Económicas
Universidad de Alicante
E-03071 Alicante, Spain
arnold@merlin.fae.ua.es

Herman Ramon
Katholieke Universiteit Leuven
Leuven B-3001, Belgium
herman.ramon@agr.kuleuven.ac.be

Christian R. Richter
Department of Economics
University of Strathclyde
Glasgow, UK
Christian.Richter@Uni-Klu.ac.at

Claudio Rocco
Universidad Central de Venezuela
Facultad de Ingeniería
Caracas 1041-A, Venezuela
rocco@neurona.ciens.ucv.ve

Simón Sosvilla-Rivero
Universidad Complutense and Fedea
Madrid, Spain
sosvilla@fedea.es

Francis E. H. Tay
Department of Mechanical Engineering
National University of Singapore
10 Kent Ridge Crescent, 119260
Singapore
mpetayeh@nus.edu.sg

Takao Terano
Graduate School of Systems Management
University of Tsukuba, Tokyo, Japan
terano@gssm.otsuka.tsukuba.ac.jp

Chunfeng Wang
Institute of Systems Engineering
Tianjin University
Tianjin 300072, P.R. China
cfwang@tju.edu.cn

Paul Wang
Department of Electrical and Computer Engineering
P.O. 90291, Duke University
Durham, NC 27708-0291, USA
ppw@ee.duke.edu

Graham J. Williams
CSIRO Mathematical and Information Sciences
GPO Box 664
Canberra ACT 2601, Australia
Graham.Williams@cmis.csiro.au

Lei Xu
Department of Computer Science and Engineering
Chinese University of Hong Kong
Shatin, N.T., Hong Kong, P.R. China
lxu@cse.cuhk.edu.hk

Xin Zhao
Institute of Systems Engineering
Tianjin University
Tianjin 300072, P.R. China
zhaoxin@hotmail.com

Constantin Zopounidis
Financial Engineering Laboratory
Dept. of Production Engineering and Management
Technical University of Crete
University Campus
73100 Chania, Greece
kostas@ergasya.tuc.gr

Part I

Introduction

1. Computational Intelligence in Economics and Finance

Shu-Heng Chen[1] and Paul P. Wang[2]

[1] AI-ECON Research Center, Department of Economics, National Chengchi University, Taipei, Taiwan 11623
email: chchen@nccu.edu.tw

[2] Department of Electrical & Computer Engineering, P.O. 90291, Duke University, Durham, NC 27708-0291, USA
email: ppw@ee.duke.edu

Computational intelligence is a consortium of data-driven methodologies which includes fuzzy logic, artificial neural networks, genetic algorithms, probabilistic belief networks and machine learning as its components. We have witnessed a phenomenal impact of this data-driven consortium of methodologies in many areas of studies, the economic and financial fields being no exception. In particular, this volume of collected works will give examples of its impact on various kinds of economic and financial modeling, prediction and forecasting, and the analysis of various phenomena which sheds new light on a fundamental understanding of the research issues. This volume is the result of the selection of high-quality papers presented at the **Second International Workshop on Computational Intelligence in Economics and Finance (CIEF'2002)**, held at the Research Triangle Park, North Carolina,United State of America, March 8–14, 2002. To complete a better picture of the landscape of this subject, some invited contributions from leading scholars were also solicited.

This introductory chapter is intended to give a structural description of the book. Not following the usual chapter-by-chapter style, our presentation shall assume that the whole book coherently addresses the following four general questions:

1. What is computational intelligence?
2. What are the computational intelligence tools?
3. What kind of economic and financial applications can best benefit from these CI tools?
4. How successful are those applications?

We believe that these four questions are very basic and of some interest to all readers, be they concerned with general themes or technical issues. It is, therefore, desirable to give an overview on what has been said on these general questions in this volume before advancing to a specific tree of the forest. With this approach in mind, this chapter is pretty much a comprehensive outline and can serve as an introduction to this subject at large.

1.1 What Is Computational Intelligence?

First, what is computational intelligence? Computational intelligence (CI) is a new development paradigm of intelligent systems which has resulted from a synergy between fuzzy sets, artificial neural networks, evolutionary computation, machine learning, etc., broadening computer science, physics, engineering, mathematics, statistics, psychology, and the social sciences alike.

Most books on CI, e.g., [1.28] and [1.66], include at least the following three main pillars of CI methodologies:

- fuzzy logic (FL),
- artificial neural nets (ANN), and
- evolutionary computation (EC).

However, given these three basic pillars of methodologies, disciplines which share some similar or related features have been developed at different stages and they may be included as well. In fact, without a tight and formal definition, CI can easily be broadened by bringing them together. This "soft membership" actually enriches each of the participating disciplines and fosters new research perspectives in the emerging field of computational intelligence. What will be presented in this book is exactly this growing, emerging and interacting picture.

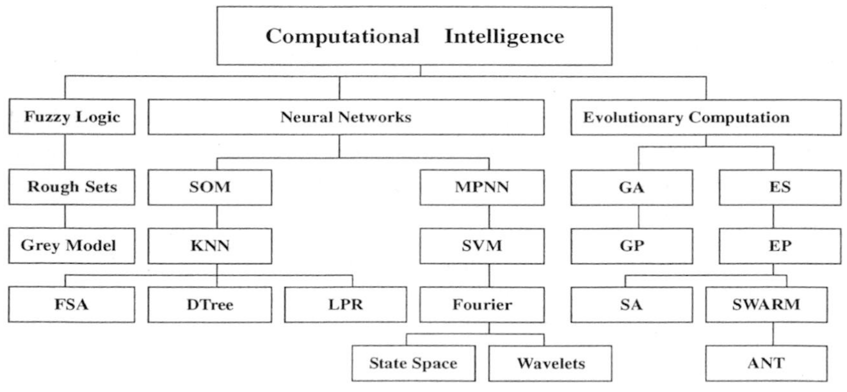

Fig. 1.1. A family tree of computational intelligence

Figure 1.1 shows a possible hierarchy of the methodologies for CI. At the top level the three main branches are: fuzzy logic, artificial neural nets, and evolutionary computation. They are arranged from left to right, which is consistent with the chronological order of their development and growth. Starting from the fuzzy logic, the tree extends down to *rough sets* and further down to the *grey model*. This branch can be considered as a response to two pursuits for the study of intelligent behavior. Figure 1.2 depicts the further evolution of Fuzzy Logic (FL). Firstly, as shown in the left half of Fig. 1.2, it is an attempt to develop a powerful notion and complementarity of uncertainty beyond the conventional probabilistic uncertainty. Secondly, as shown in the other half of Fig. 1.2, it is the acknowledgment of the significance of *natural language* in dealing with the accommodation of the uncertainty. Some psychologists have argued that our ability to process an immense amount of information efficiently is the outcome of applying fuzzy logic as part of our thought process. Fuzzy-logic-based languages and other techniques enable us to compress our perception of the complex world and economize (or simplify) our communication about it.

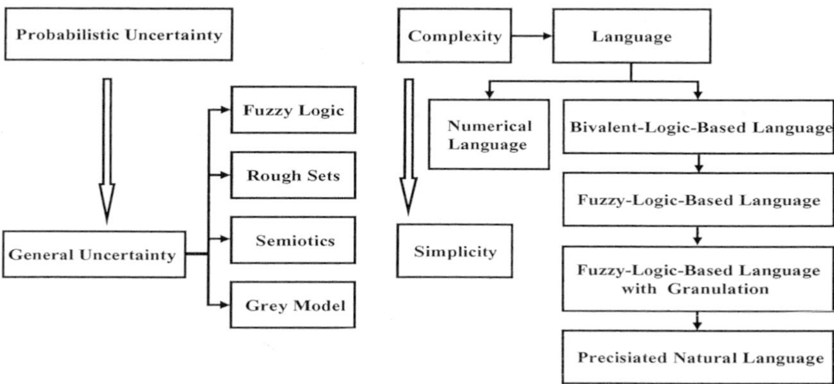

Fig. 1.2. Paradigm shifts from probabilistic uncertainty to general uncertainty: complexity reduction

Next, starting from artificial neural networks in Fig. 1.1, the path is first divided into *supervised learning*, the *multilayer perceptron neural networks* (MPNN), on the right, and *unsupervised learning*, the *Self-organizing* maps (SOM), on the left. Continuing down along the SOM branch, we encounter the *K nearest neighbors* (KNN), and further down a division into three branches: *finite state automata* (FSA), *decision trees* (DTree), and *local polynomial regressions* (LPR). The SOM branch reflects the movement from the *time domain* to the *feature domain* in time series modeling (the left panel of Fig. 1.3). In the time-domain models, extrapolation of past values into the immediate future is based on correlations among lagged observations and error terms. The feature-based models, however, select relevant prior observations based on their symbolic or geometric characteristics, rather than their location in time. Therefore, feature-based models first identify or discover features, and then act accordingly by taking advantage of these features. In that way, the movement can also be regarded as a change from the *global modeling strategies* to *local modeling strategies* (the right panel of Fig. 1.3). Consequently, one hopes that an extremely globally complex model can be decomposed into many locally simpler models.

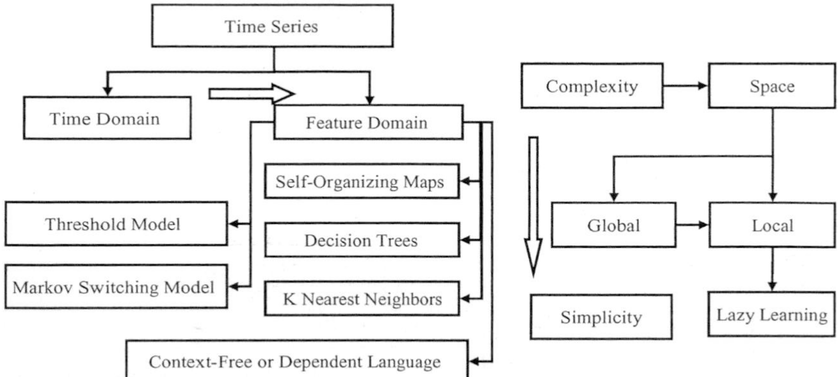

Fig. 1.3. Paradigm shifts from the time domain to the feature domain: complexity reduction

Features can be symbolic or geometric. Dealing with symbolic features, what one needs is a grammar. FSA or another even more powerful language provides the grammar required to define or to describe symbolic features, whereas SOM can tackle geometric features. Decision trees can automatically identify the symbolic features with limited logical operation, such as a limited number of ANDs and ORs. FSA, SOM and DTree classify individual objects as groups, while KNN and LPR leave individuals with a greater degree of freedom. They do not assign features to individual objects; instead, they leave each individual to select his/her own neighbors based on his/her preferences (distance metrics). FSA and SOM only carry out the task of grouping. How one should act upon each feature (group) is left for other tools. KNN, LPR and DTree can build simple models, usually linear ones, simultaneously with grouping.

Branching down in the second half of the ANN tree, we have the *support vector machine* (SVM) below MPNN. However, the latter can be treated as a generalization of the former. Further down, we have *Fourier analysis* and *wavelets*. The former is again an alternative approach to time series modeling, which places emphasis on the *frequency domain* rather than the *time domain*. Actually, time series modeling moved from the frequency domain to the time domain during the 1980s, historically speaking. Instead of reversing the direction, however, the wavelets provide an opportunity to merge both so that locality can be brought into the frequency-domain analysis as well (Fig. 1.4). As for feature-domain models, this merger is a preview of the importance of locality in time series modeling. The linkage to the Fourier analysis and wavelets could be put somewhere else within the family tree. The specific reason for leaving them here is just to point to the fact that wavelets are being used with artificial neural nets intensively nowadays.

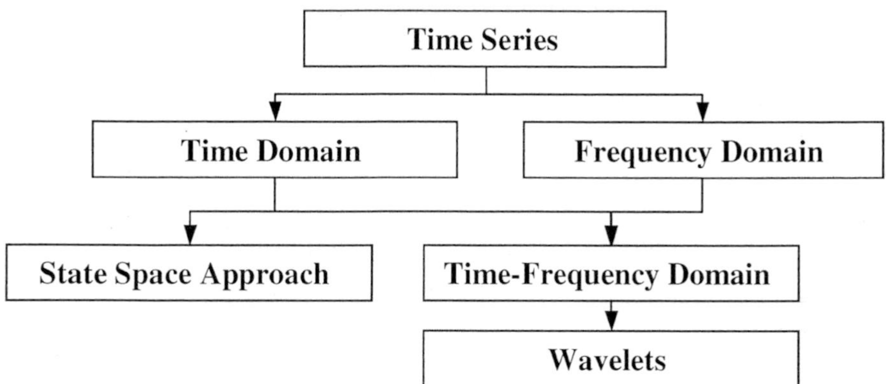

Fig. 1.4. Paradigm shifts from the time domain and the frequency domain to a synergy between the two

Finally, we focus on the rightmost part of the tree in Fig. 1.1, i.e., evolutionary computation. Based on its historical background, this node can also be further divided into two branches. *Evolutionary strategies* (ES) and *evolutionary programming* (EP) are put on the same side, whereas *genetic algorithms* (GA) is left on the other side. Historically, the operation of ES and EP has relied exclusively on *mutation*, while that of GA depends heavily on *crossover*. However this historical difference has become indistinguishable as time evolves. There is, however, another historically significant distinction existing between them, i.e., the *coding scheme*. ES and EP commonly used real coding. By contrast, GA chose to use binary coding. Again, this difference has become weaker gradually. Nonetheless, genetic programming (GP), a generalization of GA, uses the parse-tree coding (like context-free language), which establishes a distinct character in a more dramatic manner than the other three evolutionary algorithms.

The last historical difference among these four algorithms is from the viewpoint of the application orientation. ES was originally designed to deal with numerical optimization problems, whereas EP and GA were initiated for simulating intelligent behavior. This difference has also become negligible, as they have all emerged as standard numerical tools today. While considering the numerical tools for solving complex optimization problems, we also attach *simulated annealing* (SA) and *ant algorithms* (ANT) as two other alternatives. SA is included because it has frequently served as a benchmark to evaluate the performance of evolutionary algorithms. ANT has been chosen because it is a representative of *swarm intelligence* and its application to financial problems has just begun to draw the attention of researchers. This subtree

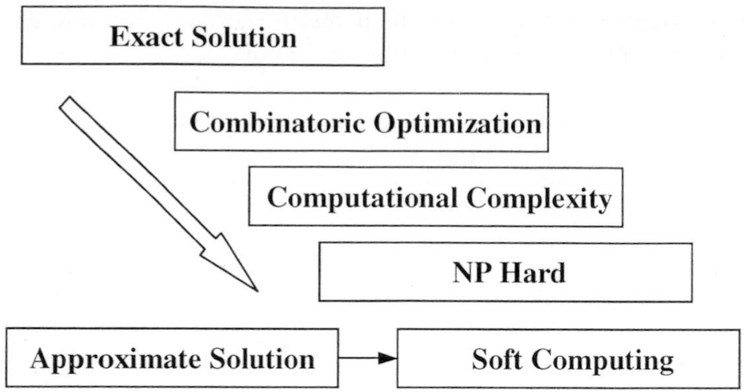

Fig. 1.5. Paradigm shifts from hard computing to soft computing

has seen rapid growth and the popularity of *soft computing*, a term coined by Prof. Lotfi Zadeh. In the domain of highly complex problems, precision is neither possible nor often desirable (Fig. 1.5). Heuristics or approximation algorithms become the only acceptable tools.

GP has a more general purpose. It was proposed to *grow* the *computer programs* which are presumably written by humans to solve specific problems. Since the users of GP do not have to know the size and shape of the solution, it becomes a very attractive tool, indeed, for nonparametric modeling. The increasing reliance by researchers on GP and artificial neural nets also reveals a stronger demand for nonlinear and nonparametric modeling (Fig. 1.6).

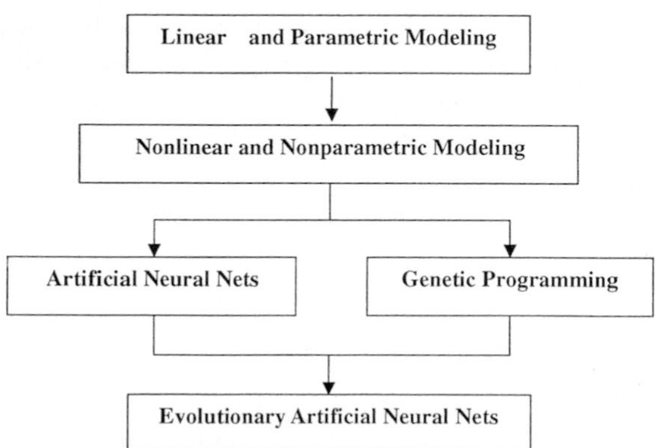

Fig. 1.6. Paradigm shifts from linear parametric modeling to nonlinear nonparametric modeling

Like many other textbooks written on this subject, we do not provide a formal definition of computational intelligence. Nevertheless, the five paradigm shifts summarized in Figs. 1.2–1.6 indicate why the era of CI has emerged and what it means for us. In addition, as a summary of the interdisciplinary nature of CI, Fig. 1.7 illustrates the relationship between computational intelligence and other fields. As the figure suggests, computational intelligence has benefited very much from physics (simulated annealing), biology (evolutionary computation and swarm intelligence), brain research (artificial neural nets), mathematical modeling of language (fuzzy logic), computer science (grammar, complexity theory and heuristics), engineering (wavelets), and machine learning (decision trees).

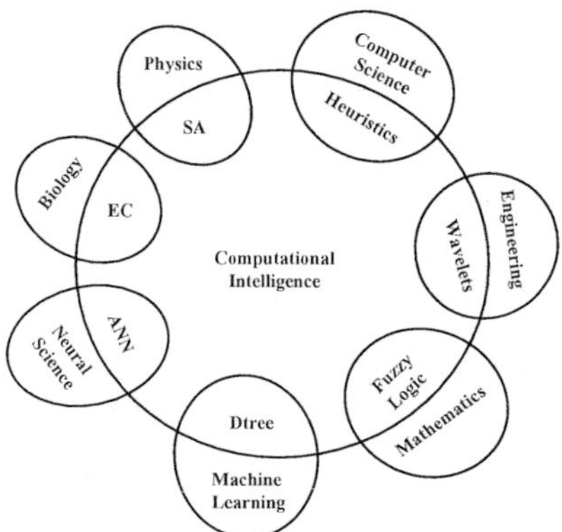

Fig. 1.7. The relationship between computational intelligence and other fields

This completes the description of Fig. 1.1. Obviously, the coverage is by no means complete. There are many more disciplines which can be added somewhere in the organization chart. The obvious reason that we do not go further for an even more detailed anatomization is inevitably due to the size of the volume. It is our hope that in the near future we will have a sequel second volume on this subject so that we can have a different summary of the evolutionary picture and the organizational chart. For the present, let us just accept Fig. 1.1 as a starting point.

All of the CI techniques introduced in this section have been applied to many disciplines, including mathematics, engineering, computer science, physiology, psychology, physics, chemistry, biology, brain research, bioinformatics, social sciences, etc. There is no need to give the general background of these in this volume. However, since this volume is written mainly for economists and finance professionals and researchers, it is imperative to know *why* these tools can be potentially relevant and effective to economic and financial modeling. Therefore, in the next few sections we shall introduce each tool by giving a brief technical review of it, and we shall indicate its relevance to economics and finance.

1.2 Fuzzy Logic and the Related Disciplines

1.2.1 Fuzzy Logic

First of all, let us begin with *fuzzy logic*. People frequently and routinely use natural language or linguistic values, such as high, low, and so on, to describe their perception, demands, anticipation, and decisions. While natural language has its ambiguities, people seem to be able to reason effectively with added vague and uncertain information, and very often the decisions they make are the outcome of their approximate reasoning. For example, consider the following economic reasoning based on natural language: "The government will adopt a loose monetary policy, if the unemployment rate is continuously high." "The economic prospects of the economy are not bright due to its low capital accumulation." "If the price spread narrows, it is high time to buy." While all these economic statements are crucial for decision making, the terms "high," "loose," "bright," "low," "narrow," and "high time" are not accurate. Fuzzy logic (FL) is a formal approach to cope with these ambiguities. Evidence on human reasoning and human thought processes support the hypothesis that at least some categories of human thought are definitely fuzzy, and the mathematical operations of fuzzy sets as prescribed by fuzzy set theory are a compatible and realistic description of how humans manipulate fuzzy concepts [1.80].

After four decades of advancement and fine tuning, the methodology of fuzzy logic has been widely used in economics, finance, and business. There are a great number of studies showing that *investment* behavior in a complex business environment is essentially fuzzy. Fuzzy logic as it turns out is the best fit for enhancing these difficult decisions. For example, [1.85] presents an operational model of decision making about advanced manufacturing technology that incorporates the mathematics of fuzzy set theory. [1.67] applies fuzzy logic to mutual fund investment. Fuzzy logic is also applied to many other real-world business activities. [1.91] and [1.12] present applications for supplier evaluation, customer targeting, scheduling, evaluating leases, choosing R&D projects, and forecasting. [1.96] deals with the marketing application. The usefulness of FL in a spectrum of applications in economics and finance is hence well established.

Fuzzy logic is not only just attractive for business practitioners, but it has also been incorporated into mainstream formal economic analysis. [1.11] introduces fuzzy general equilibrium analysis. It presents the aggregated model of microeconomics with *fuzzy behaviors*, and presents the state of the art in the *fuzzy theory of value*. [1.58] extends the application of fuzzy logic to *noncooperative oligopoly*.

Given such a huge pile of literature, this volume includes only two chapters on this subject. They represent two typical applications of fuzzy logic. Chapter 2 presents an application of *fuzzy forecasting*, whereas Chap. 3 presents one of *fuzzy investment*. Before we proceed to these two chapters, we shall first give a brief review of fuzzy logic.

Fuzzy sets are distinct from the classical sets (crisp sets) in the sense that the *membership* in the latter is *all or nothing*, whereas in the former it is a matter of *degree* (*more or less*). The degree is mathematically characterized by a *membership function*. Formally, a fuzzy set A can be denoted as follows:

$$A = \begin{cases} \sum_{x_i \in X} \mu_A(x_i)/x_i, & \text{if } X \text{ is discrete}, \\ \int_{x \in X} \mu_A(x)/x, & \text{if } X \text{ is continuous}, \end{cases} \tag{1.1}$$

where $\mu_A(x)$ is the membership function.[1] The symbols $\sum$ and $\int$ stand for the union of the membership grade. The symbol / is a marker and does not imply division. Fuzzy mathematics has been developed to define elementary *fuzzy operations* of fuzzy sets using membership functions, such as *Fuzzy OR*, *Fuzzy AND*, and *Fuzzy Complement*. Also, via some manipulation of the membership function, fuzzy mathematics can also deal with *linguistic modifiers*, such as "Very," "Extremely," "More or Less," etc., as originally suggested by its founder Lotfi A. Zadeh.

Most frequent uses of fuzzy set methodology involve the building of *fuzzy inference systems*. A fuzzy inference system is a set of *fuzzy if-then rules* that connect one or more input sets to one output set. In the literature, there are two major styles of fuzzy if-then rules, namely, the *Mamdani style* and the *Sugeno style*. In the Mamdani style both the antecedents and consequents of rules are fuzzy. In the Sugeno style only the antecedents of rules are fuzzy, whereas the consequents of rules are in general a function of the input variables. As an illustration, the Mamdani style of a fuzzy if-then rule is:

If x is "A" and y is "B," then z is "C,"

whereas the input sets "A" and "B" and the output set "C" are all fuzzy. On the other hand, the Sugeno style of a fuzzy if-then rule is:

If x is "A" and "y" is "B," then $z = f(x, y)$.

Building a fuzzy inference system consists of five main steps. First, *fuzzification*: the input variables are fuzzified by assigning a membership value with respect to different fuzzy sets. Second, a fuzzy operation, such as Fuzzy AND or Fuzzy OR, is applied in order to determine the degree to which each rule is applicable. Third, *implication*: implication methods, such as *cut*, are used to decide the membership function of the consequents of fuzzy rules. Fourth, *aggregation*: the results of the applicable rules are combined to obtain an output fuzzy set. Fifth, *defuzzification*: various methods can be used to get

[1] There are more than a dozen membership function forms used in the literature. The user of Matlab can find a wealth of built-in standard membership functions.

a number as the result of a defuzzification operation. The most common one is to use the *centroid* of the resulting fuzzy set from aggregation.

Chapter 2 should be useful for those who are interested in the details of this five-step procedure in the context of economic applications. By building a fuzzy inference system, the authors show how the judgemental forecast can be combined with the statistical forecast using the fuzzy adjustment of the gap between these forecasts. Namely, the modeling of human judgement can indeed complement the usual economic reasoning.

In addition to linguistic variables, fuzzy logic can also be applied to deal with the uncertainty in *numerical numbers*. Chapter 3 gives such an illustration. In an uncertain economic decision environment, an expert's knowledge with regard to discounting cash flows consists of a lot of vagueness instead of randomness. In this chapter, cash flow is modeled in terms of fuzzy numbers instead of random numbers. The authors show how to fuzzify all related concepts of cash flow, including fuzzy future values, fuzzy present values, fuzzy rates of return, fuzzy benefit/cost ratios, fuzzy payback periods, and fuzzy equivalent uniform annual values. Chapter 3 also reviews the applications of *fuzzy logic* to *dynamic programming* (the Bellman equation), and solves the capital budgeting issues using fuzzy dynamic programming in a synergetic manner.

1.2.2 Rough Sets

Despite the distinction in its definition, *rough set analysis*, as introduced by Zdzislaw Pawlak in 1982, is usually mentioned together with fuzzy logic. Both fuzzy logic and rough set analysis are methods for modeling uncertain, vague, or inaccurate information in a wide sense. As a matter of the fact, a "rough set" is similar to a "fuzzy set" in the sense that they are alternatives to the "crispy set," where membership in the set is a certainty. What distinguishes rough set analysis from fuzzy set analysis is that the former requires no external parameters and uses only the information presented in the given data.

Rough sets define a mathematical model of vague concepts that is used to represent the relationship of dissimilarity between objects. Rough sets are not completely specified. They can be characterized by the upper and lower approximations in the domain that define the level of uncertainty of the partial specification. The lower approximations consist of objects which belong to a concept with certainty, while the upper approximations consist of those which possibly belong. On the basis of the lower and upper approximations of a rough set, the *accuracy* of approximating the rough set can be calculated as the ratio of the cardinality of its lower approximation to the cardinality of its upper approximation. A lucid introduction to rough sets can be found in Chap. 4.

Rough sets have received much attention recently with regard to the interest in data mining techniques that combine the automation of rule extraction with domain expertise in economics and finance. [1.79] applied the rough set approach to the evaluation of bankruptcy risk. [1.78] used rough sets techniques to extract a set of rules for short-term trading on the OEX index based on the Hines indicator.[2] [1.45], [1.46], and [1.47] provide a quick grasp of the essential elements of the rough set theory for economists and finance people. [1.61] gives three exemplary economic decision problems using rough sets, namely, company evaluation, the credit policy of a bank, and the marketing strategy of a company.

While most economists and finance professionals may not be familiar with rough set analysis, the idea of *possible membership* is in fact not new to them. For a long time economists have applied probabilistic models, such as discriminate analysis, and logit and probit analysis to the issue of possible membership. Therefore, it remains an important issue as to which tool can perform better classifications, i.e., conventional probabilistic models or rough sets? Chapter 4 provides a rigorous comparison between rough sets and conventional statistical tools in classification performance.

1.2.3 Grey Forecasting

Grey system theory was first presented in 1982 by Julong Deng at Huazhong University of Science and Technology, Wuhan, China, [1.23, 1.24]. The system is named the "grey" system because the grey color lies between the white and black colors. If one considers the white system as the one with completely known information and the black system as the one with completely unknown information, then systems with partially known and partially unknown information can be called grey systems. However, this distinction is not all that not useful because all systems we encounter in the world of computational intelligence are grey. Hence, one needs to know the essential features which distinguish the "grey" concept from the "fuzzy" concept, or from the "stochastic" concept. All the differences among grey systems theory, probability, statistics, uncertainty mathematics and fuzzy mathematics are very well described in [1.23].[3]

Briefly, grey systems theory mainly studies problems with uncertainty due to the *small samples* involved. This feature makes it unique and different from probability and statistics which address problems with samples of reasonable sizes, and different from uncertainty and fuzzy mathematics which deal with problems with cognitive uncertainty. Through the grey system theory, one can utilize only a few known data by means of an *accumulated generating*

[2] The OEX or S&P 100 is a capitalization-weighted index of the 100 biggest stocks in the US market.

[3] Prof. Roger W. Brockett of Harvard University, the then editor-in-chief of the journal *Systems and Control Letters* commented on Deng's first article about grey systems as follows: "*The grey system is an initiative work*" and "*All the results are new*".

operation (*AGO*) to establish a prediction model.[4] The mathematics of AGO is introduced in Chap. 5. In the same chapter, the authors also introduce the model GM(1,1), one variable with its associated first-order difference equation, which is the most frequently used model in grey prediction.

There are already many economic and financial applications of the grey model, e.g., [1.54], [1.55], [1.56], and [1.88]. Many of them can be found in the *Journal of Grey System*. Chapter 5 applies a grey model to forecast the future price at the opening time of the spot market, i.e., 9:00 A.M. The authors choose a grey model because there is only a limited number of observations available for the future price before the opening time of the spot market, to be exact, just 5 data points.[5] This extreme situation makes the grey model an appealing choice. One distinguishing feature of this application is that the grey model is further integrated into a neural network. The authors use the predicted future price as the input to a multilayer perceptron neural network, and then forecast the spot price using the neural net. One interesting observation which one may make here is: what is the point of using the grey model as an intermediate step (preprocessing step) here? Why not directly forecast the spot price with the earlier futures price using neural nets? The answer, as evidenced by the authors, is: "Grey forecasts provide *a better initial solution* that *speeds up* the learning procedure for the neural networks, and hence give better forecasting results."

1.3 Neural Networks and Related Disciplines

1.3.1 Artificial Neural Networks

Apart from sets and logics, another major object in mathematics is *functions*. Let us consider the following general issue. We observe a time series of an economic or financial variable, such as the foreign exchange rate, $\{x_t\}$. We are interested in knowing its future, $x_{t+1}, x_{t+2}, \ldots$. For that purpose, we need to search for a function relation $f(\ )$, such that when a vector $\mathbf{x}_t$ is input into the function a prediction on $x_{t+1}, \ldots$ can be made. The question then is how to construct such a function. Tools included in this volume provide two directions to work with. They are distinguishable according to their different modeling philosophies. The first approach is based on the *universal modeling approach*, and the second one is based on the *local modeling approach*. Alternatively, we can say that the first approach is to build the function in a *time domain*, whereas the second is to work in a *feature domain* or in a *trajectory domain*.[6] We shall start with the first approach. The canonical ar-

[4] Grey theory needs as few as 4 data points to build a grey model (GM) to approximate a system.

[5] Also see Sect. 1.3.1 for a related review of this chapter.

[6] There are also other names for the local modeling approach. For example, another popular name in the literature is *guarded experts*. See [1.3].

tificial neural network (ANN) can be considered to be a representative of this paradigm.

Among all CI tools, the artificial neural network is the most widely accepted tool for economists and finance people, even though its history is much shorter than that of fuzzy logic as far as the applications to economics and finance are concerned. The earliest application of ANNs appeared in 1988, developed by Halbert White [1.92]. Since then we have witnessed an exponential growth in the number of applications. ANN is probably the only CI tool which drew serious econometricians' attention and on which many theoretical studies have been done. Both [1.93] and [1.73] gave a rigorous mathematical or statistical treatment of ANNs and hence they established ANNs with a sound foundation in the econometric field. Nowadays, ANN has already become a formal part of courses in econometrics, and, in particular, financial econometrics and financial time series. A great number of textbooks or volumes specially edited for economists and finance people are available, e.g., [1.87], [1.5], [1.6], [1.72], [1.38], [1.95], and [1.74]. Its significance to finance people can also be seen from the establishment of the journal *Neurove$t* (has referred to now *Computational Intelligence in Finance*) in 1993.

The reason why economists can embrace ANNs without any difficulties is due to the fact that an ANN can be regarded as a generalization of the already household time series model *ARMA* (autoregressive moving average). Formally, an ARMA(p,q) model presented as follows:

$$\Phi(L)x_t = \Theta(L)\epsilon_t, \tag{1.2}$$

where $\Phi(L)$ and $\Theta(L)$ are polynomials of order p and q,

$$\Phi(L) = 1 - \phi_1 L - \phi_2 L^2 - ... - \phi_p L^p, \tag{1.3a}$$

$$\Theta(L) = 1 - \theta_1 L - \theta_2 L^2 - ... - \theta_q L^q, \tag{1.3b}$$

$\{ \epsilon_t \}$ is the white noise, and L is the lag operator. ANNs can be regarded as a nonlinear generalization of these ARMA processes. In fact, more concretely, *multilayer perceptron neural networks* are nonlinear generalizations of the so-called autoregressive process

$$x_t = f(x_{t-1}, ..., x_{t-p}) + \epsilon_t, \tag{1.4}$$

whereas *recurrent neural networks* are nonlinear generalizations of the ARMA processes

$$x_t = f(x_{t-1}, ..., x_{t-p}, \epsilon_{t-1}, ..., \epsilon_{t-q}) + \epsilon_t. \tag{1.5}$$

Multilayer Perceptron Neural Networks. In terms of a multilayer perceptron neural network, (1.4) can then be represented as

$$x_t = h_2(w_0 + \sum_{j=1}^{l} w_j h_1(w_{0j} + \sum_{i=1}^{p} w_{ij} x_{t-i})) + \epsilon_t. \tag{1.6}$$

Equation (1.6) hence is a three-layer neural net (Fig. 1.8). The input layer has p inputs: $x_{t-1}, ..., x_{t-p}$. The hidden layer has l hidden nodes, and there is a single output for the output layer $\hat{x}_t$. Layers are fully connected by *weights*: w_{ij} is the weight assigned to the ith input for the jth node in the hidden layer, whereas w_j is the weight assigned to the jth node (in the hidden layer) for the output. w_0 and w_{0j} are constants, also called *biases*. h_1 and h_2 are *transfer functions*.

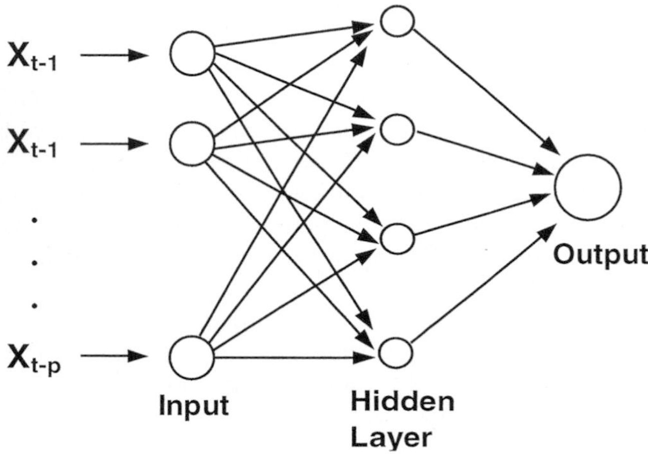

Fig. 1.8. The multilayer perceptron neural network of a nonlinear AR process

There is a rich choice of transfer functions. According to [1.18], for a multilayer perceptron network with any *Tauber-Wiener transfer functions* can be qualified as a *universal approximator*. Also, a necessary and sufficient condition for being a Tauber-Wiener function is that *it is a nonpolynomial*. In practice, a differentiable transfer function is desirable. Commonly used transfer functions for multilayer perceptron networks are the *sigmoid function*,

$$h_s(x) = \frac{1}{1+e^{-x}}, \tag{1.7}$$

and the *hyperbolic tangent function*,

$$h_t(x) = \frac{2}{1+e^{-2x}} - 1. \tag{1.8}$$

Clearly, $0 < h_s(x) < 1$ and $-1 < h_t(x) < 1$.

Chapter 5 is mainly an application of multilayer perceptron neural networks. There have been studies for an extensive period of time to see whether the prices of some financial markets can help predict the prices of other financial markets. This inquiry arises mainly because global information generation happens 24 hours a day, while the regional financial markets only operate during their *office hours.* So, during its sleeping time, the future price movement of a specific regional market may possibly be obtained by examining other, nonsleeping markets. This situation also applies to different financial markets in the same region, e.g., the futures market and the stock (spot) market. The daily trading period of an index futures contract usually *begins earlier* and *ends later* than that of its underlying spot market. In the case of Taiwan, the differences are 15 minutes earlier and 15 minutes later. The possible information content of this 30-minute gap is the motivation behind the studies in Chap. 5.

In Chap. 5, the authors experiment with two multilayer perceptron neural networks: one uses the previous day's cash closing index as the input, and the other uses the 09:00 futures, predicted by the grey model (see Sect. 1.2.3). It is found that the prediction error of the neural net with the 09:00 futures grey forecast and the previous day's closing index as inputs is at least 40% lower than that of the one that considers only the previous day's closing index. Therefore, they conclude that there is valuable information involved in the futures prices during the non-cash-trading period that can be used in forecasting the opening cash market price.

Radial Basis Network. Next to the multilayer perceptron neural network is the *radial basis network* (RBN), which is also popular in economics and finance. Radial basis function (RBF) networks are basically feedforward neural networks with a *single hidden layer*,

$$f(x) = \sum_{i}^{k} w_i \varphi(\|x - c_i\|), \tag{1.9}$$

where $\varphi(\ )$ is a *radial basis function*, c_i is the ith center, and k is the number of the center. w_i, c_i and k are to be determined by the dataset of x. Typical choices of radial basis functions are:

– the thin-plate-spline function,

$$\varphi(x) = x^2 \log^x; \tag{1.10}$$

– the Gaussian function,

$$\varphi(x) = \exp(-\frac{x^2}{\beta}); \tag{1.11}$$

– the multiquadratic function,

$$\varphi(x) = (x^2 + \beta^2)^{\frac{1}{2}}; \tag{1.12}$$

– and the inverse multiquadratic function

$$\varphi(x) = \frac{1}{(x^2 + \beta^2)^{\frac{1}{2}}}. \tag{1.13}$$

Theoretical investigation and practical results seem to show that the choice of radial basis function is not crucial to the performance of the RBF network.

It has been proved that the RBF network can indeed approximate arbitrarily well any continuous function if a sufficient number of radial-basis function units are given (the network structure is large enough) and the parameters in the network are carefully chosen. RBN also has the best approximation property in the sense of having the minimum distance from any given function under approximation. Chapter 16 applies the RBF network to solving the optimum consumption path problem in a model of economic growth.

1.3.2 Support Vector Machines

In the 1990s, based on results from *statistical learning theory* [1.89], an alternative to the artificial neural network was developed, i.e., the *support vector machine* (SVM). The SVM was founded primarily by Vladimir Vapnik, who contributed to the development of a general theory for minimizing the expected risk of losses using empirical data. Brief introductory material on the SVM can be found in [1.90], whereas [1.21] is a textbook devoted to the SVM.

Support vector machines nonlinearly map an n-dimensional input space into a high-dimensional feature space,

$$\phi : V^n \to V^m, \tag{1.14}$$

where V^n is an n-dimensional input vector space, and V^m is an m-dimensional feature vector space. Given a series of l historical observations

$$(y_1, x_1), ..., (y_l, x_l),$$

where $y_i \in V^1$ and $x_i \in V^n$, we approximate and estimate the functional relation between y_i and x_i by

$$y = f(x) = \langle w, \phi(x) \rangle + b = \sum_{i=1}^{m} w_i \phi(x)_i + b, \tag{1.15}$$

where $\langle ., . \rangle$ denotes the inner product. The vector w and the constant b are to be determined by following the *structural risk minimization principle*, borrowed from statistical learning theory. It is interesting to note some similarities between the RBN and SVM, namely, (1.9) and (1.15). However, there is also a noticeable difference. Consider an input x_i as a vector of three dimensions: $(x_{i,1}, x_{i,2}, x_{i,3})$. Then for each neuron in the hidden layer of the RBN, they all share the same form

$$(\varphi(x_{i,1}, x_{i,2}, x_{i,3}, c_1), \varphi(x_{i,1}, x_{i,2}, x_{i,3}, c_2), ...),$$

while being associated with different centers. However, each neuron in the hidden layer of the SVM may actually take different inputs. For example, the first neuron takes the first two inputs, but the second takes the last two:

$$(\phi_1(x_{i,1}, x_{i,2}), \phi_2(x_{i,2}, x_{i,3}), \ldots).$$

Also, notice that the transfer functions, $\varphi(\ )$, are the same for each neuron in the RBN, but in general are different for the SVM: $\phi_1, \phi_2, \ldots$.

In the case where the y_i are categorical, such as $y_i \in \{-1, 1\}$, the minimization process also determines a subset of $\{x_i\}_{i=1}^{l}$, called *support vectors*, and the SVM when constructed has the following form:

$$f(x) = \sum_s y_i \alpha_i^* \langle \phi(x_s), \phi(x) \rangle + b^*, \tag{1.16}$$

where α_i^* and b^* are the coefficients satisfying the structural risk minimization principle, and s is the set of all support vectors. The category assigned to the observation x, 1 or -1, will then be determined by the sign of $f(x)$:

$$y = \begin{cases} 1, & \text{if } f(x) > 0, \\ -1, & \text{if } f(x) < 0. \end{cases} \tag{1.17}$$

Equations. (1.16) and (1.17) are the SVM for the classification problem. A central concept of the SVM is that one does not need to consider the feature space in explicit form; instead, based on the Hilbert-Schmidt theory, one can use the *kernel function*, $K(x_s, x)$, where

$$K(x_s, x) = \langle \phi(x_s), \phi(x) \rangle. \tag{1.18}$$

Therefore, the SVM is also called the *kernel machine*. (1.16) can then be rewritten as

$$f(x) = \sum_s y_i \alpha_i^* K(x_s, x) + b^*. \tag{1.19}$$

Chapter 6, which applies the SVM to currency crises discrimination in Venezuela, is an application of this type. CI tools have also been applied to establish early warning systems for almost a decade. The system has been applied to bond rating, bankruptcy prediction, and financial crises detection. In Chap. 6, Claudio Rocco and José Moreno apply the support vector machine to identify and detect currency crises in Venezuela during the period from January 1980 to May 1999. This chapter is the first financial application of the SVM to the establishment of an early warning system for currency stability, to our knowledge.

Following a similar procedure, one can construct the SVM for regression problems as follows:

$$f(x) = \sum_{i=1}^{l} (\alpha_i^* - \beta_i^*) K(x, x_i) + b^*, \tag{1.20}$$

where α_i^*, β_i^* and b^* are the coefficients minimizing the corresponding objective functions. An application of (1.20) to financial time series forecasting can be found in Chap. 7.

In addition to the functional form, $f(\mathbf{x})$, the second important issue is the set of variables $\mathbf{x}$ itself, and one has to deal naturally with the problem known as *variable selection* or *feature selection*. The involvement of irrelevant variables or features may lead to a poor generalization capability. Chapter 7 addresses this issue in the framework of support vector machines (kernel machines). Their contribution here is to extend the saliency analysis from the multilayer feedforward neural network to support vector machines.

1.3.3 Self-organizing Maps

Self-organizing maps (SOMs) solve the pattern recognition problem which deals with a class of unsupervised neural networks. Basically, the SOM itself is a two-layer neural network. The input layer is composed of p cells, one for each system input variable. The output layer is composed of neurons which are placed on n-dimensional lattices (the value of n is usually 1 or 2). The SOM adopts so-called *competitive learning* among all neurons. Through competitive learning, the neurons are tuned to represent a group of input vectors in an organized manner. The mathematical background of the SOM is briefly reviewed in Chap. 8.

The SOM was invented by Teuvo Kohonen [1.53] in 1982. It has been applied with great success to many different engineering problems and to many other technical fields. [1.22] is the first volume to demonstrate the use of the SOM in finance. The applications covered in that volume include the prediction of bankruptcies, the forecasting of long-term interest rates, the trading of stock indices, the selection of mutual fund managers, the evaluation of investment opportunities, the risk analysis of countries and commercial credit, the analysis of financial statements, and real-estate property evaluations. There is no need to further justify the relevance of the SOM in finance except by adding the following remark.

In social and behavioral sciences, the ability to recognize patterns is an essential aspect of human heuristic intelligence. Herbert Simon, a Nobel Prize Laureate in economics (1978), considered pattern recognition to be critical and advocated the need to pay much more explicit attention to the teaching of pattern recognition principles. In the financial market, chartists appear to have been good at doing pattern recognition for many decades, yet little academic research has been devoted to a systematic study of these kinds of activities. On the contrary, sometimes it has been treated as nothing more than astrology, and hardly to be regarded as a rigorous science.

In the mid-1990s, researchers began to look for a mathematical foundation for charting. [1.15] and [1.63] developed an algorithm for automatically detecting geometric patterns (head-and-shoulders) in price or exchange data by looking at properly defined *local extrema*. This approach was taken further by [1.57], who applied the *kernel method* to identify local extrema and then defined five frequently used charts with these estimated local extrema:

head-and-shoulders, inverted head-and-shoulders, broadening tops and bottoms, triangle tops and bottoms, rectangle tops and bottoms, double tops and bottoms. Two statistical tests, the goodness-of-fit test and the Kolmogorov-Smirnov test, were applied to examine whether the conditional density given to the appearance of the pattern would be significantly different from the unconditional density. It was found that the conditional and unconditional distributions were significantly different for 7 of the 10 patterns in the NYSE/AMEX sample, and for all the patterns in the Nasdaq sample. This gave the first analytical result on the significance of the charts.

However, all the charts studied by [1.57] are known to us. In a sense, it is a style of supervised learning. What would happen if these charts were not all known to us? Could we still identify them? In other words, is it possible to learn these charts without supervision? It is this question that motivated Shu-Heng Chen and Hongxing He to apply SOMs as a systematic and automatic approach to *charting*, or, more generally, to *geometric pattern recognition*?

Applying a 6×6 two-dimensional to a financial time series, 36 charts are derived without supervision. Among the 36 charts, many are familiar, such as uptrends, downtrends, v-formations, rounding bottoms, rounding tops, double tops, and island reversal. Furthermore, many of these charts are able to transmit buying and selling signals. In 14 out of the 36 charts, the equity curve is monotonically increasing, featuring buying signals. In 3 out of the 36, the equity curve is monotonically decreasing, featuring selling signals. They also show that trading strategies developed from these charts may have superior profitability performance. [1.57] argued that there is a gulf between technical analysis and quantitative finance: technical analysis is primarily *visual*, whereas quantitative finance is primarily *algebraic* and *numerical*. It seems that SOMs, and many other techniques introduced in this volume, may help us to narrow the gap.

In addition to categorizing financial time series forecasting and helping trading decisions, one of the most productive applications of the SOM to economics and finance is to help us understand a series of fundamental questions concerning the causes of successes and failures. Firstly, are observation A and observation B different? Secondly, what makes them different? Thirdly, how significantly different are they? The benefits one can derive from the SOM are particularly great when we initially have no clue at all. Chapter 9 provides such an example.

In Chap. 9, the author tries to examine whether the enterprises in countries of the European Monetary Union (EMU) have shared some similar features as far as their competitiveness is concerned. If not, what are the difficulties or challenges they may encounter after embracing the Euro as their common currency? This analysis is crucial because the *convergence criteria* set by the Maastricht Treaty in 1992, also known as the Maastricht Criteria, only cover a few macroeconomic variables, such as interest rates, exchange rates and budget deficits. What the author argues in her analysis is that our understanding of the impact of the Euro on its member countries is limited, unless the diversity or similarity of enterprises' competitiveness is also taken into account. Since there is no natural way of defining the similarity of enterprises' competitiveness, it would be prudent to let the data speak for themselves, and hence the SOM, as a Self-organizing tool, provides a natural way to tackle the issue.

Using cost data and profit data for nine European countries over six years (1993–1998) from the BACH database, the author also presents *two maps of competitive advantages*, one for the cost analysis and one for the profitability analysis. Based upon the maps, countries' competitive features are Self-organized into blocks of regions. What is really interesting is that countries can be distinguished, to a large extent, by these blocks of regions via clustering. By reading the maps, one can immediately see the uniqueness of some countries, for example, Finland. Moreover, the maps can be read with time. It is also interesting to notice the movement of the same country over different blocks of regions, which may signify the progress or regression that the country experienced over the period.

1.3.4 Finite State Automata

To motivate the tool *finite state automata*, let us start from the conventional time-series modeling, and consider an n-history time series

$$X_t^n \equiv \{x_t, x_{t-1}, ..., x_{t-n}\}.$$

Supposing that we are interested in the forecasting of x_{t+1}, then conventional time series modeling would basically start by constructing a function, be it linear or nonlinear, like

$$f(x_t, x_{t-1}, ..., x_{t-k}),$$

where $k < n$. This approach basically assumes that there is a relationship existing between x_{t+1} and X_t^k. Since the time series will repeat this relationship constantly, the pattern is defined in terms of $f(\ \ )$, and the task of data mining is to discover this f by finding an estimation of it, $\hat{f}$, be it parametric or nonparametric. Moreover, the parameter k is usually exogenously given and is fixed.

Two questions arise from this modeling procedure. First, this approach implicitly assumes that if the parameter k is chosen appropriately, then

$$x_{t+1} \sim x_{t+j+1}, \quad if \quad X_t^k = X_{t+j}^k \quad \text{pointwisely.} \tag{1.21}$$

"$\sim$" here could mean that the two variables share the same distribution. Or, putting it in a broader way, it may indicate that

$$\mid x_{t+1} - x_{t+j+1} \mid < \delta, \quad \text{if} \quad \mid X_t^k - X_{t+j}^k \mid < \epsilon. \tag{1.22}$$

In plain English, it simply says that if X_t^k and X_{t+j}^k are close, then x_{t+1} and x_{t+j+1} should also be close.

The "$\epsilon - \delta$" argument can also be generalized with *categorization*: If X_t^k and X_{t+j}^k are *close* in the sense that they belong to *type* A, then x_{t+1} and x_{t+j+1} will also be close in the sense that they are related to X_t^k and X_{t+j}^k, respectively, by the same mapping f_A, where A is an index from a finite alphabet Λ, i.e., $A \in \Lambda$.[7] However, in conventional time series modeling, there exists only one global f. The existence of topological patterns, Λ, and their significance for forecasting, $\{f_A : A \in \Lambda\}$, are hence ignored.

A slight departure from the severe restriction is that one can find the famous *threshold* model, e.g., the threshold auto-regressive (TAR) model. A general form would be

$$x_{t+1} = \begin{cases} f_A(X_t^k), & \text{if } X_t^k \in A, \\ f_{\bar{A}}(X_t^k), & \text{if } X_t^k \notin A. \end{cases} \tag{1.23}$$

In this case, there are two features in the set Λ, namely A and its complement $\bar{A}$. While (1.23) is an important step for moving forward, it is still very restrictive. First, the type of the feature A is not extracted from data, but is prespecified. Second, the number of features is very small and is also exogenously given.

Some recent progress in computational intelligence can be viewed as a series of efforts made toward automatic procedures to construct the feature set Λ and the mappings $\{f_A : A \in \Lambda\}$. Earlier, we saw how this task could be done with Self-organizing maps. Chapter 8 shows how a series of 36 different trajectory patterns, $A_1, A_2, ..., A_{36}$, can be automatically discovered and grouped. However, there is a restriction: all features are constructed based on a fixed window size, i.e., a fixed number of observations. This restriction can have nontrivial effects because topological similarities usually do depend on the window size of the data.[8]

Chapter 11 signifies a breakthrough in relation to this restriction. Arnold Polanski presents an approach which does not restrict the "relevant past" or the reaction time of the system to *time windows of fixed lengths*. He proposes a language called the *pattern description language* (PDL) by which one is able to find all *frequent* subsequences, given a set of data sequences. The work is an extension of an earlier study by Norman Packard [1.65], one of the founders of the Prediction Company established in 1991 in Santa Fe, New Mexico. Those readers who have some background in the early development of financial applications of *genetic algorithms* or *genetic programming* to trading would agree that one basic difficulty in developing and evolving automated trading strategies is the *representation* and *coding* of the *current state of financial time series* on which the trading decision is based.[9] Back to our previous

---

[7] The "$\epsilon - \delta$" argument can be fuzzified: if X_t^k and X_{t+j}^k are *close*, then $\hat{x}_{t+1}$ and $\hat{x}_{t+j+1}$ will also be *close*.

[8] Consider the time series data with self-similarity.

[9] For a historical review of the progress of this work, see [1.17].

discussion, an effective solution concerns an language with a rich expression power to deal with the complexity of the set Λ.

Polanski's solutions to this problem were motivated by the work of theoretical computer scientists. He defined the patterns in terms of *regular languages*, specifically, *nondeterministic finite automata* (NFA). Abstractly, a finite state automaton (or a finite state machine) $\mathcal{M}$ is an ordered pair $(\mathcal{S}, \mathcal{A})$ of finite sets, with an action of $\mathcal{A}$ on $\mathcal{S}$. The elements of $\mathcal{S}$ are called the *states* of the automaton, $\mathcal{A}$ is called the *alphabet* of the automaton, and the elements of A are called *input symbols.* A transition table $\mathcal{T}$,

$$\mathcal{T} : \mathcal{S} \times \mathcal{A} \to \mathcal{S}, \tag{1.24}$$

associates with every $(s, a) \in \mathcal{S} \times \mathcal{A}$ a state $s' \in \mathcal{S}$. Therefore, the transition table $\mathcal{T}$ shows how the input s may change the state of the automaton from s to a state $s' = \mathcal{T}(s, a)$, also usually denoted as just sa.

If we allow ourselves to consider successive actions of inputs from $\mathcal{A}$ on states s, then in effect we extend the input set to $\mathcal{A}^*$, to be the set of all finite strings formed by concatenating input symbols from $\mathcal{A}$.[10] To see what we can do with such a machine, we designate a certain state in S to be the *initial state*, and a subset Ξ to be the *final state.* The machine then works by being presented with an input string of symbols

$$w = a_1 a_2 ... a_n$$

that it reads one by one starting at the leftmost symbol. Beginning at the initial state, s_0, the symbols determine a sequence of states,

$$s_0 a_1, (s_0 a_1) a_2, ..., (s_0 a_1 a_2 ... a_{n-1}) a_n.$$

The sequence ends when the last input symbol has been read. Furthermore, if the ending state is

$$(s_0 a_1 a_2 ... a_{n-1}) a_n \in \Xi,$$

we say that the automaton *recognizes* (or *accepts*) the string w. The *behavior* of $\mathcal{M}$ (or the *language* of $\mathcal{M}$) is defined as the set of strings in $\mathcal{A}^*$ that can be recognized by $\mathcal{M}$. The famous *Kleene's theorem* in computation theory shows that every language that can be defined by a *regular expression* can be accepted by some finite state automata.

In 1959, Michael Oser Rabin and Dana Scott provided a more general definition of an automaton by introducing the following notion of *nondeterminism* for language-recognizing finite automata:

$$\mathcal{T} : \mathcal{S} \times \mathcal{A} \to \mathrm{P}(\mathcal{A}), \tag{1.25}$$

[10] For example, if $\mathcal{A} = a, b, c$, then

$$\mathcal{A}^* = \{\aleph, a, b, c, aa, ab, ac, ba, bb, bc, ca, cb, cc, aaa, bbb, ccc, ...\},$$

where $\aleph$ is the empty string, i.e., the string of no letters.

where P($\mathcal{A}$) is the set of all subsets of $\mathcal{A}$. With (1.25), $\mathcal{T}(s,a)$ may contain more than one element of $\mathcal{A}$; therefore, the ultimate path through the machine is not determined by the input symbol alone. Human choice becomes a factor in selecting the transition path; the machine does not make all its own determinations. That is why we say this machine is nondeterministic. A string is accepted by the machine if *at least one* sequence of choices leads to a path that ends at a final state. It is important to know that nondeterminism does not increase the power of an FSA. In fact, an *equal power* theorem says that any language definable by a nondeterministic finite automaton is also definable by a deterministic finite automaton, and vice versa.

Finite state automata or other more powerful machines, such as *pushdown automata*, play an important role in the development of economic theory, in particular game theory.[11] In Chap. 11, they are applied as a language to describe financial patterns. The author supplies readers with a few examples of some useful patterns (regular expressions). In particular, a popular pattern frequently used by chartists, namely the *golden ratio rule*, is expressed in terms of regular language, and the finite state machine is implemented to search for this pattern (to recognize this pattern) in financial time series.[12] The author used the NYSE index as an illustrative example, and the pattern is found to be several times the *golden ratio matches*, with lengths ranging between 6 observations and 19 observations. From this example, it is crucial to observe that the same pattern may have different lengths, and fixing window size can then become a serious restriction.

At the end of his chapter, the author points out one direction for further research, i.e., to automatically generate useful patterns with *evolutionary algorithms*. We shall have further discussions on this in a later section. However, a few other tools useful in discovering patterns and making use of these patterns in prediction are in order.[13]

1.3.5 Decision Trees

The decision tree has become a canonical tool in machine learning. It is a *classification* procedure with a tree-like graph structure (Fig. 1.9). The data S $(=\{x_i\}_{i=1}^n)$ presented to the decision tree are of a common type, namely m *attributes* and p *decision classes*. Each attribute A_j $(j = 1, ..., m)$ partitions the n inputs into s_j distinct classes based on the attribute value,

$$A_j : S \rightarrow (a_{j,1}, a_{j,2}, ..., a_{j,s_j}). \tag{1.26}$$

[11] See [1.33] pp. 256–257 for a concrete example. Also, see [1.59] for a brief review of the application of NFA to game theory.

[12] To learn more about the golden ratio rule, see [1.31].

[13] As a continuation of Chap. 11, Chap. 12 provides a literature review on *sequence matching*. Given the rich materials presented there, the editors decided not to pursue this subject further in this introductory chapter.

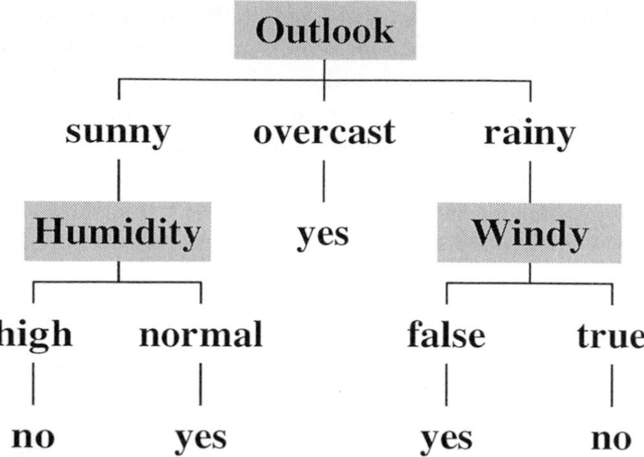

Fig. 1.9. The decision tree of the *play tennis* data

When the input x_i is presented to the tree at each node of the tree a decision is made based on a test on a value of an *attribute*, a_j. According to the result of the test, the interpretation of the tree proceeds to one of the subtrees of the node. The path will continue leading x_i to the next test until it goes through all of them, and hence reach a leaf of the tree. It is expected that all paths of the decision tree will inform us better on how different decisions are made.[14]

A decision tree is constructed based on a *top-down greedy algorithm*, known as ID3 in machine learning [1.68]. The key idea is fairly straightforward. One first finds the attribute that *best* classifies the training data, and then uses this attribute as the *root* of the decision tree. The process is repeated for each subtree. The main issue involved in this greedy algorithm is the criterion regarding the choice of the best classifying attribute. A common solution to this problem is to select the attribute with the *highest information gain*, $G(S, A)$, which is defined as the expected reduction in the entropy of the dataset S caused by knowing the value of the attribute A.

A decision tree can be converted to *if-then* rules. For the decision tree shown in Fig. 1.9 this can be represented as a decision rule as follows:

IF ([(Outlook=sunny) AND (Humidity=normal)] OR
[(Outlook=overcast)] OR
[(Outlook=rainy) AND (Windy=false)])
THEN YES (Play Tennis)

[14] In an ideal case, all inputs reaching the same branch belong to the same decision class. Of course, it is not necessarily so.

An important issue pertaining to growing decision trees is *when to stop*. Would it be desirable to grow the decision tree until it perfectly matches the data? To avoid *overfitting*, the answer is generally "no." However, in practice, a greedy algorithm will grow the full tree first, to be pruned later. There are two different types of pruning. The first one is to prune the tree directly, known as *reduced error pruning* [1.69]. The second is first to convert the tree into rules, and then to prune (generalize) each rule independently by removing preconditions that increase classification accuracy. This can be done using the famous algorithm C4.5 [1.70]. In addition to pruning, one can also use a complexity measure, such as the *minimum description length* (MDL) to halt tree growth when the MDL is found.

In contrast to artificial neural nets, *decision trees*, among all CI tools, are the least exploited tool, so to speak, in economics and finance.[15] Publications on the economic and financial applications of decision trees are nearly nonexistent. This is indeed a surprise to us, in particular considering that learning decision rules are no more difficult than many of the alternatives, e.g., rough sets.

In Chap. 13, Shu-Heng Chen and Tzu-Wen Kuo introduce a feature-based (rule-based) time series model based on the algorithm *Cubist*. Cubist provides an approximation to a nonlinear time series by partitioning the series into many subsets, and then constructing a linear regression model over each subset. By converting the decision tree into rules, one can have the time series model represented by (1.27).

$$x_{t+1} = \begin{cases} f_1(X_t^m), & \text{if } X_t^m \in A_1, \\ f_2(X_t^m), & \text{if } X_t^m \in A_2, \\ \cdot & \\ \cdot & \\ \cdot & \\ f_n(X_t^m), & \text{if } X_t^m \in A_n. \end{cases} \tag{1.27}$$

This way of representing a time series can be related to many other approaches introduced in this volume. Firstly, it is a generalization of the threshold regression model (see Sect. 1.3.4). What makes Cubist different from the threshold regression model is that the number of thresholds or gates is not fixed, and the functional form (the variables included) can change from one local model to yet another local model. This certainly endows Cubist with greater powers of expression than the threshold regression model. Secondly, Cubist can be used for comparisons with other feature-based regression models whose features are discovered by other tools, such as finite-state automata, context-free language, Self-organizing maps, and many others, as surveyed in [1.3]. Finally, the classification does not have to be crisp, it can be fuzzy as

[15] One of the earliest applications is [1.60].

well, and that leads us to the intention of the fuzzy classification and regression model methodology [1.83], [1.81]. This idea can be further enriched by the fusion of the evolutionary algorithms, and extended into the well-known *genetic fuzzy-classifier* model [1.19]. [1.84] is one of the earliest financial applications of *genetic fuzzy-classifiers.*

In Chap. 13, Cubist has been applied to both data from real markets and artificial markets. The purpose is to examine whether the efficient market hypothesis, if validated by rigorous econometrics, would also nullify the effectiveness of machine learning tools. The answer is interestingly *no.*

1.3.6 K Nearest Neighbors

In 1998, a time series prediction competition was held during the *International Workshop on Advanced Black-Box Techniques for Nonlinear Modeling.* The data to be predicted were available from November 1997 till April 1998 at Leuven. The data was generated from a generalized Chua's circuit, a well-known chaotic dynamic system. Seventeen entries had been submitted before the deadline. The winner of the competition was James McNames, and the strategy he used was the *nearest-trajectory algorithm.* By using this algorithm to speed up nearest-neighbor algorithms, McNames was able to make an accurate prediction up to 300 points in the future of the chaotic time series. At first sight, this result may be a surprise for some, because KNN is not technically demanding as opposed to many other well-known tools introduced in this volume, but it could outperform many other familiar advanced techniques, such as neural nets, wavelets, Kohonen maps, and Kalman filters in that competition.[16]

KNN can be related to*decision trees* (1.27). What makes them different is that the latter have categories $A_1, ..., A_n$ to host input variables X_t^m, while the former have X_t^m itself as a center of a hosting category, which will invite its own *neighbors*, X_s^m $(s < t)$, by ranking the *distance* $\parallel X_t^m - X_s^m \parallel$ over all $s < t$ from the closest to the farthest. Then the k closest X_s^ms will constitute the neighbors of X_t^m, $\mathcal{N}(X_t^m)$. Now, for the purpose of predicting x_{t+1}, one can first study the functional relation between x_{s+1} and X_s^m, $\forall s \in \mathcal{N}(X_t^m)$, i.e.,

$$x_{s+1} = f_t(X_s^m), s \in \mathcal{N}(X_t^m). \tag{1.28}$$

One then forecasts x_{t+1} based on $\hat{f}_t$, an estimation of f_t,

$$\hat{x}_{t+1} = \hat{f}_t(X_t^m). \tag{1.29}$$

Let us make a brief remark on what makes KNN different from the conventional time series modeling techniques. The conventional time series modeling, known as the Box-Jenkins approach, is a *global* model, which is concerned

[16] For details of the competition report, see [1.82].

with the estimation of the function, be it linear or nonlinear, in the following form

$$x_{t+1} = f(x_t, x_{t-1}, ..., x_{t-m}) + \epsilon_t = f(X_t^m) + \epsilon_t, \tag{1.30}$$

by using all of the information up to t, i.e., $X_s^m, \forall s \leq t$, and the estimated function $\hat{f}$ is assumeed to hold for every single point in time. As a result, what will affect x_{t+1} most is its immediate past $x_t, x_{t-1}, ...$ under the law of motion estimated by all available samples. For KNN, while what affects x_{t+1} most is also its immediate past, the law of motion is estimated *only* with *similar* samples, *not all* samples. The estimated function $\hat{f}_t$ is hence assumed to only hold for that specific point in time. Both KNN and SOM challenge the conventional Box-Jenkins methodology by characterizing the hidden patterns in a different form. In their formulation, hidden patterns are not characterized by time location, but by topological trajectories.

Technical issues involved here are the choice of the distance function $d(X_t^m, X_s^m)$, the choice of the functional form f_t, the choice of the number of neighbors k, and the choice of the embedding dimension m. Chapter 14 presents an excellent discussion of some of these technical choices as well as a literature review of the financial applications of KNN. Given this splendid review, there is no need for the editors to add some redundant materials here except to bring readers' attention to this lengthy, but well-worth reading, chapter.

Lazy Learning. In literature, a number of disciplines which are closely related to KNN have been grouped together with several different titles: *lazy learning*, *memory-based learning*, *instance-based learning*, *exemplar-based learning*, and *locally weighted learning*.[17] The ideas are pretty similar to what has been said on (1.28) and (1.29). Technical differences can arise from the employment of weighting functions (kernel functions) and local model structures. In some cases, the influences of neighbors are further weighted by their closeness to the host sample, and in some other cases, they are not. Also, local model can be linear regression, polynomial, or general nonlinear models. In addition to KNN, the *local polynomial regression* presented in Chap. 12 gives us another example of financial applications of lazy learning. Weiqiang Lin, Mehmet A. Orgun and Graham J. Williams give a lucid technical introduction to local polynomial regression, and it would be redundant for the editors to go over those fundamental materials again. Interested readers are, however, referred to [1.30] for more details. Furthermore, [1.94] is the first book on the application of *locally weighted regression* (*kernel regression*) to the problem of building trading systems.

[17] [1.1] is the first collection on this topic.

1.3.7 Hidden Markov

Earlier, while discussing Chap. 8, it would have been interesting to know whether the patterns identified by SOM when presented as a sequence, i.e., a sequence of SOM-derived patterns, had a *global structure*, instead of just leaving the patterns identified in sequence. For example, it may be useful to know that if the current pattern is "double top," what is the probability that we will witness a pattern of "down trend" in a month? To address questions of this type, a typical approach is to consider those patterns as states and to build transition probabilities among these states. This brings us to the familiar *Markov processes*. While Chap. 8 does not consider this Markov modeling of the patterns, Chap. 12 does conduct the estimation of the transition probabilities among different states in the context of hidden Markov model.

Chapter 12 can be connected to Chaps. 13 and 22 in a more general modeling philosophy. To see this, suppose that the series $\{x_t\}$ is generated by a random draw from a set of models (functions) $\mathcal{F}$, say, $f_1, f_2, ..., f_n$. Each time the random draw follows a distribution which is determined by the *state* (feature, pattern) of the series $\{x_t\}$, or simply just the window X_t^m. For example, if X_t^m is found to be state j, then the random draw on $\mathcal{F}$ follows the distribution $p_1^j, ..., p_n^j$ $(j = 1, ..., J)$. Furthermore, the state j $(j = 1, ..., J)$ follows a stochastic process, e.g., a first-order Markov process with a transition probability matrix $P(a, b)$, where $a, b = 1, ..., J$. The modeling task is then to identify these J states, the m models, and the distribution $p_1^j, ..., p_n^j$ with the transition probability matrix. If the state is not determined by X_t^m but by an unobserved series $\{z_t\}$, then the Markov process is also called the hidden Markov model. In the latter case, z_t has also to be estimated.

In application, this general model may be simplified in different ways. For example, (1.27) (the decision tree) is a simple version of this general model in the sense that the distribution $p_1^j, ..., p_n^j$ is degenerated for all j, and the number of states (J) is assumed to be the same as the number of equations (n). Besides, the underlying transition probability matrix $P(a, b)$ is not taken into account. The *Markov switching model*, frequently used in econometrics, provides another simple version in that J=n=2, and the distribution $p_1^j, ..., p_n^j$ is also degenerated, while the state is determined by an unobserved dummy variable $\{z_t\}$. A good example of this type of application can be found in Chap. 22. Chapter 12, on the other hand, simplifies the general model by directly defining the state of the model, which is based on the three consecutive observations of the foreign exchange rates.

1.3.8 Fourier Transforms and Wavelet Transforms

Most economic and financial time series are *nonstationary*, or, at best, *quasi stationary*. This is not surprising at all because these data are generated from complex adaptive systems. However, this nonstationary property does render many conventional statistical techniques inadequate or insufficient to cope with the problem. One example is *Fourier analysis*.

Fourier analysis or *spectral analysis* is the *frequency-domain approach*, as opposed to the *time-domain approach*, to time series. The former is associated with the *spectral*, whereas the latter is associated primarily with the *correlation*. The spectral can be defined as the discrete Fourier transform of the autocovariance function,

$$\phi_{xx}(z) = \sum_{\tau=-\infty}^{\infty} c_{xx}(\tau) z^{\tau}, \tag{1.31}$$

where $c_{xx}(\tau)$ is the autocovariance function of the zero-mean stationary time series $\{x_t\}$, and z is the complex variable $e^{-i\omega} = \cos\omega - i\sin\omega$ $(-\pi \leq \omega \leq \pi)$. Formally, $\phi_{xx}(z)$ is called the *power spectral density* or the *autocovariance generating function* of $\{x_t\}$.

Since the publication of [1.40], spectral analysis has been used in economics for almost four decades. Chapter 21 intends to make use of spectral analysis to examine the interest behavior under different international monetary regimes, and attempts to see the effect of learning agents who were adapting to these different regimes, including during the ERM (Exchange Rate Mechanism) crisis. This chapter reviews materials on input-output relations for spectral densities, including the *cross-spectral density*, *transfer function*, *cospectrum*, *quadrature spectrum*, *phase spectrum*, and *gain spectrum*. By examining the phase spectrum and the gain spectrum at different points in time, the authors are able to analyze the impact of agents' learning behavior on interest rates.

What is being presented in Chap. 21 is not a *static* spectral analysis, but an adaptive version of it. This has already led us to see a limitation of Fourier analysis, which we shall now treat in a more formal fashion. A disadvantage of Fourier analysis is that frequency information can only be extracted for the complete duration of a signal. This results in spreading frequency information over all time and, thus, the loss of frequency characteristics of a time series in the time domain. Therefore, although we might be able to determine all the frequencies present in a signal, we do not know when these frequencies are emitted. The same sine and cosine functions can represent *very different moments* in a signal because they are shifted in phase so that they amplify or cancel each other. It is then clear that Fourier analysis is very poorly suited to very brief signals, or signals that change suddenly and unpredictably. However, in the area of economics and finance, the brief changes often carry the most interesting and significant information.

In order to isolate signal discontinuities, one would like to have some very *short* basis functions. At the same time, in order to obtain detailed frequency analysis, one would like to have some very *long* basis functions. The best way to achieve this is to have *short high-frequency basis functions* and *long low-frequency ones*. This is exactly what wavelet transforms can do, as shown below.

In the wavelet transforms, a function $f(t)$ can be expressed as an additive combination of the wavelet coefficients at different resolution levels. More precisely,

$$f(t) = \sum_j \sum_k \beta_k^j \psi_k^j(t) = \sum_k \alpha_k^{j_0} \phi_k^{j_0}(t) + \sum_{j \geq j_0} \sum_k \beta_k^j \psi_k^j(t), \tag{1.32}$$

where

$$\alpha_k^{j_0} = \int f \phi_k^{j_0} dt, \quad \beta_k^j = \int f \psi_k^j dt, \tag{1.33}$$

$$\phi_k^j(t) = 2^{\frac{j}{2}} \phi(2^j t - k), \quad \psi_k^j(t) = 2^{\frac{j}{2}} \psi(2^j t - k). \tag{1.34}$$

$\alpha_k^{j_0}$ represents smooth coefficients at the coarsest resolution level j_0, and β_k^j represents detailed coefficients at the finest resolution level j, where j indicates the frequency information, and k denotes the time information. The function ϕ is also denoted as the *father wavelet* or the *scaling function*. The linear combination of $\{\phi_k^j\}$ produces the *mother wavelet function* ψ. There are different wavelet bases. Different wavelet families make different trade-offs between how compactly the basis functions are *localized in space* and how *smooth* they are. Wavelet families proposed in the literature include the Harr wavelet, the Morlet wavelet, the Coiflet wavelet, the Meyer wavelet, and the Daubechies wavelet. The Haar wavelet is the simplest one, and it is often used for educational purposes. The Daubechies wavelet transform is perhaps the most elegant one, and hence has become the cornerstone of wavelet applications today.

Wavelets are no longer an unfamiliar tool for economists and finance people. The recent publication [1.39] provides an introduction to the field for economists and is intended for financial analysts, to help them see the relevance of wavelets to economics and finance. Wavelets have made possible a new econometric tool for extracting signals from white noise or for denoising noisy data, which is an essential step in time series analysis. An obvious problem in separating noise from a signal is knowing which is which. Traditional methods require knowing or assuming something about how smooth a signal is. With wavelets, one only needs to know very little about the signal. Chapter 10 illustrates the use of wavelets in denoising noisy data. Dimitrios Moshou and Herman Ramon apply the Harr wavelet transform to the Dow-Jones stock index, and from the nondecay of the wavelet coefficients, stock market crashes can be recognized. They further illustrate the use of the wavelet transform to smooth stock prices. What has not been shown in this chapter, but has been shown elsewhere, is the advantage of *wavelet-based smoothing* over the frequently used simple moving average. For example, [1.86] has shown that the wavelet filters provide improvements over simple averaging and differencing filters, when applied to neural networks for forecasting the S&P 500 index.

In addition to denoising or smoothing noisy data, wavelets, employed jointly with other CI tools, in particular neural networks, has produced appealing results. As mentioned above, [1.86] applied wavelet-based bandpass filters to process financial data for input to neural-network-based financial predictors. A similar direction of research was also undertaken by [1.4]. In [1.4], all of the wavelet coefficients for a particular time point are taken together as a feature vector and fed into the input of a dynamical recurrent neural network.[18] Chapter 10 provides another possibility for the use of wavelets with other CI tools. In this chapter, the features of a trading period, say 32 trading days, are characterized by a vector of wavelet coefficients. The principal component analysis is applied to these vectors of wavelet coefficients, and the derived principal components are further used to train the SOM (see Sect. 1.3.3). Based on the SOM built and its associated U matrix (a distance matrix), abnormal states (isolated distant clusters), which are related to the abrupt change in the stock price, can be identified. As a result, the authors assert that by following the trajectories leading to these abnormal states, one can predict a forthcoming change in the trend of the stock value.

Wavelets have also been used to estimate *density*, and the wavelet-based density estimator can be further used in the field of forecasting. [1.64] used wavelets to estimate a regression function $M(x) = E(y \mid x)$ without having to first specify a particular form. Within the framework of density estimation, the problem can be formulated as follows:

$$M(x) = E(y \mid x) = \int y \frac{f(y,x)}{f_1(x)} dx, \tag{1.35}$$

where f_1 is the density of x marginal to the joint density $f(y,x)$. [1.64] used wavelets as the basis for density estimation.

1.4 Evolutionary Computation and the Related Disciplines

1.4.1 Evolutionary Computation: the Common Features

The last important pillar of computational intelligence is so-called *evolutionary computation* (EC). EC uses *nature* as an inspiration. While it also has a long history of utilization in economics and finance, it is, relatively speaking, a new kid in the block, as compared with neural networks, and even more so as compared to fuzzy logic. It has also drawn less attention from economists and financial analysts than the other two approaches. To gauge the comparison, there are already about a dozen books or volumes on economic and financial applications using fuzzy logic and neural nets. In the area of EC,

[18] The coefficients employed in [1.4] are *derived from* wavelet coefficients, but are not themselves wavelet coefficients.

there are only three volumes edited for economists and financiers: [1.10],[1.16], and [1.17] . Evolutionary computation is generally considered to be a consortium of *genetic algorithms* (GA), *genetic programming* (GP), *evolutionary programming* (EP), and *evolutionary strategies* (ES).

The history of evolutionary computation can be traced back to the mid-1960s, when evolutionary strategies were originated by Ingo Rechenberg [1.71], Hans-Paul Schwefel [1.76], and Peter Bienert at the Technical University of Berlin, the development of genetic algorithms was originated by John Holland of the University of Michigan, and evolutionary programming was originated by Lawrence Fogel [1.34] at the University of California at Los Angeles.[19] Despite their nontrivial differences, they share a common structure, as shown in Fig. 1.10.

```
begin
    t := 0;
    Initialize P(t);
    evaluate P(t);
    while not terminating do
    begin
        M(t) := select-mates(P(t));
        O(t) := alternation(M(t));
        evaluate(O(t));
        P(t+1) := select(O(t)∪P(t));
        t := t+1;
    end
end
```

Fig. 1.10. A pseudo program of evolutionary computation

Evolutionary computation starts with an initialization of a population of individuals (solution candidates), called $P(0)$, with a *population size* to be supplied by the users. These solutions will then be evaluated based on an *objective function* or a *fitness function* determined by the problem we encounter. The continuation of the procedure will hinge on the *termination criteria* supplied by users. If these criteria are not met, then we shall move to the next stage or *generation* by adding 1 to the time counter, say from t to $t+1$. Two major operators are conducted to form the new generation, which can be regarded as a *correspondence*, as follows,

$$F_{s_2} \circ F_a \circ F_{s_1}(P(t)) = P((t+1)), \tag{1.36}$$

where F_{s_1} and F_{s_2} denote *selection*, and F_a denotes *alteration*. The main purpose of the first-stage selection, F_{s_1}, is to make a mating pool (a collection

[19] For a description of the birth of EC, see [1.77], [1.32], and [1.28].

of parents), $M(t)$, which can in turn be used to breed the new generation:

$$F_{s_1}(P(t)) = M(t). \tag{1.37}$$

Once the mating pool is formed, F_a is applied to generate *offspring*, $O(t)$, from these parents. Two major steps (*genetic operators*) are involved here, namely, *recombination* (*crossover*), denoted by F_r, and *mutation*, denoted by F_m, which shall be detailed later.

$$F_a(M(t)) = F_m \circ F_r(M(t)) = O(t). \tag{1.38}$$

These offspring will be evaluated first, and they then enter the second-stage selection *with* or *without* their parents $P(t)$. Finally, the new generation $P(t+1)$ is formed as a result of the second-stage selection:

$$F_{s_2}(O(t) \cup P(t)) = P((t+1)). \tag{1.39}$$

After that, we shall go back to the beginning of the loop, and then check the termination criteria to see whether we shall stop or start another generation of runs. Also, see Fig. 1.11 for the evolution loop.

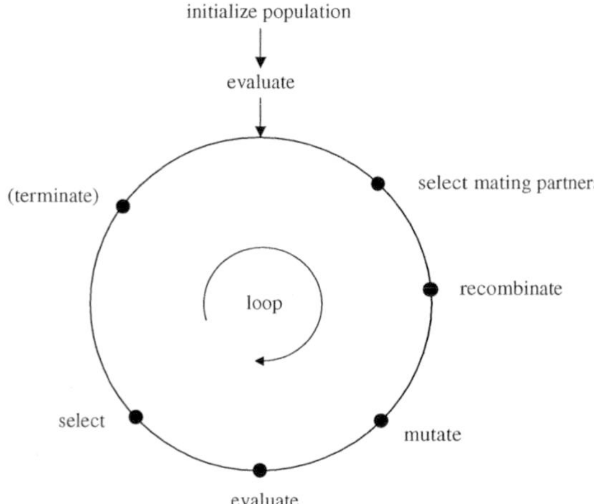

Fig. 1.11. The evolutionary loop

Based on the description above, it is perhaps beneficial to have the seven major components of evolutionary algorithms listed as follows for quick reference:

- individuals and their representations,
- initialization,
- fitness evaluation,
- selection,
- mutation,
- recombination,
- replacement.

1.4.2 Evolutionary Strategies

We shall illustrate each of these components mainly within the context of *evolutionary strategies*. Individuals are also called *chromosomes*. The individual in ES is represented as a pair of real-valued vectors $v = (x, \sigma)$, where the x represents a point in the solution space, and σ is a standard deviation vector that determines the mutation step size. Generally, σ is also called the *strategy parameter* in ES, and x is called the *object variable*.

The population size of ES is usually characterized by two parameters μ and λ. The former is the population size of $P(t)$, whereas the latter is the population size of $O(t)$. Selection F_{s_1} is much more straightforward in ES than in GA. Usually, it takes the whole $P(t)$ as the mating pool and parents are randomly selected from this mating pool. However, selection F_{s_2} in ES can be more intriguing. There are two schemes F_{s_2} in ES, known as the $(\mu+\lambda)$ scheme (the Plus scheme) and the (μ, λ) scheme (the Comma scheme). In the $(\mu + \lambda)$ scheme, μ individuals produce λ offspring, and a new population is formed by selecting μ individuals from the $\mu + \lambda$ individuals. In the (μ, λ) scheme, μ individuals produce λ offspring, and a new population is formed by selecting μ individuals from the λ offspring. There is generally no constraint for μ and λ for the $(\mu+\lambda)$ scheme, but for the (μ, λ) scheme, to make selection meaningful, μ has to be strictly less than λ; moreover, $\lambda/\mu \approx 7$ is an ideal ratio.

Mutation is considered the major ES operator for alteration on the chromosomes. Mutation is applied to this individual to perturb real-valued parameters. If we let v be the parent randomly selected from $P(t)$, then mutation on v can be described as follows:

$$v' = (x', \sigma') = (f_{m_x}(x), f_{m_\sigma}(\sigma)), \tag{1.40}$$

where

$$f_{m_x}(x) = x + N(0, (\sigma')^2), \tag{1.41}$$

and

$$f_{m_\sigma}(\sigma) = \sigma \exp(\tau N(0,1)). \tag{1.42}$$

$N(0, \sigma^2)$ denotes the normal distribution with mean 0 and variance σ^2.[20] Notice that in implementation, (1.42) has to be computed before (1.41). This is because x' is obtained by mutating x with the new standard deviation σ'.[21]

---

[20] Here, for simplicity, we assume that x is a real-valued number. In a more general setting, the variable x can be a vector. In that case, σ should be replaced by the variance-covariance matrix Σ.

[21] In (1.42), $(\sigma')^2$ is determined randomly. There is, however, some way to make it adaptive. For example, in the (1+1)-ES case, one has the famous 1/5-success rule. $(\sigma')^2$ can also be determined in a self-adaptive way. In that case, the learning rate τ can be set as a function of time. For details, see [1.77].

Recombination operators compose new chromosomes from corresponding parts of two or more chromosomes. For the binary case, two chromosomes $v_1 = (x_1, \sigma_1^2)$ and $v_2 = (x_2, \sigma_2^2)$ are to be recombined by an operator f_r. We can describe the composition of a new chromosome v' as follows:

$$v' = (x', \sigma') = (f_{r_x}(x_1, x_2), f_{r_\sigma}(\sigma_1^2, \sigma_2^2,)). \tag{1.43}$$

Each element of the object and strategy parameter is a recombination of the respective entries of v_1 and v_2. There are great varieties of f_{r_x} and f_{r_σ}. In the ES literature, they are differentiated by the terms *discrete* or *intermediate*, *dual* (sexual) or *global* (panmictic). With a *discrete* recombination function, one of the corresponding components is chosen at random and declared the new entry. With an intermediate recombination, a linear combination of the corresponding components is declared the new entry. More formally, consider x' as an example:

$$x' = \begin{cases} x_1 \;\; or \;\; x_2, & \text{discrete}, \\ \chi x_1 + (1-\chi)x_2, & \text{intermediate}, \end{cases} \tag{1.44}$$

where $\chi \in [0, 1]$ denotes a uniform random variable. So far we have only considered the one-dimensional x. An n-dimensional x can further complicate the recombination function, and that is where the terms *dual* and *global* come from. *Dual* means that two parents are chosen at random for the creation of the offspring. *Global* means that one parent is chosen anew for *each component* of the offspring:

$$x_i' = \begin{cases} x_{1,i} \;\; or \;\; x_{2,i}, & \text{discrete, dual}, \\ x_{1,i} \;\; or \;\; x_{(2),i}, & \text{discrete, global}, \\ \chi x_{1,i} + (1-\chi)x_{2,i}, & \text{intermediate, dual}, \\ \chi x_{1,i} + (1-\chi)x_{(2),i}, & \text{intermediate, global}, \end{cases} \tag{1.45}$$

where $x_{(2),i}$ indicates that parent 2 is chosen anew for each vector component i $(i = 1, 2, ..., n)$.

Both applications of the evolutionary strategies in this volume are concerned with macroeconomic issues. Chapter 16 applies ES to solving the *optimum consumption* problem in a one-sector growth model, usually known as the *policy function* problem. This type of application has interested researchers, as evidenced by EC literature, for a while [1.75, 1.27]. For example, [1.75] applied genetic programming to solve the policy function in a simple optimal growth model where a known analytical solution exists. [1.27] approached this problem with genetic algorithms and neural networks. In Chap. 16, the policy function is approximated by radial basis networks, and the number of centers and the weights of the network are evolved by evolutionary strategies. The policy function problem as addressed in [1.75] is deterministic, whereas the problem dealt with in Chap. 16 is stochastic, and hence more complicated.

One of the interesting findings as concluded in this chapter is that the authors are unable to draw a conclusion on the significance of population size. In fact, they present quite mixed results for different experiments: some are favorable to the single-member ES, while some are favorable to the multi-member ES.

Chapter 17 applies (1+1)-ES to estimate a simple linear regression model, and the authors then use the derived linear regression model to forecast the inflation rate. The forecasting performance of ES is then compared with that of NNs.

1.4.3 Evolutionary Programming

While evolutionary programming was proposed about the same time as evolutionary algorithms, their initial motives were quite different. Evolutionary strategies were developed as a method to solve *parametric optimization problems*, whereas evolutionary programming was developed as a method for simulated *intelligence behavior.* Lacking a capability to predict, an agent cannot adapt its behavior to meet the desired goals, and the success in predicting an environment is a prerequisite for intelligent behavior. As [1.36] puts it:

> Intelligent behavior is a composite ability to predict one's environment coupled with a translation of each prediction into a suitable response in the light of some objective. ([1.36], p. 11)

During the early stage, the prediction experiment can be illustrated with a sequence of symbols taken from a finite alphabet, say, a repeating sequence "(101110011101)*" from the alphabet $\{0, 1\}$. The task then is to create an algorithm that would operate on the observed indexed set of symbols and produce an output symbol that agrees with the next symbol to emerge from the environment. Lawrence Fogel took *finite state automata* (FSA) as the machine to predict the sequence. Finite state automata is a device which begins in one state and, upon receiving an input symbol, changes to another state according to its current state and the input symbol.[22] EP was first proposed to evolve a population of finite state machines that provides successively better predictions.

Earlier efforts following [1.36] mostly involved the use of finite state automata for predictions, generalization and gaming. Nonetheless, in the late 1980s, interest increased in applying EP to difficult combinatorial optimization problems, including evolving artificial neural networks. When EP was applied to continuous parameter optimization problems, its structure was very similar to ES. The main difference was that, for some reasons discussed in [1.35], EP has not relied on the recombination operator. Second, parent and offspring population sizes are usually identical in EP, i.e., $(\mu = \lambda)$. Third, because of that, EP only conducts the $(\mu + \mu)$ selection scheme, and sometimes it is a hybrid of tournament and $(\mu + \mu)$ selection, which allows less fit solutions a small chance to propagate.

---

[22] See Sect. 1.3.4 for a more detailed review.

1.4.4 Genetic Algorithms and Genetic Programming

Among the four disciplines in EC, the genetic algorithm is the most popular one used in economic and financial studies. [1.16] is the first volume devoted to the economic and financial applications of evolutionary computation. [1.17] evolves as an even sharper focus on the financial applications of genetic algorithms and genetic programming. Since these two volumes have already been made available to the readers, and they provide a pretty general landscape of the current state of GA and GP related to economics and finance problems, the editors do not intend to give a further introduction to GA and GP. Chapter 15, however, provides a brief review of the early applications of GP to financial time series with some examinations of their statistical behavior.

1.4.5 Simulated Annealing

Simulated annealing is a general optimization technique for solving combinatorial optimization problems. In practice, many large-scale combinatorial optimization problems cannot be solved to reach an optimality. This situation is caused primarily by the fact that many of their computations are NP-hard.[23] Given this situation, one is forced to use approximation algorithms or heuristics, for which there is usually no guarantee that the solution found by the algorithm indeed is optimal, but for which polynomial bounds on the computation time can be imposed.

Simulated annealing, as introduced by [1.51], is a stochastic version of *neighborhood search* or *local search*. It is also known as *probabilistic hill climbing*. It differs from the deterministic local search, say, hill climbing, in its *acceptance decision* of a new configuration (solution candidate): it will accept the transition to a new configuration which corresponds to an *increase* in the cost function (energy function) with a probability determined by the associated *cooling schedule*, e.g., the well-known *Metropolis algorithm*. This randomization design makes SA a solution for reducing the possibility of falling into a local optimum, and a benchmark with which other proposed solutions, such as evolutionary algorithms, can be compared. Some in-depth treatment of SA can be found in Chap. 19 in this volume.

Simulated annealing was known to economists more than a decade ago [1.42], [1.43], and [1.41]. In Chap. 19, Lu and Cheng apply SA to select a portfolio of 50 stocks. They relate the energy function to the excess Sharpe return so that maximizing the excess Sharpe return is equivalent to minimizing energy, and the simulated annealing is hence applied to solve the defined portfolio optimization problem.

---

[23] NP-hard problems are unlikely to be solvable using an amount of computational effort which is bounded by a *polynomial function* of the size of the problem. For details, see [1.37].

1.4.6 Ant Algorithms

While ants are practically blind, they are still able to find the *shortest paths.* [1.44] studied ants in a laboratory. An ant colony was given access to food linked to the nest by two bridges of different lengths. Ants must choose one way or the other. Experimental observations show that after a few minutes, most ants use the shortest branch. *How do they do it?* The answer is *pheromones.* Ants deposit pheromones along the trail whilst going to the food source and returning. At a decision point, they make a probabilistic choice based on the amount of pheromone along each branch. This process is *autocatalytic*, as choosing a path increases the probability it will be chosen again.

These observations inspired a new type of algorithm called *ant algorithms* or *ant colony optimization* (ACO). Formally speaking, ACO studies are artificial systems that take inspiration from the behavior of real ant colonies and which are used to solve discrete (combinatorial) optimization problems. The first ACO system was introduced by Marco Dorigo in his Ph.D. thesis [1.25], and was called the Ant System (AS). Ant algorithms have been applied to a wide range of problems including the Traveling Salesman Problem (TSP), routing, and telecommunications.

Since the ant algorithm was originally motivated by the shortest distance problem, it can be easily illustrated with the traveling salesman problem.[24] Consider a set of m ants traveling through n towns. Ants can choose their own paths by visiting each town in different orders, but each town must be visited once and once only. Each ant at time t chooses the next town where it will be at time $t+1$ (time is discrete). When an ant decides which town to move to next, it does so with a probability, called the *transition probability*, to be explained later. When the ant completes its tour, called a *cycle*, it lays down a substance called the *pheromone trail* on each edge (i, j) visited, where (i, j) is the direct path connecting towns i and j.[25] The amount of the pheromone trail the kth ant lays on the edge (i, j), $\Delta\tau_{ij}^k$, can be computed:

$$\Delta\tau_{ij}^k = \frac{Q}{L_k}, \tag{1.46}$$

[24] This brief introduction to the ant algorithms is based on [1.26].

[25] Depending on how ants drop pheromones on the trail as they pass, there are variants of the ant algorithm. The one to be introduced here is called the *ant-cycle algorithm.* In this model, each ant lays its trail till the end of the tour. This is also called the *delayed pheromone update.* Alternatively, in the *ant-quantity algorithm* and the *ant-density algorithm*, each ant lays its trail at each step, without waiting for the end of the tour. This is called the *online step-by-step pheromone update.* The ant-cycle algorithm uses *global information*, that is, its ants lay an amount of trail which is proportional to how good the solution produced was. In fact, ants producing shorter paths contribute a higher amount of trail marker than ants whose tour was poor. On the contrary, both the ant-quantity and ant-density algorithms use *local information.* Their search is not directed by any measure of the final result achieved.

if the kth ant uses the edge (i, j) in its tour during a cycle (i.e., between t and $t+n$); otherwise, $\Delta\tau_{ij}^k = 0$. L_k is the tour length of the kth ant, and Q is a constant.

After each ant lays its trail, the trail intensity of each edge (i, j) can be updated using the following formula:

$$\tau_{ij}(t+n) = \rho \cdot \tau_{ij}(t) + \Delta\tau_{ij}, \tag{1.47}$$

where

$$\Delta\tau_{ij} = \sum_{k=1}^{m} \Delta\tau_{ij}^k. \tag{1.48}$$

In (1.47), in order to avoid unbounded accumulation of the pheromone trail, ρ ($\rho < 1$) is a coefficient that represents the *evaporation* of the pheromone trail between t and $t+n$. The inclusion of the parameter ρ implements a form of *forgetting*. It favors exploration of new areas of the search space, and hence its purpose is to avoid a too rapid convergence of the algorithm to a suboptimal region.

The trail density, τ_{ij}, forms an important basis on which an ant decides which town to move to next during his next tour, as the transition probability from town i to j is given as follows:

$$p_{ij}^k(t) = \begin{cases} \frac{[\tau_{ij}(t)]^\alpha \cdot [\eta_{ij}]^\beta}{\sum_{k \in \text{allowed}_k} [\tau_{ik}(t)]^\alpha \cdot [\eta_{ik}]^\beta}, & if \;\; j \in \text{allowed}_k, \\ 0, & \text{otherwise}, \end{cases} \tag{1.49}$$

where $\eta_{ij} = 1/d_{ij}$, the inverse of the distance between the city i and j, denotes *visibility*. Equation (1.49) says that the probability of an ant following a certain route is a function not only of the *pheromone intensity* but also a function of what the ant can see (*visibility*). The parameters α and β control the relative importance of the pheromone intensity versus visibility. A high value for α means that the trail is very important and therefore ants tend to choose edges chosen by other ants in the past. On the other hand, low values of α make the algorithm very similar to a stochastic multigreedy algorithm. A positive value introduces a positive feedback effect, known as autocatalysis, to the ant system.

With time, it is likely that the amount of pheromone the ants deposit will increase on the shorter paths and decrease on the longer paths. In the limit, the shortest path may be found as the emergent property of the ant system. Of course, like other soft computing algorithms, there is no guarantee that the shortest path will always be found. In fact, [1.26] showed the significance of the parameters α, β, ρ and m, the initial distribution of ants over different towns, and the number of elitists. In the literature, the use of genetic algorithms has also been suggested to optimize the ant system.

Ant systems are clearly a *population-based* approach. There is a population of ants, with each ant finding a solution and then communicating with the other ants via the pheromone intensity and the transition probability. In this respect it is similar to evolutionary algorithms.

Among all CI tools presented in this volume, the ant algorithm perhaps is the newest kid on the block for economists. The economic and financial applications of the ant algorithm are still in their infancy. Nevertheless, a few applications are known to us. [1.14] cited business advantages to using ant algorithms intelligently and summarized many examples of real businesses that are using ant algorithms in order to solve problems or even to improve efficiency. For example, making use of ant algorithms, Hewlett-Packard has developed software programs that can find the most efficient way to route phone traffic over a telecommunications network. Southwest Airlines has used a similar model to efficiently route their cargos. [1.50] uses *ant programming* to derive analytical approximations for determining the implied volatility of European-style options as well as American call options. Another example can be found in Chap. 18, which is an interesting example of the application of the ant algorithm to business failure prediction. Business failure prediction is a research subject which has been intensively treated by economists, statisticians, and AI researchers alike. Various classification techniques ranging from statistics to AI have been applied to attack this problem. Chapter 18 gives a short but nice summary of its past. It shows readers how the ant algorithm represents a new possible alternative for future research in economics and finance.

Chapter 18 makes some changes to the standard ant algorithm as presented above. The most dramatic change is made to the transition probability. The original multiplication form of (1.49) is now modified:

$$p_{ij}^k(t) = \begin{cases} \frac{[\tau_{ij}(t)]^\alpha + [\eta_{ij}]^\beta}{\sum_{k \in \text{allowed}_k} [\tau_{ik}(t)]^\alpha + [\eta_{ik}]^\beta}, & if \quad j \in \text{allowed}_k, \\ 0, & \text{otherwise.} \end{cases} \tag{1.50}$$

Furthermore, as an effective way to fine-tune the control parameters α and β, the authors introduce a functional relation between the coefficients α and β. To avoid stagnation and nonconvergence, they also revise the pheromone trail update equation, (1.48), by adopting a dynamic update mechanism of the pheromone trail. Since the ant algorithm was originally motivated by the shortest path problem, some essential elements such as "towns" and "visibility" need to be appropriately defined when one wants to apply the algorithm to classification problems. In particular, it is not clear what the role of *physical distance* in classification applications is. Chapter 18 proposes some adjustments which justify the relevance of the ant algorithm to a larger class of problems.

1.5 State-Space Approach

The last discipline included in this volume reminds us all about the state-space approach of so-called modern control theory developed by control engineers from the 1970s up to the present day [1.9, 1.13]. The relevance of the optimal control theory to economics and the financial world was recognized by Richard Bellman and others at an early stage. Unfortunately, economists no longer consider that controlling the dynamic system was very useful or relevant for managing the real economy. The reason for this reality has very much to do with the fact that the mathematical modeling of the economics seldom fits into the very restricted canonical form of the linear control systems used by control engineers. One obvious exception seems to be the problem of predicting dynamic systems, which still motivates economists to borrow some techniques familiar to control engineers, in particular, the *state-space approach to time series modeling* [1.62, 1.8]. Starting from the late 1980s, Kalman filtering and the recursive estimation technique found their way to become standard tools for a large population of econometricians [1.2, 1.48].

A state-space description of a time-invariant linear discrete dynamical system can be written as

$$x_{t+1} = Ax_t + B\mu_t, \tag{1.51a}$$

$$y_t = Cx_t + D\nu_t, \tag{1.51b}$$

where x_t is an n-dimensional state vector, y_t is an m-dimensional output vector, and μ_t and ν_t are, respectively, n- and m-dimensional Gaussian white noise. Equation (1.51a) is called the *state vector equation*, and (1.51b) is called the *observation vector equation*. In control theory, the series $\{x_t\}$ is unobservable and can only be estimated from the observation series $\{y_t\}$. The well-known observer theory addresses this very issue. In economics, this theoretical representation can have useful applications. For example, x_t can be the (unobservable) real wage, whereas y_t is the (observable) nominal wage. Another example is x_t as the fundamental value of a stock, with y_t as its market value counterpart.

In Chap. 20, x_t is regarded as a k-dimensional vector of hidden *factors* in the context of *arbitrage pricing theory* (APT). There the k factors are not directly observable, but can be extracted from the observations, y_t, which are taken as stock returns in this application. The number of k needs to be determined simultaneously with the estimation of the parameters A and C, and the covariance matrix of state and observation noise has also to be estimated. All these are done by means of the *temporal factor analysis* (TFA) invented by the authors. Using TFA, the stock returns of m companies are determined linearly using these k common factors and the idiosyncratic factors which are uncorrelated between different companies. The authors also show that these k hidden common factors, originally extracted from stock returns, can also be the hidden common factors for the economic variables, such as interest rates.

As we already encountered in Sect. 1.3, many disciplines and chapters are included in this volume mainly for the purpose of dealing with *unstable relationships* among economic and financial variables. In addition to the disciplines introduced in Sect. 1.3, the state-space model can be an alternative for dealing with this issue. For an illustration, a basic linear regression model with a *changing coefficient vector* β_t is represented by

$$y_t = x_t\beta_t + \epsilon_t \quad (t = 1, ..., T), \tag{1.52a}$$

$$\beta_t = M\beta_{t-1} + \eta_t \quad (t = 1, ..., T), \tag{1.52b}$$

where x_t is a row vector of k fixed explanatory variables, and ϵ_t and η_t are Gaussian noise. In this state-space representation, β_t is the state variable, whereas y_t is the observation variable. x_t, corresponding to C in (1.51b), is known in this case, and M is also assumed to be known (usually it is taken as the identity matrix). Then the main econometric issue here is to estimate β_t, i.e., the constantly changing coefficients. A technique used here is the well-known *Kalman filter* for linear time-varying cases [1.48]. Chapter 21 applies the Kalman filter to estimate the unstable relationship in the UK short-term interest rate regression equation. The time-domain adaptive regression is then converted to a corresponding time-variant analysis in the frequency domain (see also Sect. 1.3.8). The authors have shown that while the frequency-domain analysis can detect a change in the risk premium preference during the sample period, the time-domain analysis does not present this feature.

While Chaps. 20 and 21 apply the state-space approach to *linear Gauss-Markov* time series, Chap. 22 generalizes the application to a *nonlinear* and *non-Gaussian* time series. Given such a general nonlinear system, the estimation problem concerning tracking the state variables becomes more difficult. The simple linear Kalman filter can no longer be applied. Under this situation, the frequently used approach is the method known to econometricians as the *extended Kalman filter*, which linearizes the state and observation equation using a Taylor series. However, that is not the way the authors use. Instead, they introduce the alternative known method of *particle filters* or *sequential Monte Carlo methods.*

Monte Carlo methods are very flexible in the sense that they do not require any assumptions about the probability distributions of the data. Moreover, experimental evidence suggests that these methods lead to improved results. From a Bayesian perspective, sequential Monte Carlo methods allow one to compute the posterior probability distributions of interest on-line. Yet, the methods can also be applied within a maximum likelihood context. As a result, they have been applied to solve a large number of interesting real problems, including econometrics problems, as shown in [1.52].

Particle filtering approximates the posterior densities using swarms of points in the sample space of nonlinear non-Gaussian dynamic models. These points are called "particles." The posterior distribution can then be approximated by a discrete distribution which has support on each of the particles. A particle cloud representation of the filtering distribution evolves through time using *importance sampling* and *resampling* ideas. Chapter 22 presents a brief introduction to particle filters, and more advanced materials can also be found in [1.20].

The state-space approach is used to model the international business cycle transmission between Germany and the US in Chap. 22. The authors use the *random walk plus noise* model to characterize the industrial production of these two countries. Furthermore, to be able to capture the mechanism of the business cycle, the authors further add to this model a *Markov switching process* with two states, one bullish and the other bearish. This is an interesting state-constrained problem. However, the state variables are not independent between the two countries. In fact, the transmission between them is represented by an interdependent first-order Markov transition matrix. The observation variable in this model is industrial production, whereas the state variables are the hidden trend (random walk), the state of the economy (boom or recession) and many unknown parameters. The purpose here is to keep track of the joint likelihood of the state of the two economies and their associated transmission mechanism. Particle filters are applied to serve this very objective.

1.6 Application Domains

The third issue addressed in this volume encompasses the application domains. What are the concrete economic and financial applications of CI? In which specific areas would economists or finance people benefit from CI tools the most? Based on what has been reviewed in Sect. 1.2 – 1.5, the chapters included in this volume cover four major areas of economic and financial applications, namely, forecasting, trading, classifications, and optimization; these can be concisely tabulated. In fact, while the table of contents of this volume is organized based on the tools employed, for some readers it may be even more useful to reorganize the materials according to the application domains. The seven application areas are tabulated as follows:

1. **Data Preprocessing:** Chaps. 5, 10.
2. **Features Selection and Classification:** Chaps. 4, 7, 8, 9, 10, 11, 18.
3. **Forecasting:** Chaps. 2, 5, 7, 12, 13, 14, 15, 17.
4. **Trading:** Chaps. 8, 11, 14, 19.
5. **Early Warning Systems:** Chaps. 6, 10, 18.
6. **Econometric Analysis:** Chapts. 15, 20, 21, 22.
7. **Optimization:** Chaps. 3, 16, 17, 19.

Another way of classifying, enabling readers to have a quick grasp of the economic and financial applications of CI, is to directly give a menu containing the real data used in the chapters, organized as follows:

1. **Stock Markets:**
   - the daily five-minute transaction data of futures and cash prices from October 1, 1998 to December 31, 1999 in Taiwan (Chap. 5)
   - stock index futures and government bonds from the Chicago Mercantile Market (Chap. 7)
   - daily closing prices of Taiwan stock indices (TAIEX) (Chap. 8)
   - the closing prices of the Dow Jones Stock Index, S&P 500, IBM and Microsoft (Chap. 10)
   - New York Stock Exchange index (Chap. 11)
   - the 1-minute Hang Seng stock index (Chap. 13)
   - stocks in the Russell 2000 and S&P 500 SmallCap indices plus all domestic stocks between the high and low market capitalization ranges of these two indices (Chap. 19)
   - closing prices of 86 trading stocks of Hong Kong from HSI, HSCCI, and HSCEI (Chap. 20)
2. **Foreign Exchange Markets:**
   - daily exchange rates between the US dollar and the Australian dollar (Chap. 12)
   - the tick-by-tick $ECU/$US exchange rate (Chap. 13)
   - exchange-rate data of nine currencies (Chap. 14)
3. **Interest Rates:**
   - the UK 2-year interest rate, the UK 10-year interest rate, the UK base rate, the German 2-year interest rate, and the US 2-year interest rate over the period 1982–1998 (Chap. 21)
4. **Industries and Companies:**
   - the balance sheet and income and loss statement data on the enterprises of nine EMU countries from the BACH (Bank for the Accounts of Companies Harmonized) project database (Chap. 9)
   - companies' financial ratios from Moody's industrial manuals (Chap. 18)
5. **Macroeconomic Data:**
   - domestic credit in real terms, money supply (M2), international reserves, inflation, oil prices, stock indices, exchange rates (Chap. 6)
   - money supply (M2), divisia, consumer price index (Chap. 17)
   - the monthly industrial production indices (seasonally adjusted) of the US and Germany from January 1961 to December 2000 (Chap. 22)

1.7 Performance Evaluation

Fourth, do we see the successful applications of Computational Intelligence (CI) in economics and finance? To what extent can we rightfully claim such a success? Specifically, how can we evaluate the performance or the effectiveness of CI in economic and financial modeling?

1.7.1 Prediction

Chapter 14 specifically provides a basic framework to evaluate the forecasting and trading performance of CI tools. The framework consists of the choices of *benchmarks* and *evaluation criteria*. The usual benchmark chosen to play against CI forecasting tools is the *random walk*. Nonetheless, other simple or canonical models can also be included as well, such as the linear time series model (the ARIMA model) or a vanilla version of an artificial neural network.

Depending on the application, forecasting or trading, performance criteria are also different. For forecasting, what concerns us is the *prediction accuracy*, which is measured using error statistics, such as the mean square error (MSE), root mean square error (RMSE), mean absolute percentage error (MAPE), root mean squared percentage error (RMSPE), Theil's U, and the hit ratio. In addition to these primitive statistics, more advanced statistical tests have been proposed in recent years to give a more rigorous measure of forecasting performance. Two of them are seen in Chap. 14, the *Diebold-Mariano Test* and the *Pesaran-Timmerman Test*. The first test is motivated by a loss-function-independent measure, whereas the second test is triggered by the issue of data snooping. To enhance the credibility of CI tools, the use of these advanced statistical tests is certainly very crucial.

1.7.2 Classification

For classification problems, the usual benchmarks are discriminate analysis, logit, and probit. Criteria involved are the typical Type-I error (α), the Type-II error (β), the overall accuracy rate (or the total probability of misclassification), and some mathematical combination of all of these criteria. Sometimes, it is also important to see whether the classification rules are *comprehensive* or easy to understand, rather than being just a black box.

Chapter 4 in this volume conducts a rigorous experimental procedure to evaluate the merits of rough sets over the conventional statistical tools in classification. The authors consider two types of discriminate analysis, both linear and quadratic, and logit analysis as the benchmarks. What makes this chapter particularly interesting is that the data applied to the performance evaluation are all *artificial*. The data are generated from four types of *probabilistic models*, including multivariate normal, uniform, exponential, and log-normal models. The fundamental design together with the control of other parameters covers five important aspects of multivariate data in the real applications, namely, symmetry, normality, dispersion, correlation, and overlap. Furthermore, the authors have further complicated the task with two other factors: different sizes of training samples and different numbers of groups.

Given the seven factors, a seven-way analysis of variance was conducted, and it was found that all seven factors are important in accounting for the misclassification rates. Using Tukey's tes, the authors are able to show that the general performance of the rough sets approach is superior to that of the other three statistical procedures. Detailed analysis from Tukey's test further shows that the rough sets approach provides consistently lower error rates both when small and larger training samples are employed. The rough sets approach also has some advantages when the data originate from asymmetric distributions. In conclusion, even though the rough sets approach does not perform uniformly well over all combinations of the seven factors, its performance seems to be robust under different scenarios. This feature is particularly attractive because it implies that the rough sets approach can work with a larger class of data environments.

A similar evaluation procedure is also seen in Chap. 18, where the authors examine the performance of the ant algorithm in business failure prediction. In addition to the conventional benchmarks, the authors also include genetic programming and the recursive partition algorithm as competitors. The results support the superiority of ANT over the alternatives.

In Chap. 6, a criterion called the *noise/signal ratio* (NSR) is also used. The NSR is a function of α and β and is defined as:

$$NSR = \frac{\alpha}{1-\beta}. \tag{1.53}$$

Based on this criterion, it is found that the Support Vector Machine (SVM) has a lower NSR than the multivariate probit and the signal approach had. With this evidence, it is claimed that SVM is claimed to be superior to the conventional tools.

1.7.3 Trading

As to the evaluation of the CI trading tools, the buy-and-hold strategy is still considered to be a highly competitive alternative. This benchmark is used in Chap. 8. At a more advanced level, technical trading rules, such as the moving average rules and the filter rules, can also be included.

As for criteria, what concerns users are profits or, more precisely, *risk-adjusted profits*. A primitive criterion is the mean annual total return. This criterion can be strengthened with a series of considerations, such as transaction costs and risk. The two criteria used in Chap. 14, which take these factors into account, are the mean annual net return and the Sharpe ratio. In addition, the ideal profit ratio is also employed by them to see how far away the CI tool is from the best that one may possibly have. The *excess Sharpe ratio* is used in Chap, 19. The issue dealt with in this chapter is not only the market-timing decision but also the portfolio optimization problem. Criteria used to evaluate the performance of portfolios can be even more complicated than those used to evaluate market-timing decisions. This is because of the involvement of regulatory constraints as added to the portfolio construction. In practice, one has also to take into account individual investor requirements, and nontrivial risk and subjective quality measures.[26] Chapter 19 is a good case in point.

Unlike the previous criteria, which basically provide just an evaluation based on a snapshot, the equity curve allows users to trace the whole history of a trading performance. This criterion is used in Chap, 8. Based on the analysis of various equity curves, the authors find that the SOM-derived charts transmit profit signals, and simple trading strategies built from these charts can beat the buy-and-hold strategy.

1.7.4 Function Optimization

In the application domain of function optimization, the general evaluation procedure is to select testbeds with the best-so-far or some frequently used algorithms, and then evaluate the proposed algorithm accordingly. This is exactly what Chap. 16 does. Given the policy function problem in a one-sector growth model, the authors compared their proposed evolutionary strategy neural networks (ESNNs) with two popular approaches used in the same problem, namely, the *linear quadratic approximation* (LQA) method and the *parameterized expectation* (PE) method. The respective performances are based on the utility attained from the policy function, and the deviation from the first-order condition (the Euler equation). It is found that in all cases, be they deterministic or stochastic ones, ESNNs perform the best among the three methods.

In Chap. 19, the simulated annealing is evaluated with the benchmark, the gradient descent method. It is found that the portfolio selection based on the former outperforms the gradient maximization method as well as the market index in excess returns, namely, the Sharpe ratio and the turnover ratio.

1.8 Concluding Remarks: Agent-Based Models

We do not intend to leave the readers with the impression that this book merely contains just *many* CI tools in the tool box. One recent research trend in machine learning and computational intelligence is the extensive use of the hybrid system, in which many tools work synergetically together as a *multi-agent system*. Examples abound in this volume as well. For example, grey models are useful to help preprocess the data used as the input to the neural nets. Another example is that the evolutionary strategies can indeed

[26] For example, [1.7] has shown how financial regulation can complicate the objective function and constraints in the case of a pension fund; [1.49] exemplified how the short-run fluctuation in excess returns and volatility can modify the standard problem into a more difficult tactical asset allocation problem. Both suggested the use of GAs to tackle the optimization problem.

enhance the determination of a network topology. The genetic algorithms can be applied to optimizing the ant system. The Self-organizing maps can help us to read wavelets. Finally, the fuzzy logic and rough sets can reduce the complexity of the decision tree while enhancing its power of expression. Furthermore, there are also some potential techniques that remain unexploited in this volume; nevertheless, they are worth a try! For example, fuzzy logic can be applied to the discrete time series model by fuzzifying the real-valued time series, and transforming the real-value time series into a discretized fuzzy time series.

Building an integrated framework of CI tools can be achieved in two ways. The first one is motivated by the need to manage the data warehouse. It is, therefore, desirable to organize all CI tools as a team, as a multi-agent system, for effective knowledge discovery and data mining. Chapter 19 provides a nice view on how these systems look. It would never be an overstatement to explain the significance of CI tools in the development of information technology. Chapter 23 goes further and introduces a 3C-DRIVE framework to address the business value created by the adoption of information technology in an era of electronic commerce reform.

Alternatively, an integrated treatment of CI tools can also be motivated by the *agent-based simulation*, currently known as *agent-based computational economics* and *agent-based computational finance*, or, more broadly, *agent-based computational social sciences*. Chapter 23 presents an agent-based simulation of information technology management in a competitive industry. Agents (company managers) differ in their risk attitudes in information technology adoption. The agent-based simulation developed by the authors, Yasuo Kadono and Takao Terano, is very similar to the well-known *Sugarscape* introduced by Joshua Epstein and Robert Axtell in the early stage of agent-based computational social sciences [1.29].

It was a temptation to include more agent-based simulation studies in this volume. In fact, the conferences **CIEF'2000** and **CIEF'2002** received many submissions on this very subject. However, due to the pages constraint and the focus of the book, the editors were forced to make an unpopular choice. Instead, there will be a special issue of the journal *Information Sciences* that will present more on the remarkable influence of computational intelligence on agent-based computational economics and finance. The special issue of *Information Sciences* and the 23 chapters in this volume viewed aggregately indeed represent a pretty good picture of the subject "computational intelligence in economics and finance."

Even so, the editors were hesitant to "stop" here mainly because there are still many important CI disciplines yet to be covered. The future realization of their potential should not be overlooked. Among them, the editors would like to single out in particular the methodologies of *independent component analysis*, *reinforcement learning*, *inductive logical programming*, *classifier systems*, and *Bayesian networks*, not to mention many ongoing and highly fascinating hybrid systems. A way to make up for their omission is to revisit this subject later. We certainly hope that we can do so in the near future. We shall close this introductory tour by welcoming from our prospective readers any comments and suggestions for further improvements.

References

1.1 Aha, D. (1997): Lazy Learning. Kluwer

1.2 Aoki, M. (1987): State Space Modeling of Time Series. Springer-Verlag

1.3 Armano, G., Murru A., Marchesi, M. (2002): NXCS—A Hybrid Approach to Stock Indexes Forecasting. In: Chen, S.-H. (Ed.), Genetic Algorithms and Genetic Programming in Computational Finance, Kluwer

1.4 Aussem, A., Campbell, J., Murtagh F. (1998): Wavelet-Based Feature Extraction and Decomposition Strategies for Financial Forecasting. Computational Intelligence in Economics and Finance, Vol. 6, No. 2, 5–12

1.5 Azoff, M. (1994): Neural Network Time Series: Forecasting of Financial Markets. Wiley

1.6 Baestaens, D.-E., Van Den Bergh, W., Wood D. (1994): Neural Network Solutions for Trading in Financial Markets, Pitman

1.7 Baglioni, S., Sorbello, D., Pereira, C., Tettamanzi, A. G. B. (2000): "Evolutionary Multiperiod Asset Allocation," In: Whitley, D., Goldberg, D., Cantú-Paz, E., Spector, L., Parmee, I., Beyer, H. -G. (Eds.), Proceedings of the Genetic and Evolutionary Computation Conference, 597–604. Morgan Kaufmann

1.8 Balakrishnan, A. (1987): Kalman Filtering Theory. Optimization Software

1.9 Balakrishnan, A. (1988): State Space Theory of Systems. Optimization Software

1.10 Bauer, R., Jr. (1994): Genetic Algorithms and Investment Strategies. Wiley

1.11 Billot, A. (1995): Economic Theory of Fuzzy Equilibria: an Axiomatic Analysis. Springer-Verlag, 2nd edition

1.12 Bojadziev, G., Bojadziev, M., Zadeh, L. (1997): Fuzzy Logic for Business, Finance, and Management. World Scientific

1.13 Boguslavskij, I. (1988): Filtering and Control. Optimization Software

1.14 Bonabeau, E., Meyer, C. (2002): Swarm Intelligence: a Whole New Way to Think About Business. Harvard Business School Press

1.15 Chang, K., Osler, C. (1994), "Evaluating Chart-Based Technical Analysis: the Head-and-Shoulders Pattern in Foreign Exchange Markets," Working Paper, Federal Reserve Bank of New York

1.16 Chen, S.-H. (2002a): Evolutionary Computation in Economics and Finance, Physica-Verlag

1.17 Chen, S.-H. (2002b): Genetic Algorithms and Genetic Programming in Computational Finance, Kluwer

1.18 Chen, T., Chen, H. (1995): Universal Approximation to Nonlinear Operators by Neural Networks with Arbitrary Activation Functions and Its Application to Dynamical Systems. IEEE Transactions on Neural Netwroks, Vol. 6, 911–917

1.19 Cordón, O., Herrera, F., Hoffmann, F., Magdalena, L. (2001): Genetic Fuzzy Systems. World Scientific

1.20 Crisan, D. (2001): Particle Filters—a Theoretical Perspective. In: Doucet, A., Freitas, N., Gordon, N. (2001) (Eds.), Sequential Monte Carlo Methods in Practice, Springer-Verlag, 17–41

1.21 Cristianini, N., Shawe-Taylor, J. (2000): An Introduction to Support Vector Machines and Other Kernel-Based Learning Methods. Cambridge University Press
1.22 Deboeck, G., Kohonen, T. (1998), Visual Explorations in Finance with Self-organizing Maps, Springer-Verlag
1.23 Deng, J. (1982): Control Problems of Grey System. System and Control Letters, No. 5, 288–294
1.24 Deng, J. (1989): Introduction to Grey System Theory, Journal of Grey System, Vol. 1, No. 1, 1–24
1.25 Dorigo, M. (1992): Optimization, Learning and Natural Algorithms, Ph.D. Thesis, Politecnico di Milano, Italy, in Italian
1.26 Dorigo, M., Maniezzo, V., Colorni, A. (1996): The Ant System: Optimization by a Colony of Cooperating Agents, IEEE Transactions on Systems, Man, and Cybernetics, Part B, Vol. 26, No. 1, 29–41
1.27 Duffy, J., McNelis, P. D. (2001): Approximating and Simulating the Stochastic Growth Model: Parameterized Expectations, Neural Networks, and the Genetic Algorithm. Journal of Economic Dynamics and Control **25(9)**, 1273–1303
1.28 Eberhart, R., Simpson, P., Dobbins, R. (1996): Computational Intelligence PC Tools. AP Professional
1.29 Epstein, J., Axtell, R. (1996): Growing Artificial Societies: Social Science from the Bottom Up. MIT Press
1.30 Fan, J., Gijbels, I. (1996): Local Polynomial Modeling and Its Applications. Chapman & Hall.
1.31 Fischer, R. (2001): The New Fibonacci Trader: Tools and Strategies for Trading Success. Wiley
1.32 Fogel, D. (1995): Evolutionary Computation—Toward a New Philosophy of Machine Intelligence. IEEE Press
1.33 Fogel, D., Chellapilla, K., Angeline, P. (2002): Evolutionary Computation and Economic Models: Sensitivity and Unintended Consequences. In: Chen, S. H. (Ed.), Evolutionary Computation in Economics and Finance, Physica-Verlag, 245–269
1.34 Fogel, L. (1964): On the Organization of Intellect, Ph.D. Thesis, UCLA
1.35 Fogel, L. (1997): A Retrospective View and Outlook on Evolutionary Algorithms. In: Reusch, B. (Ed.), Computational Intelligence: Theory and Applications, 5th Fuzzy Days, Springer-Verlag, Berlin, 337–342
1.36 Fogel, L. J., Owens, A. J., Walsh, M. J. (1966): Artificial Intelligence through Simulated Evolution, Wiley
1.37 Garey, M., Johnson, D. (1979): Computers and Intractability, a Guide to the Theory of NP-Completeness. Freeman
1.38 Gately, E. (1996): Neural Networks for Financial Forecasting. Wiley
1.39 Gencay, R., Selcuk, F., Whitcher, B. (2001): an Introduction to Wavelets and Other Filtering Methods in Finance and Economics, Academic Press
1.40 Granger, C., Hatanaka, M. (1964): Spectral Analysis of Economic Time Series. Princeton
1.41 Goffe, W. (1996): SIMANN: A Global Optimization Alogorithm Using Simulated Annealing. Studies in Nonlinear Dynamics and Econometrics. Vol. 1, No. 3
1.42 Goffe, W., Ferrier, G., Rogers, J. (1992): Simulated Annealing: an Initial Application in Econometrics. Computer Science in Economics and Management. Vol. 5
1.43 Goffe, W., Ferrier, G., Rogers, J. (1994): Global Optimization of Statistical Functions with Simulated Annealing. Journal of Econometrics. Vol. 60, No. 1/2. January/February. 65–99

1.44 Goss. S., Aron. S., Deneubourg, J. L., Pasteels, J. M. (1989): Self-organized Shortcuts in the Argentine Ant. Naturwissenschaften 76, 579–581
1.45 Hampton, J. (1997): Rough Set Theory—The Basics (Part 1). Journal of Computational Intelligence in Finance, Vol. 5, No. 6, 25–29
1.46 Hampton, J. (1998): Rough Set Theory—The Basics (Part 2). Journal of Computational Intelligence in Finance, Vol. 6, No. 1, 40–42
1.47 Hampton, J. (1998): Rough Set Theory—The Basics (Part 3). Journal of Computational Intelligence in Finance, Vol. 6, No. 2, 35–37
1.48 Harvey, A. (1989): Forecasting Structural Time Series Models and the Kalman Filter. Cambridge
1.49 Hiemstra, Y. (1996): Applying Neural Networks and Genetic Algorithms to Tactical Asset Allocation, Neurove$t Journal, 4(3), 8–15
1.50 Keber, C. (2002): Collective Intelligence in Option Pricing—Determining Black-Scholes Implied Volatilities with the Ant Programming Approach, working paper, University of Vienna
1.51 Kirkpatrick, S., Gelatt, C., Vecchi, M. (1983), Optimization by Simulated Annealing, Science, Vol. 220, 671–680
1.52 Kitagawa, G., Sata, S. (2001): Monte Carlo Smoothing and Self-organizing State Space Model. In: Doucet, A., Freitas, N., Gordon, N. (2001) (Eds.), Sequential Monte Carlo Methods in Practice, Springer-Verlag, 177–196
1.53 Kohonen, T. (1982): Self-organized Foundation of Topologically Correct Feature Maps. Biological Cybernetics **43**, 59–69
1.54 Lin, C. -T., Yang, S. -Y. (1999): Selection of Home Mortgage Loans Using Grey Relational Analysis. Journal of Grey System, Vol. 11, No. 4, 359–368
1.55 Lin, C. -T., Chen, L. -H. (1999): A Grey Analysis of Bank Re-decreasing the Required Reserve Ratio. Journal of Grey System, Vol. 11, No. 2, 119–132
1.56 Lin, C. -T., Chang, P. -C. (2001): Forecast the Output Value of Taiwan's Machinery Industry Using the Grey Forecasting. Journal of Grey System, Vol. 13, No. 3, 259–268
1.57 Lo, A., Mamaysky, H., Wang, J. (2000): Foundations of Technical Analysis: Computational Algorithms, Statistical Inference, and Empirical Implementation, Journal of Finance, Vol. LV, No. 4, 1705–1765
1.58 Mansur, Y. (1995): Fuzzy Sets and Economics: Applications of Fuzzy Mathematics to Non-cooperative Oligopoly. Edward Elgar
1.59 Marks, R. (2002): Playing Games with Genetic Algorithms. In: Chen, S. H. (Ed.), Evolutionary Computation in Economics and Finance, Physica-Verlag, 31–44
1.60 Messier, W. F., Hansen, J. V. (1988): Inducing Rule for Expert System Development: an Example Using Default and Bankruptcy Data. Management Science **34**, 1403–1415
1.61 Moózek, A., Skabek, K. (1998): Rough Sets in Economic Applications. In: Polkowski, L., Skowron, A. (Eds.), Rough Sets in Knowledge Discovery 2: Applications, Case Studies and Software Systems, Physica-Verlag. Chap. 13
1.62 Mortensen, R. (1987): Random Signals and Systems. Wiley
1.63 Osler, C., Chang, K. (1995): Head and Shoulder: Not Just a Flaky Pattern. Staff Report No. 4, Federal Researve Bank of New York
1.64 Pan, Z., Wang, X. (1998): Wavelet-Based Density Estimator Model for Forecasting. Computational Intelligence in Economics and Finance, Vol. 6, No. 1, 6–13
1.65 Packard, N. (1990): A Genetic Learning Algorithm for the Analysis of Complex Data, Complex Systems 4, No. 5, 543–572
1.66 Pedrycz, W. (1997): Computational Intelligence: An Introduction. CRC Press
1.67 Peray, K. (1999): Investing in Mutual Funds Using Fuzzy Logic. CRC Press

1.68 Quinlan, R. (1986): Induction of Decision Trees. Machine Learning 1(1), 81–106
1.69 Quinlan, R. (1987): Simplifying Decision Trees. International Journal of Man-Machine Studies 27(3), 221–234
1.70 Quinlan, R. (1993): C4.5: Programs for Machine Learning. Morgan Kaufmann
1.71 Rechenberg, I. (1965): Cybernetic Solution Path of an Experimental Problem. Royal Aircraft Establishment, Library Translation No. 1122, August, Farnborough, UK
1.72 Refenes, A. -P. (1995): Neural Networks in the Capital Markets. Wiley
1.73 Refenes, A. -P., Zapranis, A. (1999): Principles of Neural Model Identification, Selection and Adequacy: with Applications in Financial Econometrics. Springer
1.74 Shadbolt, J., Taylor, J. (2002): Neural Networks and the Financial Markets—Predicting, Combining, and Portfolio Optimisation. Springer
1.75 Schmertmann, C. P. (1996): Functional Search in Economics Using Genetic Programming. Computational Economics **9(4)**, 275–298
1.76 Schwefel, H. -P. (1965): Kybernetische Evolution als Strategies der Experimentellen Forschung in der Strömungstechnik. Diploma Thesis, Technical University of Berlin
1.77 Schewfel, H. -P. (1995): Evolution and Optimum Seeking, Wiley
1.78 Skalkoz, C. (1996): Rough Sets Help Time the OEX. Neuralve$t Journal, Nov./Dec., 20–27
1.79 Slowinski, R., Zopounidis, C. (1995): Applications of the Rough Set Approach to Evaluation of Bankruptcy Risk. International Journal of Intelligent Systems in Accounting, Finance and Management **4**, 27–41
1.80 Smithson, M. J. (1987): Fuzzy Set Analysis for Behavioral and Social Sciences. Springer-Verlag, New York
1.81 Sugeno, M., Yasukawa, T. (1993): A Fuuzy-Logic-Based Approach to Qualitative Modeling. IEEE Transactions on Fuzzy Systems, Vol. 1, 7–31
1.82 Suykens, J., Vandewalle, J. (1998): The K.U. Leuven Time Series Prediction Competition. In: Suykens, J., Vandewalle, J. (Eds.), Nonlinear Modeling: Advanced Black-Box Techniques, Kluwer, 241–253
1.83 Takagi, T., Sugeno, M. (1985): Fuzzy Identification of Systems and Its Applications to Modeling and Control. IEEE Transactions on Systems, Man, and Cybernetics, Vol. 15, 116–132
1.84 Tay, N., Linn, S. (2001): Fuzzy Inductive Reasoning, Expectation Formation and the Behavior of Security Prices. Journal of Economic Dynamics and Control, Vol. 25, 321–361
1.85 Taylor, P., Abdel-Kader, M., Dugdale, D. (1998): Investment Decisions in Advanced Manufacturing Technology—A Fuzzy Set Theory Approach. Ashgate
1.86 Thomason, M. R. (1997): Financial Forecasting with Wavelet Filters and Neural Networks. Computational Intelligence in Economics and Finance, Vol. 5, No. 2, 27–32
1.87 Trippi, R. R., Turban, E. (1993): Neural Networks in Finance and Investing. Irwin
1.88 Tu, Y.-C., Lin, C.-T., Tsai, H.-J. (2001): The Performance Evaluation Model of Stock-Listed Banks in Taiwan—By Grey Relational Analysis and Factor Analysis. The Journal of Grey System, Vol. 13, No. 2, 153–164
1.89 Vapnik, V. (1998a): Statistical Learning Theory. Wiley
1.90 Vapnik, V. (1998b): The Support Vector Method of Function Estimation. In: Suykens, J., Vandewalle, J. (Eds.), Nonlinear Modeling: advanced Black-Box Techniques, Kluwer, Boston, 55–85
1.91 Von Altrock, C. (1996): Fuzzy Logic and Neurofuzzy Applications in Business and Finance. Prentice Hall

1.92 White, H. (1988): Economic Prediction Using Neural Networks, the Case of IBM Daily Stock Returns. Proceedings of IEEE International Conference on Neural Networks, Vol. 2, IEEE, New York, 451–458

1.93 White, H. (1992): Artificial Neural Networks—Approximation Theory and Learning Theory. Blackwell

1.94 Wolberg, J. (2000): Expert Trading Systems: Modeling Financial Markets with Kernel Regression. Wiley

1.95 Zirilli, J. (1996): Financial Prediction Using Neural Networks. International Thomson Publishing

1.96 Zopounidis, C., Pardalos, P., Baourakis, G. (2002): Fuzzy Sets in Management, Economy & Marketing. World Scientific

Part II

Fuzzy Logic and Rough Sets

2. Intelligent System to Support Judgmental Business Forecasting: the Case of Estimating Hotel Room Demand

Mounir Ben Ghalia[1] and Paul P. Wang[2]

[1] Engineering Department, University of Texas-Pan American, Edinburg, TX 78539 USA
email: benghalia@panam.edu

[2] Electrical and Computer Engineering Department, Duke University, Durham, NC 27708 USA
email: ppw@ee.duke.edu

Forecasting is an instrumental tool for strategic decision- making in any business activity. Good forecasts can reduce the uncertainty about the future and, hence, help managers make better decisions. Virtually all statistical forecasting techniques depend on the continuity of historical data and time series and may not predict a discontinuous change in the business environment. Often times, this discontinuity is known to managers who then must rely on their judgment to make forecast adjustments. In this paper, we discuss the role of judgmental forecasting and take the problem of estimating future hotel room demand as a practical business application. Next, we propose IS-JFK: an intelligent system to support judgmental forecasting and knowledge of managers. To account for vagueness in the knowledge elicited from managers and the approximate nature of their reasoning, the system is built around fuzzy IF-THEN rules and uses fuzzy logic for decision inference. IS-JFK supports two methods for forecast adjustments: 1) a direct approach and 2) an approach based on fuzzy intervention analysis. Actual data from a hotel property are used in some case-scenario simulations to illustrate the merits of the intelligent support system.

2.1 Introduction

Judgment has received considerable attention by cognitive psychologists interested in human reasoning and decision-making [2.19] [2.21] [2.32] [2.35] [2.2]. In statistics and economics, human judgment of subjective probabilities has been studied in relation to decision theory [2.24] [2.25]. This paper investigates the role of human judgment in business forecasting and takes the problem of estimating future hotel room demand as a practical example.

All statistical techniques used for forecasting require a series of historical data that can be used in computing the forecast [2.16]. These techniques depend on the continuity of historical data and may not predict a discontinuous change in the business environment; for example, forecasting the room demand for a period that coincides with the opening of a new competitor in the same area. In such situations, while historical data might not reflect the impact of the new competitor on room demand, the hotel manager can come up with a forecast based on subjective judgments.

Many researchers and practitioners agree that judgment should play "an" important role in forecasting [2.30]. However, there is still a debate about how this role should be played and to what extent [2.4]. Butler et al. [2.7] suggest that statistical approaches ought to be used as tools to provide a first approximation of forecasts using historical data. The role of judgment is to "massage" these forecasts to take into account influencing factors not included in the historical data. Makridakis and Wheelwright [2.26] argue that it is difficult to say which method provides the best forecast, a quantitative/statistical method, or a method based on judgment. However, they recognize that human forecasters are capable of processing much more information than most of the quantitative methods. The advantage of human forecasters is that most often they know about specific events that are likely to affect the forecasts.

On the same subject, Makridakis et al.[2.27] point out that each method has its own limitations. Judgmental forecasting approaches are prone to suffer from biases, while quantitative methods cannot cope with significant changes in the environment of organization. They discussed the possibility of combining judgmental and statistical approaches in order to obtain more accurate forecasts. One way of combining the two approaches is to give more weight to formal quantitative methods when there are no changes in the environment and more weight to judgmental methods when changes do occur.

A review of the published work reveals that there are still problems in deciding how the combination of statistical and judgmental approaches should be achieved. In particular, "the" role of judgment in the combined forecasting has not been made clear and issues of implementation have not been carefully addressed.

Human experts are expected to disagree with forecasts produced by statistical techniques for the following two main reasons: 1) when the events affecting the environment of the organization do not seem to be taken into account in the generated forecasts or 2) when the forecasts just do not "feel" right. Reason 1) implies that the experts know about specific events that are very likely to affect forecasts, but not incorporated in the model or the historical data used by the statistical methods. In such situations, experts exercise their judgment based on knowledge, inputs, and years of experience. Reason 2), on the other hand, suggests a judgment based on intuition, beliefs and expectation (and possibly based on experience as well). However, psychologists may argue that people have a tendency to generate forecasts which are consistent with their prior beliefs and expectations resulting in biases [2.13].

While nearly all previous studies have focused on "subjective" And "intuitive" judgment [Reason 2]], inadequate attention has been paid to "informed" judgment in forecasting [Reason 1)] and how it could be implemented in real-world applications. This study focuses on the role of informed

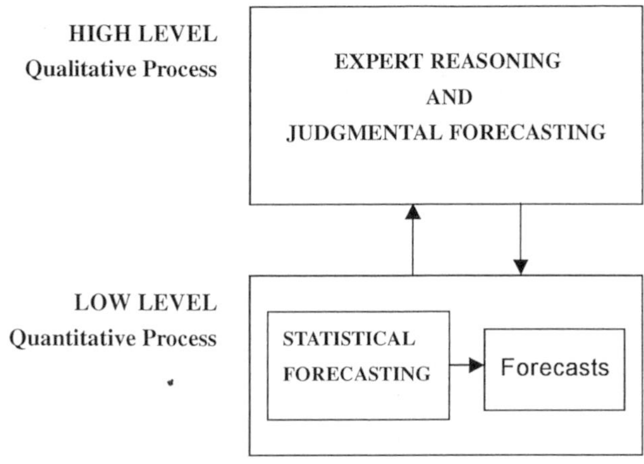

Fig. 2.1. Hierarchical structure for forecasting

judgmental forecasting and looks at issues of implementation for a real-world business problem.

In the business world where statistical approaches are used to generate forecasts, human experts (or managers) should be assigned a supervisory role to exercise their informed judgment in a two-level hierarchical structure as depicted in Fig. 2.1. They should be given the possibility of overriding the predictions of statistical and quantitative models. In this structure, statistical forecasting occupies the lower level and its output is monitored by a higher "intelligent" level representing expert reasoning and judgmental forecasting.

While this intelligent hierarchical structure could be used in almost any application where judgmental forecasting might play an important role, the specific details of implementation could differ from one problem to another. In this paper, we discuss the implementation of such structure to the problem of forecasting future hotel room demand. However, we hope that this study helps to inspire other real-world applications on the possibilities of considering judgmental forecasting.

Based on the hierarchical structure for forecasting shown in Fig. 2.1, this paper discusses the development of intelligent system to support judgmental forecasting and knowledge IS-JFK. The intelligent support system is not intended to eliminate the statistical techniques nor to completely automate the forecasting task. Instead, it is intended to aid managers in making systematic and consistent adjustments to demand forecast produced by statistical techniques.

To better understand the importance of forecasting in the hotel business, Section 2.2 reviews the hotel revenue management system and emphasizes the role of forecasting. In Section 2.3, we review the statistical techniques and discuss their limitations. In particular, we explain their use in forecasting hotel room demand. Since the intelligent system is intended to support hotel managers' judgmental forecasting, it is important to 1) explain the nature of judgment in relation to demand forecast; 2) show why supporting judgmental forecasting is important; and 3) explain how it can improve hotel revenue. We address these points in Section 2.4. IS-JFK is a knowledge-based system and an important part of this study has focused on knowledge engineering. The different methods used for knowledge acquisition are discussed in Section 2.5. When working on forecast problems, hotel managers are often faced with uncertainty and often have to work with incomplete knowledge and heuristics. Also, the decisions they make do not usually lead to conclusions which are absolutely certain. Section 2.5 addresses this point and discusses the motivation for representing the knowledge via fuzzy *IF-THEN* rules as well as choosing fuzzy logic as a tool for approximate reasoning. The structure of IS-JFK is then presented in Section 2.6. There are different ways of solving forecast problems in the hotel business. Hence, the knowledge has to be structured and represented to imitate the way it is structured and represented in the minds of hotel managers. In this study, two different approaches are developed to support managers in adjusting demand forecast. The first approach is referred to as the direct approach and is discussed in Section 2.7. The second approach makes use of *fuzzy intervention analysis* developed in this study and presented in Section 2.8. Both approaches are backed up by simulation results based on actual hotel data and presented in the respective sections. Finally, concluding remarks are presented in Section 2.9.

2.2 Revenue Management System for Hotels

Hotels belong to the category of capacity-constrained service firms that include airlines, car rental companies, and cruise line industries among others [2.22]. The units of capacity are the rooms forming the products to be sold. The demand for each room is an uncertain process and before any arrival date there is only a finite horizon over which the hotel can sell its rooms.

The demand market for hotel rooms is commonly split into two major segments: business travelers and leisure travelers. Leisure travelers tend to book early as vacations are planned ahead. They are usually price sensitive and may be flexible in the days of travel in the sake of a good bargain. Business travelers, on the other hand, tend to book late, often a few days before arrival. They are usually time sensitive and willing to pay higher rates. Within these two major market segments, hotels have introduced additional segments such as corporate travelers, government employees, American Automobile Association (AAA), American Association of Retired Persons (AARP), etc. Each market segment is associated with a specific discounted rate that is some percentage lower than the rack rate. A walk-in guest pays rack rate if he or she does not qualify for any of the available discounted rates. Convention guests usually pay a discounted rate negotiated in advance between the meeting planner and the hotel sales manager. The availability of a rate for each arrival date depends primarily on the specific market demand for that

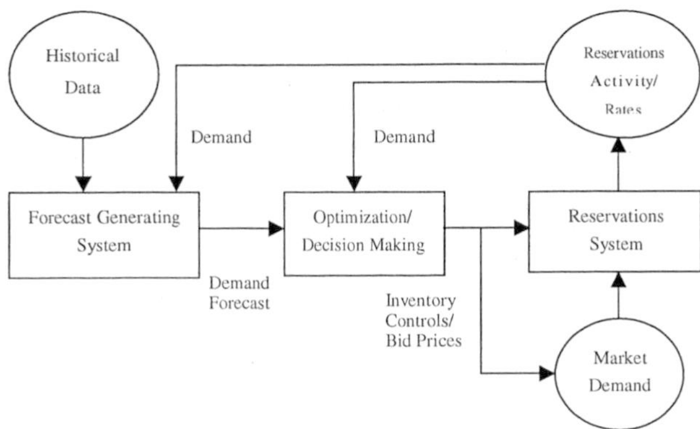

Fig. 2.2. Revenue management system

specific date. If high late-room demand is anticipated for an arrival date, then an appropriate number of rooms can be offered only to higher rate paying customers by eliminating the availability of all other lower discounted rates. This market segmentation results in hotel guests paying different rates for staying in similar rooms but seen as different products. This fact can also be seen in the airlines industry [2.1] and car rental industry [2.8]. However, one can argue that this is a business fact where all parties can be seen as winners. In fact, a hurried business traveler finds a room on short notice for a peak busy day, the vacation traveler finds a time when deep discounts are available, and the hotel manager achieves both higher occupancy and higher average rate.

Hotel managers often have to decide the best use of the fixed capacity of rooms and how to allocate room products to available demand from different market segments in such a way as to maximize total revenue over the finite sales horizon. Thus, the problem is to determine when to sell, how much to sell at what price, and to which market segment. Financial success is often coined with the management's ability to make an intelligent use of the fixed capacity.

A technique for maximizing revenues is known as revenue management, also referred to as yield management in the literature [2.22] [2.12]. A revenue management system applies basic economic principles to pricing and controlling rooms inventory with the objective of maximizing revenues. As shown in Fig. 2.2, a revenue management system (RMS) executes two main functions: forecasting and optimization. The forecasting system attempts to derive future demand using historical data and current reservations activity provided by the reservations system. Based on the demand forecast and the remaining capacity, the optimization function sets the appropriate controls on the rooms inventory. In many hotels, rooms inventory is controlled via bid prices. For each arrival date in the planning horizon, RMS computes an appropriate bid price. A reservation request can be accepted only if the rate requested is at least the bid price, otherwise it is denied.

The bid prices are designed to encourage maximum utilization of the hotel capacity while indicating a minimum revenue for each arrival day. Therefore, if demand exceeds available capacity, the optimization module increases the bid price to extract high-valued customers (higher rate paying customers) of the total demand. If demand is below available capacity, the optimization module reduces the bid price to stimulate demand market. Thus, prices at which hotel rooms are offered affect the demand for those rooms (Fig. 2.2).

2.2.1 Demand Forecasting as Part of RMS

Revenue management depends on accurate demand forecasts. In Fig. 2.2, forecasting is a part of a closed-loop RMS. Current forecasts affect the decisions made by the optimization stage and in turn those decisions are likely to change the forecasts. For example, if the decision made is to lower bid prices, then this may stimulate the demand forecasts. But if the decision made is to increase the bid prices, then the hotel may see a decline in demand from price-sensitive customers. Failing to see the forecasting as part of a closed-loop system may hinder the success of such system. A good forecast will result in better inventory optimization and management. Thus, the objective is to be able to obtain a good forecast consistently for all future days, rather than having an exact prediction of the room demand on some days.

2.2.2 Optimization

Demand forecasts for each market segment and remaining capacity provide inputs to a mathematical programming model that decides the bid price for each arrival day. Several researchers have developed different mathematical programming approaches to determine the optimal pricing and inventory controls [2.14] [2.15] [2.33]. Such methods include deterministic linear programming, probabilistic linear programming, and stochastic dynamic programming.

2.3 Forecasting Using Statistical Techniques

2.3.1 Statistical Forecasting Techniques

Several methods have been used for the purpose of forecasting data in a variety of business applications [2.29]. Different methods vary in the manner in which the historical data is modeled. Regressional methods seek to explain the data with one or more input variables and the model relates the data to the inputs with a set of coefficients [2.29]. A very popular forecast method is the Box-Jenkins approach to time-series modeling and forecasting [2.6]. Future values are forecasted as a function of past observations and errors. However, these models are very complicated and difficult to implement.

The simple exponential smoothing procedure forecasts future data based on past observations [2.10]. In this method, previous observations are discounted such that recent observations are given more weights while observations further in the past are given less weight.

A more general variation of the simple exponential smoothing procedure is the Holt-Winters method [2.34]. This method owes its popularity to the fact that it is very simple to implement and is comparable with any other univariate forecasting procedure in terms of accuracy [2.9]. The Holt-Winters method is an extension of the exponentially weighted moving average (EWMA) procedure [2.18]. The distinctive feature of the Holt-Winters procedure is that it incorporates linear trend and seasonality into the simple exponential smoothing algorithm [2.18]. The trend represents the direction in which the time series is moving, while the seasonality explains the effects of different seasons in the data.

2.3.2 Application to the Estimation of Hotel Room Demand

In [2.31], the Holt-Winters procedure was applied to the forecast of hotel room demand. The forecasted value of demand is comprised of two components: the long term and the short term forecasts (STFs). The long-term forecast (LTF) estimates the final demand for the different arrival dates/market segment combinations well in advance of the arrival dates. The demands for future arrival dates are estimated based on historical data. The forecast may be made as much as a year ahead of the arrival date. The STF on the other hand, estimates the final demand only after the hotel property starts receiving booking for an arrival day. Typically, most of the advance booking requests are received during the 60 days before the arrival day. The STF is an estimate of demand for future dates based on the actual advance booking activity. This is in contrast to the LTF, which estimates the demand entirely on the basis of the historical demand pattern. In estimating the demand, the STF uses the current reservations held and a booking profile that gives the historical fraction of demand that has been booked by this time [2.31].

The final forecast is a weighted combination of LTF and STF. LTF is the dominant component of the final forecast when the arrival day is far away from the processing day. Conversely, as the arrival day approaches, the STF becomes the dominant component, since the STF depends on the actual booking rate. The corresponding weights associated with each of the two forecasts also change as the arrival day approaches. Initially, when the arrival day is far away, the LTF weight is near unity, while the STF weight is almost zero. As the arrival day gets closer, the STF weight increases and the LTF weight decreases. Eventually, when the arrival day is very close, the STF weight will approach unity while the LTF weight will be near zero.

2.3.3 Limitations of Statistical Forecasting

The simulation results showed that while good forecasts are obtained from some days, there are occasions on which the algorithm does not perform satisfactorily [2.31]. The primary reason for this is that the forecast is based entirely on a quantitative model and hard data. It does not take into account external/non-random effects which may have influenced the demand, but known to the hotel manager. Also, the algorithm is unable to distinguish nonrecurring events.

Some interviews have been conducted with hotel managers at different hotels. One hotel manager gave an example of when statistical demand forecast can deviate considerably from the actual demand. One convention had been held at the hotel for several consecutive years and during the same time of the year, only to move the following year to a different location. When the convention was taking place at the hotel, there used to be a very high demand during the convention period. The hotel manager knows that the convention is not taking place at the hotel this year and, thus, a lower demand than the previous years is expected. Nevertheless, the computer system was still showing high demand for the convention period. The hotel manager explained the negative impact that this has had on his hotel revenue since the optimization stage was computing the values of its decision variables based on bad forecast. The only time the forecast starts to catch up with the new level of demand is during the short-term horizon when actual bookings start to accumulate in the reservations systems.

Several researchers have recognized the limitations of statistical forecasting techniques and that historical data can be a poor guide. To quote from Jenkins [2.20]: "The use of model presupposes that the statistical behavior of the future will be similar to the statistical behavior of the past. However, if it is believed that some future events are likely to be untypical of past behavior, then some scope is needed for making adjustments to the forecasts obtained from the model. This tuning is a judgmental matter and depends on "intelligence" not available to the model.

All the statistical techniques used for forecasting require a series of historical data that can be used in computing the forecast. These techniques depend on the continuity of historical data and may not predict a discontinuous change in the business environment as in the example described by the hotel manager. Often times, this discontinuity is known to managers whose forecast is likely to outperform the one generated by the statistical model. The hotel managers interviewed insisted that their experience and knowledge of the citywide events are valuable inputs to the forecast system. Over their years of experience, they have noticed that in many situations their forecast is more accurate than what the system suggests.

Although, hotel managers have the capability to override the system forecast by changing the forecast values, this action is not systematic. The override may have to be done continuously until the system catches up with the demand. In addition, a pure statistical forecast model cannot accommodate the wealth of information that the hotel staff possess about the impact of future events on room demand, which, for the hotel business, could represent a fine line between gain and loss of revenue.

2.4 Judgmental Forecasting in the Hotel Industry

2.4.1 What Do Hotel Managers Know?

A hotel manager who has worked at one same hotel for several years is able to give a rough estimate of future room demand for a particular day of the year (for example, room demand for a Tuesday of the third week of the year) based on his or her knowledge of what the demand was for the same day during the previous years. It is interesting to note that as in the case with statistical forecasting techniques, hotel managers also base their judgment on their knowledge of historical demand when they try to forecast future room demand. However, they process this information differently. There is an association between each day of the year and the room demand for that particular day. Each association constitutes a part of the knowledge stored in the hotel manager's memory. Studies in cognitive psychology suggest that knowledge is stored in memory forming an associative network [2.3] [2.11]. An outsider to the hotel business might be overwhelmed by the huge amount of information of which hotel managers are knowledgeable. In fact, hotel managers have to know the room demand not only at the aggregate level, but also at the market segment level. We asked one hotel manager how he possibly could remember the demand for rooms for a specific day of the year during the past years and how he uses that information to estimate what the demand is going to be for the same day this year. He explained that for every hotel, the year is divided into a certain number of seasons. His hotel, for example, has three main seasons. Each season is composed of several weeks. The mean average demand for rooms is almost similar for all weeks within one same season. The demand pattern for weekdays and weekends are also comparable from one week to week within the same season. Thus, there is less information to remember than what it might first seem. His knowledge of this demand pattern and the demand trend in his hotel coupled with his years of experience are his main inputs when making a judgment on what the future demand for rooms is likely going to be.

But historical data are not the only information that hotel managers rely upon when forecasting demand. In fact, the problem is more complicated than that. Beyond historical data, there are several mechanisms that influence the room demand forecast. Fig. 2.3 depicts a conceptual model that describes the environment of a hotel property and shows the variables that influence room demand. This model was developed after a series of discussions with

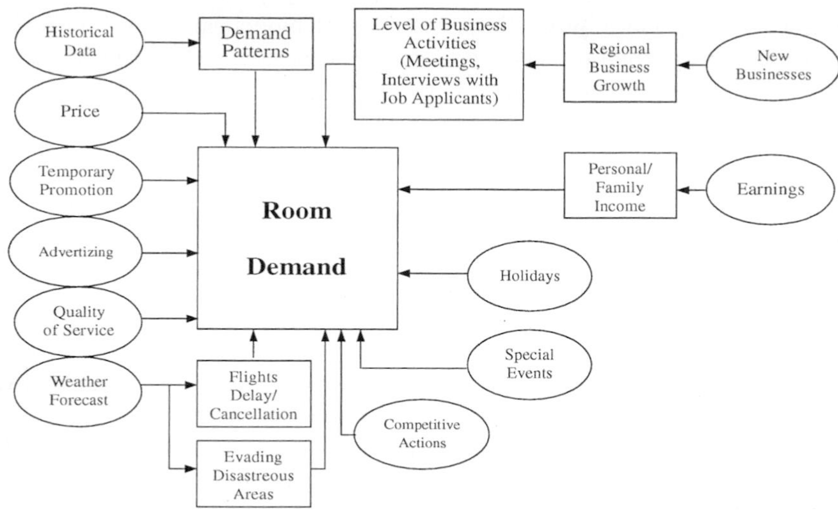

Fig. 2.3. Conceptual model describing mechanisms influencing hotel room demand forecasts

hotel managers who, in qualitative terms, explained to us the causal impact of changing events on room demand to be forecast.

The influencing variables can affect the long-term or the short-term room demand. Advertising and promotional activities, for example, have been seen to stimulate room demand. Advertising can be conducted at the national level for major hotel chains or at the regional level for smaller hotels. The hotel managers involved in this study expressed that when they run some promotions and advertise about their property at the regional level, they experience some increase in room demand. The adjustment of price also can stimulate demand. Special events such as the Super Bowl affect the demand for rooms for the period during which event takes place. The actions of competitors also affect room demand in many different ways. When a competitor lowers its room prices below the competition several weeks before a target period, then neighboring hotels may experience lower demand early in the planning horizon, but they also may expect an increase in room demand near the target period when the competitor is filled at capacity. Weather also has an effect on room demand. Bad weather can cause the demand for rooms to drop at resort hotels. Inclement weather in other regions may push residents to evade their cities and shelter in safer regions. For hotels near airports, weather causing delay and cancellation in flight may force people to stay at a hotel or check out a day or two later. The hotel sees this as an increase in demand. Hotels in regions with blooming businesses also expect the demand for their rooms to increase year after year as the increase in business activities leads to more meetings and job applicants coming for interviews.

For resort hotels, the national economy and the earnings of the populations is very likely to affect the demand for room (in negative or positive way). A rising economy encourages people to spend money and go on vacations in resort hotels.

Hotel managers are knowledgeable about the influence that these variables may have on room demand for their hotels during the different seasons. They also know when and how these variables may affect their businesses and to what extent. These variables are not available to the time-series models used by the statistical forecasting techniques. When hotel managers estimate future demand, they are already informed about the major variables that affect the demand. Their judgment is, therefore, very instrumental in reducing the uncertainties about the forecasts. While these variables are difficult to incorporate in statistical techniques, they seem to be easily incorporated in hotel managers' mind and used intelligently in judgmental forecasting. It is fair to say that unlike statistical forecasting techniques, hotel managers look at the big picture when deriving their own forecasts.

2.4.2 How Can Informed Judgmental Forecasting Help Increase Revenue?

The planning horizon for an arrival date in the future can extend to several months before that date. If accurate forecasts of room demand at the market segment level can be obtained earlier in the planning horizon, then optimal decisions can be made by selling the right number of rooms to the right market segment at the right time. Such optimal decisions ensure that the total revenue is maximized.

If there are unprecedented events that are expected to affect room demand, then accurate forecasts can be produced by taking into account the effect of such events. Forecasting based on historical data and statistical techniques cannot predict the occurrence of unprecedented events earlier in the planning horizon. The effect of such events may become apparent to statistical techniques only after a significant delay which can result in revenue opportunities being lost.

A hotel manager can exercise his judgment on what the demand forecast should be based on his knowledge of all the different influencing factors. Such factors can become apparent to the manager early in the planning horizon and, therefore, a positive impact on revenue can be made by adjusting forecasts earlier instead of waiting for the statistical forecasting techniques to catch up after a significant delay.

2.5 Knowledge Engineering for IS-JFK

In this study, much attention has been given to knowledge acquisition and representation (see Fig. 2.4). The knowledge acquisition can be defined as *the*

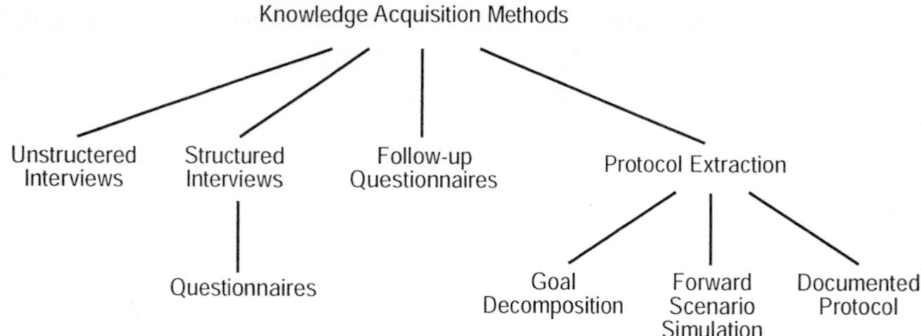

Fig. 2.4. Knowledge acquisition methods used in the study

transfer and transformation of problem solving expertise from some knowledge source to a program. Sources of knowledge include human experts (hotel managers and field managers (experienced personnel working with different hotel managers at different hotels)), textbooks, articles, and so forth. In this study, the effort has been directed toward identifying and interpreting expert knowledge from interviews with hotel managers followed directly from verbal and documented data to rules as a form of knowledge representation in IS-JFK. It is important to recognize that trying to elicit all these different forms of knowledge required a strong cooperation on the part of the hotel managers whose contribution has been instrumental for the success of this study. The knowledge acquisition technique employed for this project consisted of the steps discussed next.

2.5.1 Unstructured Interviews

Unstructured interviews conducted during first contacts with hotel managers helped to establish an initial map of the forecast and revenue management domains. In particular, the unstructured interviews provided a basis for preparing two types of questionnaires. The first questionnaire was designated to field managers and the second one to hotel managers. It was interesting to notice that a person who has been a manager for 20 years knows how to do things, but he may encounter difficulty relating the hows and whys of his knowledge to others. This is why several methods have been employed in this study to better elicit expert knowledge.

2.5.2 Structured Interviews

A structured interview with a field manager was conducted first. The purpose of this interview was threefold. First, the outcome of the interview helped to gain more insight into the revenue management domain, the different categories of hotels, and an overall account of the experience of the field manager as well as hotel managers with existing systems. Second, this interview provided information about a certain number of hotels that were designated as interview sites. Finally, the outcome of this interview helped to embellish the questionnaire designated for hotel managers. The embellished questionnaire was then used during structured interview sessions with two hotel managers. During the interview sessions, the hotel managers were guided to focus on one specific aspect of the revenue management system at a time.

2.5.3 Follow-up Questionnaires

After taking notes and analyzing the answers given by the hotel managers, a follow-up revised questionnaire was sent to each of the hotel managers to seek clarification on some of the answers and to obtain answers for some of the questions that were not asked during the structured interview sessions due to time constraint.

2.5.4 Protocol Extraction

We asked hotel managers to explicitly explain how some specific problems are solved. The main objective was to get insight into the reasoning and decision-making process of experienced hotel managers. The explication of hotel managers' problem-solving protocol helped to: 1) identify the important factors needed to be taken into consideration while reasoning about a specific situation and 2) understand the strategies of how these factors are applied in decision making. The following protocol extraction methods have been applied.

1. *Goal Decomposition and Forward Scenario Simulation:* We formulated several different problem scenarios and asked the managers to describe the chain of steps to be followed as the problems are being solved. The managers were also asked to specify the important decision factors and to explain their strategy.
2. *Documented Problem Solving:* In addition to the hypothetical problem scenarios, we wanted to examine the hotel managers' problem-solving strategies in actual situations and without our intervention. A technique for knowledge elicitation (called the observational approach) has been suggested in the literature[2.17]. The observational technique consists in observing and recording the behavior of the experts as they work on real problems in their normal working environment, in as unobtrusive a way as possible. However, this approach is time consuming and was deemed not effective for our study. As an alternative solution, we asked the hotel managers to participate in *documented problem solving weeks* during which they record their problem-solving steps and their daily strategic decisions on paper while working on actual problems. In particular, the

managers were asked to provide 1) a print out of the statistical forecasting results; 2) the nature of override, if any, made by the hotel managers; and 3) an explanation of why such an override was made. This form of knowledge that we tried to elicit is interesting and powerful because it provides us with *documented* knowledge of how hotel managers carry out their tasks and how they make their strategic decisions in real-world situations. This information was then sent to us for analysis.

2.5.5 Knowledge Analysis and Representation

The analysis of the knowledge acquired from the hotel managers revealed the following important points.

1. There are different ways of solving forecast problems in the hotel business. Therefore, the knowledge has to be carefully structured and represented in order to imitate the way it is structured and represented in the minds of hotel managers.
2. Managers frequently use linguistic values, such as high, low, and so on, to describe demand forecasts.
3. When working on forecast problems, hotel managers are often faced with uncertainty and often have to work with incomplete knowledge and heuristics. Nevertheless, they seem to be able to reason with vague and uncertain information and very often the decisions they make are the outcome of their approximate reasoning. They always strive to exercise their best judgment given the information that is available to them.

In light of this, the knowledge is represented by fuzzy *IF-THEN* rules. This is a natural representation of the knowledge acquired from hotel managers. In order to manage the uncertainty problem, the inference engine of the intelligent system uses a fuzzy logic-based approximate reasoning mechanism. The motivation for our approach comes from the fact that the theory of fuzzy sets offers an effective methodological framework to represent knowledge together with a reasoning mechanism when decisions need to be made based on uncertain and vague information[2.37].

Two different approaches are developed to support managers in adjusting demand forecast (Fig. 2.5): a direct approach for forecast adjustment and forecast adjustment via fuzzy intervention analysis. These approaches are discussed in the following two sections.

2.6 Intelligent System to Support Judgmental Forecast and Knowledge (IS-JFK)

The role of the intelligent system is to support managers to incorporate their knowledge and judgment in the forecast decision-making process. The system is not intended to eliminate the statistical techniques nor to completely

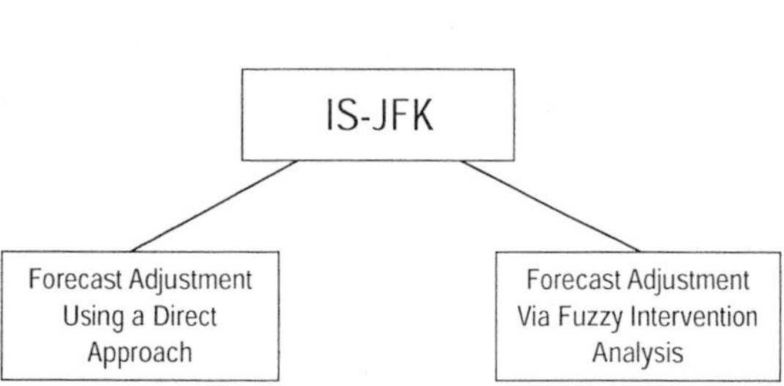

Fig. 2.5. The intelligent system IS-JFK supports two different approaches to adjust demand forecast. (1) direct approach. (b) approach based on fuzzy intervention analysis. The two approaches model the way hotel managers reason about the forecast problem in real-world situations

automate the forecasting task. Instead, the intelligent support system is intended to aid managers to systematically and consistently make adjustments to forecasts produced by statistical techniques. Without such support system, the only override to forecasts that managers can exercise is by changing the numerical values of forecasts. However, there are several serious limitations to such method. First, hotel managers feel uncomfortable working with what appears to be precise numbers. Their day-to-day operation involves working with qualitative reasoning and analysis and, therefore, overriding numbers with numbers does not come natural. Furthermore, often times hotel managers might know that some specific events are likely to cause an increase in room demand for certain days in the future, but cannot give a precise number. For instance, they can describe the effect of some events on demand as causing a slight increase, a huge increase in demand, a moderate increase, etc. In addition, there might be several events happening all at the same time and affecting room demand in different directions; for instance, some events may cause an increase in demand while some others may cause a decrease in demand. With a support system, hotel managers can input their knowledge about the different events affecting future arrival days and their judgment on what their effects are expected to be. The system will then compile all the entered information, process it and produce a final judgment of what the demand forecast should be.

For less experienced hotel managers, this IS-JFK can serve as a teaching tool for novice managers by showing them how to reason with different events and how to recognize their effect. A block diagram of IS-JFK is given in Fig. 2.6. The different components of the system are described next.

2.6.1 Fuzzification

Numerical forecasts are generated by statistical techniques. For the hotel revenue management application, managers can view demand forecasts for up to one year ahead. In the IS-JFK, the linguistic variable *forecast* takes on linguistic values. In the fuzzification stage, numerical values of forecasts are mapped into linguistic values described by fuzzy sets. This procedure is described in Section 2.7.

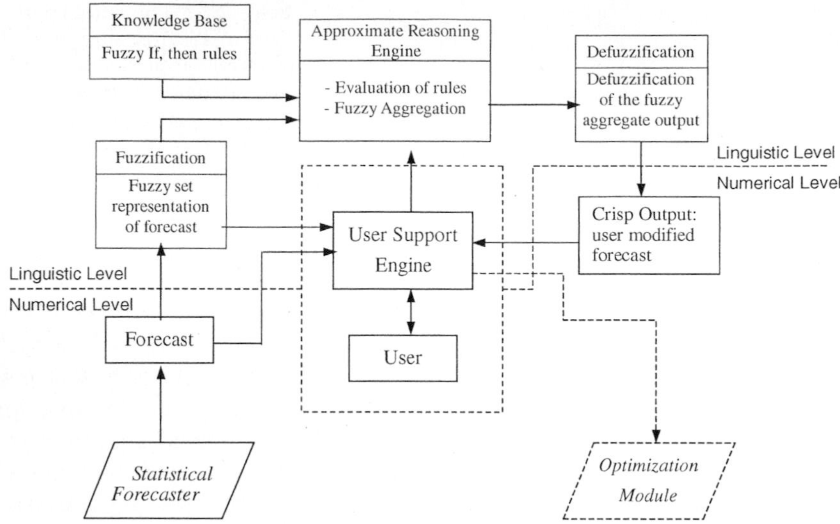

Fig. 2.6. Block diagram of the intelligent support system IS-JFK

2.6.2 Knowledge Base

The knowledge base consists of a set of fuzzy IF-THEN rules. These rules are acquired from experienced hotel managers through knowledge engineering as discussed in the previous section.

2.6.3 Database

Different hotel properties have different parameters (for example, hotel capacity). Also, statements like "*the forecast is high*" can mean different things for different hotels. To guarantee a consistency in all forecast adjustment at the same hotel property, a calibration session is required before the use of the system. The calibration parameters are stored in the database, which can be used by the reasoning engine and can be updated by the hotel manager.

2.6.4 User Support Engine

The user support engine (USE) interacts with the user (the manager). The manager is guided on how to enter his or her knowledge and can viewthe outcomeof the session through USE.

2.6.5 Approximate Reasoning Engine

Approximate reasoning engine (ARE) uses information from three different modules: the knowledge base, the inputs of the user via USE, and the fuzzified forecast generated by the statistical forecaster. The first function of ARE is to evaluate the production rules. Its second function is to aggregate the fuzzy outputs resulting from all firing rules into one fuzzy set. The flow of information between USE and ARE is unidirectional: inputs of the user are transferred to ARE via USE in order to be processed along with other information inputs to ARE.

2.6.6 Defuzzification

The defuzzification stage maps the fuzzy output into a numerical value, which reflects the modified forecast. If the user is satisfied with the adjusted forecast, then he or she can ask for a reoptimization based on the newly generated numerical forecast data. In this study, we look only at the task of forecasting.

2.7 Forecast Adjustment Using a Direct Approach

For situations in which the manager has a knowledge of what the demand for rooms is very likely to be for a future arrival date, then he or she can input that knowledge directly into the system. For instance, the manager could input "very high" or "average" to describe the demand forecast for the first Tuesday of the month of March. The knowledge of future demand is usually a compilation of knowledge about specific events expected to affect a future date, knowledge of room demand for that date during the past years, and manager's own experience and intuition. Based on the manager's input for that particular future date, IS-JFK monitors the forecast generated by the statistical method and makes any necessary adjustment to it before it is sent to the optimization module.

2.7.1 Fuzzy IF-THEN Rules for the Direct Approach

For the direct approach, there are two input variables and one output variable. One of the input variables is the hotel manager's judgmental forecast given as a linguistic value as described earlier. We will refer to this input as the manager demand forecast (MDF). The second input is a statistical demand forecast (SDF) generated by the statistical forecaster. When supplied with these two inputs, IS-JFK decides how SDF needs to be adjusted so that it conforms to the manager's input or MDF. This decision is the output of IS-JFK and consists of a demand forecast adjustment (DFA) action which suggests either to increase or decrease SDF or no action is necessary. The input variables are modeled by sets containing five linguistic values

$$\text{Input 1: Statistical Demand Forecast (SDF)} = \text{FS} = \{\text{L, AL, A, AH, H}\}$$

$$\text{Input 2: Manager Demand Forecast (MDF)} = \text{FM} = \{\text{L, AL, A, AH, H}\}$$

Statistical Demand Forecast \ Manager Demand Forecast	Low L	About Low AL	Average A	About High AH	High H
Low L	0 (1)	PS (2)	PM (3)	PL (4)	PVL (5)
About Low AL	NS (6)	0 (7)	PS (8)	PM (9)	PL (10)
Average A	NM (11)	NS (12)	0 (13)	PS (14)	PM (15)
About High AH	NL (16)	NM (17)	NS (18)	0 (19)	PS (20)
High H	NVL (21)	NL (22)	NM (23)	NS (24)	0 (25)

Fig. 2.7. Fuzzy *IF-THEN* rules for forecast adjustment using the direct approach

where

L low;
AL about low;
A average;
AH about high;
H high.

The output variable is modeled by a set containing nine linguistic values: Demand Forecast Adjustment Action (DFA) $= \mathcal{F}_A =$ {NVL, NL, NM, NS, ZE, PS, PM, PL, PVL}, where NVL = negative very large, NL = negative large, NM = negative medium, NS = negative small, ZE = zero, PS = positive small, PM = positive medium, PL = positive large, and PVL = positive very large. The linguistic values of DFA mean corresponding changes to SDF: negative stands for lowering SDF, positive for increasing SDF, and zero for no action. The fuzzy rule base for forecast adjustment using the direct approach is shown in Fig. 2.7.

There are 25 fuzzy *IF-THEN* rules for forecast adjustment using the direct approach (Fig. 2.7). The fuzzy rules are

$$If\ SDF\ is\ F_s\ and\ MDF\ is\ F_m\ then\ DFA\ is\ F_a \tag{2.1}$$

or compactly

$$(F_s\ and\ F_m) \rightarrow F_a \tag{2.2}$$

where $F_a \in \mathcal{F}_S,\ F_m \in \mathcal{F}_M,\ and\ F_a \in \mathcal{F}_A$. For example, Rule 4 says that: if the statistical demand forecast is "low" and manager thinks that the demand forecast should be "about high," then the statistical demand forecast needs to be increased by a "large quantity." This rule can simply be written as: Rule 4: *If SDF is L and MDF is AH , then DFA is PL*, or compactly as: $(L\ and\ AH) \rightarrow PL$.

2.7.2 Example 1

Through an example, we illustrate the direct approach forecast adjustment by showing the reasoning process and the computation taking place in IS-JFK, but which are transparent to the user. The parameters for this example are taken from an actual hotel and the fuzzy sets and their corresponding membership functions, describing the values of the linguistic variables involved, have been developed with consultation of two hotel managers. The hotel managers were asked questions like: "How would you describe an *average demand*?" We received two types of responses: 1) "When the demand corresponds to an occupancy percentage of *about 50%*, then the demand is said to be average" and 2) "When the demand corresponds to an occupancy percentage between 40 and 60%, then the demand is said to be average." Both types of responses are vague and uncertain. Fuzzy sets are used to represent and handle such vagueness and uncertainty. The membership functions presented in Figs. 2.8–2.10 were obtained after certain iterations and discussions with the hotel managers. During this process, we explained to the managers the concept of fuzzy sets. This has helped us to determine the support and the shape of each membership function.

The hotel managers who participated in this study suggested that the universe of discourse should reflect the occupancy fraction (or occupancy percentage) instead of the total number of demand for rooms. The occupancy fraction is computed as the number of room demand divided by the hotel capacity (i.e., total number of rooms in the hotel)

$$occupancy\ fraction = \frac{total\ room\ demand}{hotel\ capacity} \tag{2.3}$$

The hotel managers explained that during their daily activities they often use forecast occupancy fractions (or percentage numbers) because it is easier to reason with such quantities than with the total room demand. Thus, in this study, the linguistic variables SDF, MDF, and DFA are described in terms of occupancy fractions.

The universe of discourse of occupancy fractions is $\mathcal{X} = [0, 1]$ and an element of $\mathcal{X}$ is denoted by x . The universe of discourse of forecast adjustment actions is $\mathcal{Z} = [-1, 1]$ and an element of $\mathcal{Z}$ is denoted by z . The fuzzy set describing the linguistic values of SDF, MDF, and DFA are reported in Figs. 2.8–2.9, and 2.10, respectively.

Now assume that for an arrival day a , the room demand forecast generated by the statistical forecaster is 182. For a hotel with 289 room capacity,

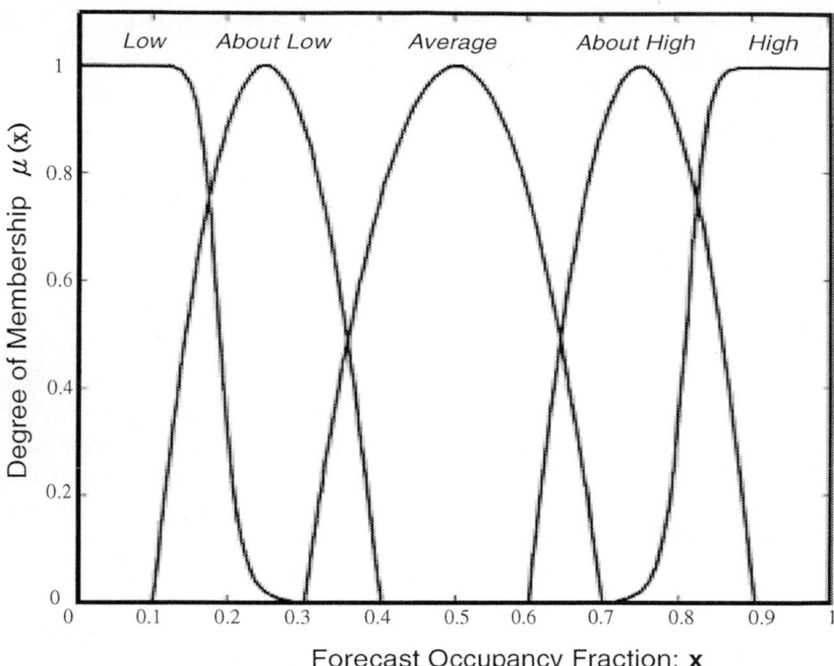

Fig. 2.8. Membership functions describing the input SDF

the statistical forecast occupancy fraction is $x_a = 182/289 = 0.63$. The numerical value $x_a = 0.63$ is described by a fuzzy set singleton S_a with the membership function $\mu_{S_a}(x)$ satisfying $\mu_{S_a}(x) = 1\ if\ x = x_a = 0.63$, and $\mu_{S_a}(x) = 0,\ if\ x \neq x_a = 0.63$.

For that same arrival date a, the manager is expecting the room demand to be "very high." The fuzzy modifier "high" is modeled by the fuzzy set $VH = H^2$

$$Very\ High = VH = H^2 = \int_X [\mu_H(x)]^2/x. \tag{2.4}$$

Based on these two inputs, the observation (or fact) $\mathcal{O}$ is formulated as

$$\mathcal{O} : SDF\ is\ S_a\ and\ MDF\ is\ VH \tag{2.5}$$

or simply as

$$\mathcal{O} : (S_a\ and\ VH). \tag{2.6}$$

In order to determine which rules will fire as a result of the observation $\mathcal{O}$, the inference process will determine the degree of match of the observation $\mathcal{O}$ with the antecedents of all the IF-THEN forecast adjustment rules (Rules 1 through 25).

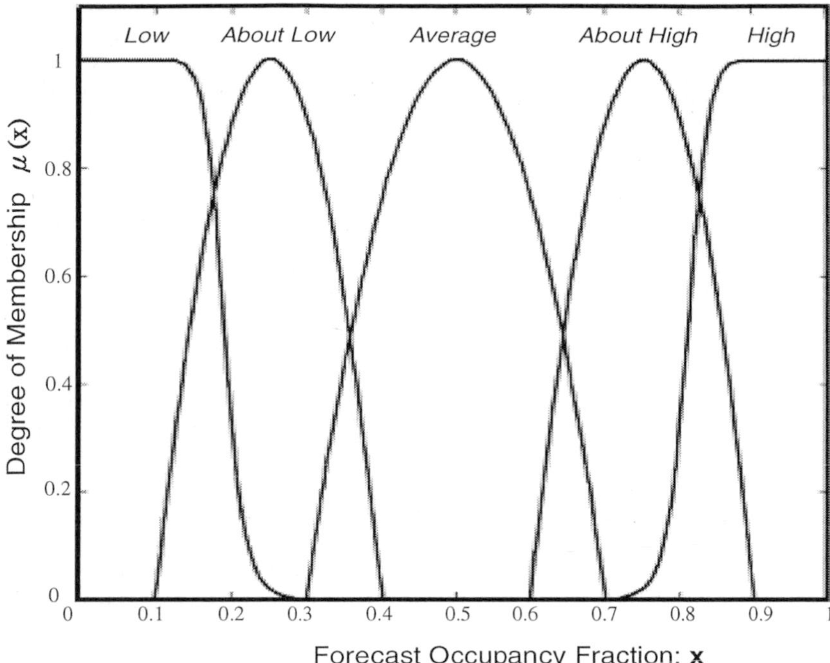

Fig. 2.9. Membership functions describing the input MDF

The intersection of the fuzzy singleton S_a with all the SDF membership functions results in a zero membership degree for all fuzzy sets of $\mathcal{F}_S$ except for fuzzy sets and A and AH. The second attribute of the observation $\mathcal{O}$ is VH, which is a subset of (i.e., $VH \subset H$). Hence, the rules that will fire are those whose antecedents have as a first attribute A or AH and as a second attribute H. These rules are Rules 15 and 20, defined as

$$\text{Rule 15: } (A \text{ and } H) \rightarrow PM$$
$$\text{Rule 20: } (AH \text{ and } H) \rightarrow PS$$

If the connective "and" were to be represented by the minimum operator ($\wedge$), then in the statement (L and H) one would have $\mu_{LandH}(x) = \mu_L(x) \wedge \mu_H(x) = 0$, for $x \in \mathcal{X}$. This means that if the statistical demand forecast is low (L) and the manager demand forecast is high (H), then the level of firing of Rule 5 [$(L\ and\ H) \rightarrow PVL$] is zero and, consequently, no action would be taken. However, this does not reflect the way managers merge these linguistic values in order to infer the appropriate action needed, where in this case the demand forecast needs to be adjusted with a very large positive quantity.

To imitate the way managers process linguistic information, the "and" connective in the observation or the premise of the *IF-THEN* rules should be represented by a more general operator [2.38]. This general operator is

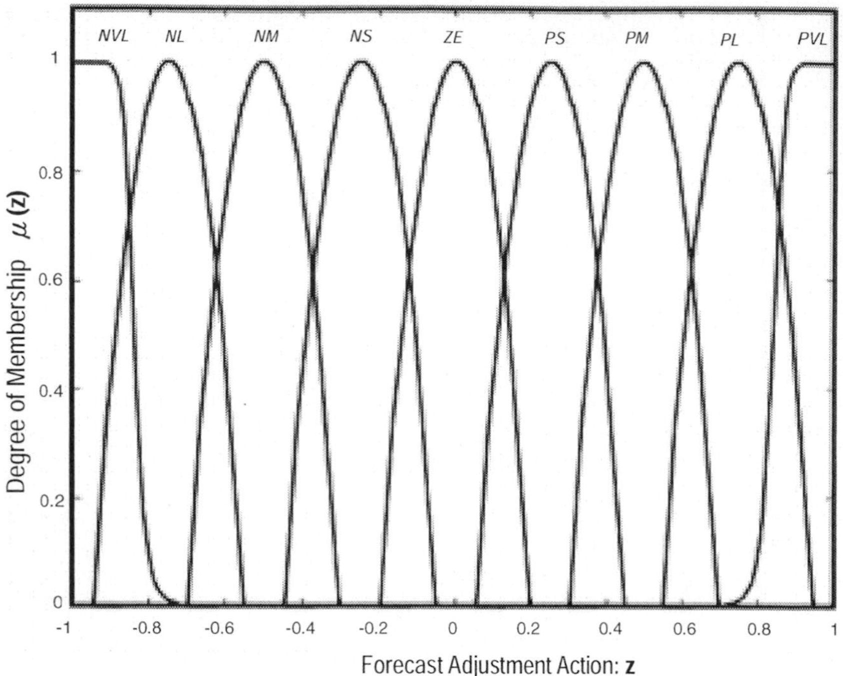

Fig. 2.10. Membership functions describing DFA

denoted by $\bar{\wedge}$ and is defined as

$$\mu_{F_s and F_m}(x) = \mu_{F_s \bar{\wedge} F_m}(x) \tag{2.7}$$

$$= (1-\xi)\mu_{F_s \cap F_m}(x) + \xi \mu_{F_s \cup F_m}(x) \tag{2.8}$$

$$= (1-\xi)(\mu_{F_s}(x) \wedge \mu_{F_m}(x)) + \xi(\mu_{F_s}(x) \vee \mu_{F_m}(x)) \tag{2.9}$$

$$for x \in \mathcal{X}$$

where $\xi \in [0,1]$ is the grade of compensation. When $\xi = 0$, the operator $\bar{\wedge}$ is reduced to the classical minimum operator $\wedge$ associated with the classical intersection between the connected fuzzy sets (i.e., there is no compensation). When $\xi = 1$, the operator $\bar{\wedge}$ is reduced to the classical maximum operator $\vee$ associated with the classical union between the connected fuzzy sets (i.e., there is a full compensation). Here we take $\xi = 0.4$.

Remarks:

1. The general operator overcomes the limitations of the classical minimum operator for the current forecast problem.
2. After testing with different values for the grade of compensation, the value $\xi = 0.4$ has led to acceptable level of rules firing for the problem at hand, but is not necessarily the optimal value to choose.

The operator $\bar{\wedge}$ as defined in (2.7)–(2.9) may not lead to a normal fuzzy set. The fuzzy set describing the observation $\mathcal{O}$ results in the firing of the appropriate forecast adjustment action fuzzy rules. If the fuzzy set is not in normal form, it will dampen the fuzzy implication, thus producing ineffective forecast adjustments. In order to achieve meaningful inference, the membership function of the observation $\mathcal{O}$ is normalized as

$$\mu_{\mathcal{O}}(x) = \frac{\mu_{F_s and F_m}(x)}{sup(F_s \ and \ F_m)}, \qquad for x \ \in \mathcal{X} \tag{2.10}$$

where "$sup(F_s \ and \ F_m)$" is the maximum membership value attained by the fuzzy set $(F_s \ and \ F_m)$ computed as in (2.7)–(2.9).

Now, we define the fuzzy relation matrix $R(x, z)$ as

$$\begin{aligned} R(x,z) &= R_{F_s and F_m \to F_a}(x,z) \\ &= \mu_{F_s and F_m}(x) \wedge \mu_{F_a}(z), \qquad for \ x \in \mathcal{X} \ and \ z \in \mathcal{Z}. \end{aligned} \tag{2.11}$$

Note that the fuzzy implication in (2.11) is chosen to be of the Mamdani type [2.28], which is widely used in fuzzy systems.

The fuzzy relation $R(x, z)$ would have an infinite dimension if $\mathcal{X}$ and $\mathcal{Z}$ were considered as continuum sets. However, this would not be a practical result. For this reason, discrete sets $\mathcal{X}$ and $\mathcal{Z}$ are used instead and are defined, respectively, as

$$\mathcal{X} \supset X = \{x \in \mathcal{X} | x = k\rho_x, k = 0, 1, 2, \cdots, K_x \ and \ \rho_x > 0\} \tag{2.12}$$

and

$$\begin{aligned} \mathcal{Z} \supset Z = \{ \ z \in \mathcal{Z} | z = l\rho_z, l = -L_z, -L_z + 1, \cdots, 0, 1, 2, \\ \cdots, L_z \ and \ \rho_z > 0\} \end{aligned} \tag{2.13}$$

where ρ_x and ρ_z are positive constants that set the resolutions with which $\mathcal{X}$ and $\mathcal{Z}$ are discretized to obtain the support sets X and Z , respectively. For this study, we chose $\rho_x = 0.01$ and $\rho_z = 0.01$ and, thus. $K_x = 100$ and $L_z = 200$. These values are reasonable enough to generate meaningful results in terms of representing occupancy fractions.

The demand forecast adjustment action is inferred by applying Zadeh sup-min compositional rule of inference [2.36], which is described by the following inference expression:

Observation $\mathcal{O}$:	$(S_a \ and \ VH)$
Existing Knowledge Rule 15 (R_{15}):	$(A \ and \ H) \to PM$
Existing Knowledge Rule 20 (R_{20}):	$(AH \ and \ H) \to PS$
Conclusion DFA:	$(\mathcal{O} \circ R_{15}) \cup (\mathcal{O} \circ R_{20})$

where the operator $\circ$ denotes the sup-min composition defined by

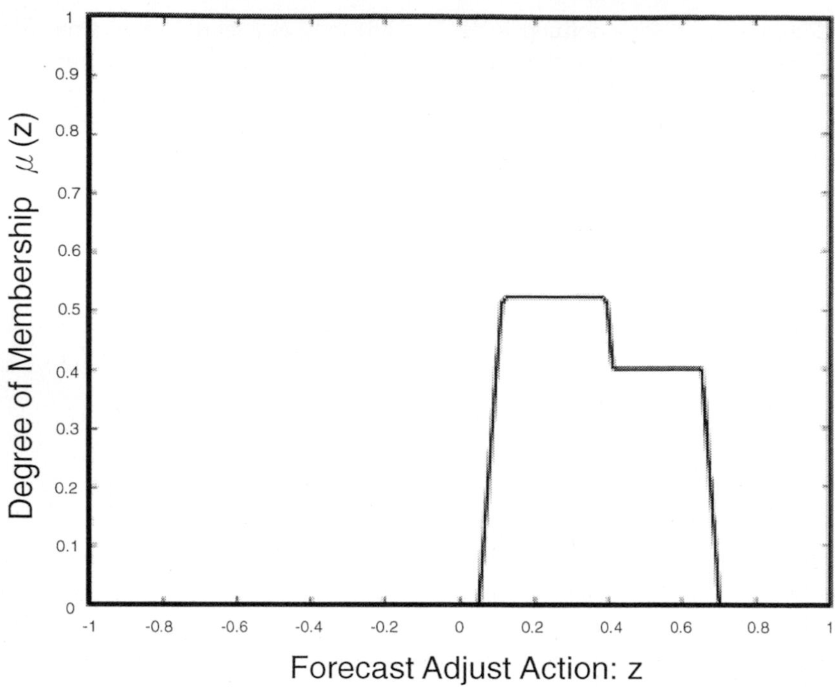

Fig. 2.11. The fuzzy set output $DFA = (\mathcal{O} \circ R_{15}) \cup (\mathcal{O} \circ R_{20})$ for Example 1

$$\mathcal{O} \circ R_i = \sup_{x \in X}(\mathcal{O} \cap R_i) = \sup_{x \in X}(\mu_{\mathcal{O}}(x) \wedge \mu_{R_i}(x, z)) \qquad i = 15, 20. \tag{2.14}$$

The membership function describing the fuzzy output DFA is computed as

$$\mu_{FDA}(z) = (\sup_{x \in X}(\mu_{\mathcal{O}}(x) \wedge \mu_{R_{15}}(x, z))) \vee (\sup_{x \in X}(\mu_{\mathcal{O}}(x) \wedge \mu_{R_{20}}(x, z))), \quad for\ z \in Z. \tag{2.15}$$

The fuzzy set output DFA is shown in Fig. 2.11. For the forecast problem, we need a numerical value that indicates how much the statistical demand forecast needs to be adjusted for an arrival day a . Here, we use *center-of-area* defuzzification method as follows:

$$\hat{z}_a = Defuzz(DFA) = \frac{\sum_{z \in Z} z \mu_{DFA}(z)}{\sum_{z \in Z} \mu_{DFA}(z)} = 0.36. \tag{2.16}$$

The new occupancy fraction forecast for arrival day a is $\hat{x}_a = x_a + \hat{z}_a = 0.63 + 0.36 = 0.99$. Thus, the adjusted room demand forecast is $\hat{x}_a \times hotel\ capacity = \hat{x}_a \times 289 = 286$.

Remark: The computed value of adjustment $\hat{z}_a$ may produce a demand greater than 100% of capacity. However, this is not a problem. In fact, the system is intended to forecast the unconstrained hotel room demand. By unconstrained demand, we mean the requests for reservations received without regard to capacity, pricing, or any other constraint. This is important, since the forecast is intended to serve as an input to the optimization module as it was explained in Section 2.2.

2.8 Fuzzy Adjustment via Fuzzy Intervention Analysis

The direct approach discussed in the previous section does not support the explicit consideration of the effects of external and special events on room demand. In fact, with the direct approach, the effects of such events are implicitly considered in the final demand forecast supplied by hotel managers. However, in some situations hotel managers may not be able to supply the final demand forecast for an arrival day, but only an estimate of an increase or decrease in demand due to a particular event affecting that particular arrival day. Also, one single external event could exhibit a dynamic effect, that is, it could affect room demand not only for one single arrival day but for several consecutive arrival days.

Experienced hotel managers are well positioned to provide a qualitative description of the mechanisms of the effects of special events on room demand. Such description is given in terms of qualitative amounts of increase or decrease in room demand with respect to the normally expected demand if no such events were taking place.

To support the reasoning process of hotel managers in analyzing the nature of effects that some events are likely to have on room demand for future arrival days, a special IS-JFK module is designed for that purpose. The module is intended to allow managers to incorporate their knowledge about future events as well as the nature and extent of their effects on room demand for future arrival days. The mechanisms of the effects of special events are described via fuzzy intervention models developed in this study as an extension to classical mathematical intervention models. First let us start by briefly reviewing classical intervention models and their use in forecast.

2.8.1 Intervention Models

An arrival day during which a special event takes place could be characterized by an *intervention variable* in the form of an *impulse variable* taking on the value "1." Those arrival days during which there are no special events taking place would then be characterized by zeros. Estimated variation (i.e., increase or decrease) in room demand is related to the event by a transfer function model. This transfer function model is used to describe the dynamic relationship between the event and the variation in room demand. Such dynamic models, involving the use of intervention variables as input variables, are called intervention models [2.5]. Fig. 2.3 shows several different factors and events that can influence room demand forecast. Examples for which intervention models could be used in the room demand problem include:

1. the effect of a city-wide event on room demand;
2. the effect of bad weather on room demand;
3. the effect of promotions and discounted rates on room demand;
4. the effect of a competitor's strategy and actions on room demand.

Consider the intervention shown in Fig. 2.12(a). The intervention model is described by

$$\eta_a = \omega_0 \zeta_a, \qquad where\ \zeta_a(t) = \begin{cases} 1 & if\ \ t = a \\ 0 & if\ \ t \neq a \end{cases} \tag{2.17}$$

where ω_0 denotes the height of the single period pulse modeling the intervention (or event) during arrival day a and η_a denotes the increase (or decrease) in room demand for arrival day a due to the intervention.

Let X_a denote the room demand forecast for an arrival day a generated by the statistical forecaster. Because of the intervention, the demand forecast is adjusted and a new demand forecast is computed as

$$\hat{x}_a = x_a + \eta_a = x_a + \omega_0 \zeta_a. \tag{2.18}$$

Intervention models are in general described by the Koyck type transfer function [2.23] given by

$$v(B) = \frac{\omega(B)}{\delta(B)} \tag{2.19}$$

where

$$\omega(B) = \omega_0 + \omega_1 B + \omega_2 B^2 + \cdots + \omega_h B^h \tag{2.20}$$

and

$$\delta(B) = 1 - \delta_1 B - \delta_2 B^2 - \cdots - \delta_r B^r \tag{2.21}$$

where B is the back shift operator (i.e., $B\zeta_a(t) = \zeta_a(t-1)$, $B^2\zeta_a(t) = \zeta_a(t-2)$, etc.) If an intervention (or event) occurs during arrival day a (ζ_a), then it may have an effect not only on day a but also on neighboring days $a+1$ $[B\zeta_a(t)]$, $a+2[B^2\zeta_a(t)]$, etc. Different values for the parameters h and r lead to different intervention models. Examples of intervention models based on the Koyck type transfer function are shown in Fig. 2.12.

Remarks:

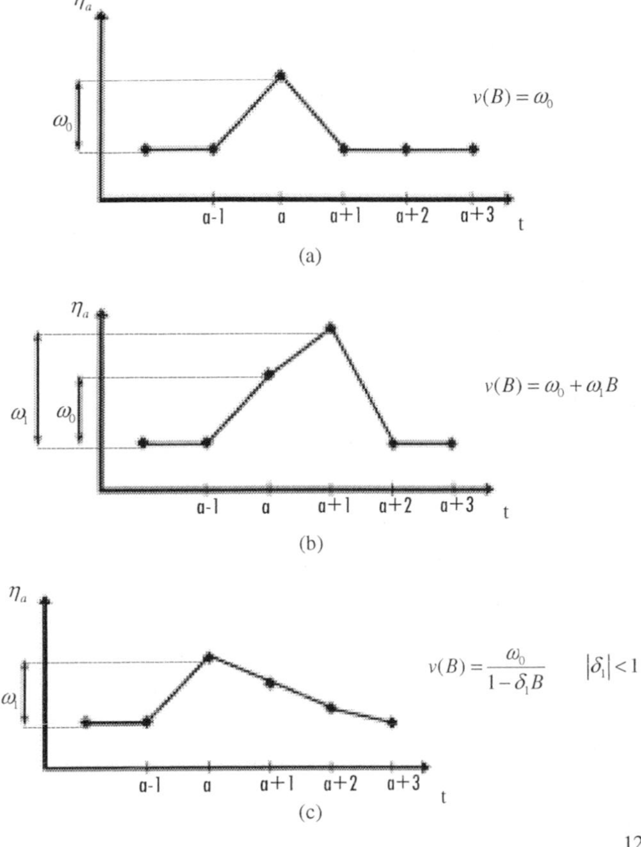

12

Fig. 2.12. Examples of interventions. (a)Type I where $v(b) = \omega_0$. (b)Type II where $v(b) = \omega_0 + \omega_1 B$. (c)Type III where $v(B) = \omega_0/(1 - \delta_1 B)$ and $|\delta_1| < 1$

1. Intervention model type II can be seen as a finite impulse response model where the effect of an intervention is limited to a finite number of future arrival days.
2. Intervention model type III can bee seen as an infinite impulse response model where the effect of an intervention can (theoretically) be spread over an infinite number of future arrival days.

Intervention models help to analyze the mechanisms of the effect of special events in order to adjust demand forecast accordingly. A major limitation of classical mathematical intervention models is that they assume precise quantitative information about the mechanisms of the effects of external events. However, the events affecting hotel room demand are difficult to quantify and, therefore, the use of mathematical intervention models may not be practical. Thus, the objective is to develop intervention models that imitate the way managers describe the mechanisms of special events. In the following section, we introduce a more realistic solution that describes the mechanisms of external events via fuzzy intervention models.

2.8.2 Fuzzy Intervention Models

By definition, a fuzzy intervention model is a fuzzy *IF-THEN* representation of a classic intervention model. Consider the following example: "*if* an important convention is being held on the first Wednesday of the month of April, *then* it is expected that room demand will increase for Tuesday by about 75% above the normal for that day." In this example, convention attendees are assumed to arrive the night before the convention day and to leave at the end of the convention day. Using the classical intervention analysis, Tuesday will be characterized by an impulse and the Koyck type transfer function is $v(B) = \omega_0 = 75\%$. However, in order to account for the intervention in the intelligent system, a fuzzy intervention model needs to be generated. To achieve this, the hotel manager needs to specify the values of the following variables: `arrival_day` and `intervention_effect` , where `arrival_day` is the date for which the event starts to having an effect on demand and is the corresponding demand forecast adjustment action (DFA) required for `arrival_day`. For this example, `arrival_day` is Tuesday and `intervention_effect` is $PL \in \mathcal{F}_A$ (refer to Fig. 2.10). The fuzzy intervention model can then be represented by the following fuzzy rule:

$$If \;\; arrival\_day \;\; is \;\; Tuesday, \; then \; DFA \;\; is \;\; PL. \tag{2.22}$$

In general, the fuzzy intervention model of type I can be generated for any arrival day and represented by the fuzzy rule

$$Type \; I: \;\; IF \;\; arrival\_day \; is \; a, \; then \; DFA \;\; is \; DFA_a \tag{2.23}$$

where $DFA_a \in \mathcal{F}_A$ is the intervention_effect corresponding to a demand forecast adjustment action for arrival day a. Fuzzy intervention model of type I is an elementary model that can be used to build fuzzy intervention models of other types. For instance, referring to Fig. 2.12, fuzzy intervention models of types II and III can be generated and represented as follows:

$Type \; II:$

$$If \;\; arrival\_day \; is \; a, \; then \; DFA \;\; is \; DFA_a \tag{2.24}$$

$$If \;\; arrival\_day \; is \; a+1, \; then \; DFA \;\; is \; DFA_{a+1} \tag{2.25}$$

$Type \; III:$

$$If \;\; arrival\_day \; is \; a, \; then \; DFA \;\; is \; DFA_a \tag{2.26}$$

$$If \;\; arrival\_day \; is \; a+1, \; then \; DFA \;\; is \; DFA_{a+1} \tag{2.27}$$

$$If \;\; arrival\_day \; is \; a+2, \; then \; DFA \;\; is \; DFA_{a+2} \tag{2.28}$$

$$If \;\; arrival\_day \; is \; a+3, \; then \; DFA \;\; isDFA_{a+3} \tag{2.29}$$

where DFA_a, DFA_{a+1}, DFA_{a+2}, and DFA_{a+3} are elements of $\mathcal{F}_A$.

Early in the planning horizon, the statistical forecaster uses historical data to generate LTFs of room demand for future arrival days. As it is discussed in Section 2.3.1, such forecasts may be made as early as one year before the arrival day. If there is a new external event that is expected to affect the demand for a future arrival day, then early in the planning horizon the statistical forecast may not reflect the variation in demand due to that event. In fact, early in the planning horizon more weight is put on LTF and less weight on STF as no actual reservations are being registered in the reservation system [2.31]. A few weeks before the arrival day and as a relatively important number of reservations start to build up in the system, more weight is put on the STF. This time frame is referred to as the short term planning horizon during which the statistical forecaster is expected to catch up with a more accurate final demand forecast. Depending on the type of hotel (e.g., hotels catering to business travelers, hotels catering to leisure travelers, etc.), short term planning horizon can extend from a few weeks to several months.

In their practice, hotel managers look out for any special events that are likely to affect the demand for future arrival days. They often mark the affected days on a calendar and discuss the mechanisms of the effects of the events with their staff. However, the existing forecast systems do not allow managers to systematically input their knowledge in order to properly adjust the forecast. They mostly rely on daily overrides of system forecasts till the system catches up. The managers we talked to expressed that this practice can be frustrating, whereas with the system IS-JFK, hotel managers are given the opportunity to systematically adjust the demand forecast early in the planning horizon by choosing the appropriate fuzzy intervention model for a particular event affecting the demand for a future arrival day and, thereby, are able to set optimal inventory controls leading to a revenue increase. Since it is expected that the statistical forecaster to catch up with the final demand as the arrival day approaches, the fuzzy intervention models have to be switched off gradually during the short term horizon. The rule base used in forecast adjustment via fuzzy intervention analysis is shown in Fig. 2.13.

For a hotel catering to business travelers, the fuzzy sets describing the planning types (PT) long term horizon (LT) and short term (ST) horizon are shown in Fig. 2.14. The figure shows that for this type of hotel, most of the bookings occur during the last four weeks before arrival (Day 0) and, hence, fuzzy set ST is constructed accordingly. Whereas, the fuzzy set LT characterizes the long term horizon which extends from about three weeks to 60 days before the arrival day. The fuzzy sets describing the estimated increase or decrease in final room demand due to interventions are the same as those used to describe the fuzzy sets describing forecast adjustment action (DFA) of the direct approach (refer to Fig. 2.10).

Planning Type (PT)	Demand Forecast Adjustment Action (DFA)								
	NVL	NL	NM	NS	0	PS	PM	PL	PVL
Long Term Horizo LT	NVL	NL	NM	NS	0	PS	PM	PL	PVL
Short Term Horizo ST	0	0	0	0	0	0	0	0	0

Fig. 2.13. Fuzzy *If-THEN* rule base for forecast adjustment *via* fuzzy intervention analysis

2.8.3 Example 2

This example shows that IS-JFK allows hotel managers to incorporate their judgmental forecast early in the planning horizon. Consider an intervention of the type I shown in Fig. 2.12(a). The hotel manager believes that due to a specific future event, there will be a medium increase in room demand for arrival day. The hotel manager inputs are

`arrival_day` a
`intervention_effect` PM

From the rule-base in Fig. 2.13, the following *IF-THEN* rules are formulated for this situation:

Rule 1: *If* `arrival_day` is a and PT is LT, *then* DFA is PM
Rule 2: *If* `arrival_day` is a and PT is ST, *then* DFA is 0

Simulation results are shown in Fig. 2.15. Early in the planning horizon (i.e., during LT), the statistical forecaster estimates that the occupancy fraction for arrival day (Day 0 in Fig. 2.15) will be around 0.41. When bookings start to build up in the reservation system, statistical forecaster starts to catch up to finally converge to the actual final demand about two weeks before arrival. It can be seen that the input of the hotel manager helps to adjust the forecast and to obtain a more accurate demand forecast early in the planning horizon. As the arrival day approaches, the fuzzy intervention model is gradually switched off. This is done according to the fuzzy *IF-THEN* rules for forecast adjustment (Fig. 2.13) and the definitions of the fuzzy sets long term horizon and short term horizon in Fig. 2.14. During the long term horizon more weight is given to the forecast adjustment action. That weight decreases gradually as the arrival day approaches and the hotel enters the short term planning period (Fig. 2.13).

This example forms the basis for any forecast adjustment based on fuzzy intervention analysis. If one arrival day is affected by more than one intervention, the same fuzzy rule base given in Fig. 2.13 will be used for each

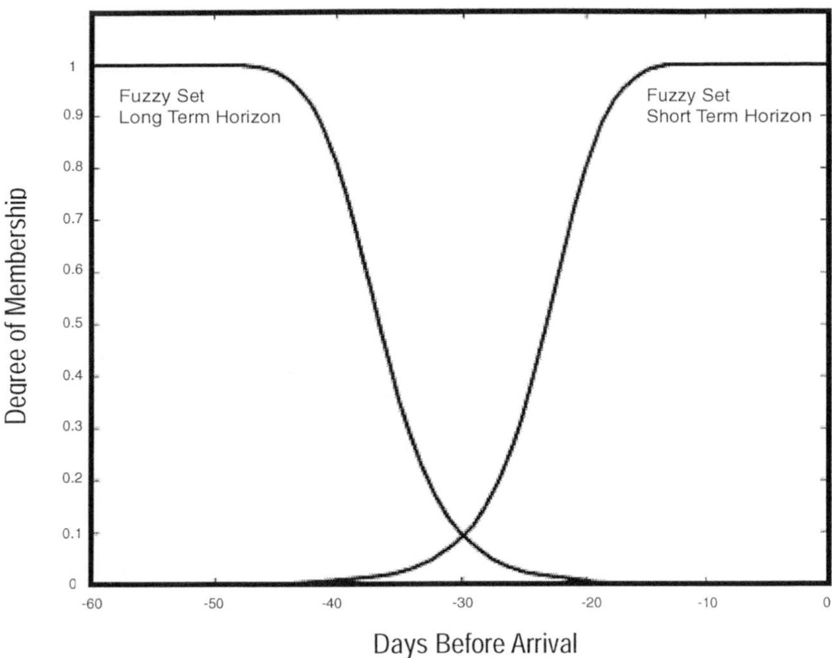

Fig. 2.14. Fuzzy sets describing planning horizons LT and ST for a hotel catering to business travelers

event. The final forecast adjustment will be the aggregation of the different adjustments.

2.9 Conclusion

In this paper, we presented IS-JFK which is a knowledge-based system that supports the judgmental forecasting and knowledge of hotel managers. IS-JFK allows managers to adjust demand forecasts for future arrival days when there is a discontinuous change in the business environment that statistical techniques fail to detect. The exogenous events that affect room demand can be numerous for a hotel property, which makes numerical overrides of demand forecasts difficult and inconsistent. Whereas with the intelligent system, hotel managers can incorporate their knowledge about future events and their judgment about the extent of effect of each event on room demand. This information is processed by the intelligent system and appropriate adjustments to statistical forecasts are made.

The knowledge in IS-JFK has been represented by fuzzy IF-THEN rules. This is a natural representation of the knowledge acquired from hotel managers. Fuzzy set theory provided an effective methodological framework to

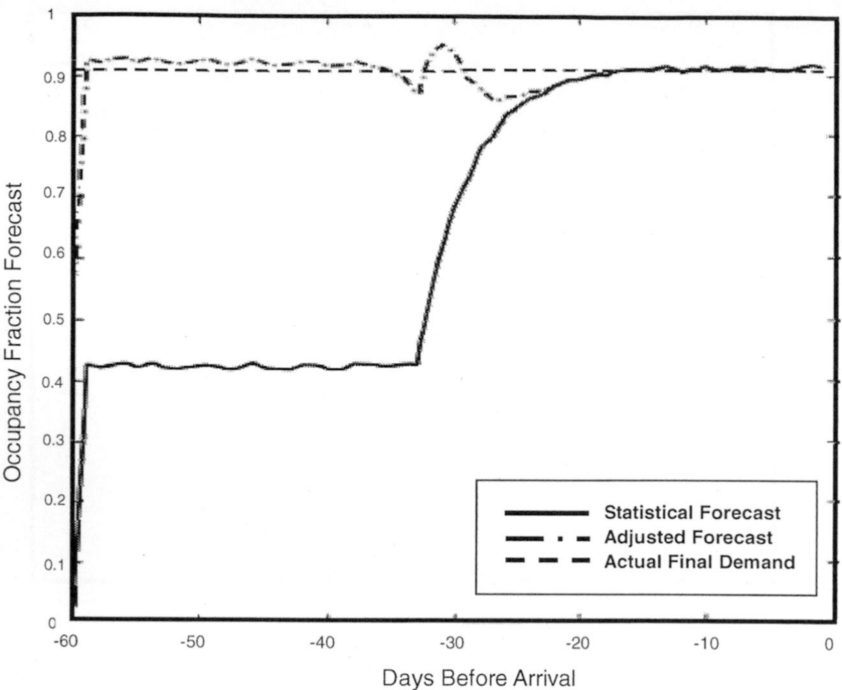

Fig. 2.15. Forecast adjustment via fuzzy intervention analysis for Example 2 (Day 0 is the arrival day)

represent the acquired knowledge together with a reasoning mechanism in order to manage uncertainty and vagueness of information based on which hotel business decisions were made.

Two different approaches have been developed to support managers in adjusting demand forecast: 1) a direct approach for forecast adjustment and 2) forecast adjustment via fuzzy intervention analysis. Problems scenarios and simulation results based on actual hotel data have demonstrated the effectiveness of IS-JFK.

In this study, the cooperation of hotel managers was very instrumental. In addition to acquiring their knowledge, we chose to involve the hotel managers in all aspects of the conceptualization and development of the intelligent system. Their input was important to define and characterize the fuzzy sets used in the system.

Further research work needs to be done to make IS-JFK mature and implementable in different types of hotel properties (such as business traveler hotels, business/convention center hotels, leisure traveler hotels, and so on). Different types of hotels are affected by different types of external events. In this study, we focused only on hotels catering to business travelers. Another aspect of future research is to make IS-JFK a training tool for novices in hotel management.

Acknowledgements

The authors would like to thank the hotel managers and the field manager, who participated in this study, for their cooperation and valuable input. The authors are also grateful for the great efforts made by Chung-Chin Tai to transformed the original document into the Latex format.

References

2.1 Alstrup, J., Boas, S., Madsen, O., Vidal, R. (1986): Booking Policy for Flights with Two Types of Passengers. Eur. J. Operations Res., Vol. 27, 274–288

2.2 Anderson, N. H. (1986): A Cognitive Theory of Judgment and Decision. In: Brehmer, B., Jungermann, H., Lourens, P., Sevons, G. (Eds.), New Directions in Research on Decision-Making. Elsevier

2.3 Anderson, J. R., Bower, G. H. (1973): Human Associative Memory. Winston

2.4 Beach, L. R., Christensen-Szalanski, J., Barnes, V. (1987): Assessing Human Judgment: Has it been done, can it be done, should it be done? In: Wright, G., Ayton, P. (Eds.), Judgmental Forecasting. Wiley, 49–62

2.5 Box, G. E. P., Yiao, G. F. C. (1975): Intervention Analysis with Applications to Economic and Environmental Problems. J. Amer. Statistical Assoc, Vol. 70, 70–79

2.6 Box, G. E. P., Jenkins, G. M. (1976): Time Series Analysis, Forecasting and Control. Holden-Day

2.7 Butler, W. F., Kavesh, R. A., Platt, R. B. (1974): Methods and Techniques of Business Forecasting. PrenticeHall

2.8 Carrol, W. J., Grimes, R. C. (1995): Evolutionary Change in Product Management: Experiences in the Car Rental Industry. Interfaces, Vol. 25, No. 5, 84–104

2.9 Chatfield, C. (1978): The Holt-winters Forecasting Procedure. Appl. Statistics, Vol. 27, No. 3, 264–279

2.10 Chatfield, C. (1980): The Analysis of Time Series: An Introduction, 2nd ed., Chapman and Hall

2.11 Collins, A. M., Quillian, M. R. (1969): Retrieval Time from Semantic Memory. J. Verbal Learning Verbal Behavior, Vol. 8, 241–248

2.12 Cross, R. G. (1997): Revenue Management. Broadway Books

2.13 Evans, J. B. T. (1987): Beliefs and Expectations as Causes of Judgmental Bias. In: Wright, G., Ayton, P. (Eds.), Judgmental Forecasting, Wiley, 31-47

2.14 Gallego, G., van Ryzin, G. (1994): Optimal Dynamic Pricing of Inventories with Stochastic Demand Over Finite Horizons. Management Sci., Vol. 40, 999–1020

2.15 Gallego, G., van Ryzin, G. (1997): A Multiproduct Dynamic Pricing Problem and Its Applications to Network Yield Management. Operations Res., Vol. 45, No. 1, 24–41

2.16 Gilchrist, W. (1976): Statistical Forecasting. Wiley

2.17 Hart, A. (1992): Knowledge Acquisition for Expert Systems, 2nd ed. McGraw-Hill

2.18 Harvey, A. C. (1993): Time Series Models, 2nd ed. MIT Press
2.19 Hogarth, R. M. (1987): Judgment and Choice: The Psychology of Decision. Wiley
2.20 Jenkins, G. M. (1980): Practical Experiences with Modeling and Forecasting Time Series. In: Anderson O. D. (Ed.), Forecasting Amsterdam, 43–166
2.21 Kaplan, M. F., Shwartz, S. (1975): Human Judgment and Decision Processes. Academic
2.22 Kimes, S. E. (1989): Yield Management: A Tool for Capacity-constrained Ser-vice Firms. J. Operations Management, Vol. 8, No. 4, 348–363
2.23 Koyck, L. M. (1954): Distributed Lags and Investment Analysis. In: Amsterdam ,Contributions to Economic Analysis, Vol. 4, North-Hol-land
2.24 Lee, W. (1971): Decision Theory and Human Behavior. Wiley
2.25 Lindley, D. V. (1985): Making Decisions. Wiley
2.26 Makridakis, S., Wheelwright, S. C. (1979): Forecasting the Future and the Future of Forecasting. In Amsterdam ,Studies in the Managerial Sciences, Vol. 12 Elsevier
2.27 Makridakis, S., Wheelwright, S. C., McGhee, V. E. (1983): Forecasting: Methods and Applications. Wiley
2.28 Mamdani, E. H., Assilian, S. (1977): An Experiment in Linguistic Synthesis with A Fuzzy Logic Controller. Int. J. Man-Machine Studies, Vol. 7, 1–13
2.29 Montgomery, D. C., Johnson, L. A. (1976): Forecasting and Time Series Analysis. McGraw-Hill
2.30 Phillips, L. D. (1987): On the Adequacy of Judgmental Forecasts. In: Wright, G. and Ayton, P., (Eds.), Judgmental Forecasting, Wiley, 11–30
2.31 Rajopadhye, M., Ben Ghalia, M., Wang, P. P., Baker, T., Eister, C. (1999): Forecasting Uncertain Hotel Room Demand. In: Proc. Amer. Contr. Conf.
2.32 Wallsten, T. S., Budescu, B. V. (1983): Encoding Subjective Probabilities: A Psychological and Psychometric Review. Management Sci., Vol. 29, 151–173
2.33 Weatherford, L. R., Bodily, S. E. (1992): A Taxonomy and Research Overview of Perishable Asset Revenue Management: Yield Management, Overbooking and Pricing. Operations Res., Vol. 40, 831–844
2.34 Winters, P. R. (1960): Forecasting Sales by Exponentially Weighted Moving Averages. Management Sci., Vol. 6, No. 3, 324–342
2.35 Wright, G., Phillips, L. D. (1979): Personality and Probabilistic Thinking: An Explanatory Study. British J. Psychol., Vol. 70, 295–303
2.36 Zadeh, L. A. (1973): Outline of a New Approach to the Analysis of Complex and Decision Processes. IEEE Trans. Syst., Man, Cybern., Vol. 3, 28–44
2.37 Zadeh, L. A. (1983): The Role of Fuzzy Logic in the Management of Uncertainty in Expert Systems. Fuzzy Sets Syst., Vol. 11, 199–227
2.38 Zimmermann, H. J., Zysnos, P. (1980): Latent Connectives in Human Decision Making. Fuzzy Sets Syst., Vol. 4, 37–51

3. Fuzzy Investment Analysis Using Capital Budgeting and Dynamic Programming Techniques

Cengiz Kahraman[1] and Cafer Erhan Bozdağ[2]

[1] Istanbul Technical University, Industrial Engineering Department, 80680, Maçka, Istanbul, Turkey
email: kahramanc@itu.edu.tr

[2] Istanbul Technical University, Industrial Engineering Department, 80680, Maçka, Istanbul, Turkey
email: bozdagc@itu.edu.tr

In an uncertain economic decision environment, an expert's knowledge about discounting cash flows consists of a lot of vagueness instead of randomness. Cash amounts and interest rates are usually estimated by using educated guesses based on expected values or other statistical techniques to obtain them. Fuzzy numbers can capture the difficulties in estimating these parameters. In this chapter, the formulas for the analysis of fuzzy present value, fuzzy equivalent uniform annual value, fuzzy future value, fuzzy benefit-cost ratio, and fuzzy payback period are developed and given some numeric examples. Then the examined cash flows are expanded to geometric and trigonometric cash flows and using these cash flows fuzzy present value, fuzzy future value, and fuzzy annual value formulas are developed for both discrete compounding and continuous compounding. The fuzzy dynamic programming is applied to the situation where each investment in the set has the following characteristics: the amount to be invested has several possible values, and the rate of return varies with the amount invested. Each sum that may be invested represents a distinct level of investment, and the investment therefore has multiple levels. A fuzzy present worth based dynamic programming approach is used. A numeric example for a multilevel investment with fuzzy geometric cash flows is given. A computer software named FUZDYN is developed for various problems such as alternatives having different lives, different uniform cash flows, and different ranking methods.

3.1 Introduction

The purpose of this chapter is to develop the fuzzy capital budgeting techniques and a fuzzy dynamic programming method for multilevel investments. The analysis of fuzzy future value, fuzzy present value, fuzzy rate of return, fuzzy benefit/cost ratio, fuzzy payback period, fuzzy equivalent uniform annual value are examined for the case of discrete compounding and continuous compounding.

To deal with vagueness of human thought, Zadeh [3.37] first introduced the fuzzy set theory, which was based on the rationality of uncertainty due to imprecision or vagueness. A major contribution of fuzzy set theory is its capability of representing vague knowledge. The theory also allows mathematical operators and programming to apply to the fuzzy domain.

A fuzzy number is a normal and convex fuzzy set with membership function $\mu_A(x)$ which both satisfies normality: $\mu_A(x)=1$, for at least one $x \in R$ and convexity: $\mu_A(x') \geq \mu_A(x_1) \Lambda \mu_A(x_2)$, where $\mu_A(x) \in [0,1]$ and $\forall x' \in [x_1, x_2]$. '$\Lambda$' stands for the minimization operator.

Quite often in finance future cash amounts and interest rates are estimated. One usually employs educated guesses, based on expected values or other statistical techniques, to obtain future cash flows and interest rates. Statements like *approximately between \$ 12,000 and \$ 16,000* or *approximately between 10% and 15%* must be translated into an exact amount, such as *\$ 14,000* or *12.5%* respectively. Appropriate fuzzy numbers can be used to capture the vagueness of those statements.

A tilde will be placed above a symbol if the symbol represents a fuzzy set. Therefore, $\tilde{P}, \tilde{F}, \tilde{G}, \tilde{A}, \tilde{i}, \tilde{r}$ are all fuzzy sets. The membership functions for these fuzzy sets will be denoted by $\mu(x \,|\, \tilde{P}), \mu(x \,|\, \tilde{F}), \mu(x \,|\, \tilde{G})$, etc. A fuzzy number is a special fuzzy subset of the real numbers. The extended operations of fuzzy numbers are given in the appendix. A triangular fuzzy number (TFN) is shown in Fig. 3.1 The membership function of a TFN ($\tilde{M}$)defined by

$$\mu(x \,|\, \tilde{M}) = (m_1, f_1(y \,|\, \tilde{M})/m_2, m_2/f_2(y \,|\, \tilde{M}), m_3) \tag{3.1}$$

where $m_1 \prec m_2 \prec m_3$, $f_1(y \,|\, \tilde{M})$ is a continuous monotone increasing function of y for $0 \leq y \leq 1$ with $f_1(0 \,|\, \tilde{M}) = m_1$ and $f_1(1 \,|\, \tilde{M}) = m_2$ and $f_2(y \,|\, \tilde{M})$ is a continuous monotone decreasing function of y for $0 \leq y \leq 1$ with $f_2(1 \,|\, \tilde{M}) = m_2$ and $f_2(0 \,|\, \tilde{M}) = m_3$. $\mu(x \,|\, \tilde{M})$ is denoted simply as $(m_1/m_2, m_2/m_3)$.

A flat fuzzy number (FFN) is shown in Fig. 3.2 The membership function of a FFN, $\tilde{V}$ is defined by

$$\mu(x \,|\, \tilde{V}) = (m_1, f_1(y \,|\, \tilde{V})/m_2, m_3/f_2(y \,|\, \tilde{V}), m_4) \tag{3.2}$$

where $m_1 \prec m_2 \prec m_3 \prec m_4$, $f_1(y \,|\, \tilde{V})$ is a continuous monotone increasing function of y for $0 \leq y \leq 1$ with $f_1(0 \,|\, \tilde{V}) = m_1$ and $f_1(1 \,|\, \tilde{V}) = m_2$ and $f_2(y \,|\, \tilde{V})$ is a continuous monotone decreasing function of y for $0 \leq y \leq 1$ with $f_2(1 \,|\, \tilde{V}) = m_3$ and $f_2(0 \,|\, \tilde{V}) = m_4$. $\mu(y \,|\, \tilde{V})$ is denoted simply as $(m_1/m_2, m_3/m_4)$.

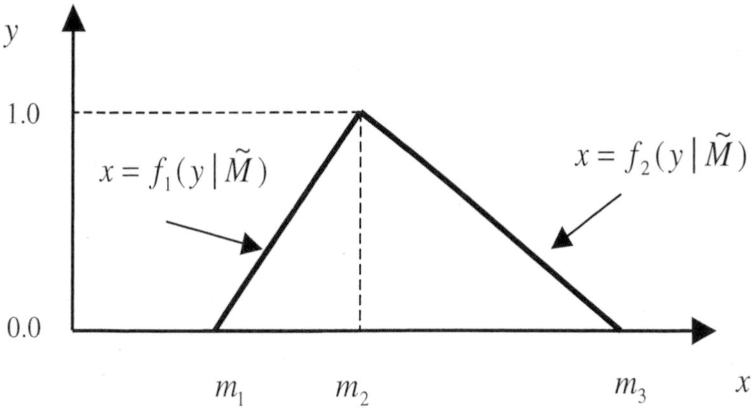

Fig. 3.1. A triangular fuzzy number, $\tilde{M}$

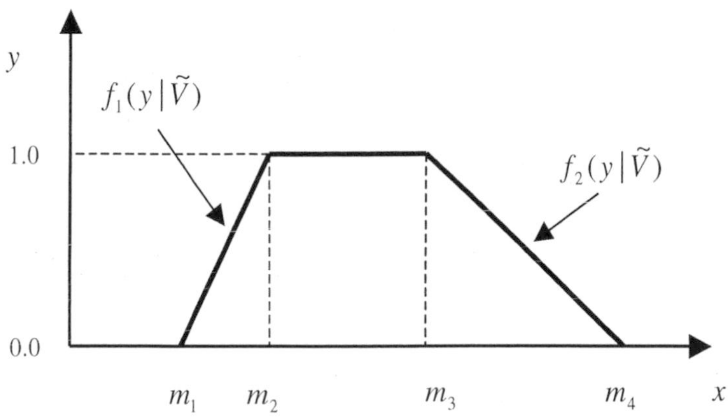

Fig. 3.2. A flat fuzzy number, $\tilde{V}$

The fuzzy sets $\tilde{P}, \tilde{F}, \tilde{G}, \tilde{A}, \tilde{i}, \tilde{r}$ are usually fuzzy numbers but n will be discrete positive fuzzy subset of the real numbers [3.5]. The membership function $\mu(x\,|\tilde{n}\,)$ is defined by a collection of positive integers n_i, $1 \leq i \leq K$, where

$$\mu(x\,|\tilde{n}\,) = \begin{cases} \mu(n_i\,|\tilde{n}\,) = \lambda_i, & 0 \leq \lambda_i \leq 1 \\ 0 \quad , & \text{otherwise} \end{cases} \tag{3.3}$$

Karsak [3.26] develops some measures of liquidity risk supplementing fuzzy discounted cash flow analysis. Iwamura and Liu [3.16] develop chance constrained integer programming models for capital budgeting in fuzzy environments. Boussabaine and Elhag [3.4] examine the possible application of the fuzzy set theory to the cash flow analysis in construction projects. Dimitrovski and Matos [3.9] present an approach to including nonstatistical uncertainties in utility economic analysis by modelling uncertain variables with fuzzy numbers. Kuchta [3.28] proposes fuzzy equivalents of all the classical capital budgeting methods.

3.2 Fuzzy Present Value (PV) Method

The present-value method of alternative evaluation is very popular because future expenditures or receipts are transformed into equivalent dollars now. That is, all of the future cash flows associated with an alternative are converted into present dollars. If the alternatives have different lives, the alternatives must be compared over the same number of years.

Chiu and Park [3.8] propose a present value formulation of a fuzzy cash flow. The result of the present value is also a fuzzy number with nonlinear membership function. The present value can be approximated by a TFN. Chiu and Park [3.8]'s formulation is

$$\begin{aligned} P\tilde{V} = & \left[\sum_{t=0}^{n}\left(\frac{\max(P_t^{l(y)},0)}{\prod_{t'=0}^{t}(1+r_{t'}^{r(y)})}+\frac{\min(P_t^{l(y)},0)}{\prod_{t'=0}^{t}(1+r_{t'}^{l(y)})}\right),\right. \\ & \left.\sum_{t=0}^{n}\left(\frac{\max(P_t^{r(y)},0)}{\prod_{t'=0}^{t}(1+r_{t'}^{l(y)})}+\frac{\min(P_t^{r(y)},0)}{\prod_{t'=0}^{t}(1+r_{t'}^{r(y)})}\right)\right] \end{aligned} \tag{3.4}$$

where$P_t^{l(y)}$: the left representation of the cash at time t, $P_t^{r(y)}$: the right representation of the cash at time t, $r_t^{l(y)}$: the left representation of the interest rate at time t, $r_t^{r(y)}$: the right representation of the interest rate at time t.

Buckley's [3.5] membership function for $\tilde{P}_n$,

$$\mu(x\left|\tilde{P}_n\right.) = (p_{n1}, f_{n1}(y\left|\tilde{P}_n\right.)/p_{n2}, p_{n2}/f_{n2}(y\left|\tilde{P}_n\right.), p_{n3}) \tag{3.5}$$

is determined by

$$f_{ni}(y|\tilde{P}_n) = f_i(y|\tilde{F})(1+f_k(y|\tilde{r}))^{-n} \tag{3.6}$$

for i = 1,2 where $k = i$ for negative $\tilde{F}$ and $k = 3 - i$ for positive $\tilde{F}$. Ward [3.34] gives the fuzzy present value function as

$$P\tilde{V} = (1+r)^{-n}(a,b,c,d) \tag{3.7}$$

where (a, b, c, d) is a trapezoidal fuzzy number.

3.3 Fuzzy Capitalized Value Method

A specialized type of cash flow series is perpetuity, a uniform series of cash flows that continues indefinitely. An infinite cash flow series may be appropriate for such very long-term investment projects as bridges, highways, forest harvesting, or the establishment of endowment funds where the estimated life is 50 years or more.

In the nonfuzzy case, if a present value P is deposited into a fund at interest rate r per period so that a payment of size A may be withdrawn each and every period forever, then the following relation holds between P, A, and r:

$$P = \frac{A}{r} \tag{3.8}$$

In the fuzzy case, lets assume all the parameters as triangular fuzzy numbers: $\tilde{P} = (p_1, p_2, p_3)$ or $\tilde{P} = (((p_2 - p_1)y + p_1), (p_2 - p_3)y + p_3)$ and $\tilde{A} = (a_1, a_2, a_3)$ or $\tilde{A} = (((a_2 - a_1)y + a_1), (a_2 - a_3)y + a_3)$ and $\tilde{r} = (r_1, r_2, r_3)$ or $\tilde{r} = ((r_2 - r_1)y + r_1, (r_2 - r_3)y + r_3)$, where y is the membership degree of a certain point of A and r axis. If $\tilde{A}$ and $\tilde{r}$ are both positive,

$$\tilde{P} = \tilde{A} \emptyset \tilde{r} = (a_1/r_3, a_2/r_2, a_3/r_1) \tag{3.9}$$

or

$$\tilde{P} = (((a_2-a_1)y+a_1)/((r_2-r_3)y+r_3), ((a_2-a_3)y+a_3)/((r_2-r_1)y+r_1)) \tag{3.10}$$

If $\tilde{A}$ is negative and $\tilde{r}$ is positive,

$$\tilde{P} = \tilde{A} \emptyset \tilde{r} = (a_1/r_1, a_2/r_2, a_3/r_3) \tag{3.11}$$

or

$$\tilde{P} = (((a_2-a_1)y+a_1)/((r_2-r_1)y+r_1), ((a_2-a_3)y+a_3)/((r_2-r_3)y+r_3)) \tag{3.12}$$

Now, let $\tilde{A}$ be an expense every nth period forever, with the first expense occurring at n. For example, an expense of $(\$5,000, \$7,000, \$9,000)$ every third year forever, with the first expense occurring at t=3. In this case, the fuzzy effective rate $\tilde{e}$ may be used as in the following:

$$f_i(y|\,\tilde{e}) = (1 + (1/m) f_i(y|\,\tilde{r}'))^m - 1 \tag{3.13}$$

where $i = 1, 2; f_1(y|\,\tilde{e})$:a continuous monotone increasing function of y; $f_2(y|\,\tilde{e})$:a continuous monotone decreasing function of y; m: the number of compounding per period; $\tilde{r}'$: the fuzzy nominal interest rate per period. The membership function of $\tilde{e}$ may be given as

$$\mu(x|\,\tilde{e}) = (e_1, f_1(y|\,\tilde{e})/e_2, e_2/f_2(y|\,\tilde{e}), e_3) \tag{3.14}$$

If $\tilde{A}$ and $f_i(y|\,\tilde{e})$ are both positive,

$$\tilde{P} = \tilde{A}\emptyset\tilde{e} = [((a_2 - a_1)y + a_1)/f_2(y|\,\tilde{e}), ((a_2 - a_3)y + a_3)/f_1(y|\,\tilde{e})] \quad (3.15)$$

If $\tilde{A}$ is negative and $f_i(y|\,\tilde{e})$ is positive,

$$\tilde{P} = \tilde{A}\emptyset\tilde{e} = [((a_2 - a_1)y + a_1)/f_1(y|\,\tilde{e}), ((a_2 - a_3)y + a_3/f_2(y|\,\tilde{e}))] \quad (3.16)$$

$(a_2 - a_1)y + a_1$and $(a_2 - a_3)y + a_3$ can be symbolized as $f_1(y|\,\tilde{a})$ and $f_2(y|\,\tilde{a})$ respectively.

3.4 Fuzzy Future Value Method

The future value (FV) of an investment alternative can be determined using the relationship

$$FV(r) = \sum_{t=0}^{n} P_t(1+r)^{n-t} \quad (3.17)$$

where *FV(r)* is defined as the future value of the investment using a minimum attractive rate of return *(MARR)* of $r\%$. The future value method is equivalent to the present value method and the annual value method.

Chiu and Park's [3.8] formulation for the fuzzy future value has the same logic of fuzzy present value formulation:

$$\{\sum_{t=0}^{n-1} [\max(P_t^{l(y)},0) \prod_{t'=t+1}^{n} (1+r_{t'}^{l(y)}) + \min(P_t^{l(y)},0) \prod_{t'=t+1}^{n} (1+r_{t'}^{r(y)})] + P_n^{l(y)},$$

$$\sum_{t=0}^{n-1} [\max(P_t^{r(y)},0) \prod_{t'=t+1}^{n} (1+r_{t'}^{r(y)}) + \min(P_t^{r(y)},0) \prod_{t'=t+1}^{n} (1+r_{t'}^{l(y)})] + P_n^{r(y)}\}$$

(3.18)

Buckley's [3.5] membership function $\mu(x|\,\tilde{F})$ is determined by

$$f_i(y \mid \tilde{F}_n) = f_i(y \mid \tilde{P})(1 + f_i(y \mid \tilde{r}))^n \quad (3.19)$$

For the uniform cash flow series, $\mu(x|\,\tilde{F})$ is determined by

$$f_{ni}(y \mid \tilde{F}) = f_i(y \mid \tilde{A})\beta(n, f_i(y|\,\tilde{r})) \quad (3.20)$$

where i =1,2 and $\beta(n,r) = (((1+r)^n - 1)/r)$ and $\tilde{A} \succ 0$ and $\tilde{r} \succ 0$.

3.5 Fuzzy Benefit/Cost Ratio Method

The benefit/cost ratio (BCR) is often used to assess the value of a municipal project in relation to its cost; it is defined as

$$BCR = \frac{B - D}{C} \tag{3.21}$$

where B represents the equivalent value of the benefits associated with the project, D represents the equivalent value of the disbenefits, and C represents the project's net cost. A BCR greater than 1.0 indicates that the project evaluated is economically advantageous. In BCR analysis, costs are not preceded by a minus sign.

When only one alternative must be selected from two or more mutually exclusive (stand-alone) alternatives, a multiple alternative evaluation is required. In this case, it is necessary to conduct an analysis on the incremental benefits and costs. While calculating $\Delta B_{2-1}/\Delta C_{2-1}$ ratio, the costs and benefits of the alternative with higher first cost are subtracted from the costs and benefits of the alternative with smaller first cost. Suppose that there are two mutually exclusive alternatives. In this case, for the incremental BCR analysis ignoring disbenefits the following ratios must be used:

$$\Delta B_{2-1}/\Delta C_{2-1} = \Delta PVB_{2-1}/\Delta PVC_{2-1} \tag{3.22}$$

where PVB: present value of benefits, PVC: present value of costs.
If $\Delta B_{2-1}/\Delta C_{2-1} \geq 1.0$, the alternative 2 is preferred.

In the case of fuzziness, first, it will be assumed that the largest possible value of Alternative 1 for the cash in year t is less than the least possible value of Alternative 2 for the cash in year t. The fuzzy incremental BCR is

$$\Delta\tilde{B}\big/\Delta\tilde{C} =$$

$$\left(\frac{\sum\limits_{t=0}^{n}(B_{2t}^{l(y)} - B_{1t}^{r(y)})(1+r^{r(y)})^{-t}}{\sum\limits_{t=0}^{n}(C_{2t}^{r(y)} - C_{1t}^{l(y)})(1+r^{l(y)})^{-t}}, \frac{\sum\limits_{t=0}^{n}(B_{2t}^{r(y)} - B_{1t}^{l(y)})(1+r^{l(y)})^{-t}}{\sum\limits_{t=0}^{n}(C_{2t}^{l(y)} - C_{1t}^{r(y)})(1+r^{r(y)})^{-t}}\right) \tag{3.23}$$

If $\Delta\tilde{B}\big/\Delta\tilde{C}$ is equal or greater than (1, 1, 1), Alternative 2 is preferred.

In the case of a regular annuity, the fuzzy $\tilde{B}\big/\tilde{C}$ ratio of a single investment alternative is

$$\tilde{B}/\tilde{C} = \left(\frac{A^{l(y)}\gamma(n, r^{r(y)})}{C^{r(y)}}, \frac{A^{r(y)}\gamma(n, r^{l(y)})}{C^{l(y)}}\right) \tag{3.24}$$

where $\tilde{C}$ is the first cost and $\tilde{A}$ is the net annual benefit, and $\gamma(n, r) = (((1+r)^n - 1)/(1+r)^n r)$.

The $\Delta\tilde{B}/\Delta\tilde{C}$ ratio in the case of a regular annuity is

$$\Delta\tilde{B}/\Delta\tilde{C} = (\frac{(A_2^{l(y)} - A_1^{r(y)})\gamma(n, r^{r(y)})}{C_2^{r(y)} - C_1^{l(y)}}, \frac{(A_2^{r(y)} - A_1^{l(y)})\gamma(n, r^{l(y)})}{C_2^{l(y)} - C_1^{r(y)}}) \quad (3.25)$$

3.6 Fuzzy Equivalent Uniform Annual Value (EUAV) Method

The *EUAV* means that all incomes and disbursements (irregular and uniform) must be converted into an equivalent uniform annual amount, which is the same each period. The major advantage of this method over all the other methods is that it does not require making the comparison over the least common multiple of years when the alternatives have different lives [3.3]. The general equation for this method is

$$EUAV = A = NPV\gamma^{-1}(n, r) = NPV[\frac{(1+r)^n r}{(1+r)^n - 1}] \quad (3.26)$$

where *NPV* is the net present value. In the case of fuzziness, $N\tilde{P}V$ will be calculated and then the fuzzy $EU\tilde{A}V$ ($\tilde{A}_n$) will be found. The membership function $\mu(x \mid \tilde{A}_n)$ for $\tilde{A}_n$ is determined by

$$f_{ni}(y| \tilde{A}_n) = f_i(y| N\tilde{P}V)\gamma^{-1}(n, f_i(y \mid \tilde{r})) \quad (3.27)$$

and *TFN(y)* for fuzzy *EUAV* is

$$\tilde{A}_n(y) = (\frac{NPV^{l(y)}}{\gamma(n, r^{l(y)})}, \frac{NPV^{r(y)}}{\gamma(n, r^{r(y)})}) \quad (3.28)$$

3.7 Fuzzy Payback Period (FPP) Method

The payback period method involves the determination of the length of time required to recover the initial cost of investment based on a zero interest rate ignoring the time value of money or a certain interest rate recognizing the time value of money. Let C_{j0} denote the initial cost of investment alternative j, and R_{jt} denote the net revenue received from investment j during period t. Assuming no other negative net cash flows occur, the smallest value of m_j ignoring the time value of money such that

$$\sum_{t=1}^{m_j} R_{jt} \geq C_{j0} \quad (3.29)$$

or the smallest value of m_j recognizing the time value of money such that

$$\sum_{t=1}^{m_j} R_{jt}(1+r)^{-t} \geq C_{j0} \quad (3.30)$$

defines the payback period for the investment j. The investment alternative having the smallest payback period is the preferred alternative. In the case of fuzziness, the smallest value of m_j ignoring the time value of money such that

$$(\sum_{t=1}^{m_j} r_{1jt}, \sum_{t=1}^{m_j} r_{2jt}, \sum_{t=1}^{m_j} r_{3jt}) \geq (C_{1j0}, C_{2jo}, C_{3j0}) \tag{3.31}$$

and the smallest value of m_j recognizing the time value of money such that

$$(\sum_{t=1}^{m_j} \frac{R_{jt}^{l(y)}}{(1+r^{r(y)})^t}, \sum_{t=1}^{m_j} \frac{R_{jt}^{r(y)}}{(1+r^{l(y)})^t}) \geq ((C_{2j0} - C_{1j0})y + C_{1j0}, (C_{2j0} - C_{3j0})y + C_{3j0}) \tag{3.32}$$

defines the payback period for investment j, where r_{kjt} : the kth parameter of a triangular fuzzy R_{jt}; C_{kj0}: the kth parameter of a triangular fuzzy C_{j0}; $R_{jt}^{l(y)}$: the left representation of a triangular fuzzy R_{jt}; $R_{jt}^{r(y)}$: the right representation of a triangular fuzzy R_{jt}. If it is assumed that the discount rate changes from one period to another, $(1+r^{r(y)})^t$ and $(1+r^{l(y)})^t$ will be $\prod_{t'=1}^{t}(1+r_{t'}^{r(y)})$ and $\prod_{t'=1}^{t}(1+r_{t'}^{l(y)})$ respectively.

3.8 Ranking Fuzzy Numbers

It is now necessary to use a ranking method to rank the TFNs such as Chiu and Park's [3.8], Chang's [3.6] method, Dubois and Prade's [3.10] method, Jain's [3.17] method, Kaufmann and Gupta's [3.27] method, Yager's [3.36] method. These methods may give different ranking results and most methods are tedious in graphic manipulation requiring complex mathematical calculation. In the following, three of the methods, which do not require graphical representations, are given.

Kaufmann and Gupta [3.27] suggest three criteria for ranking TFNs with parameters (a,b,c). The dominance sequence is determined according to priority of:

1. Comparing the ordinary number (a+2b+c)/4
2. Comparing the mode, (the corresponding most promise value), b, of each TFN.
3. Comparing the range, c-a, of each TFN.

The preference of projects is determined by the amount of their ordinary numbers. The project with the larger ordinary number is preferred. If the ordinary numbers are equal, the project with the larger corresponding most promising value is preferred. If projects have the same ordinary number and most promising value, the project with the larger range is preferred.

Liou and Wang [3.32] propose the total integral value method with an index of optimism $\omega \in [0,1]$. Let $\tilde{A}$ be a fuzzy number with left membership function $f_{\tilde{A}}^{L}$ and right membership function $f_{\tilde{A}}^{R}$. Then the total integral value is defined as:

$$E_{\omega}\left(\tilde{A}\right) = \omega E_{R}\left(\tilde{A}\right) + (1-\omega) E_{L}\left(\tilde{A}\right) \tag{3.33}$$

where

$$E_{R}\left(\tilde{A}\right) = \int_{\alpha}^{\beta} x f_{\tilde{A}}^{R}(x)\, dx \tag{3.34}$$

$$E_{L}\left(\tilde{A}\right) = \int_{\gamma}^{\delta} x f_{\tilde{A}}^{L}(x)\, dx \tag{3.35}$$

where $-\infty \leq \alpha \leq \beta \leq \gamma \leq \delta \geq +\infty$ and a trapezoidal fuzzy number is denoted by $(\alpha, \beta, \gamma, \delta)$. For a triangular fuzzy number,$\tilde{A} = (a, b, c)$,

$$E_{\omega}\left(\tilde{A}\right) = \frac{1}{2}\left[\omega (a+b) + (1-\omega)(b+c)\right] \tag{3.36}$$

and for a trapezoidal fuzzy number, $\tilde{B} = (\alpha, \beta, \gamma, \delta)$,

$$E_{\omega}\left(\tilde{B}\right) = \frac{1}{2}\left[\omega (\gamma+\delta) + (1-\omega)(\alpha+\beta)\right] \tag{3.37}$$

Chiu and Park's [3.8] weighted method for ranking TFNs with parameters (a, b, c) is formulated as

$$((a+b+c)/3) + wb$$

where w is a value determined by the nature and the magnitude of the most promising value. The preference of projects is determined by the magnitude of this sum.

The computer software developed by the authors, FUZDYN, has the ability to use many ranking methods that are tedious in graphic manipulation requiring complex mathematical calculation. To select the ranking method required by the decision maker, the following form in Fig. 3.3 is used:

3.9 Fuzzy Internal Rate of Return (IRR) Method

The *IRR* method is referred to in the economic analysis literature as the discounted cash flow rate of return, internal rate of return, and the true rate of return. The internal rate of return on an investment is defined as the rate

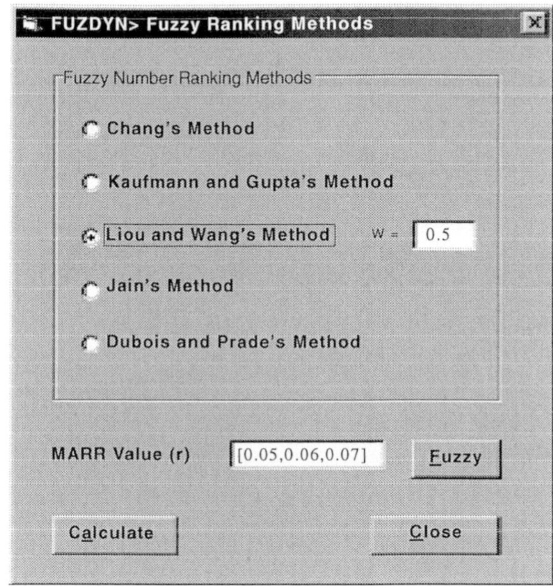

Fig. 3.3. The form of fuzzy ranking methods

of interest earned on the unrecovered balance of an investment. Letting r^* denote the rate of return, the equation for obtaining r^* is

$$\sum_{t=1}^{n} P_t(1+r^*)^{-t} - FC = 0 \tag{3.38}$$

where P_t is the net cash flow at the end of period t.

Assume the cash flow $\tilde{F} = \tilde{F}_0, \tilde{F}_1, ..., \tilde{F}_N$ is fuzzy. $\tilde{F}_n$ is a negative fuzzy number and the other $\tilde{F}_i$ may be positive or negative fuzzy numbers. The fuzzy $IRR(\tilde{F}, n)$ is a fuzzy interest rate $\tilde{r}$ that makes the present value of all future cash amounts equal to the initial cash outlay. Therefore, the fuzzy number $\tilde{r}$ satisfies

$$\sum_{i=1}^{n} PV_{k(i)}(\tilde{F}_i, \tilde{r}) = -\tilde{F}_0 \tag{3.39}$$

where $\sum$ is fuzzy addition, $k(i)$=1 if $\tilde{F}_i$ is negative and $k(i)$=2 if $\tilde{F}_i$ is positive.

Buckley [3.5] shows that such simple fuzzy cash flows may not have a fuzzy *IRR* and concludes that the *IRR* technique does not extend to fuzzy cash flows. Ward [3.34] considers Eq. (3.38) and explains that such a procedure can not be applied for the fuzzy case because the right hand side of Eq. (3.38) is fuzzy, 0 is crisp, and an equality is impossible.

3.10 An Expansion to Geometric and Trigonometric Cash Flows

When the value of a given cash flow differs from the value of the previous cash flow by a constant percentage, $g\%$, then the series is referred to as a *geometric series*. If the value of a given cash flow differs from the value of the previous cash flow by a sinusoidal wave or a cosinusoidal wave, then the series is referred to as a *trigonometric series* .

3.10.1 Geometric Series–Fuzzy Cash Flows in Discrete Compounding

The present value of a crisp geometric series is given by

$$P = \sum_{n=1}^{N} F_1(1+g)^{n-1}(1+i)^{-n} = \frac{F_1}{1+g}\sum_{n=1}^{N}(\frac{1+g}{1+i})^n \tag{3.40}$$

where F_1 is the first cash at the end of the first year. When this sum is made, the following present value equation is obtained:

$$P = \begin{cases} F_1[\frac{1-(1+g)^N(1+i)^{-N}}{i-g}], & i \neq g \\ \frac{NF_1}{1+i} \quad , & i = g \end{cases} \tag{3.41}$$

and the future value is

$$F = \begin{cases} F_1[\frac{(1+i)^N-(1+g)^N}{i-g}], & i \neq g \\ NF_1(1+i)^{N-1} \quad , & i = g \end{cases} \tag{3.42}$$

In the case of fuzziness, the parameters used in Eq.(3.40) will be assumed to be fuzzy numbers, except project life. Let $\gamma(i,g,N) = [\frac{1-(1+g)^N(1+i)^{-N}}{i-g}], i \neq g$. As it is in Fig. 3.1 and Fig. 3.2, when k=1, the left side representation will be depicted and when k=2, the right side representation will be depicted. In this case, for $i \neq g$

$$f_{Nk}(y \mid \tilde{P}_N) = f_k(y \mid \tilde{F}_1)\gamma(f_{3-k}(y|\tilde{i}), f_{3-k}(y \mid \tilde{g}), N) \tag{3.43}$$

In Eq.(3.43), the least possible value is calculated for $k=1$ and $y=0$; the largest possible value is calculated for $k=2$ and $y=0$; the most promising value is calculated for k= 1 or $k=2$ and $y=1$.

To calculate the future value of a fuzzy geometric cash flow, let $\zeta(i,g,N) = [\frac{(1+i)^N-(1+g)^N}{i-g}], i \neq g$. Then the fuzzy future value is

$$f_{Nk}(y|\tilde{F}_N) = f_k(y \mid \tilde{F}_1)\zeta(f_k(y \mid \tilde{i}), f_k(y \mid \tilde{g}), N) \tag{3.44}$$

In Eq. (3.44), the least possible value is calculated for $k = 1$ and $y = 0$; the largest possible value is calculated for $k = 2$ and $y = 0$; the most promising value is calculated for $k = 1$ or $k = 2$ and $y = 1$. This is also valid for the formulas developed at the rest of the paper.

The fuzzy uniform equivalent annual value can be calculated by using Eq. (3.45):

$$f_{Nk}(y \mid \tilde{A}) = f_k(y \mid \tilde{P}_N)\vartheta(f_k(y \mid \tilde{i}), N) \tag{3.45}$$

where $\vartheta(i, N) = [\frac{(1+i)^N i}{(1+i)^N - 1}]$ and $f(y \mid \tilde{P}_N)$ is the fuzzy present value of the fuzzy geometric cash flows.

3.10.2 Geometric Series–Fuzzy Cash Flows in Continuous Compounding

In the case of crisp sets, the present and future values of discrete payments are given by Eq.(3.46) and Eq.(3.47) respectively:

$$P = \begin{cases} F_1[\frac{1-e^{(g-r)N}}{e^r - e^g}], & r \neq g \\ \frac{NF_1}{e^r}, & g = e^r - 1 \end{cases} \tag{3.46}$$

$$F = \begin{cases} F_1[\frac{e^{rN} - e^{gN}}{e^r - e^g}], & r \neq g \\ NF_1 e^{r(N-1)}, & g = e^r - 1 \end{cases} \tag{3.47}$$

and the present and future values of continuous payments are given by Eq.(3.48) and Eq.(3.49) respectively:

$$P = \begin{cases} F_1[\frac{1-e^{N(g-r)}}{r-g}], & r \neq g \\ \frac{NF_1}{1+r}, & r = g \end{cases} \tag{3.48}$$

$$F = \begin{cases} F_1[\frac{e^{rN} - e^{gN}}{r-g}], & r \neq g \\ \frac{NF_1 e^{rN}}{1+r}, & r = g \end{cases} \tag{3.49}$$

The fuzzy present and future values of the fuzzy geometric discrete cash flows in continuous compounding can be given as in Eq.(3.50) and Eq.(3.51) respectively:

$$f_{Nk}(y \mid \tilde{P}_N) = f_k(y \mid \tilde{F}_1)\beta(f_{3-k}(y \mid \tilde{r}), f_{3-k}(y \mid \tilde{g}), N) \tag{3.50}$$

$$f_{Nk}(y \mid \tilde{F}) = f_k(y \mid \tilde{F}_1)\tau(f_k(y \mid \tilde{r}), f_k(y \mid \tilde{g}), N) \tag{3.51}$$

where $\beta(r, g, N) = \frac{1-e^{(g-r)N}}{e^r - e^g}, r \neq g$ for present value and $\tau(r, g, N) = \frac{e^{rN} - e^{gN}}{e^r - e^g}, r \neq g$ for future value.

The fuzzy present and future values of the fuzzy geometric continuous cash flows in continuous compounding can be given as in Eq.(3.52) and Eq.(3.53) respectively:

$$f_{Nk}(y|\tilde{P}_N) = f_k(y|\tilde{F}_1)\eta(f_{3-k}(y \mid \tilde{r}), f_{3-k}(y \mid \tilde{g}), N) \tag{3.52}$$

$$f_{Nk}(y \mid \tilde{F}_N) = f_k(y \mid \tilde{F}_1)\upsilon(f_k(y \mid \tilde{r}), f_k(y \mid \tilde{g}), N) \tag{3.53}$$

where $\eta(r, g, N) = \frac{1-e^{(g-r)N}}{r-g}$, $\upsilon(r, g, N) = \frac{e^{rN} - e^{gN}}{r-g}, r \neq g$

3.10.3 Trigonometric Series–Fuzzy Continuous Cash Flows

In Fig. 3.4, the function of the semi-sinusoidal wave cash flows is depicted. This function, $h(t)$, is given by Eq.(3.54) in the crisp case:

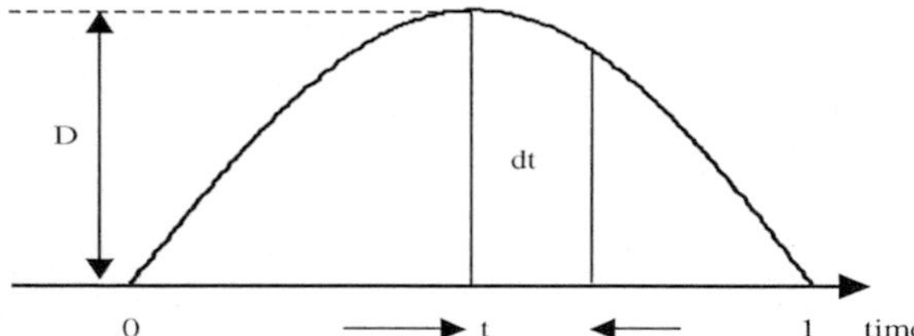

Fig. 3.4. Semi sinusoidal wave cash flow function

$$h(t) = \begin{cases} D\sin(\pi\, t), & 0 \le t \le 1 \\ 0 \quad , & \text{otherwise} \end{cases} \tag{3.54}$$

The future value of a semi-sinusoidal cash flow for T=1 and g is defined by Eq. (3.55) :

$$V(g, 1) = D\int_0^1 e^{r(1-t)} \sin(\pi\, t)\, dt = D[\frac{\pi(2+g)}{r^2 + \pi^2}] \tag{3.55}$$

Fig. 3.5 shows the function of a cosinusoidal wave cash flow. This function, $h(t)$, is given by Eq.(3.56):

$$h(t) = \begin{cases} D(\cos(2\pi\, t) + 1), & 0 \le t \le 1 \\ 0 \quad , & \text{otherwise} \end{cases} \tag{3.56}$$

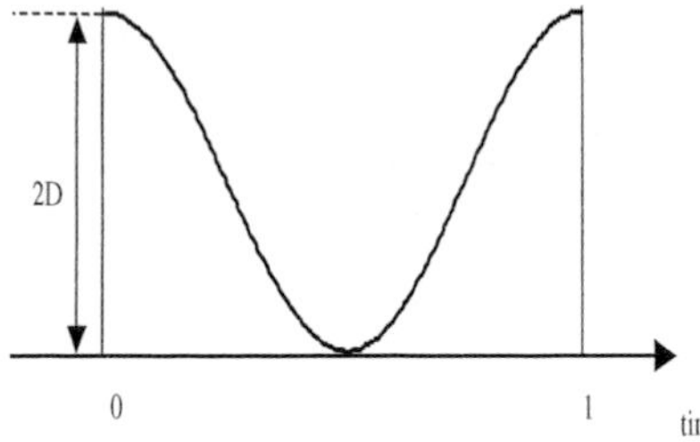

Fig. 3.5. Cosinusoidal wave cash flow function

The future value of a cosinusoidal cash flow for T=1 and g is defined as

$$V(g,1) = D\int_0^1 e^{r(1-t)}(\cos(2\pi\, t)+1)dt = D[\frac{gr}{r^2+4\pi^2}+\frac{g}{r}] \quad (3.57)$$

Let the parameters in Eq. (3.55), r and g, be fuzzy numbers. The future value of the semi-sinusoidal cash flows as in Fig. 3.6 is given by

$$f_{Nk}(y \mid \tilde{F}_N) = f_k(y \mid \tilde{D})\phi(f_{3-k}(y \mid \tilde{r}), f_k(y \mid \tilde{g}))\varphi(f_k(y \mid \tilde{r}), N) \quad (3.58)$$

where $\phi(r,g) = \pi(2+g)/(r^2+\pi^2)$, $\varphi(r,N) = (e^{rN}-1)/(e^r-1)$.

The present value of the semi-sinusoidal cash flows is given by Eq. (3.59):

$$f_{Nk}(y \mid \tilde{P}_N) = f_k(y \mid \tilde{D})\phi(f_{3-k}(y \mid \tilde{r}), f_k(y \mid \tilde{g}))\psi(f_{3-k}(y \mid \tilde{r}), N) \quad (3.59)$$

where $\psi(r,N) = (e^{rN}-1)/((e^r-1)e^{rN})$.

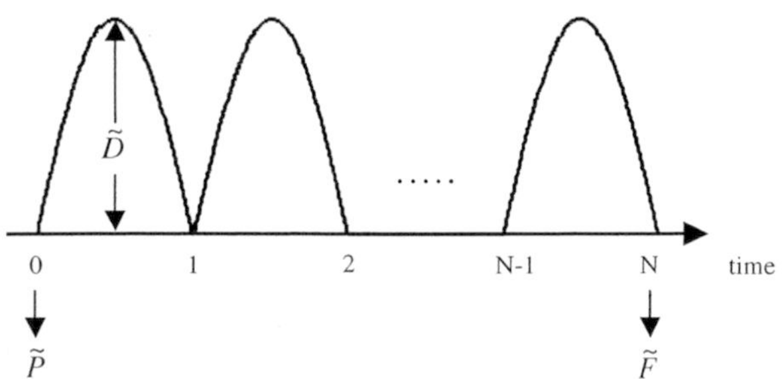

Fig. 3.6. Fuzzy sinusoidal cash flow diagram

3.10.4 Numeric Example I

The continuous profit function of a firm producing ice cream during a year is similar to semi-sinusoidal wave cash flows whose g is around 4%. The maximum level in $ of the ice-cream sales is between the end of June and the beginning of July. The profit amount obtained on this point is around

\$120,000. The firm manager uses a minimum attractive rate of return of around 10%, compounded continuously and he wants to know the present worth of the 10-year profit and the possibility of having a present worth of \$1,500,000.

'Around \$120,000' can be represented by a TFN, (\$100,000;\$120,000;\$130,000). 'Around 10%' can be represented by a TFN, (9%;10%;12%). 'Around 4%' can be represented by a TFN, (3%;4%;6%)

$f_2(y|\tilde{r}) = 0.12 - 0.02y$ $\quad f_1(y|\tilde{r}) = 0.09 + 0.01y$

$f_1(y|\tilde{D}) = 100,000 + 20,000y$ $\quad f_2(y|\tilde{D}) = 130,000 - 10,000y$

$f_{10,1}(y|\tilde{P_{10}}) = f_1(y|\tilde{D})\Phi(f_2(y|\tilde{r}), f_1(y|\tilde{g}))\Psi(f_2(y|\tilde{r}), 10)$

$f_{10,2}(y|\tilde{P_{10}}) = f_2(y|\tilde{D})\Phi(f_1(y|\tilde{r}), f_2(y|\tilde{g}))\Psi(f_1(y|\tilde{r}), 10)$

$f_1(y|\tilde{g}) = 0.03 + 0.01y$ $\quad f_2(y|\tilde{g}) = 0.06 - 0.02y$

$$f_{10,1}(y|\tilde{P}_{10}) = (100,000 + 20,000y) \times \left[\frac{\pi(2.03 + 0.01y)}{[(0.12 - 0.02y)^2 + \pi^2]}\right] \left[\frac{e^{(0.12-0.02y)10} - 1}{e^{0.12-0.02y} - 1}\right] \frac{1}{e^{(0.12-0.02y)10}}$$

$$f_{10,2}(y|\tilde{P}_{10}) = (130,000 - 10,000y) \times \left[\frac{\pi(2.06 - 0.02y)}{[(0.09 + 0.01y)^2 + \pi^2]}\right] \left[\frac{e^{(0.09+0.01y)10} - 1}{e^{0.09+0.01y} - 1}\right] \frac{1}{e^{(0.09+0.01y)10}}$$

For $y = 1$, the most possible value is $f_{10,1}(y|\tilde{P}_{10}) = f_{10,2}(y|\tilde{P}_{10}) = \$467,870.9$.
For $y = 0$, the smallest possible value is $f_{10,1}(y|\tilde{P}_{10}) = \$353,647.1$.
For $y = 0$, the largest possible value is $f_{10,2}(y|\tilde{P}_{10}) = \$536,712.8$.

It seems to be impossible to have a present worth of \$1,500,000.

The present and future values of the fuzzy cosinusoidal cash flows as in Fig. 3.7 can be given by Eq. (3.60) and Eq. (3.61) respectively:

$$f_{Nk}(y \mid \tilde{P}_N) = f_k(y \mid \tilde{D})\xi(f_{3-k}(y \mid \tilde{r}), f_k(y \mid \tilde{g}))\Psi(f_{3-k}(y \mid \tilde{r}), N) \quad (3.60)$$

where $\xi(r, g) = [\frac{gr}{r^2+4\pi^2} + \frac{g}{r}]$ and the fuzzy future value is

$$f_{Nk}(y \mid \tilde{F}_N) = f_k(y \mid \tilde{D})\xi(f_{3-k}(y \mid \tilde{r}), f_k(y \mid \tilde{g}))\varphi(f_k(y \mid \tilde{r}), N) \quad (3.61)$$

3.10.5 Numeric Example II

The continuous cash flows of a firm is similar to cosinusoidal cash flows. The maximum level of the cash flows during a year is around \$780,000. The fuzzy nominal cost of capital is around 8% per year. The fuzzy geometric growth rate of the cash flows is around 4% per year. Let us compute the future worth of a 10 year working period.

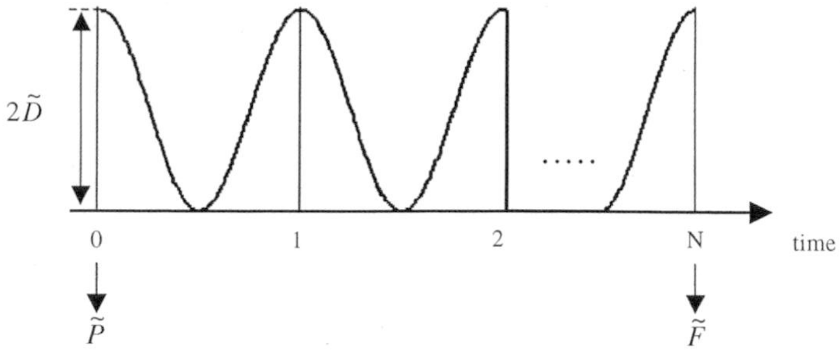

Fig. 3.7. Fuzzy sinusoidal cash flow diagram

Let us define
$\tilde{D} = (\$300,000; \$390,000; \$420,000), f_1(y|\tilde{D}) = 350,000 + 40,000y$,
$f_2(y|\tilde{D}) = 420,000 - 30,000y$, $\tilde{r} = (6\%; 8\%; 10\%)$, $f_1(y|\tilde{r}) = 0.06 + 0.02y$,
$f_2(y|\tilde{r}) = 0.10 - 0.02y$, $\tilde{g} = (3\%; 4\%; 5\%)$, $f_1(y|\tilde{g}) = 0.03 + 0.01y$,
$f_2(y|\tilde{g}) = 0.05 - 0.01y$

$$f_{10,1}(y|\tilde{F}_{10}) = (350,000 + 40,000y) \times \left[\frac{(0.03 + 0.01y)(0.10 - 0.02y)}{(0.10 - 0.02y)^2 + 4\pi^2} + \frac{0.03 + 0.01y}{0.10 - 0.02y}\right]\left[\frac{e^{(0.06+0.02y)10} - 1}{e^{0.06+0.02y} - 1}\right]$$

$$f_{10,2}(y|\tilde{F}_{10}) = (420,000 - 30,000y) \times \left[\frac{(0.05 - 0.01y)(0.06 + 0.02y)}{(0.06 + 0.02y)^2 + 4\pi^2} + \frac{0.05 - 0.01y}{0.06 + 0.02y}\right]\left[\frac{e^{(0.10-0.02y)10} - 1}{e^{0.10-0.02y} - 1}\right]$$

For $y = 1$, the most possible value is $f_{10,1}(y|\tilde{F}_{10}) = \$2,869,823.5$.
For $y = 0$, the smallest possible value is $f_{10,1}(y|\tilde{F}_{10}) = \$1,396,331.5$.
For $y = 0$, the largest possible value is $f_{10,2}(y|\tilde{F}_{10}) = \$5,718,818.9$.

3.11 Dynamic Programming for Multilevel Investment Analysis

Dynamic programming is a technique that can be used to solve many optimization problems. In most applications, dynamic programming obtains solutions by working backward from the end of a problem toward the beginning, thus breaking up a large, unwieldy problem into a series of smaller, more tractable problems. The characteristics of dynamic programming applications are [3.35]

- The problem can be divided into stages with a decision required at each stage.

- Each stage has a number of states associated with it.
- The decision chosen at any stage describes how the state at the current stage is transformed into the state at the next stage.
- Given the current state, the optimal decision for each of the remaining stages must not depend on previously reached states or previously chosen decisions.
- If the states for the problem have been classified into one of T stages, there must be a recursion that relates the cost or reward earned during stages t, $t+1,\ldots,T$ to the cost or reward earned from stages $t+1$, $t+2,\ldots,T$.

The dynamic programming recursion can often be written in the following form. For a min problem with fixed output:

$$\begin{aligned} f_t\left(i\right) = \;& \min\{\text{ (cost during stage t)} + \\ & f_{t+1}\left(\text{new state at stage t}+1\right)\} \end{aligned} \tag{3.62}$$

and for a max problem with fixed input, it is

$$\begin{aligned} f_t\left(i\right) = \;& \max\{\text{ (benefits during state t)} + \\ & f_{t+1}\left(\text{new state at stage t}+1\right)\} \end{aligned} \tag{3.63}$$

or for a max problem neither input nor output fixed, it is

$$\begin{aligned} f_t\left(i\right) = \max\{(\text{'benefits} - \text{costs' during state t}) + \\ f_{t+1}\left(\text{new state at stage t}+1\right)\} \end{aligned} \tag{3.64}$$

where the minimum in Eq. (3.62) or maximum in Eq. (3.63) and Eq. (3.64) is over all decisions that are allowable, or feasible, when the state at stage t is i. In Eq. (3.62), $f_t\left(i\right)$ is the minimum cost and in Eq. (3.63) the maximum benefit incurred from stage t to the end of the problem, given that at stage t the state is i.

In deterministic dynamic programming, a specification of the current state and current decision is enough to tell us with certainty the new state and costs during the current stage. In many practical problems, these factors may not be known with certainty, even if the current state and decision are known. When we use dynamic programming to solve problems in which the current period's cost or the next period's state is random, we call these problems *probabilistic dynamic programming problems (PDPs)*. In a *PDP*, the decision-maker's goal is usually to minimize expected cost incurred or to maximize expected reward earned over a given time horizon.

Many *PDPs* can be solved using recursions of the following forms. For min problems:

$$\begin{aligned} f_t\left(i\right) = \;& \min_a\Big[\left(\text{ expected cost during stage t}|\,\text{i, a }\right) + \\ & \sum_j p\left(j|\,i,a,t\right) f_{t+1}\left(j\right)\Big] \end{aligned} \tag{3.65}$$

and for max problems:

$$f_t(i) = \max_a \Big[(\text{expected reward during stage } \mathrm{t} | \mathrm{i,a}) + \sum_j p(j|i,a,t) f_{t+1}(j) \Big] \tag{3.66}$$

where

i : the state at the beginning of stage t.

a: all actions that are feasible when the state at the beginning of stage t is i.

$p(j|i,a,t)$:the probability that the next period's state will be j, given that the current state is i and action a is chosen.

In the formulations above, we assume that benefits and costs received during later years are weighted the same as benefits and costs received during earlier years. But later benefits and costs should be weighted less than earlier benefits and costs. We can incorporate the time value of money into the dynamic programming recursion in the following way. For a max problem with neither input nor output fixed,

$$\begin{aligned} f_t(i) = \max \{ & (\text{'benefits} - \text{costs' during state t}) + \\ & \tfrac{1}{(1+r)} f_{t+1} (\text{new state at stage t} + 1) \} \end{aligned} \tag{3.67}$$

where r is the time value of money.

Many capital budgeting problems allow of a *dynamic* formulation. There may actually be several decision points, but even if this is not so, if the decision problem can be divided up into *stages* then a discrete dynamic expression is possible. Many problems allow of either static or dynamic expression. The choice of form would be up to the problem solver. Characteristically, a dynamic economizing model allocates scarce resources between alternative uses between initial and terminal times. In the case of equal-life multilevel investments, each investment in the set has the following characteristic: the amount to be invested has several *possible* values, and the rate of return varies with the amount invested. Each sum that may be invested represents a distinct *level* of investment, and the investment therefore has multiple levels. Examples of multilevel investments may be the purchase of labor-saving equipment where several types of equipment are available and each type has a unique cost. The level of investment in labor-saving equipment depends on the type of equipment selected. Another example is the construction and rental of an office building, where the owner-builder has a choice concerning the number of stories the building is to contain [3.29].

3.11.1 Fuzzy Dynamic Programming: Literature Review

Fuzzy dynamic programming has found many applications to real-world problems: Health care, flexible manufacturing systems, integrated regional development, transportation networks and transportation of hazardous waste, chemical engineering, power and energy systems, water resource systems.

Li and Lai [3.31] develop a new fuzzy dynamic programming approach to solve hybrid multi-objective multistage decision-making problems. They present a methodology of fuzzy evaluation and fuzzy optimization for hybrid multi-objective systems, in which the qualitative and quantitative objectives are synthetically considered. Esogbue [3.11] presents the essential elements of fuzzy dynamic programming and computational aspects as well as various key real world applications. Fu and Wang [3.13] establish a model in the framework of fuzzy project network by team approach under the consideration of uncertain resource demand and the budget limit. The model is transformed into a classical linear program formula and its results show that the cause-effect relations of insufficient resources or over due of the project is identified for better management. Lai and Li [3.30] develop a new approach using dynamic programming to solve the multiple-objective resource allocation problem. There are two key issues being addressed in the approach. The first one is to develop a methodology of fuzzy evaluation and fuzzy optimization for multiple-objective systems. The second one is to design a dynamic optimization algorithm by incorporating the method of fuzzy evaluation and fuzzy optimization with the conventional dynamic programming technique. Esogbue [3.12] considers both time and space complexity problems associated with the fuzzy dynamic programming model. Kahraman *et al.* [3.23] use fuzzy dynamic programmingto combine equal-life multilevel investments. Huang *et al.*[3.14] develop a fuzzy dynamic programmingapproach to solve the direct load control problem of the air conditioner loads. Kacprzyk and Esogbue [3.18] survey major developments and applications of fuzzy dynamic programming which is advocated as a promising attempt at making dynamic programming models more realistic by a relaxation of often artificial assumptions of precision as to the constraints, goals, states and their transitions, termination time, etc. Chin [3.7] proposes a new approach using fuzzy dynamic programmingto decide the optimal location and size of compensation shunt capacitors for distribution systems with harmonic distortion. The problem is formulated as a fuzzy dynamic programming of the minimization of real power loss and capacitor cost under the constraints of voltage limits and total harmonic distortion. Hussein and Abo-Sinna [3.15] propose a new approach using fuzzy dynamic programmingto solve the multiple criteria resource allocation problems. They conclude that solutions obtained by the approach are always efficient; hence an "optimal" compromise solution can be introduced. Berenji [3.2] develops a new algorithm called Fuzzy Q-Learning, which extends Watkin's Q-Learning method. It is used for decision processes in which the goals and/or the constraints, but not necessarily the system under control, are fuzzy in nature. He shows that fuzzy Q-Learning provides an alternative solution simpler than the Bellman-Zadeh's[3.1] fuzzy dynamic programming approach.

3.11.2 Crisp Dynamic Programming for Multilevel Investments

The solution of a dynamic programming problem of multilevel investments consists of the following steps:

1. Devise all possible investments that encompass plans A and B alone, applying an upper limit of $\$L$ to the amount invested. Compound the corresponding annual dividends. Let Q denote the amount of capital to be allocated to the combination of plans A and B, where Q can range from \$ X to \$ kX where k=1, 2, 3... Although both plans A and B fall within our purview in this step, it is understood that Q can be allocated to A alone or to B alone.
2. Identify the most lucrative combination of Plans A and B corresponding to every possible value of Q.
3. Devise all possible investments that encompass plans A, B, and C, and identify the most lucrative one.

Now let's consider the selection among multilevel investments when crisp cash flows are known. In other words, let's deal with the problem from capital budgeting viewpoint.

Newnan [3.33] shows that independent proposals competing for funding should be picked according to their IRR values- monotonically from highest to lowest. Ranking on present-worth values (computed at a specified MARR) may not give the same results. Given a specified minimum attractive rate of return(MARR) value, Newnan [3.33] suggests that proposals be ranked on the basis of

$$\text{Ranking ratio} = \frac{\text{Proposal PW(MARR)}}{\text{Proposal first cost}} \tag{3.68}$$

where PW is the present worth of a proposal. The larger ratio indicates the better proposal.

Now assume that cash flows for l independent proposals that have passed a screening based on a MARR of $r\%$ are given in Table 3.1 and we have a $\$L$ capital limitation. The problem is which combination of proposals should be funded. The solution consists of the following steps:

1. Devise all possible investments that encompass proposals 1 and 2 alone, applying an upper limit of $\$L$ to the amount invested. Compute the present worth of each proposal in the possible combinations using the discounted cash flow techniques. $\$L$ can be allocated to proposal 1 alone or to proposal 2 alone or to any other combination.
2. Identify the most lucrative combination of proposals 1 and 2 corresponding to every possible value of $\$L$, using the ranking ratio in Eq.(3.68).
3. Devise all possible investments that encompass proposals 1, 2, and 3, and identify the most lucrative one as in step 2.

Table 3.1. Cash flows for l independent proposals

End-of-period-cash-flow, $						
Proposal	Investment,$	Period 1	Period 2	Period 3	...	Period n
1	$X	CF_{11}^1	CF_{12}^1	CF_{13}^1	...	CF_{1n}^1
	$2X	CF_{11}^2	CF_{12}^2	CF_{13}^2	...	CF_{1n}^2
	$3X	CF_{11}^3	CF_{12}^3	CF_{13}^3	...	CF_{1n}^3
	...	...	...	...	...	...
	$$k$X	CF_{11}^k	CF_{12}^k	CF_{13}^k	...	CF_{1n}^k
2	$X	CF_{21}^1	CF_{22}^1	CF_{23}^1	...	CF_{2n}^1
	$2X	CF_{21}^2	CF_{22}^2	CF_{23}^2	...	CF_{2n}^2
	$3X	CF_{21}^3	CF_{22}^3	CF_{23}^3	...	CF_{2n}^3
	...	...	...	...	...	...
	$$k$X	CF_{21}^k	CF_{22}^k	CF_{23}^k	...	CF_{2n}^k
...	...	...	...	...	...	...
l	$X	CF_{l1}^1	CF_{l2}^1	CF_{l3}^1	...	CF_{ln}^1
	$2X	CF_{l1}^2	CF_{l2}^2	CF_{l3}^2	...	CF_{ln}^2
	$3X	CF_{l1}^3	CF_{l2}^3	CF_{l3}^3	...	CF_{ln}^3
...	...	...	...	...	...	...
	$$k$X	CF_{l1}^k	CF_{l2}^k	CF_{l3}^k	...	CF_{ln}^k

4. Continue increasing the number of proposals in the combination until the number is l and identify the most lucrative combination.

In Table 3.1, CF_{lt}^k indicates the cash flow of proposal l in period t at kth level of investment.

3.11.3 Fuzzy Dynamic Programming for Multilevel Investments

Assume that we know the fuzzy cash flows of multilevel investments and we deal with the problem from capital budgeting viewpoint. Given a fuzzy specified (MARR) value, proposals can be ranked on the basis of

$$\text{Ranking ratio} = \frac{\text{Proposal fuzzy PW(MARR)}}{\text{Proposal fuzzy first cost}} \tag{3.69}$$

where PW is the present worth of a proposal. The larger ratio indicates the better proposal. Kahraman *et al.* [3.22] and Kahraman [3.19] use fuzzy present worth and fuzzy benefit/cost ratio analysis for the justification of manufacturing technologies and for public work projects.

Now assume that cash flows for l independent proposals that have passed a screening based on a MARR of $\tilde{r}\%$ are given in Table 3.2 and we have a $\$\tilde{L}$ capital limitation. In Table 3.2, $\tilde{CF}_{lt}^k$ indicates the fuzzy cash flow of proposal l in period t at kth level of investment. The problem is which combination of proposals should be funded. The solution consists of the following steps:

1. Devise all possible investments that encompass proposals 1 and 2 alone, applying an upper limit of $\$\tilde{L}$ to the fuzzy amount invested. Compute the fuzzy present worth of each proposal in the possible combinations using the fuzzy discounted cash flow techniques [3.20], [3.21]. $\$\tilde{L}$ can be allocated to proposal 1 alone or to proposal 2 alone or to any other combination.
2. Identify the most lucrative combination of proposals 1 and 2 corresponding to every possible value of $\$\tilde{L}$, using the ranking ratio in Eq. (3.69). Use a ranking method of fuzzy numbers to identify the most lucrative combination.
3. Devise all possible investments that encompass proposals 1, 2, and 3, and identify the most lucrative one as in step 2. Use a ranking method of fuzzy numbers to identify the most lucrative combination.
4. Continue increasing the number of proposals in the combination until the number is l and identify the most lucrative combination. Use a ranking method of fuzzy numbers to identify the most lucrative combination.

Table 3.2. Fuzzy cash flows for l independent proposals

End-of-period-cash-flow, $						
Proposal	Investment,$	Period 1	Period 2	Period 3	...	Period n
1	$\$\tilde{X}$	$\tilde{CF}_{11}^{1}$	$\tilde{CF}_{12}^{1}$	$\tilde{CF}_{13}^{1}$	...	$\tilde{CF}_{1n}^{1}$
	$\$2\tilde{X}$	$\tilde{CF}_{11}^{2}$	$\tilde{CF}_{12}^{2}$	$\tilde{CF}_{13}^{2}$	...	$\tilde{CF}_{1n}^{2}$
	$\$3\tilde{X}$	$\tilde{CF}_{11}^{3}$	$\tilde{CF}_{12}^{3}$	$\tilde{CF}_{13}^{3}$	...	$\tilde{CF}_{1n}^{3}$
	...	...	...	...	...	...
	$\$k\tilde{X}$	$\tilde{CF}_{11}^{k}$	$\tilde{CF}_{12}^{k}$	$\tilde{CF}_{13}^{k}$	...	$\tilde{CF}_{1n}^{k}$
2	$\$\tilde{X}$	$\tilde{CF}_{21}^{1}$	$\tilde{CF}_{22}^{1}$	$\tilde{CF}_{23}^{1}$	...	$\tilde{CF}_{2n}^{1}$
	$\$2\tilde{X}$	$\tilde{CF}_{21}^{2}$	$\tilde{CF}_{22}^{2}$	$\tilde{CF}_{23}^{2}$	...	$\tilde{CF}_{2n}^{2}$
	$\$3\tilde{X}$	$\tilde{CF}_{21}^{3}$	$\tilde{CF}_{22}^{3}$	$\tilde{CF}_{23}^{3}$	...	$\tilde{CF}_{2n}^{3}$
	...	...	...	...	...	...
	$\$k\tilde{X}$	$\tilde{CF}_{21}^{k}$	$\tilde{CF}_{22}^{k}$	$\tilde{CF}_{23}^{k}$	...	$\tilde{CF}_{2n}^{k}$
...	...	...	...	...	...	...
l	$\$\tilde{X}$	$\tilde{CF}_{l1}^{1}$	$\tilde{CF}_{l2}^{1}$	$\tilde{CF}_{l3}^{1}$	...	$\tilde{CF}_{ln}^{1}$
	$\$2\tilde{X}$	$\tilde{CF}_{l1}^{2}$	$\tilde{CF}_{l2}^{2}$	$\tilde{CF}_{l3}^{2}$	...	$\tilde{CF}_{ln}^{2}$
	$\$3\tilde{X}$	$\tilde{CF}_{l1}^{3}$	$\tilde{CF}_{l2}^{3}$	$\tilde{CF}_{l3}^{3}$	...	$\tilde{CF}_{ln}^{3}$
	...	...	...	...	...	...
	$\$k\tilde{X}$	$\tilde{CF}_{l1}^{k}$	$\tilde{CF}_{l2}^{k}$	$\tilde{CF}_{l3}^{k}$	...	$\tilde{CF}_{ln}^{k}$

3.11.4 A Numeric Example

A firm has $(15000, 21000, 27000) available for investment, and three investment proposals are under consideration. Each proposal has these features:

the amount that can be invested is a multiple of \$(5000, 7000, 9000); the investors receive annual unequal receipts; each proposal has a useful life of three years. Table 3.3 lists the annual geometric receipts corresponding to the various fuzzy levels of investment. Devise the most lucrative composite investment using fuzzy dynamic programming. The company-specified MARR value,$\tilde{r}\%$, is (5%, 6%, 7%) per year.

Table 3.3. Fuzzy cash flows for three independent proposals ($\times$ \$1,000)

Proposal	Investment, \$	Year 1	Year 2	Year 3
1	\$(5, 7, 9)	(3, 4, 5)	(3.3, 4.4, 5.5)	(3.63, 4.84, 6.05)
	\$(10, 14, 18)	(5, 6, 7)	(5.6, 6.72, 7.84)	(6.272, 7.526, 8.78)
	\$(15, 21, 27)	(8, 9, 10)	(9.12, 10.26, 11.4)	(10.396, 11.696, 12.996)
2	\$(5, 7, 9)	(3, 4, 6)	(3.3, 4.4, 6.6)	(3.63, 4.84, 7.392)
	\$(10, 14, 18)	(4, 6, 7)	(4.48, 6.72, 7.84)	(5.017, 7.526, 8.78)
	\$(15, 21, 27)	(5, 9, 10)	(5.7, 10.26, 11.4)	(6.498, 11.696, 12.996)
3	\$(5, 7, 9)	(3, 3, 4)	(3.3, 3.3, 4.4)	(3.630, 3.63, 4.84)
	\$(10, 14, 18)	(5, 7, 7)	(5.6, 7.84, 7.84)	(6.272, 7.526, 7.526)
	\$(15, 21, 27)	(8, 9, 12)	(9.12, 10.26, 13.68)	(10.396, 11.696, 15.595)

In FUZDYN, the project definition is as in Fig. 3.8.

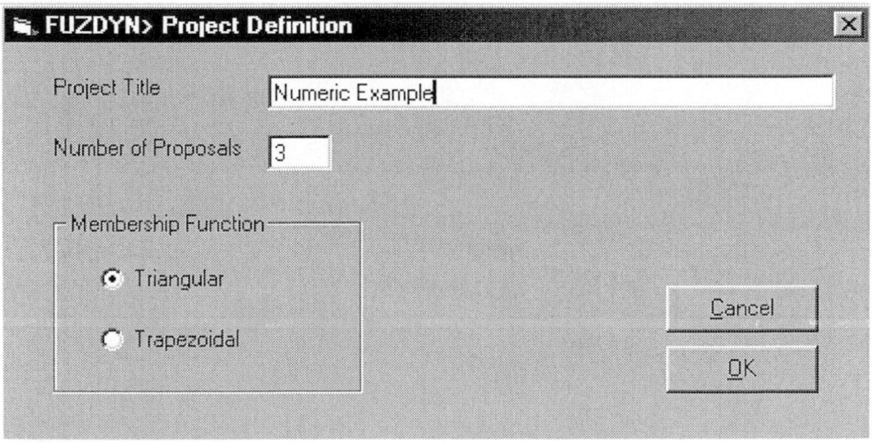

Fig. 3.8. Project definition

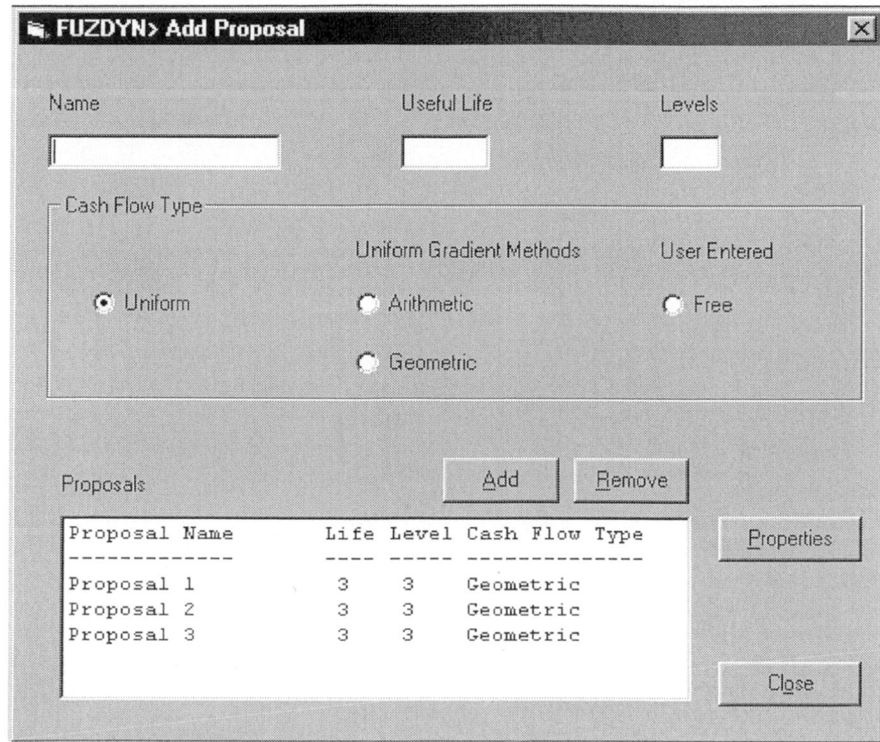

Fig. 3.9. The form of parameter input for proposals

As it can be seen from Table 3.3, the geometric growth rates (g) for the annual receipts at the investment levels are 10%, 12%, and 14% respectively and they are given as crisp rates in the problem and $f_1\left(y|\,\tilde{r}\right) = 0.05 + 0.01y$, $f_2\left(y|\,\tilde{r}\right) = 0.07 - 0.01y$, $\gamma\left(f_{3-k}\left(y|\,\tilde{r}\right), g, n\right)$, k=1,2.

In FUZDYN, data input for proposals is shown in Fig. 3.9. In Fig. 3.10 the data regarding fuzzy investment cost, fuzzy growth rate, and the benefit of the first year are entered and in Fig. 3.11, it is shown how a fuzzy number is entered.

For the total investment of \$(15000, 21000, 27000) in proposals 1 and 2:

– Investment in proposal 1: \$ (15000, 21000, 27000) and proposal 2: \$ 0

We find $f_1\left(y|\,\tilde{F}_1\right) = 1000y + 8000$, $f_2\left(y|\,\tilde{F}_1\right) = 10000 - 1000y$. For k= 1, $f_{3,1}\left(y|\,\tilde{P}\right) = (1000y + 8000)\left[\frac{(1.14)^3(1.07-0.01y)^{-3}-1}{0.07+0.01y}\right]$ and for $y = 0$, $f_{3,1}\left(y|\,\tilde{P}\right)$ =\$ 23,929 and for $y = 1$, $f_{3,1}\left(y|\,\tilde{P}\right)$ =\$ 27,442. For $k = 2$, $f_{3,2}\left(y|\,\tilde{P}\right) = (10000 - 1000y)\left[\frac{1-(1.14)^3(1.05+0.01y)^{-3}}{0.01y-0.09}\right]$ and for $y = 0$, $f_{3,2}\left(y|\,\tilde{P}\right)$ =\$ 31,090.

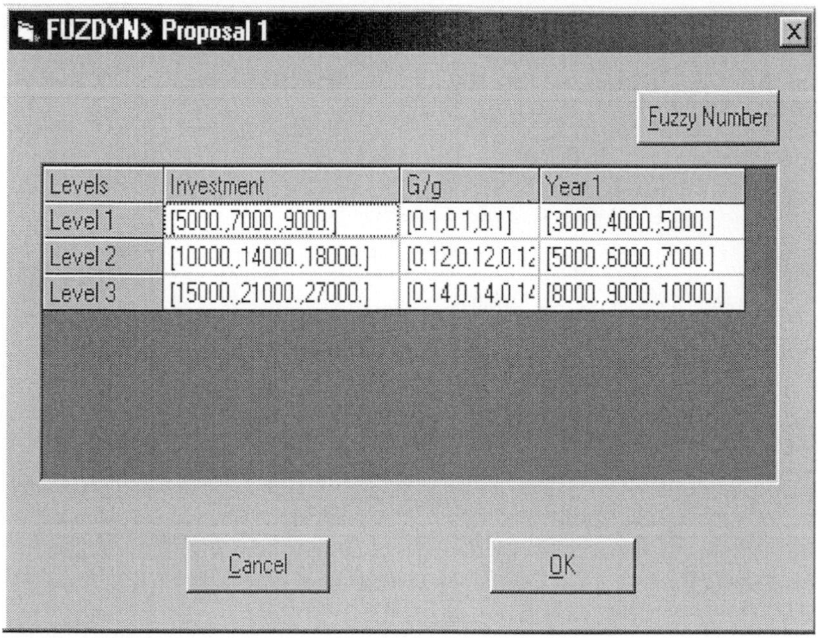

FUZDYN> Proposal 1

Fuzzy Number

Levels	Investment	G/g	Year 1
Level 1	[5000.,7000.,9000.]	[0.1,0.1,0.1]	[3000.,4000.,5000.]
Level 2	[10000.,14000.,18000.]	[0.12,0.12,0.1:	[5000.,6000.,7000.]
Level 3	[15000.,21000.,27000.]	[0.14,0.14,0.1	[8000.,9000.,10000.]

Cancel OK

Fig. 3.10. The forms related to data input for proposals

Now we can calculate the net PW and the fuzzy ranking ratio:

$$\begin{aligned} NPW_1 &= \$\,(23,929; 27,442; 31,090) - \$\,(15,000; 21,000; 27,000) \\ &= \$\,(-3,071; +6,442; +16,090) \end{aligned}$$

$$\begin{aligned} \text{Ranking ratio} &= \frac{\text{Proposal fuzzy PW(MARR)}}{\text{Proposal fuzzy first cost}} \\ &= \frac{\$\,(-3,071; +6,442; +16,090)}{\$\,(15,000; 21,000; 27,000)} \\ &= (-0.114; +0,307; +1,073) \end{aligned}$$

– Investment in proposal 1: \$ (10000, 14000, 18000) and proposal 2: \$ (5000, 7000, 9000)

For proposal 1:
$f_1\left(y|\tilde{F}_1\right) = 1000y + 5000,\ \ f_2\left(y|\tilde{F}_1\right) = 7000 - 1000y.$
For $k = 1$, $f_{3,1}\left(y|\tilde{P}\right) = (1000y + 5000)\left[\frac{(1.12)^3(1.07-0.01y)^{-3}-1}{0.05+0.01y}\right]$ and for $y = 0$, $f_{3,1}\left(y|\tilde{P}\right)$ =\$ 14,684 and for $y = 1$, $f_{3,1}\left(y|\tilde{P}\right)$ =\$ 17,960.

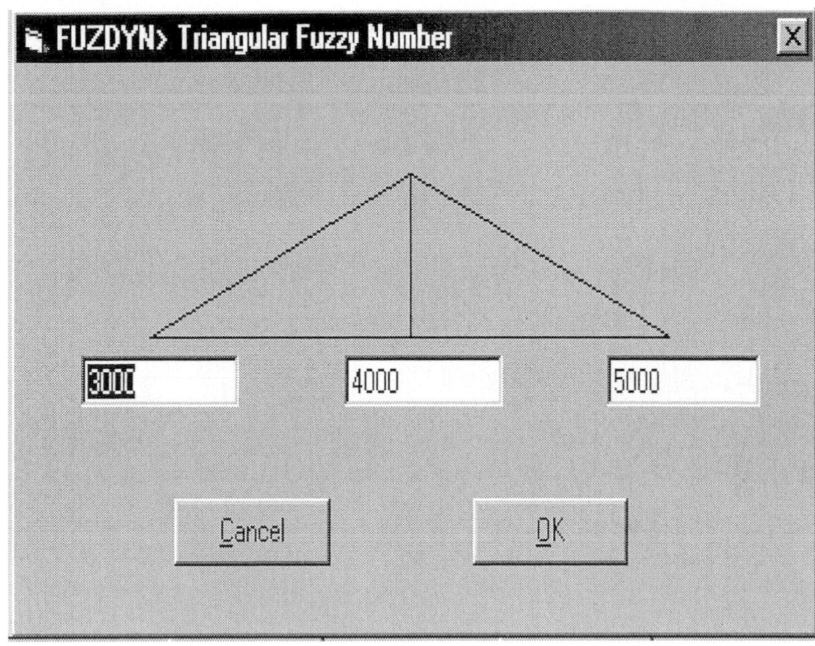

Fig. 3.11. The forms related to data input for proposals

For $k = 2$, $f_{3,2}\left(y|\,\tilde{P}\right) = (7000 - 1000y)\left[\frac{1-(1.12)^3(1.05+0.01y)^{-3}}{0.01y-0.07}\right]$ and for $y = 0$, $f_{3,2}\left(y|\,\tilde{P}\right) =$\$ 21,363

For proposal 2:

$f_1\left(y|\,\tilde{F}_1\right) = 1000y + 3000, \quad f_2\left(y|\,\tilde{F}_1\right) = 6000 - 2000y$

For k= 1, $f_{3,1}\left(y|\,\tilde{P}\right) = (1000y + 3000)\left[\frac{(1.10)^3(1.07-0.01y)^{-3}-1}{0.03+0.01y}\right]$ and for $y = 0$, $f_{3,1}\left(y|\,\tilde{P}\right) =$\$ 8,649 and for $y = 1$, $f_{3,1}\left(y|\,\tilde{P}\right) =$\$ 11,753.

For $k = 2$, $f_{3,2}\left(y|\,\tilde{P}\right) = (6000 - 2000y)\left[\frac{1-(1.10)^3(1.05+0.01y)^{-3}}{0.01y-0.05}\right]$ and for $y = 0$, $f_{3,2}\left(y|\,\tilde{P}\right) =$\$ 17,972.

Now we can calculate the net PW and the fuzzy ranking ratio:

$$\begin{aligned} PW_{1,2} &= PW_1 + PW_2 \\ &= \$\,(14,684; 17,690; 21,393) + \$\,(8,649; 11,753; 17,972) \\ &= \ \$\,(23,333; 29,443; 39,365) \end{aligned}$$

$$\begin{aligned} NPW_{1,2} &= \ \$\,(23,333; 29,443; 39,365) - \$\,(15,000; 21,000; 27,000) \\ &= \$\,(-3,667; +8,443; +24,365) \end{aligned}$$

$$\text{Ranking ratio} = \frac{\$\,(-3,667;+8,443;+24,365)}{\$\,(15,000;21,000;27,000)} = (-0.136;+0.402;+1.624)$$

– Investment in proposal 1: \$ (5000, 7000, 9000) and proposal 2: \$ (10000, 14000, 18000)

For proposal 1:
$f_1\left(y|\tilde{F}_1\right) = 1000y + 3000, \quad f_2\left(y|\tilde{F}_1\right) = 5000 - 1000y.$

For $k = 1$, $f_{3,1}\left(y|\tilde{P}\right) = (1000y + 3000)\left[\frac{(1.10)^3(1.07-0.01y)^{-3}-1}{0.03+0.01y}\right]$ and for $y = 0$, $f_{3,1}\left(y|\tilde{P}\right)$ =\$ 8,649 and for $y = 1$, $f_{3,1}\left(y|\tilde{P}\right)$ =\$ 11,753.
For $k = 2$, $f_{3,2}\left(y|\tilde{P}\right) = (5000 - 1000y)\left[\frac{1-(1.10)^3(1.05+0.01y)^{-3}}{0.01y-0.05}\right]$ and for $y = 0$, $f_{3,2}\left(y|\tilde{P}\right)$ =\$ 14,977.

For proposal 2:
$f_1\left(y|\tilde{F}_1\right) = 2000y + 4000, \quad f_2\left(y|\tilde{F}_1\right) = 7000 - 1000y$
For $k = 1$, $f_{3,1}\left(y|\tilde{P}\right) = (2000y + 4000)\left[\frac{(1.12)^3(1.07-0.01y)^{-3}-1}{0.05+0.01y}\right]$ and for $y= 0$, $f_{3,1}\left(y|\tilde{P}\right)$ =\$ 11,747 and for $y = 1$, $f_{3,1}\left(y|\tilde{P}\right)$ =\$ 17,960.
For k =2, $f_{3,2}\left(y|\tilde{P}\right) = (7000 - 1000y)\left[\frac{1-(1.12)^3(1.05+0.01y)^{-3}}{0.01y-0.07}\right]$ and for $y = 0$, $f_{3,2}\left(y|\tilde{P}\right)$ =\$ 21,363.

Now we can calculate the net PW and the fuzzy ranking ratio:

$$\begin{aligned} PW_{1,2} &= PW_1 + PW_2 \\ &= \$\,(8,649;11,753;14,977) + \$\,(11,747;17,960;21,363) \\ &= \$\,(20,396;29,713;36,340) \end{aligned}$$

$$\begin{aligned} NPW_{1,2} &= \$\,(20,396;29,713;36,340) - \$\,(15,000;21,000;27,000) \\ &= \$\,(-6,604;+8,713;+21,340) \end{aligned}$$

$$\text{Ranking ratio} = \frac{\$\,(-6,604;+8,713;+21,340)}{\$\,(15,000;21,000;27,000)} = (-0.025;+0.415;+1.423)$$

– Investment in proposal 1: \$ 0 and proposal 2: \$ (15000, 21000, 27000)

We find $f_1\left(y|\tilde{F}_1\right) = 4000y + 5000$, $f_2\left(y|\tilde{F}_1\right) = 10000 - 1000y$.
For $k=1$, $f_{3,1}\left(y|\tilde{P}\right) = (4000y + 5000)\left[\frac{(1.14)^3(1.07-0.01y)^{-3}-1}{0.07+0.01y}\right]$ and for $y=0$, $f_{3,1}\left(y|\tilde{P}\right) =\$\ 14{,}956$ and for $y = 1$, $f_{3,1}\left(y|\tilde{P}\right) =\$\ 27{,}442$.
For $k=2$, $f_{3,2}\left(y|\tilde{P}\right) = (10000 - 1000y)\left[\frac{1-(1.14)^3(1.05+0.01y)^{-3}}{0.01y-0.09}\right]$ and for $y = 0$, $f_{3,2}\left(y|\tilde{P}\right) =\$\ 31{,}090$.
Now we can calculate the net PW and the fuzzy ranking ratio:

$$\begin{aligned} NPW_2 &= \$\,(14,956; 27,442; 31,090) - \$\,(15,000; 21,000; 27,000) \\ &= \$\,(-12,044; +6,442; +16,090) \end{aligned}$$

$$\begin{aligned} \text{Ranking ratio} &= \frac{\$\,(-12,044; +6,442; +16,090)}{\$\,(15,000; 21,000; 27,000)} \\ &= (-0.446; +0,307; +1,073) \end{aligned}$$

To select the most lucrative combination of an investment of \$ (15,000; 21,000; 27,000), we will use Liou and Wang's [3.32] method. For a moderately optimistic decision-maker, $\omega = 0.5$.

Table 3.4. Identifying the most lucrative combination of \$ (15,000; 21,000; 27,000) for the first stage

Ranking ratio, $\tilde{A}$	$E_\omega(\tilde{A}) = \frac{1}{2}\left[\omega\,(a+b) + (1-\omega)\,(b+c)\right]$
(−0.114; +0.307; +1.073)	0.393
(−0.136; +0.402; +1.624)	0.573*
(−0.025; +0.415; +1.423)	0.557
(−0.446; +0.307; +1.073)	0.310

As it can be seen from Table 3.4, the most lucrative combination is to invest \$ (10,000; 14,000; 18,000) in proposal 1 and invest \$ (5,000; 7,000; 9,000) in proposal 2.

For the total investment of \$ (10000, 14000, 18000) in proposals 1 and 2:

– Investment in proposal 1: (10000, 14000, 18000) and proposal 2: \$ 0

$$\begin{aligned} \text{Ranking ratio} &= \frac{\$\,(-3,316; +3,960; +11,363)}{\$\,(10,000; 14,000; 18,000)} \\ &= (-0.184; +0.283; +1.136) \end{aligned}$$

– Investment in proposal 1: \$ (5000, 7000, 9000) and proposal 2: \$ (5000, 7000, 9000)

$$\text{Ranking ratio} = \frac{\$\,(-702; +9,506; +22,949)}{\$\,(10,000; 14,000; 18,000)} = (-0.039; +0.679; +2.295)$$

– Investment in proposal 1: \$ 0 and proposal 2: \$ (10000, 14000, 18000)

$$\text{Ranking ratio} = \frac{\$\,(-6,253; +3,960; 11,363)}{\$\ (10000,\ 14000,\ 18000)} = (-0.347; +0.283; +1.136)$$

To select the most lucrative combination of an investment of \$ (10,000; 14,000; 18,000), we will again use Liou and Wang's [3.32] method. For a moderately optimistic decision-maker, $\omega = 0.5$.

Table 3.5. Identifying the most lucrative combination of \$ (10,000; 14,000; 18,000) for the second stage

Ranking ratio, $\tilde{A}$	$E_\omega\left(\tilde{A}\right) = \frac{1}{2}\left[\omega\,(a+b) + (1-\omega)\,(b+c)\right]$
(−0.184; +0.282; +1.136)	0.379
(−0.039; +0.679; +2.295)	0.904*
(−0.347; +0.283; +1.136)	0.339

As it can be seen from Table 3.5, the most lucrative combination is to invest \$ (5,000; 7,000; 9,000) in proposal 1 and invest \$ (5,000; 7,000; 9,000) in proposal 2.

For the total investment of \$ (5000, 7000, 9000) in proposals 1 and 2:

– Investment in proposal 1: \$ (5000, 7000, 9000) and proposal 2: \$ 0

$$\text{Ranking ratio} = \frac{\$\,(-351; +4,753; +9,977)}{\$\ (5000,\ 7000,\ 9000)} = (-0.039; +0.679; +1.995)$$

– Investment in proposal 1: \$ 0 and proposal 2: \$ (5000, 7000, 9000)

$$\text{Ranking ratio} = \frac{\$\,(-351; +4,753; +12,972)}{\$\,(5,000; 7,000; 9,000)} = (-0.039; +0.679; +2.594)$$

It is obvious that the most lucrative combination of an investment of \$ (5,000; 7,000; 9,000) is to invest \$ (5,000; 7,000; 9,000) in proposal 2.

Now we will devise all possible investments that encompass proposals 1, 2, and 3, and identify the most lucrative one.

– Investment in proposals 1+2: \$ (15000, 21000, 27000) and proposal 3: \$ 0

$$\text{Ranking ratio} = \frac{\$\,(-3,667; +8,443; +24,365)}{\$\,(15,000; 21,000; 27,000)} = (-0.136; +0.402; +1.624)$$

- Investment in proposals 1+2: \$ (10000, 14000, 18000) and proposal 3: \$ (5000, 7000, 9000)

$$\text{Ranking ratio} = \frac{\$\,(-1,053; +11,321; +29,930)}{\$\,(15,000; 21,000; 27,000)} = (-0.039; +0.539; +1.995)$$

- Investment in proposals 1+2: \$ (5000, 7000, 9000) and proposal 3: \$ (10000, 14000, 18000)

$$\text{Ranking ratio} = \frac{\$\,(-3,667; +11,707; +24,335)}{\$\,(15,000; 21,000; 27,000)} = (-0.136; +0.557; +1.622)$$

- Investment in proposals 1+2: \$ 0 and proposal 3: \$ (15000, 21000, 27000)

$$\text{Ranking ratio} = \frac{\$\,(-3,071; +6,442; +10,308)}{\$\,(15,000; 21,000; 27,000)} = (-0.114; +0.307; +0.687)$$

To select the most lucrative combination of an investment of \$ (15,000; 21,000; 27,000), we will again use Liou and Wang's [3.32] method. For a moderately optimistic decision-maker, $\omega = 0.5$.

Table 3.6. Identifying the most lucrative combination of \$ (15,000; 21,000; 27,000) for the last stage

Ranking ratio, $\tilde{A}$	$E_\omega(\tilde{A}) = \frac{1}{2}\left[\omega\,(a+b) + (1-\omega)\,(b+c)\right]$
(−0.136; +0.402; +1.624)	0.573
(−0.039; +0.539; +1.995)	0.759*
(−0.136; +0.557; +1.622)	0.650
(−0.114; +0.307; +0.687)	0.328

As it can be seen from Table 3.6, the most lucrative combination is to invest \$ (10,000; 14,000; 18,000) in proposal 1 and proposal 2 and invest \$ (5,000; 7,000; 9,000) in proposal 3. Then the final solution is to invest \$ (5000, 7000, 9000) in proposal 1 and \$ (5000, 7000, 9000) in proposal 2, and \$ (5,000; 7,000; 9,000) in proposal 3.

The final solution found by FUZDYN is given in Fig. 3.12:

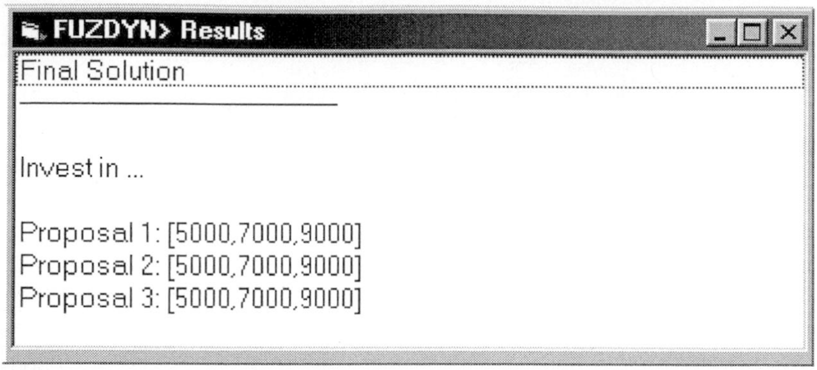

FUZDYN> Results

Final Solution

Invest in ...

Proposal 1: [5000,7000,9000]
Proposal 2: [5000,7000,9000]
Proposal 3: [5000,7000,9000]

Fig. 3.12. Final solution

3.12 Conclusions

In this chapter, capital budgeting techniques in the case of fuzziness and discrete compounding have been studied. The cash flow profile of some investments projects may be geometric or trigonometric. For these kind of projects, the fuzzy present, future, and annual value formulas have been also developed under discrete and continuous compounding in this chapter. Fuzzy set theory is a powerful tool in the area of management when sufficient objective data has not been obtained. Appropriate fuzzy numbers can capture the vagueness of knowledge. The other financial subjects such as replacement analysis, income tax considerations; continuous compounding in the case of fuzziness can be also applied [3.24], [3.25]. Comparing projects with unequal lives has not been considered in this paper. This will also be a new area for a further study. Dynamic programming is a powerful optimization technique that is particularly applicable many complex problems requiring a sequence of interrelated decisions. In the paper, we presented a fuzzy dynamic programming application for the selection of independent multi level investments. This method should be used when imprecise or fuzzy input data or parameters exist. In multilevel mathematical programming, input data or parameters are often imprecise or fuzzy in a wide variety of hierarchical optimization problems such as defense problems, transportation network designs, economical analysis, financial control, energy planning, government regulation, equipment scheduling, organizational management, quality assurance, conflict resolution and so on. Developing methodologies and new concepts for solving fuzzy and possibilistic multilevel programming problems is a practical and interesting direction for future studies.

Acknowledgements

The paper was carefully refereed by Yachi Hwang. The author is grateful to her for many useful corrections and valuable suggestions made to the original draft.

References

3.1 Bellman, R., Zadeh, L. A. (1970): Decision-Making in a Fuzzy Environment. Management Science, Vol. 14, B-141–164
3.2 Berenji, H. R. (1994): Fuzzy Q-Learning: A New Approach for Fuzzy Dynamic Programming. Proceedings of the 3rd IEEE International Conference on Fuzzy Systems, Vol. 1, 486–491
3.3 Blank, L. T., Tarquin, J. A. (1987): Engineering Economy, Third Edition, McGraw-Hill
3.4 Boussabaine, A. H., Elhag, T. (1999): Applying fuzzy techniques to cash flow analysis. Construction Management and Economics, Vol. 17, No. 6, 745–755
3.5 Buckley, J. U. (1987): The Fuzzy Mathematics of Finance, Fuzzy Sets and Systems, Vol. 21, 257–273
3.6 Chang, W. (1981): Ranking of Fuzzy Utilities with Triangular Membership Functions, Proc. Int. Conf. of Policy Anal. and Inf. Systems, 263-272
3.7 Chin, H. -C. (1995): Optimal Shunt Capacitor Allocation by Fuzzy Dynamic Programming. Electric Power Systems Research, Vol. 35, No. 2, 133–139
3.8 Chiu, C. Y. (1994): Park, C. S., Fuzzy Cash Flow Analysis Using Present Worth Criterion. The Engineering Economist, Vol. 39, No. 2, 113–138
3.9 Dimitrovski, A. D., Matos, M. A. (2000): Fuzzy Engineering Economic Analysis. IEEE Transactions on Power Systems, Vol. 15, No. 1, 283–289
3.10 Dubois, D., Prade, H. (1983): Ranking Fuzzy Numbers in the Setting of Possibility Theory. Information Sciences, Vol. 30, 183–224
3.11 Esogbue, A. O. (1999): Fuzzy Dynamic Programming: Theory and Applications to Decision and Control. Proceedings of the 18th International Conference of the North American Fuzzy Information Processing Society (NAFIPS'99), 18–22
3.12 Esogbue, A. O. (1999): Computational Complexity of Some Fuzzy Dynamic Programs. Computers and Mathematics with Applications 11, 47–51
3.13 Fu, C. -C., Wang, H. -F. (1999): Fuzzy Project Management by Team Approach. Proceedings of the IEEE International Fuzzy Systems Conference (FUZZ-IEEE'99), Vol. 3, 1487–1492
3.14 Huang, K. Y., Yang, H. T., Liao, C. C., Huang, C. -L. (1998): Fuzzy Dynamic Programming for Robust Direct Load Control. Proceedings of the 2nd International Conference on Energy Management and Power Delivery (EMPD'98), Vol. 2, 564–569
3.15 Hussein, M. L., Abo-Sinna, M. A. (1995): Fuzzy Dynamic Approach to the Multicriterion Resource Allocation Problem. Fuzzy Sets and Systems 2, 115–124
3.16 Iwamura, K., Liu, B. (1998): Chance Constrained Integer Programming Models for Capital Budgeting in Fuzzy Environments. Journal of the Operational Research Society, Vol. 49, No. 8, 854–860
3.17 Jain, R. (1976): Decision Making in the Presence of Fuzzy Variables. IEEE Trans. on Systems Man Cybernet, Vol. 6, 698–703

3.18 Kacprzyk, J., Esogbue, A. O. (1996) Fuzzy Dynamic Programming: Main Developments and Applications. Fuzzy Sets and Systems, Vol. 1, No. 81, 31–45
3.19 Kahraman, C. (2001): Capital Budgeting Techniques Using Discounted Fuzzy Cash Flows, in Soft Computing for Risk Evaluation and Management: Applications in Technology Environment and Finance, Ruan, D., Kacprzyk, J., Fedrizzi, M. (Eds.), Physica-Verlag, 375–396
3.20 Kahraman, C. (2001): Fuzzy Versus Probabilistic Benefit/Cost Ratio Analysis for Public Works Projects. International Journal of Applied Mathematics and Computer Science 3, 101–114
3.21 Kahraman, C., Ruan, D., Tolga, E. (2001): Capital Budgeting Techniques Using Discounted Fuzzy Versus Probabilistic Cash Flows, Information Sciences, forthcoming
3.22 Kahraman, C., Tolga, E., Ulukan, Z. (2000): Justification of Manufacturing Technologies Using Fuzzy Benefit/Cost Ratio Analysis. International Journal of Production Economics 1, 5–52
3.23 Kahraman, C., Ulukan, Z., Tolga, E. (1998): Combining Equal-Life Multilevel Investments Using Fuzzy Dynamic Programming. Proceedings of the Third Asian Fuzzy Systems Symposium (AFSS'98), 347–351
3.24 Kahraman, C., Ulukan, Z. (1997): Continuous Compounding in Capital Budgeting Using Fuzzy Concept. In: Proceedings of 6th IEEE International Conference on Fuzzy Systems (FUZZ-IEEE'97), Bellaterra-Spain, 1451–1455
3.25 Kahraman, C., Ulukan, Z. (1997): Fuzzy Cash Flows Under Inflation. In: Proceedings of Seventh International Fuzzy Systems Association World Congress (IFSA'97), University of Economics, Vol. IV, 104–108
3.26 Karsak, E. E. (1998): Measures of Liquidity Risk Supplementing Fuzzy Discounted Cash Flow Analysi. Engineering Economist, Vol. 43, No. 4, 331–344
3.27 Kaufmann, A., Gupta, M. M. (1988): Fuzzy Mathematical Models in Engineering and Management Science, Elsevier Science Publishers B.V.
3.28 Kuchta, D. (2000): Fuzzy capital budgeting, Fuzzy Sets and Systems, Vol. 111, 367–385
3.29 Kurtz, M. (1995): Calculations for Engineering Economic Analysis, McGraw-Hill
3.30 Lai, K. K., Li, L. (1999): A Dynamic Approach to Multiple-Objective Resource Allocation Problem, European Journal of Operational Research, Vol. 117, No. 2, 293–309
3.31 Li, L., Lai, K. K. (2001): Fuzzy Dynamic Programming Approach to Hybrid Multi-objective Multistage Decision-Making Problems, Fuzzy Sets and Systems, Vol. 117, No. 1, 13–25
3.32 Liou, T. -S., Wang, M. -J. (1992): Ranking Fuzzy Numbers with Integral Value, Fuzzy Sets and Systems, 247–255
3.33 Newnan, D. G. (1988): Engineering Economic Analysis, 3rd Edition, Engineering Press
3.34 Ward, T. L. (1985): Discounted Fuzzy Cash Flow Analysis, in 1985 Fall Industrial Engineering Conference Proceedings, 476-481
3.35 Winston, W. L. (1994): Operations Research: Applications and Algorithms, Duxbury Press
3.36 Yager, R. R. (1980): On Choosing Between Fuzzy Subsets, Kybernetes, 151–154
3.37 Zadeh, L. A. (1965): Fuzzy Sets, Information and Control, Vol. 8, 338–353

Appendix

One of the most basic concepts of fuzzy set theory which can be used to generalize crisp mathematical concepts to fuzzy sets is the extension principle. Let X be a cartesian product of universes $X = X_1 \dots X_r$, and $\tilde{A}_1, ..., \tilde{A}_r$ be r fuzzy sets in $X_1 ,\dots, X_r$, respectively. f is a mapping from X to a universe Y, $y = f(x_1 ,\dots, x_r)$. Then the extension principle allows us to define a fuzzy set $\tilde{B}$ in Y by

$$\tilde{B} = \{(y, \mu_{\tilde{B}}(y)) | \, y = f(x_1, ..., x_r), (x_1, ..., x_r) \in X\} \tag{A.1}$$

where

$$\mu_{\tilde{B}}(y) = \begin{cases} \sup\limits_{(x_1,\dots,x_r)\in f^{-1}(y)} \min\{\mu_{\tilde{A}_1}(x_1), ..., \mu_{\tilde{A}_r}(x_r)\}, \; if \; f^{-1}(y) \neq \emptyset \\ 0 \qquad , \quad \text{otherwise} \end{cases} \tag{A.2}$$

where f^{-1} is the inverse of f.

Assume $\tilde{P}$=(a, b, c) and $\tilde{Q}$=(d, e, f). a, b, c, d, e, f are all positive numbers. With this notation and by the extension principle, some of the extended algebraic operations of triangular fuzzy numbers are expressed in the following.

Changing Sign

$$-(a, b, c) = (-c, -b, -a) \tag{A.3}$$

or

$$-(d, e, f) = (-f, -e, -d) \tag{A.4}$$

Addition

$$\tilde{P} \oplus \tilde{Q} = (a + d, b + e, c + f) \tag{A.5}$$

and

$$k \oplus (a, b, c) = (k + a, k + b, k + c) \tag{A.6}$$

or

$$k \oplus (d, e, f) = (k + d, k + e, k + f) \tag{A.7}$$

if k is an ordinary number (a constant).

Subtraction

$$\tilde{P} - \tilde{Q} = (a - f, b - e, c - d) \tag{A.8}$$

and

$$(a, b, c) - k = (a - k, b - k, c - k) \tag{A.9}$$

or

$$(d, e, f) - k = (d - k, e - k, f - k) \quad (A.10)$$

if k is an ordinary number.

Multiplication

$$\tilde{P} \otimes \tilde{Q} = (ad, be, cf) \quad (A.11)$$

and

$$k \otimes (a, b, c) = (ka, kb, kc) \quad (A.12)$$

or

$$k \otimes (d, e, f) = (kd, ke, kf) \quad (A.13)$$

if k is an ordinary number.

Division

$$\tilde{P} \oslash \tilde{Q} = (a/f, b/e, c/d) \quad (A.14)$$

4. Rough Sets Theory and Multivariate Data Analysis in Classification Problems: a Simulation Study

Michael Doumpos and Constantin Zopounidis

Financial Engineering Laboratory, Department of Production Engineering and Management, Technical University of Crete, University Campus, 73100 Chania, Greece
email: {dmichael, kostas}@ergasya.tuc.gr

The classification of a set of objects into predefined homogenous groups is a problem with major practical interest in many fields. Over the past two decades several non-parametric approaches have been developed to address the classification problem, originating from several scientific fields. This paper is focused on a rule induction approach based on the rough sets theory and the investigation of its performance as opposed to traditional multivariate statistical classification procedures, namely the linear discriminant analysis, the quadratic discriminant analysis and the logit analysis. For this purpose an extensive Monte Carlo simulation is conducted to examine the performance of these methods under different data conditions.

4.1 Introduction

The classification problem involves the assignment of a set of observations (objects, alternatives) described over some attributes or criteria into predefined homogeneous classes. Such problems are often encountered in many fields, including finance, marketing, agricultural management, human recourses management, environmental management, medicine, pattern recognition, etc. This major practical interest of the classification problem has motivated researchers in developing an arsenal of methods in order to develop quantitative models for classification purposes. The linear discriminant analysis (LDA) was the first method developed to address the classification problem from a multidimensional perspective. LDA has been used for decades as the main classification technique and it is still being used at least as a reference point for comparing the performance of new techniques that are developed. Other widely used parametric classification techniques, developed to overcome LDA's restrictive assumptions (multivariate normality, equality of dispersion matrices between groups), include quadratic discriminant analysis (QDA), logit and probit analysis and the linear probability model (for two-group classification problems).

During the last two decades several alternative non-parametric classification techniques have been developed, including mathematical programming techniques [4.8], multicriteria decision aid methods [4.7], neural networks

[4.24], and machine learning approaches [4.22], [4.28], [4.29], [4.3], [4.4], [4.18]. Among these techniques, the rough sets theory developed following the concepts of machine learning, has several distinguishing and attractive features, including data reduction, handling of uncertainty, ease of interpretation of the developed classification model, etc. The purpose of this paper is to explore the performance of a rule induction approach developed on the concepts of rough sets, as opposed to traditional statistical classification procedures. For this purpose an extensive Monte Carlo simulation is conducted. Prior research on the comparison of the rough set approach with statistical classification techniques (mainly with LDA) has been focused on limited experiments conducted on real-world data sets (mainly medical data sets). Such comparisons have been performed by Krusinska et al. [4.19], [4.20] as well as by Teghem and Benjelloun [4.43]. Stefanowski [4.36] provides a general discussion of the comparison of the rough sets theory and statistical methods in classification problems. Consequently, the present study significantly extends the research in this area, through the consideration of a wide range of simulated data sets having different properties.

The rest of the paper is organized as follows. Section 2 presents a brief outline of the basic concepts of the rough sets approach and the associated rule induction methodology considered in this study. Section 3 provides details on the design of the Monte Carlo simulation and the factors considered in the conducted experiments. Section 4 discusses the obtained results for the rough sets approach and compares them with the results of LDA, QDA and logit analysis. Finally, Section 5 concludes the paper and outlines some further research directions.

4.2 Outline of the Rough Sets Approach

Pawlak [4.25] introduced the rough sets theory as a tool to describe dependencies between attributes, to evaluate the significance of attributes and to deal with inconsistent data. As an approach to handle imperfect data (uncertainty and vagueness), it complements other theories that deal with data uncertainty, such as probability theory, evidence theory, fuzzy set theory, etc. Generally, the rough set approach is a very useful tool in the study of classification problems, regarding the assignment of a set of objects into prespecified classes. Recently however, there have been several advances in this field to allow the application of the rough set theory to choice and ranking problems as well [4.9].

The rough set philosophy is founded on the assumption that with every object some information (data, knowledge) is associated. This information involves two types of attributes; condition and decision attributes. Condition attributes are those used to describe the characteristics of the objects. For instance the set of condition attributes describing a firm can be its size, its financial characteristics (profitability, solvency, liquidity ratios), its organization, its market position, etc. The decision attributes define a partition of the objects into classes according to the condition attributes (for example, good firms, bad firms).

On the basis of these two types of attributes an information table is formed such as the one presented in Table 4.1. The rows of the information table represent the objects and its columns represent the condition and decision attributes. In the example of Table 4.1, the decision attribute d defines a dichotomic classification of the set of objects: $d = A$ means acceptance, $d = R$ means rejection. The classification $Y = \{Y_A, Y_R\}$ is composed of two decision classes: $Y_A = \{x_1, x_4, x_6, x_7, x_{10}\}$ and $Y_R = \{x_2, x_3, x_5, x_8, x_9\}$ corresponding to the objects accepted and rejected by the decision maker, respectively. The set of condition attributes is denoted by $A = \{a_1, a_2, a_3, a_4\}$.

Table 4.1. An example of an information table (Source: Slowinski, et al. 1997)

Objects	Condition attributes				Decision
	a_1	a_2	a_3	a_4	d
x_1	1	2	1	3	A
x_2	1	1	1	1	R
x_3	2	1	1	2	R
x_4	3	3	1	1	A
x_5	3	2	1	1	R
x_6	3	3	1	3	A
x_7	1	3	0	2	A
x_8	2	1	0	3	R
x_9	1	1	1	1	R
x_{10}	3	2	1	1	A

Objects characterized by the same information are considered to be indiscernible. This indiscernibility relation constitutes the main mathematical basis of the rough set theory. Any set of all indiscernible objects is called an elementary set and forms a basic granule of knowledge about the universe. In the example of Table 4.1, there are eight A-elementary sets: the couples of indiscernible objects $\{x_2, x_9\}$, $\{x_5, x_{10}\}$ and six remaining discernible objects.

Any set of objects being a union of some elementary sets is referred to as crisp (precise) otherwise the set is rough (imprecise, vague). Consequently, each rough set has a boundary-line consisting of cases (objects) which cannot be classified with certainty as members of the set or of its complement. Therefore, a pair of crisp sets, called the lower and the upper approximation can represent a rough set. The lower approximation consists of all objects that certainly belong to the set and the upper approximation contains objects that possibly belong to the set. The difference between the upper and the lower approximation defines the doubtful region, which includes all objects that cannot be certainly classified into the set. The lower and the upper approximations as well as the doubtful region of the decision classes Y_A and Y_R in the example of Table 4.1, are the following:

– Decision class Y_A:
  Lower approximation: $\{x_1, x_4, x_6, x_7\}$,
  Upper approximation: $\{x_1, x_4, x_5, x_6, x_7, x_{10}\}$,
  Doubtful region: $\{x_5, x_{10}\}$,
– Decision class Y_R:
  Lower approximation: $\{x_2, x_3, x_8, x_9\}$,
  Upper approximation: $\{x_2, x_3, x_5, x_8, x_9, x_{10}\}$,
  Doubtful region: $\{x_5, x_{10}\}$.

On the basis of the lower and upper approximations of a rough set, the accuracy of approximating the rough set can be calculated as the ratio of the cardinality of its lower approximation to the cardinality of its upper approximation. In the example used, the accuracy of approximating the sets A and R is equal in both cases to 0.67 (4/6). The overall quality of classification, defined as the ratio of the sum of the cardinalities of all lower approximations to the total number of objects is 0.8 (8/10).

On the basis of these approximations, the first major capability that the rough set theory provides is to reduce the available information, so as to retain only what is absolutely necessary for the description and classification of the objects being studied. This is achieved by discovering subsets of the attributes' set that describes the considered objects, which can provide the same quality of classification as the whole attributes' set. Such subsets of attributes are called reducts. Generally, the reducts are more than one. In such a case the intersection of all reducts is called the core. The core is the collection of the most relevant attributes, which cannot be excluded from the analysis without reducing the quality of the obtained description (classification). In the example used, there are two reducts. The first consists of the condition attributes a_1 and a_2, while the second one consists of the condition attributes a_2 and a_4. Thus, the core includes only the condition attribute a_2.

The decision maker can examine all obtained reducts and proceed to the further analysis of the considered problem according to the reduct that best describes reality. Heuristic procedures can also be used to identify an appropriate reduct [4.33].

The subsequent steps of the analysis involve the development of a set of rules for the classification of the objects. The rules developed through the rough set approach have the following form:

IF *conjunction of elementary conditions*
THEN *disjunction of elementary decisions*

In the example of Table 4.1, two sets of minimal decision rules can be developed on the basis of the two reducts mentioned above. These rules are the following (the numbers in brackets indicate the strength of the rules):

– Rules corresponding to the reduct $\{a_1, a_2\}$:
 1. **if**$a_2 = 3$ **then** $d = A$ [3]
 2. **if**$a_1 = 1$ **and** $a_2 = 2$ **then** $d = A$ [1]
 3. **if**$a_2 = 1$ **then** $d = R$ [4]
 4. **if**$a_1 = 3$ **and** $a_2 = 2$ **then** $d = A$ **or** $d = R$ [1/1]

– Rules corresponding to the reduct $\{a_2, a_4\}$:
 1. **if**$a_2 = 3$ **then** $d = A$ [3]
 2. **if**$a_2 = 2$ **and** $a_4 = 3$ **then** $d = A$ [1]
 3. **if**$a_2 = 1$ **then** $d = R$ [4]
 4. **if**$a_2 = 2$ **and** $a_4 = 2$ **then** $d = A$ **or** $d = R$ [1/1]

The developed rules can be consistent if they include only one decision in their conclusion part, or approximate if their conclusion involves a disjunction of elementary decisions. Approximate rules are consequences of an approximate description of decision classes in terms of blocks of objects (granules) indiscernible by condition attributes. Such a situation indicates that using the available knowledge, one is unable to decide whether some objects belong to a given decision class or not. In the two sets of rules presented above there are three exact decision rules (rules 1-3) and one approximate decision rule (rule 4). Each rule is associated with a strength, that represents thenumber of objects satisfying the condition part of the rule and belonging to the class that the conclusion of the rule suggests (for approximate rules, the strength is calculated for each class separately). The higher the strength of the rules, the more general the rule is considered to be (its condition part is shorter and less specialized).

Generally, the rule induction process within the rough sets theory can be performed in a variety of ways. Rule induction is a significant area of the machine learning field. Well-known rule induction algorithms include the ID3 algorithm [4.28], the C4.5 algorithm [4.29], the algorithms of the AQ family [4.22], the CN2 algorithm [4.4] and the CART algorithm [4.3]. Rule induction based on the concepts of the rough sets theory has also been a focal point interest for researchers working on the rough set approach. Appropriate algorithms have been developed by Grzymala-Busse [4.14], Slowinski and Stefanowski [4.31], Skowron [4.30], Ziarko et al. [4.47], Stefanowski and Vanderpooten [4.40], Mienko et al. [4.23], Stefanowski [4.38], Greco et al. [4.11]. Reviews on rough set based rule induction algorithms are provided in the works of Greco et al. [4.13], Stefanowski [4.39] and Slowinski et al. [4.34].

The rule induction approach employed in this study is based on the MODLEM algorithm of Stefanowski [4.38]. This algorithm is an extension of the LEM2 algorithms [4.14], which is among the most popular rule induction approaches within the context of the rough sets theory. The MODLEM algorithm is a machine learning inspired algorithm that uses elements of rough sets theory in one phase - i.e. if the learning examples are inconsistent, then decision rules are induced from approximations of decision classes. So, in this algorithm the rough sets are used only to calculate approximations but the rule induction phase follows the typical machine learning paradigm of sequential covering [4.38]. The major feature of the MODLEM algorithm is that it does not require the realization of a discretization of the attributes values prior to its application. Therefore, it is well suited to problems involving continuous attributes such as the ones studied through the Monte Carlo simulation that is employed in this study. Furthermore, the algorithm leads to the development of a compact set of rules with high strength [4.15]. The algorithm is applied in combination with the LERS classification system [4.14], which enables the handling of cases where an object's description matches more than one conflicting rules, or cases where an object does not match any rule. Alternatively to the LERS system other classification schemes can also be employed in combination with the MODLEM algorithm. The most popular classification scheme among researchers working in the field of rough sets is the use of the value closeness relation [4.32], [4.37]. However, the use of the value closeness relation necessitates the specification of a number of parameters by the decision maker. Since this is a simulation study, the specification of the parameters involved in the development of the value closeness relation would entail a significant degree of arbitrariness. Therefore, instead of the value closeness relation approach the classification scheme of the LERS system is used in the present simulation study.

More details on the rough sets theory can be found in Pawlak and Slowinski [4.26], [4.27] as well as in the recent extensive review paper of Komorowski et al. [4.17].

4.3 Experimental Design

In order to perform a thorough examination of the classification performance of the MODLEM algorithm for rough set based rule induction, as opposed to traditional classification procedures, an extensive experimental study is conducted using several different data conditions. Except for the MODLEM algorithm, three other statistical and econometric procedures are considered: linear discriminant analysis (LDA), quadratic discriminant analysis (QDA), and logit analysis (LA). The factors considered in this experimental design are illustrated in Table 4.2. The simulation was conducted on a Pentium III 600Mhz PC, using Matlab 5.2 for data generation as well as for the application of LA and QDA. Appropriate codes for the application of LDA and the MODLEM algorithm were written by the authors in the Visual Basic 6 programming environment. The results of the simulation have been analyzed using the SPSS 10 statistical package.

The first of the factors involving the properties of the data investigated in the conducted experimental design involve their distributional form (F_2).

Table 4.2. Factors investigated in the experimental design

Factors	Levels
F_1: Classification procedures	1. Linear discriminant analysis (LDA)
	2. Quadratic discriminant analysis (QDA)
	3. Logit analysis
	4. MODLEM algorithm
F_2: Statistical distribution	1. Mulrivariate normal
	2. Mulrivariate uniform
	3. Mulrivariate exponential
	4. Mulrivariate log–normal
F_3: Number of groups	1. Two groups
	2. Three groups
F_4: Size of the training sample	1. 36 objects, 5 attributes
	2. 72 objects, 5 attributes
	3. 108 objects, 5 attributes
F_5: Correlation coefficient	1. Low correlation $r \in [0, 0.1]$
	2. Higher correlation $r \in [0.2, 0.5]$
F_6: Group dispersion matrices	1. Equal
	2. Unequal
F_7: Group overlap	1. Low overlap
	2. High overlap

While many studies conducting similar experiments have been concentrated on univariate distributions to consider multivariate non-normality, in this study multivariate distributions are considered. This specification enables the investigation of additional factor in the experiment, such as the correlation of the attributes and the homogeneity of the group dispersion matrices. Actually, using a univariate distribution implies that the attributes are independent, a case that is hardly the situation encountered in real-world problems. The first two of the multivariate distributions that are considered (normal and uniform) are symmetric, while the exponential (this is actually a multivariate distribution that resembles the exponential distribution in terms of its skewness and kurtosis) and log-normal distributions are asymmetric, thus leading to a significant violation of multivariate normality. The generation of the multivariate non-normal data is based on the methodology presented by Vale and Maurelli [4.45].

Factor F_3 defines the number of groups into which the classification of the objects is made. In this experimental design the two-group and the three-group classification problems are considered. This specification enables the derivation of useful conclusions on the performance of the methods investigated, in a wide range of situations that are often met in practice (many real-world classification problems involve three groups).

The fourth factor is used to define the size of the training sample, and in particular the number of objects that it includes (henceforth this number is denoted as m). The factor has three levels corresponding to 36, 72 and 108 objects, distributed equally to the groups defined factor F_3. In all three cases the objects are described along five attributes. Generally, small training samples contain limited information about the classification problem being examined, but the corresponding complexity of the problem is also limited. On the other hand, larger samples provide richer information, but the complexity is also increased. Thus, the examination of the three levels of the factor enables the investigation of the performance of the classification procedures under all these cases.

The specified correlation coefficients for every pair of attributes define the off-diagonal elements of the dispersion matrices of the groups. The elements in the diagonal of the dispersion matrices, representing the variance of the attributes are specified by the sixth factor, which is considered in two levels. In the first level, the variances of the attributes are equal for all groups, whereas in the second level the variances differ. Denoting the variance of attribute x_i for group j as σ_{ij}^2, the realization of these two situations regarding the homogeneity of the group dispersion matrices is performed as follows:

1. For the multivariate normal, uniform and exponential distributions:
   - Level 1: $\sigma_{i1}^2=\sigma_{i2}^2=\sigma_{i3}^2=1$, $\forall i=1, 2, ..., 5$,
   - Level 2: $\sigma_{i1}^2=1$, $\sigma_{i2}^2=4$, $\sigma_{i3}^2=16$, $\forall i=1, 2, ..., 5$.
2. For the multivariate log-normal distribution, the variances are specified so as to assure that the kurtosis of the data ranges within reasonable levels,[1] as follows:
   - In the case of two groups:

$$\text{Level 1: } \sigma_{i1}^2 = \sigma_{i2}^2 = \begin{cases} 12, \text{ if } m = 36 \\ 14, \text{ if } m = 72 \\ 16, \text{ if } m = 108 \end{cases}, \forall i = 1, 2, ..., 5,$$

$$\text{Level 2: } \sigma_{i1}^2 = \begin{cases} 12, \text{ if } m = 36 \\ 14, \text{ if } m = 72 \\ 16, \text{ if } m = 108 \end{cases}, \sigma_{i2}^2 = 1.5\sigma_{i1}^2, \forall i = 1, 2, ..., 5.$$

   - In the case of three groups:

$$\text{Level 1: } \sigma_{i1}^2 = \sigma_{i2}^2 = \sigma_{i3}^2 \begin{cases} 4, \ \text{ if } m = 36 \\ 7, \ \text{ if } m = 72 \\ 10, \text{ if } m = 108 \end{cases}, \forall i = 1, 2, ..., 5,$$

[1] In contrast to the other distributions considered in this experimental design where the coefficients of skewness and kurtosis are fixed, in the log-normal distribution the skewness and kurtosis are defined by the mean and the variance of the attributes for each group. The procedures for generating multivariate non-normal data are able to replicate satisfactory the prespecified values of the first three moments (mean, standard deviation and skewness) of a statistical distribution. However, the error is higher for the fourth moment (kurtosis). Therefore, in order to reduce this error and consequently to have better control of the generated data, both the mean and the variance of the attributes for each group in the case of the multivariate log-normal distribution are specified so as the coefficient of kurtosis is lower than 40.

$$\text{Level 2: } \sigma_{i1}^2 = \begin{cases} 2, \text{ if } m = 36 \\ 4, \text{ if } m = 72 \\ 6, \text{ if } m = 108 \end{cases}, \sigma_{i2}^2 = 1.5\sigma_{i1}^2, \sigma_{i3}^2 = 1.5\sigma_{i2}^2, \forall i = 1, 2, ..., 5.$$

The final factor defines the degree of group overlap. The higher the overlapping is between each pair of groups, the higher is the complexity of the classification problem due to the difficulty in discriminating between the objects of each group. The degree of group overlap in this experimental design is considered using the Hotelling T^2 statistic. This is a multivariate measure of difference between the means of two groups, assuming that the attributes are multivariate normal and that the group dispersion matrices are equal. Studies conducted on the first of these assumptions (multivariate normality) have shown that actually the Hotelling T^2 is quite robust to departures from multivariate normality even for small samples [4.21]. Therefore, using the Hotelling T^2 in the non-multivariate distributions considered in this experimental design does not pose a significant problem. To overcome the second assumption regarding the homogeneity of the group dispersion matrices, the modified version of the Hotelling T^2 defined by Anderson [4.1] is employed in the case where the dispersion matrices are not equal. The use of these measures of group overlap in the conducted experimental design is performed as follows: Initially, the means of all five attributes for the first group is fixed to a specific value (1 for the case of multivariate normal, uniform and exponential distribution, and 8 in the case of the log-normal distribution). Then, the means of the attributes for the second group are specified so as the Hotelling T^2 (or its modified version) between the means of groups 1 and 2 is significant at the 1% and the 10% significance level, corresponding to low and high degree of group overlap. Similarly, the means of the third group are specified so as the Hotelling T^2 (or its modified version) between the means of groups 2 and 3 is significant at the 1% and the 10% significance level.

For each combination of the factors F_2-F_7 (192 combinations) a training sample and a validation sample are generated, having all the properties that these factors specify. The size of the training sample is defined by the factor F_4, while the size of the validation sample (holdout sample) is fixed at 216 in all cases. For each factor combination 20 replications are performed. Therefore, during this experiment the number of samples considered is 7,680 ($192 \times 20 = 3{,}840$ training samples matched to 3,840 validation samples). Overall, the conducted experiment involves a $4 \times 4 \times 2 \times 3 \times 2 \times 2 \times 2$ full-level factorial design consisting of 768 treatments (factor combinations).

4.4 Results

The analysis of the results obtained in this experimental design, is focused only in classification errors for the validation samples. The rough sets approach correctly classify all objects in the training sample and consequently there is no point of conducting a comparison with the statistical classification procedures with regard to their performance in the training sample.

The examination of the results obtained from the experimental design is performed through a seven-way analysis of variance, using the transformed misclassification rates of the methods, on the basis of the transformation:

$$2 \times \arcsin \sqrt{\text{error rate}} \,. \tag{4.1}$$

This transformation has been proposed by several researchers in order to stabilize the variance of the misclassification rates [4.2], [4.16]. The ANOVA results presented in Table 4.3, indicate that the seven main effects (the considered factors), three two-way interaction effects and one three-way interaction effect explain more than 76% of the total variance measured using the Hays ω^2 statistic. None of the remaining interaction effects explains more than 1% of the total variance, and consequently they are not reported.

Table 4.3. Major explanatory effects regarding the classification performance of the methods (seven-way ANOVA results)

Effects	df	Sum of squares	Mean squares	F	ω^2
F_6	1	198.249	198.249	17111.30	17.76
$F_1 \times F_6$	3	132.835	44.278	3821.75	11.90
$F_1 \times F_2$	9	122.608	13.623	1175.84	10.97
F_1	3	84.637	28.212	2435.07	7.58
F_3	1	83.150	83.150	7176.84	7.45
F_2	3	71.409	23.803	2054.48	6.39
F_4	2	47.897	23.949	2067.06	4.29
F_7	1	42.719	42.719	3687.14	3.83
$F_1 \times F_4$	6	29.572	4.929	425.41	2.64
F_5	1	24.603	24.603	2123.58	2.20
$F_1 \times F_2 \times F_6$	9	12.452	1.384	119.42	1.11

In these results the interaction effects are of major interest. All four interaction effects that are found to explain a high proportion of the total variance in the obtained results, involve the interaction of the factor F_1 (classification procedures) with other factors, in particular the homogeneity of the group dispersion matrices (F_6), the distributional form of the data (F_2), and the training sample size (F_4). Table 4.4 summarizes the results of all methods throughout all experiments, while Tables 4.5-4.8 provide further details on the comparison of the methods in terms of the aforementioned two and three-way interaction effects that are found significant through the analysis of variance. Each of these tables reports the average transformed error rate, the true error rate (in parentheses) and the grouping obtained through the Tukey's test[2] on

[2] Tukey's honestly significantly different test is a post-hoc comparison technique that follows the results of ANOVA enabling the identification of the means that

the average transformed error rates [cf. Eq. (4.1)]. The homogeneous groups of classification techniques formed by the Tukey's test are denoted A (classification techniques with the lower error rate) to D (classification techniques with the higher error rate).

Table 4.4. Overall average error rates

	Average error	Tukey's grouping
LDA	1.200 (32.20%)	C
QDA	1.0671 (26.99%)	B
Logit	1.1891 (31.69%)	C
Rough sets	1.0302 (25.65%)	A

Table 4.5. Average error rates for different structures of the group dispersion matrices (factor: F_6)

	Homogeneity of dispersion matrices			
	Equal		Unequal	
	Average error	Tukey's grouping	Average error	Tukey's grouping
LDA	1.2391 (33.98%)	C	1.1610 (30.42%)	C
QDA	1.3283 (38.17%)	D	0.8059 (15.80%)	A
Logit	1.2183 (33.01%)	B	1.1599 (30.37%)	C
Rough sets	1.1552 (30.89%)	A	0.9051 (20.42%)	B

The results indicate that, overall, the rough sets approach outperforms all statistical procedures, followed by QDA, while the performances of LDA

most contribute to the considered effect. In this simulation study the Tukey's test is used to perform all pairwise comparisons among average classification error rates (transformed error rates) of each pair of methods to form homogenous sets of methods according to their classification error rate. Each set includes methods that do not present statistically significant differences with respect to their classification error rates (see Yandell [4.46] for additional details).

Table 4.6. Average error rates for different distributions (factor: F_2)

	Distribution			
	Normal		Uniform	
	Average error	Tukey's grouping	Average error	Tukey's grouping
LDA	1.1900 (31.69%)	B	1.2312 (33.57%)	C
QDA	1.0917 (27.79%)	A	1.0634 (27.02%)	A
Logit	1.1827 (31.36%)	B	1.2200 (33.03%)	C
Rough sets	1.2795 (35.97%)	C	1.1219 (29.30%)	B
	Exponential		Log-normal	
	Average error	Tukey's grouping	Average error	Tukey's grouping
LDA	1.1642 (30.66%)	C	1.2147 (32.86%)	C
QDA	1.0392 (26.97%)	B	1.0740 (27.16%)	B
Logit	1.1517 (30.11%)	C	1.2020 (32.25%)	C
Rough sets	0.6769 (11.83%)	A	1.0422 (25.51%)	A

and LA are similar. The further analysis indicates that when the homogeneity of the dispersion matrices is considered, the rough sets approach is the best classifier in the cases where the dispersion matrices are equal, while in the opposite case the QDA outperforms all the other procedures. However, the results of Table 4.5 clearly indicate that QDA is very sensitive to departures from its assumption regarding the heterogeneity of the group dispersion matrices.

The distributional form of the data is also a significant factor in explaining the differences in the error rates of the methods. The results of Table 4.6 show that the rough sets approach provides significantly lower error rates than all the statistical procedures, when the data come from non-symmetric distributions, such as the exponential and the log-normal distribution. On the contrary, when the underlying distribution of the data is symmetric then QDA is found to be the best classifier, even in the case of multivariate non-normality (uniform distribution). This is not surprising; Clarke et al. [4.5] investigated the performance of QDA for departures from non-normality and they concluded that QDA is quite robust to non-normal data, expect for the

Table 4.7. Average error rates for different training sample sizes (factor: F_4)

	Training sample size					
	36		72		108	
	Average error	Tukey's grouping	Average error	Tukey's grouping	Average error	Tukey's grouping
LDA	1.1092 (26.04%)	B	1.2342 (33.68%)	C	1.3017 (36.87%)	B
QDA	1.0585 (26.26%)	B	1.0686 (27.11%)	B	1.0741 (27.58%)	A
Logit	1.0555 (25.29%)	B	1.2225 (33.11%)	C	1.2894 (36.27%)	B
Rough sets	1.0099 (24.51%)	A	1.0205 (25.27%)	A	1.0601 (27.19%)	A

case where skewness is high. This justifies the results of this experimental design.

The consideration of the interaction between the homogeneity of the group dispersion matrices and the distributional form of the data (Table 4.8) clarifies further the above results. According to the results of Table 4.8, the LDA and the LA outperform both rough sets and QDA when the distribution is symmetric and the group dispersion matrices are equal, whereas in the case of homogeneous group dispersion matrices QDA outperforms all procedures, followed by the rough sets approach. In the cases of the exponential and the log-normal distributions, which are asymmetric, the rough sets approach provides consistently lower error rates than the statistical procedures, except for the case of the log-normal distribution with unequal group dispersion matrices where the QDA provides the best performance. However, once again the high sensitivity of the QDA to departures from its assumption regarding the heterogeneity of the group dispersion matrices, becomes apparent. For all distributional forms with equal group dispersion matrices, the QDA performs worse than the other classification procedures, expect for the multivariate normal distribution where it outperforms rough sets, whereas its performance when the group dispersion matrices are equal is greatly improved.

Finally, the size of the training sample is also found to be a significant factor. The results of Table 4.7 indicate that the rough sets approach provides consistently lower error rates both when small and larger training samples are employed. The differences between the rough sets and the statistical classification procedures are significant for small and medium size training samples (36 and 72 objects), while in the case of a larger training sample the difference between the rough sets and the QDA is not significant. The performances of LDA and LA deteriorates significantly as the training sample size increases, whereas the rough sets approach and QDA provide more robust results. In

Table 4.8. Average error rates for different distributions and group dispersion matrices (factors: F_2, F_6)

	Normal distribution			
	Group dispersion matrices			
	Equal		Unequal	
	Average error	Tukey's grouping	Average error	Tukey's grouping
LDA	1.2200 (33.08%)	A	1.1601 (30.30%)	B
QDA	1.3164 (37.60%)	B	0.8671 (17.99%)	A
Logit	1.2033 (32.34%)	A	1.1622 (30.39%)	B
Rough sets	1.4090 (42.08%)	C	1.1500 (29.86%)	B
	Uniform distribution			
	Group dispersion matrices			
	Equal		Unequal	
	Average error	Tukey's grouping	Average error	Tukey's grouping
LDA	1.2603 (34.94%)	A	1.2021 (32.20%)	C
QDA	1.3460 (39.00%)	B	0.7808 (15.04%)	A
Logit	1.2403 (33.98%)	A	1.1996 (32.07%)	C
Rough sets	1.3386 (38.07%)	B	0.9053 (19.90%)	B
	Exponential distribution			
	Group dispersion matrices			
	Equal		Unequal	
	Average error	Tukey's grouping	Average error	Tukey's grouping
LDA	1.2429 (34.18%)	C	1.0855 (27.14%)	C
QDA	1.3255 (38.07%)	D	0.7529 (13.88%)	B
Logit	1.2107 (32.71%)	B	1.0927 (27.51%)	C
Rough sets	0.7631 (14.54%)	A	0.5908 (9.12%)	A
	Log-normal distribution			
	Group dispersion matrices			
	Equal		Unequal	
	Average error	Tukey's grouping	Average error	Tukey's grouping
LDA	1.2332 (33.70%)	B	1.1961 (32.02%)	C
QDA	1.3251 (38.02%)	C	0.8229 (16.30%)	A
Logit	1.2189 (33.02%)	B	1.1851 (31.49%)	C
Rough sets	1.1100 (28.24%)	A	0.9744 (22.78%)	B

the case of the QDA, its improvement relative to the other procedures when the training sample size increases can be justified on the basis that larger samples provide a better representation of the true form of the group dispersion matrices, thus enabling the QDA to take full advantage of its assumption regarding the heterogeneity of the group dispersion matrices.

4.5 Conclusions

The aim of this study was to explore the performance of a rough sets based rule induction approach (the MODLEM algorithm) in classification problems. Over the past two decades, the rough sets theory has emerged as a significant methodological tool to address complex decision-making problems, where a classification of the considered objects is involved.

A thorough comparison was performed with traditional statistical classification approaches, on the basis of an extensive experimental design involving several factors regarding the properties of the data involved. The results indicate that the rough sets based rule induction approach could be considered as a promising classification methodology compared to well-established existing procedures, at least in the cases where the data originate from asymmetric distributions. This finding fully confirms the results of previous studies conducted on real-world data [4.19], [4.20]. The major "rival" of the rough sets, the QDA also performs well in many cases, but the homogeneity of the group dispersion matrices heavily affects its performance.

These results are encouraging with regard to the performance of the rough sets based rule induction in addressing classification problems. Of course, any experimental study is subject to limitations in terms of the factors that it considers and the issues that it addresses. In that regard, the analysis and the corresponding findings of this study could be further extended to address some additional significant issues, including:

- The analysis of additional data conditions including the case of qualitative data and outliers. Especially the former case (qualitative data) is of major interest to real-world decision-making. Handling qualitative attributes with traditional statistical techniques has several problems derived by the requirement that the qualitative measurement scale of such attributes should be transformed into a quantitative one. On the other hand, the rough sets approach is quite flexible in handling qualitative attributes and consequently it would be interesting to investigate its performance compared to statistical techniques in the presence of qualitative attributes.
- A comparison with other rule induction algorithms from the machine learning paradigm. Such a comparison will contribute to the understanding of the relative advantages of rough sets based rule induction over other similar approaches (e.g., ID3, C4.5, CART, CN2) that are widely used by researchers in the machine learning community.

- The consideration of alternative ways to induce decision rules within the context of the rough sets theory. The MODLEM algorithm considered in this study leads to the construction of a minimal set of rules (i.e. if one rule is deleted, the set of rules is no more able to cover all the objects in the training set). However, it would be interesting to consider the development of a satisfying set of rules [4.40], [4.23], [4.39]. In that regard the developed set of decision rules would respect some predetermined conditions. For instance the rule induction procedure could be directed towards developing a set of rules covering a fixed percentage of objects, rules that cover a limited number of negative examples, rules of a specific maximal length of their conditions part, rules that include some specific important attributes in their conditions, etc. The investigation of the impact of such conditions to the classification accuracy of the developed rules is an issue of major interest in terms of the functionality of rough sets based rule induction approaches.
- The consideration of multicriteria classification (sorting) problems. Very often in classification problems, criteria (i.e. attributes with preferentially ordered domains) and preferentially ordered classes are considered. Such problems are referred to as sorting problems (within the multicriteria decision making paradigm; [4.10], [4.11]). Bond rating, credit risk assessment, stock evaluation, country risk assessment, etc., are some very common examples of sorting problems from the field of finance. The rough sets approach has recently been extended to take into consideration the specific characteristics of sorting problems. In particular Greco et al. [4.9], [4.12] proposed the use of the dominance relation instead of the indiscernibility relation as the founding concept of the rough set theory when criteria are involved instead of attributes. This approach is well suited to the analysis of sorting problems, but it is also suitable in addressing choice and ranking problems within the rough sets theory framework. Bearing in mind the fact that sorting problems are often encountered in real-world decision making, it would be quite interesting to investigate the performance of the aforementioned recent advances in the rough sets theory compared to statistical, machine learning or multicriteria techniques.

However, despite these issues that need further research and the findings of this extensive experimental study, there are two significant features of the rough sets theory and the associated rule induction approaches that should be emphasized, with regard to providing meaningful decision support.

- They provide a sound mechanism for data reduction thus minimizing the information required in making decisions. This reduces the time and cost of data gathering and management, as well as the time and cost of the whole decision making process. Towards this direction it would be interesting to extent the scope of the simulation study presented in this paper to cope not only with the classification accuracy of rough sets based rule induction techniques, but also with their data reduction capabilities. This

will enable the derivation of useful results with regard to the way that data conditions are related to the data reduction capabilities of rough sets based rule induction.
- They enable the development of rule-based classification models that are easy to understand. This second feature is very significant in terms of decision support, and it is hardly shared by the statistical approaches explored in this study. These features have been the main reasons for the widespread application of rough sets in addressing numerous real-world problems from the fields of finance [4.6], engineering [4.41], automatic control [4.42], medicine [4.44], etc.

Acknowledgements

The paper was refereed by Chueh-Yung Tsao and Chia-Lin Chang. The authors are grateful to the referees for their careful reading of the paper.

References

4.1 Anderson, T. W. (1958): An Introduction to Multivariate Statistical Analysis. Wiley
4.2 Bajgier, S. M., Hill, A. V. (1982): An Experimental Comparison of Statistical and Linear Programming Approaches to the Discriminant Problem. Decision Sciences 13, 604–618
4.3 Breiman, L., Friedman, J. H., Olsen, R. A., Stone, C. J. (1984): Classification and Regression Trees. Pacific Grove
4.4 Clark, P., Niblett, T. (1989): The CN2 Induction Algorithm. Machine Learning, Vol. 3, 261–283
4.5 Clarke, W. R, Lachenbruch, P. A, Broffitt, B. (1979): How Non-normality Affects the Quadratic Discriminant Function. Communications in Statistics: Theory and Methods, Vol. 8, No. 13, 1285–1301
4.6 Dimitras, A. I., Slowinski, R., Susmaga, R., Zopounidis, C. (1999): Business Failure Prediction Using Rough Sets. European Journal of Operation Research, Vol. 114, 49–66
4.7 Doumpos, M., Zopounidis, C., Pardalos, P. M. (2000): Multicriteria Sorting Methodology: Application to Financial Decision Problems. Parallel Algorithms and Applications, Vol. 15, No. 1-2, 113–129
4.8 Freed, N., Glover, F. (1981): Simple But Powerful Goal Programming Models for Discriminant Problems. European Journal of Operational Research, Vol. 7, 44–60
4.9 Greco, S., Matarazzo, B., Slowinski, R. (1997): Rough Set Approach to Multiattribute Choice and Ranking Problems. In: Fandel, G., Gal, T. (Eds.), Multiple Criteria Decision Making. Springer-Verlag, 318–329
4.10 Greco, S., Matarazzo, B., Slowinski, R. (1998): A New Rough Set Approach to Evaluation of Bankruptcy Risk. In: Zopounidis, C. (Ed.), Operational Tools in the Management of Financial Risks. Kluwer Academic Publishers, 121–136
4.11 Greco, S., Matarazzo, B., Slowinski, R. (1998): The Use of Rough Sets and Fuzzy Sets in MCDM. In: Gal, T., Hanne, T., Stewart, T. (Eds.), Advances in Multiple Criteria Decision Making. Kluwer Academic Publishers, 14.1–14.59

4.12 Greco, S., Matarazzo, B., Slowinski, R. (1998): Extension of the Rough Set Approach to Multicriteria Decision Support. INFOR, Vol. 38, No. 3, 161–196
4.13 Greco, S., Matarazzo, B., Slowinski, R. (2001): Rough Sets Theory for Multicriteria Decision Analysis. European Journal of Operational Research, Vol. 129, No. 3, 1–47
4.14 Grzymala-Busse, J. W. (1992): LERS - A system for Learning from Examples Based on Rough Sets. In: Slowinski, R. (Ed.), Intelligent Decision Support: Handbook of Applications and Advances of the Rough Sets Theory. Kluwer Academic Publishers, 3–18
4.15 Grzymala-Busse, J. W., Stefanowski, J. (2001): Three Discretization Methods for Rule Induction. International Journal of Intelligent Systems, Vol. 16, 29–38
4.16 Joachimsthaler, E. A., Stam, A. (1988): Four Approaches to the Classification Problem in Discriminant Analysis: An Experimental Study. Decision Sciences, Vol. 19, 322–333
4.17 Komorowski, J., Pawlak Z., Polkowski L., Skowron A. (1999): Rough Sets: Tutorial. In: Pal,S. K., Skowron, A. (Eds.), Rough Fuzzy Hybridization: A New Trend in Decision-Making. Springer, 3–98
4.18 Kordatoff, Y., Michalski, R. S. (1990): Machine Learning: An Artificial Intelligence Approach (Vol. II). Morgan Kaufmann Publishers
4.19 Krusinska, E., Slowinski, R., Stefanowski, J. (1992): Discriminant Versus Rough Sets Approach to Vague Data Analysis. Journal Applied Stochastic Models and Data Analysis, Vol. 8, 43–56
4.20 Krusinska, E., Stefanowski, J., Stromberg, J. E. (1994): Comparability and Usefulness of Newer and Classical Data Analysis Techniques: Application in Medical Domain Classification. In: E. Didey et al. (Eds.), New Approaches in Classification and Data Analysis. Springer Verlag, 644–652
4.21 Mardia, K. V. (1975): Assessment of Multinormality and the Robustness of Hotelling's T^2 Test. Applied Statistics, Vol. 24, 163–171
4.22 Michalski, R. S. (1969): On the Quasi-minimal Solution of the General Covering Problem. Proceedings of the 5th International Federation on Automatic Control, Vol. 27, 109–129
4.23 Mienko, R., Stefanowski, J., Toumi, K., Vanderpooten, D. (1996): Discovery Oriented Induction of Decision Rules. Cahier du LAMSADE 141, Universite de Paris Dauphine
4.24 Patuwo, E., Hu, M. Y., Hung, M. S. (1993): Two-group Classification Using Neural Networks. Decision Sciences, Vol. 24, 825–845
4.25 Pawlak, Z. (1982): Rough Sets. International Journal of Information and Computer Sciences, Vol. 11, 341–356
4.26 Pawlak, Z., Slowinski, R. (1994): Decision Analysis Using Rough Sets. International Transactions in Operational Research, Vol. 1, No. 1, 107–114
4.27 Pawlak, Z., Slowinski, R. (1994): Rough Set Approach to Multi-attribute Decision Analysis. European Journal of Operational Research, Vol. 72, 443–459
4.28 Quinlan, J. R. (1986): Induction of Decision Trees. Machine Learning, Vol. 1, 81–106
4.29 Quinlan, J. R. (1993): C4.5: Programs for Machine Learning. Morgan Kaufmann Publishers
4.30 Skowron, A. (1993): Boolean Reasoning for Decision Rules Generation. In: Komorowski, J., Ras, Z. W. (Eds.), Methodologies for Intelligent Systems (Lecture Notes in Artificial Intelligence Vol. 689). Springer-Verlag, 295–305
4.31 Slowinski, R., Stefanowski, J. (1992): RoughDAS and RoughClass Software Implementations of the Rough Sets Approach. In: Slowinski, R. (Ed.), Intelligent Decision Support: Handbook of Applications and Advances of the Rough Sets Theory. Kluwer Academic Publishers, 445–456

4.32 Slowinski, R., Stefanowski, J. (1994): Rough Classification with Valued Closeness Relation. In: E. Diday et al. (Eds.), New Approaches in Classification and Data Analysis. Springer-Verlag, 482–488

4.33 Slowinski, R., Zopounidis, C. (1995): Application of the Rough Set Approach to Evaluation of Bankruptcy Risk. International Journal of Intelligent Systems in Accounting, Finance and Management, Vol. 4, 27–41

4.34 Slowinski, R., Stefanowski, J., Greco, S., Matarazzo, B. (2000): Rough Sets Processing of Inconsistent Information. Control and Cybernetics, Vol. 29, No. 1, 379–404

4.35 Slowinski, R., Zopounidis, C., Dimitras, A. I. (1997): Prediction of Company Acquisition in Greece by Means of the Rough Set Approach. European Journal of Operational Research, Vol. 100, 1–15

4.36 Stefanowski, J. (1992): Rough Sets Theory and Discriminant Methods in an Analysis of Information Systems. Foundations of Computing and Decision Sciences, Vol. 17, No. 2, 81–89

4.37 Stefanowski, J. (1995): Using Valued Closeness Relation in Classification Support of New Objects. In: Lin, T. Y., Wildberg, A. M. (Eds.), Soft Computing. Simulation Council Inc., 324–327

4.38 Stefanowski, J. (1998): Rough Set Based Rule Induction Techniques for Classification Problems. Proceedings of the 6th European Congress on Intelligent Techniques and Soft Computing (Vol. 1). Aachen Sept. 7–10, 109–113

4.39 Stefanowski, J. (1998): On Rough Set Based Approaches to Induction of Decision Rules. In: Polkowski, L., Skowron, A. (Eds.), Rough Sets in Knowledge Discovery. Physica-Verlag, 500–529

4.40 Stefanowski, J., Vanderpooten, D. (1994): A General Two-stage Approach to Inducing Rules from Examples. In: Ziarko, W. (Ed.), Rough Sets, Fuzzy Sets and Knowledge Discovery. Springer-Verlag, 317–325

4.41 Stefanowski, J., Slowinski, R., Nowicki, R. (1992): The Rough Sets Approach to Knowledge Analysis for Classification Support in Technical Diagnostics of Mechanical Objects. In: Belli, F., Radermacher, F. J. (Eds). Industrial & Engineering Applications of Artificial Intelligence and Expert Systems (Lecture Notes in Economics and Mathematical Systems 604). Springer-Verlag, 324–334

4.42 Szladow, A. J., Ziarko, W. P. (1992): Knowledge-based Process Control Using Rough Sets. In: Slowinski, R., (Ed.), Intelligent Decision Support: Handbook of Applications and Advances of the Rough Sets Theory. Kluwer Academic Publishers, 49–60

4.43 Teghem, J., Benjelloun, M. (1992): Some Experiments to Compare Rough Sets Theory and Ordinal Statistical Methods. In: R. Slowinski (Ed.), Intelligent Decision Support: Handbook of Applications and Advances of the Rough Sets Theory. Kluwer Academic Publishers, 267–286

4.44 Tsumoto, S. (1998): Automated Extraction of Medical Expert System Rules from Clinical Databases Based on Rough Set Theory. Information Sciences, Vol. 112, 67–84

4.45 Vale, D. C., Maurelli, V. A. (1983): Simulating Multivariate Nonnormal Distributions. Psychometrika, Vol. 48, No. 3, 465–471

4.46 Yandell, B. S. (1977): Practical Data Analysis for Designed Experiments. Chapman & Hall

4.47 Ziarko, W., Golan, D., Edwards, D. (1993): An application of DATALOGIC/R Knowledge Discovery Tool to Identify Strong Predictive Rules in Stock Market Data. In: Proceedings of the AAAI Workshop on Knowledge Discovery in Databases, Washington D.C., 89–101

Part III

Artificial Neural Networks and Support Vector Machines

5. Forecasting the Opening Cash Price Index in Integrating Grey Forecasting and Neural Networks: Evidence from the SGX-DT MSCI Taiwan Index Futures Contracts

Tian-Shyug Lee[1], Nen-Jing Chen[2], and Chih-Chou Chiu[3]

[1] Department of Business Administration, Fu-Jen Catholic University, Hsin-Chuang, Taiwan
email: badm1004@mails.fju.edu.tw

[2] Department of Economics, Fu-Jen Catholic University, Hsin-Chuang, Taiwan

[3] Institute of Commerce Automation and Management, National Taipei University of Technology, Taipei, Taiwan

This chapter investigates the information content of SGX-DT (Singapore Exchange-Derivatives Trading Limited) MSCI (Morgan Stanley Capital International) Taiwan futures contracts and its underlying cash market during the non-cash-trading (NCT) period. Previous day's cash market closing index and the grey forecasts by using the futures during the NCT period are used to forecast the 09:00 AM opening cash price index by the neural networks model. To demonstrate the effectiveness of our proposed method, the five-minute intraday data of spot and futures index from October 1, 1998 to December 31, 1999 was evaluated using the special neural network modeling. Analytic results demonstrate that the proposed model of integrating grey forecasts and neural networks outperforms the neural network model with previous day's closing index as the input variable, the random walk model and ARIMA forecasting. It therefore indicates that there is valuable information involved in the futures prices during the NCT period in forecasting the opening cash price index. Besides, grey forecasts provide a better initial solution that speeds up the learning procedure for the neural networks which in turn give better forecasting results.

5.1 Introduction

It's a common practice that the daily trading period of an index futures contract ends later and begins earlier than which of its underlying spot market. In this study, the time period between daily close and the subsequent opening of the cash index trading is defined as non-cash-trading (NCT) period. Valuable information should be obtained through the analysis of the NCT period of the futures market and should contribute to the success of implementing proper investment decisions for the underlying spot market. For SGX-DT (Singapore Exchange-Derivatives Trading Limited) MSCI Taiwan futures contracts and its underlying cash market, the trading sessions are from 08:45 AM to 12:15 PM and from 09:00 AM to 12:00 PM, respectively. Therefore, the information contents of NCT (from 12:00 to 12:15 in each trading day and from 08:45 to 09:00 in the following trading day) futures prices are analyzed in this study. And since the world's leading European and American stock markets start the daily trading after the MSCI Taiwan futures and cash markets close and end the daily trading before the MSCI Taiwan markets open, [5.8] therefore used the ARCH model to analyze the intraday return and volatility spill over effect among the New York, London, Tokyo, Hong Kong, and Taiwan stock markets. She found that there exists spill over effect from one market to the opening of the market that opens subsequently, no matter whether the trading hours of two markets overlap or not. She also found that the New York stock market has a significant spill over effect on the opening of all other markets under investigation. It is therefore hypothesized that the new information of world's leading markets will be reflected in the MSCI Taiwan index futures prices first before it reaches the cash market.

Following the above argument, this research investigates the information content of SGX-DT MSCI Taiwan index futures prices during the NCT period. And the neural networks approach is proposed to build the model in forecasting the opening cash market price index. Previous day's cash market closing index at 12:00 and the NCT futures are used to forecast the 09:00 opening cash market price index using the backpropagation neural network (BPNN) model. It is noted that the popular ARIMA approach is inappropriate in constructing the model in this research since the futures index is utilized in predicting the opening cash price index. And the neural networks are adopted in building the forecasting model since one of its distinguishing features is the ability to capture subtle functional relationships among the empirical data even though the underlying relationships are unknown or hard to describe. Besides, no strong model assumptions, like variation homogeneity and system stationarity, are required to build the model. In addition, the literature on neural networks for forecasting problems is vast and fruitful [5.33]. As to the determination of the appropriate input variables of the neural networks, the grey model (GM) is used to forecast the 09:00 futures price to account for the effect of the five 5-minute futures prices during the NCT period. And hence only previous day's cash market closing index and the GM forecast of the 09:00 futures price will be used as the input variables of the designed neural networks. In this study, BPNN with various numbers of nodes in the hidden layer, and different learning rates are extensively studied to address and solve the issue of finding the appropriate setup of the topology of the networks. Further studies are performed on the robustness of the constructed neural network in terms of different training and testing sample sizes. To evaluate the effectiveness of the proposed neural network model, the daily five-minute transaction data of index futures and cash prices from October 1, 1998 to December 31, 1999 are used as an illustrative example.

Finally, analytic results compared with neural network model with previous day's closing index as the input variable, neural network model with grey forecast of the futures price at 09:00 as the input variable, the random walk model and ARIMA forecasts will also being discussed.

The rest of the paper is organized as follows. We will briefly review the literature of lead-lag relationships and market opening and closing trading characteristics in section 5.2. Section 5.3 gives a brief introduction about neural networks and grey theory model. The development of neural network's topology in using the futures during the NCT period and previous day's cash closing index to predict the opening cash price index is presented in section 5.4. To verify the robustness of the designed neural network model, the prediction efficiency is summarized in terms of different training and testing sample sizes in section 5.5. Section 5.6 addresses the conclusion and possible future research areas.

5.2 Literature Review

5.2.1 Lead-Lag Relationship between Index Futures and Cash Markets

Removing the nonsynchronization effect in the cash index has became a standard process under a research topic of lead-lag relationship between index cash and futures markets since 1990. Though there exist researches which indicate only a feedback relationship such as [5.1], most researches find that index futures price changes lead price changes of the underlying spot market more often and more significantly than the other way. They include [5.30], [5.5], [5.22], [5.18], [5.25], [5.24] and [5.20], to name a few. Generally speaking, past studies suggest that derivative markets are informationally more efficient than the underlying spot market ([5.16]).

5.2.2 Market Trading Characteristics at the Opening and Closing Stage

A number of researches investigate the security and derivative markets' intraday trading characteristics such as trading volume, return volatility, return, and bid-ask spread, especially at the open and close of the market. In general, empirical studies report that an U shape is among one or more of the above trading characteristics. [5.13], [5.23], [5.11], and [5.6] are some examples of them.

To explain the U shape trading volume and return volatility, [5.2] develop a theoretical information model. Based on their model, periodic concentrations of trading volume are associated with high return volatility and are results of cost reducing trading strategies by both informed and liquidity traders. They state: "If the liquidity-trading volume is higher at the end of the trading day, then more informed traders will trade at this time. As a result, the prices quoted at the end of the trading day will reflect more of the information that will be released publicly during the following nontrading hours."

In contrast to [5.2], [5.4] suggest that observed volume concentrations and high return volatility at the open and close are associated with high trading cost in terms of bid-ask spread and are due to the demand for portfolio reformation. The reasons behind the U shape phenomenon are still subject to empirical research.

[5.16] investigate the information content of the end-of-the-day (EOD) Osaka Nikkei index futures returns by studying the relationship between EOD futures returns and overnight spot returns. They find that the unexpected component of EOD futures returns is positively related to the overnight spot returns as well as the trading period spot returns over the next two trading days. According to [5.12], private information affects prices for more than one trading day. [5.16] therefore suggest that EOD futures returns reflected the information of informed traders.

The aim of the current research is to learn the information content of the NCT SGX-DT MSCI Taiwan futures prices by comparing the cash opening price forecast including the NCT futures and the random walk model assumption. If the former forecasts outperform the latter then the information of the NCT futures price is considered valuable and therefore this underlying work will provide an indirect support of [5.2] information model.

5.3 Neural Networks and Grey Theory

5.3.1 Neural Networks

A neural network is a computationally intensive, algorithmic procedure for transforming inputs into desired outputs using highly inter-connected networks of relatively simple processing elements (often termed neuron cells, units or nodes-we will use nodes thereafter). Neural networks are modeled following the neural electrical-physiological activity in the human brain. The three essential features of a neural network are the nodes, the network architecture describing the connections among the nodes, and the training algorithm used to find values of the network parameters (weights) for performing a particular task. The nodes are connected to one another in the sense that the output from one node can be served as the input to yet another node. Each node transforms an input to an output using some specified function that is typically monotone, but otherwise arbitrary. This function depends on constants (parameters) whose values must be determined with a training set of inputs and outputs. Network architecture is the organization of nodes and the types of connections permitted. The nodes are arranged in a series of layers with connections between nodes in different layers, but not between nodes in the same layer. The layer receiving the inputs is the input layer. The final layer provides the target output signal is the output layer. Any layers between the input and output layers are hidden layers. A simple representation of a neural network is shown in Fig. 5.1 ([5.28]).

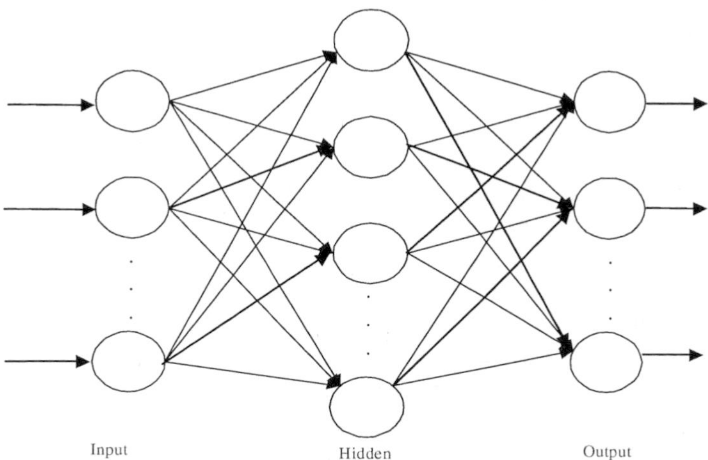

Fig. 5.1. A three-layer backpropagation neural network

Neural networks can be classified into two categories, feedforward and feedback networks. The feedback networks contain nodes that can be connected to themselves, enabling a node to influence other nodes as well. Kohonen self-organizing and the Hopfield networks are examples of this type of network. The nodes in feedforward networks can take inputs only from the previous layer and send outputs to the next layer. The ADALINE and BPNN are two typical examples of this kind of network.

BPNN is essentially a gradient steepest descent training algorithm and has been the most often utilized paradigm to date. For the gradient descent algorithm, the step size, called the learning rate, must be specified first. The learning rate is crucial for BPNN since smaller learning rates tend to slow down the learning process before convergence while larger ones may cause network to oscillate and hence unable to converge.

Neural networks are increasingly found to be useful in modeling nonstationary processes due to its associated memory characteristics and generalization capability ([5.29]). Computer scientists and statisticians have interests in the computational potentials of neural network algorithms is on the rise. [5.14] wrote a comprehensive reference on artificial neural networks. [5.3] edited a collection of papers that chronicled the major developments in neural network modeling. [5.7], [5.27], and [5.29] provided surveys describing the relevance of neural networks to the statistics community.

Despite the many satisfactory characteristics of neural networks, building a neural network model for a particular forecasting problem is by no means a trivial task. One crucial decision for a designer is to determine the appropriate network topology, the number of layers, and the number of nodes in each layer, and the appropriate learning rates. Sometimes, the parameters associated with the nonlinear sigmoid function are also an important consideration in the design peocess. The selection of these parameters is basically problem dependent. To date, there is no simple, clear-cut guidelines one can lean on it for the determination of these parameters. Guidelines are either heuristic results from statistical analysis or based on simulations from limited experiments. Hence the design of neural networks is more of an art rather than a science. Basically, the number of nodes in the input layer corresponds to the number of variables in the input vector used to forecast future target values. Ideally, we desire a small number of essential nodes for unveiling the unique features embedded in the data. Too few or too many input nodes will either affect the learning or the predicting capability of the network. Influenced by theoretical works which show that one hidden layer is sufficient for neural networks to approximate any complex nonlinear function with any desired accuracy ([5.9]; [5.17]), most researchers use only one hidden layer for forecasting purposes. As to the issue of determining the optimal number of hidden nodes is a crucial yet complicated one. In general, artificial neural network with fewer hidden nodes are preferable as they usually have better generalization ability and less overtraining problem while network with too few hidden nodes may not have enough power to model and learn the data. There is no theoretical basis for selecting this parameter although a few systematic approaches are reported. So far the most commonly approach in determining the number of hidden nodes is via experiments or trial-and-error. Several rules of thumb have also been proposed by [5.21], [5.15], [5.32], [5.31], and [5.19] for the case of the also popular one hidden layer network.

5.3.2 Grey Model and Grey Forecasting

(1) Grey Model

Grey theory model was first introduced by [5.10]. From the perspective of the infromation availability, the system which lacks information, such as structure message, operation mechanism and behavior document, are referred to as a grey system. Lacking of information is the main distinctive property of grey system, and this simply implies that the available information about a system is very limited. The problem then is how to make the best use of the limited information to approximate the system dynamic behavior. The goal of grey system and its applications is to bridge the gap existing between grey or uncertain to white or certain. Hence, the aim of grey system theory is to provide the theory, techniques, notions and ideas for resolving the latent and intricate systems. Unlike the traditional mathematics and/or statistical system method which requires a lot of sample points to construct a system dynamics, grey theory needs as few as 4 data points to build a grey model (GM) to approximate a system. In the following, basic mathematical preliminaries needed ([5.10] in building a GM are introduced.

Definition 1 *For a nonnegative time series*

$$x^{(0)} = [x^{(0)}(1), x^{(0)}(2), \ldots, x^{(0)}(n)],$$

define

$$x^{(1)} = [x^{(1)}(1), x^{(1)}(2), \ldots, x^{(1)}(n)],$$

where

$$x^{(1)}(k) = \sum_{m=1}^{k} x^{(0)}(m)$$

then

$$\begin{Bmatrix} x^{(0)}(k) = x^{(1)}(k) - x^{(1)}(k-1), & for\ k \geq 2 \\ x^{(0)}(1) = x^{(1)}(1), & for\ k = 1 \end{Bmatrix} \tag{5.1}$$

It is convenient to denote $x^{(1)} = \text{AGO}\ x^{(0)}$ *and* $x^{(0)} = \text{IAGO}\ x^{(1)}$ *where* AGO *and* IAGO *are referred to accumulated generating operation and inverse accumulated generating operation, respectively.*

Definition 2 *A first order grey differential equation having one variable is defined as* GM(1,1) *and can be written as follows:*

$$x^{(0)}(k) + az^{(1)}(k) = b, \quad k = 1, 2, \ldots, n$$

where

$$z^{(1)}(k) = 0.5[x^{(1)}(k) + x^{(1)}(k-1)], \quad k = 2, 3, \ldots, n$$

and

$$x^{(1)}(k) = \sum_{m=1}^{k} x^{(0)}(m), \quad k = 1, 2, \ldots, n$$

Note that a and b are called the developing coefficient and grey input, respectively.

The parameters a and b have to be estimated first before we can proceed to grey forecasting. The following proposition outlines the procedure for obtaining the popular least squared estimates for a and b.

Proposition 1 *Consider a* GM(1,1) *model as follows:*

$$x^{(0)}(k) + az^{(1)}(k) = b, \quad k = 1, 2, \ldots, n$$

where

$$z^{(1)}(k) = 0.5[x^{(1)}(k) + x^{(1)}(k-1)], \quad k = 2, 3, \ldots, n$$

and

$$x^{(1)} = \text{AGO } x^{(0)}$$

Then we can have

$$Y_N = B\theta$$

with

$$\theta = [a, \ b]^T$$

$$B = \begin{bmatrix} -z^{(1)}(2) & 1 \\ -z^{(1)}(3) & 1 \\ \vdots & \vdots \\ -z^{(1)}(n) & 1 \end{bmatrix}$$

and

$$Y_N = \begin{bmatrix} x^{(0)}(2) \\ x^{(0)}(3) \\ \vdots \\ x^{(0)}(n) \end{bmatrix}$$

Applying the least squared method, we can have

$$\hat{\theta} = \begin{bmatrix} a \\ b \end{bmatrix} = (B^T B)^{-1} B^T Y_N.$$

(2) Grey Forecasting

The GM(1,1) is well built after successfully estimating the values of a and b. We can then have

$$x^{(0)}(k) + 0.5a[x^{(1)}(k) + x^{(1)}(k-1)] = b, \quad k = 2, 3, \ldots, n$$

with

$$x^{(1)}(k) = x^{(1)}(k-1) + x^{(0)}(k)$$

Rearranging terms we can have

$$x^{(0)}(k) = \frac{b - ax^{(1)}(k-1)}{1+0.5a}$$

And hence the predicted value of $x^{(0)}(n+1)$ can be obtained by the following equation:

$$\hat{x}^{(0)}(n+1) = \frac{b - ax^{(1)}(n)}{1+0.5a}$$

5.4 Empirical Results and Discussion

The daily five-minute transaction data of futures and cash prices from October 1, 1998 to December 31, 1999 provided by Capital Futures Corporation, Taipei is used in this study. The 12:00 cash market closing index and the five 5-minute futures prices during the NCT period (12:05, 12:10, 12:15 in each trading day and 08:50, 08:55 in the following trading day) are used to forecast the 09:00 opening cash price index by the BPNN model. Originally all the five 5-minute futures prices are used as the input variables of the BPNN models. However, the training process is slow and the network is unable to converge when the training iterations are not big enough. Therefore the five 5-minute futures prices are first used to build a GM(1,1) model in forecasting the 09:00 futures price.

5.4.1 Grey Forecasting Model

We will be using the NCT futures prices from 1999/11/09 to 1999/11/10 in explaining the procedure of building the GM(1,1) forecasting model. The NCT futures prices during the studying period can be found in the following table.

Table 5.1. The NCT futures prices from 1999/11/09 to 1999/11/10

date time	11/09 12:05	11/09 12:10	11/09 12:15	11/10 08:50	11/10 08:55	11/10 09:00
futures price	335.1	334.5	334.5	334	333.8	334

Since

$$x^{(0)} = (335.1, 334.5, 334.5, 334, 333.8)$$

So we have

$$x^{(1)} = AGOx^{(0)} = (335.1, 669.6, 1004.1, 1338.1, 1671.9)$$

Therefore the GM(1,1) model is as follows:

$$x^{(0)}(k) + az^{(1)}(k) = b, \quad k = 1, 2, \ldots, 5$$

$$z^{(1)}(k) = 0.5[x^{(1)}(k) + x^{(1)}(k-1)], \quad k = 2, 3, \ldots, 5$$

From $x^{(0)}$ and $x^{(1)}$ we can have

$$z^{(1)} = (502.35, 836.85, 1171.1, 1505)$$

By applying the least squared techniques, we can get

$$\hat{\theta} = \begin{bmatrix} a \\ b \end{bmatrix} = (B^T B)^{-1} B^T Y_N = \begin{bmatrix} 0.000778 \\ 334.9809 \end{bmatrix}$$

With

$$B = \begin{bmatrix} -502.35 & 1 \\ -836.85 & 1 \\ -1171.1 & 1 \\ -1505 & 1 \end{bmatrix}, \quad Y_N = \begin{bmatrix} 334.5 \\ 334.5 \\ 334.0 \\ 333.8 \end{bmatrix}$$

Then we can put the estimates of a and b into the GM(1,1) model and have:

$$x^{(0)}(k) + 0.000778 * z^{(1)}(k) = 334.9809$$

or equivalently $x^{(0)}(k) + 0.5 * 0.000778 * [x^{(1)}(k) + x^{(1)}(k-1)] = 334.9809$
And since

$$x^{(1)}(k) = x^{(1)}(k-1) + x^{(0)}(k)$$

We can have

$$x^{(0)}(k) = \frac{334.9809 - 0.000778 * x^{(1)}(k-1)}{1 + 0.5 * 0.000778}$$

Therefore the forecasts of $x^{(0)}(6)$ is

$$\hat{x}^{(0)}(6) = \frac{334.9809 - 0.000778 * 1671.9}{1 + 0.5 * 0.000778} = 333.55$$

Following the same procedure, the forecasting results of all the 09:00 futures prices can be computed.

5.4.2 Neural Networks Forecasting Model

The obtained 09:00 futures grey forecasts and previous day's cash closing index are then used as the input variables of the designed BPNN model. A three-layer BPNN will be trained to build the forecasting model and the neural network simulator [5.26] developed by Vesta was used to develop the modeling network. And since there are only 2 input nodes in the input layer, the initial number of hidden nodes to be tested was chosen to be 3, 4, 5, 6, and 7. And the network has only one output node, the forecasted 09:00 opening cash price index. As [5.28] concluded that lower learning rates tended to give the best network results, learning rates 0.01, 0.05, 0.08, and 0.1 are tested during the training process.

Fig. 5.2 is a plot of the daily opening cash price index data versus time, connected by straight lines. From the data pattern revealed in Fig. 5.2, with the 307 observations during the period studied, the first 246 (80% of the total sample points) are used to train the network and the remaining 61 are retained for testing. The convergence criteria used for training are a root mean squared error (RMSE) less than or equal to 0.0001 or a maximum of 1,000 iterations. The network topology with the minimum testing RMSE is considered as the optimal network topology.

Fig. 5.2. MSCI Taiwan opening cash index from 09/01/1998 to 12/31/999

The prediction results of the neural network with combinations of different hidden nodes and learning rates are summarized in Table 5.2. From Table 5.2, the 2-6-1 topology with a learning rate of 0.10 gives the best forecasting result (minimum testing RMSE). Here 2-6-1 stands for two neurons in the input layer, six neurons in the hidden layer, and one neuron in the output layer. To examine the convergence characteristics of the proposed neural network model, the RMSE during the training process for the 2-6-1 network with the learning rate of 0.10 are depicted in Fig. 5.3. The excellent convergence characteristics of the proposed network can easily be observed. The BPNN model prediction results with previous day's cash closing index as input variable and the BPNN model with the 09:00 futures grey forecast as input variable are summarized in Tables 5.5 and 5.6. For comparison, the results of opening cash price index forecasts by the random walk, ARIMA and BPNN models are also included in Table 5.7.

Table 5.2. BPNN model prediction results

Number of nodes in the hidden layer	Learning rate	Training RMSE	Testing RMSE
3	0.01	0.014190	0.011926
	0.05	0.013107	0.010806
	0.08	0.012496	0.010220
	0.1	0.013346	0.011179
4	0.01	0.013784	0.011704
	0.05	0.012679	0.010428
	0.08	0.012559	0.010293
	0.1	0.012838	0.010667
5	0.01	0.013990	0.012044
	0.05	0.012589	0.010337
	0.08	0.012396	0.010128
	0.1	0.012419	0.010132
6	0.01	0.013758	0.011619
	0.05	0.012739	0.010485
	0.08	0.012362	0.010088
	0.1	0.012425	0.009426
7	0.01	0.014232	0.012358
	0.05	0.012512	0.010240
	0.08	0.012514	0.010337
	0.1	0.012054	0.009822

From Table 5.7, we can see that the neural network model with the 09:00 futures grey forecasts and previous day's cash closing index as input variables accurately forecasts the opening cash price being up or down in 48 out of 61 days (about 79%) in the testing sample. Fig. 5.4 depicts the forecasting results of the testing sample of the designed neural network model. It is observed that the built model provides well forecasting results. Besides, the RMSE, mean absolute deviation (MAD), mean absolute percentage error (MAPE), and the root mean squared percentage error (RMSPE) of the forecasts by the above five models can be computed and summarized in Table 5.3.

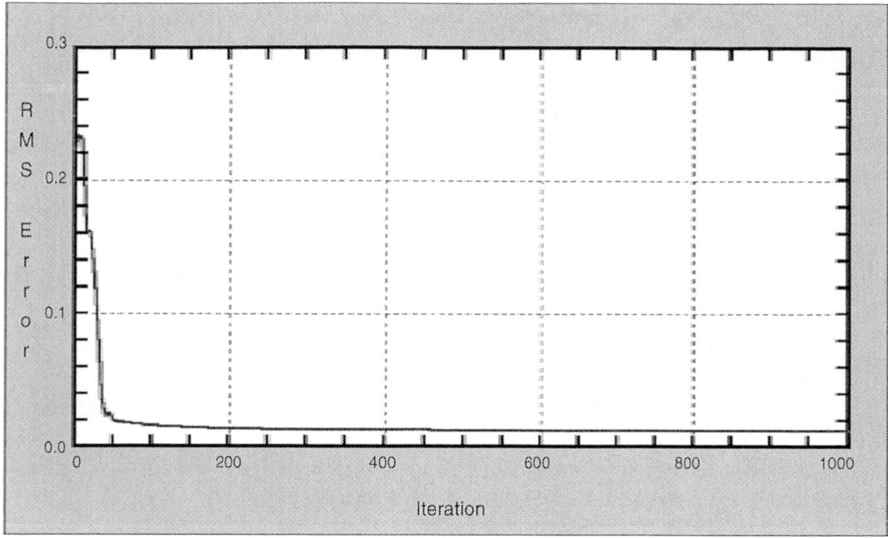

Fig. 5.3. The RMSE history in the learning process for the proposed network

Table 5.3. Error measures for five different models

Model	Testing RMSE	Testing MAD	Testing MAPE	Testing RMSPE
Random walk model	3.36	2.57	0.73%	0.95%
ARIMA model	5.35	4.51	1.28%	1.52%
BPNN with previous day's closing index as input	3.16	2.55	0.72%	0.89%
BPNN with the 09:00 futures grey forecast as input	2.51	2.04	0.58%	0.71%
BPNN with the 09:00 futures grey forecast and previous day's closing index as inputs	2.07	1.77	0.50%	0.59%

It is observed that the prediction error of the BPNN model with the 09:00 futures grey forecasts and previous day's cash closing index as input variables is at least 40% lower than those predictions by considering only previous day's closing index in the cash market. It therefore indicates that there is valuable information involved in the futures prices during the NCT period that can be used in forecasting the opening cash market price.

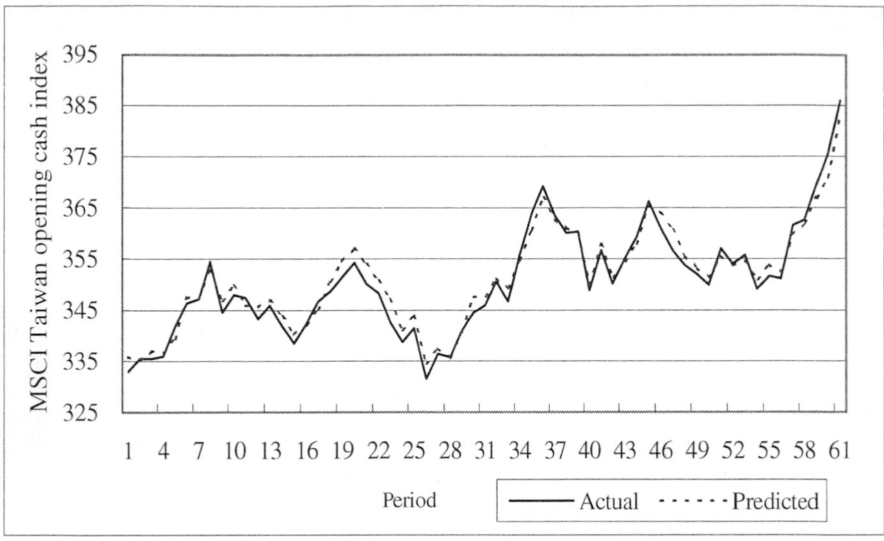

Fig. 5.4. MSCI Taiwan opening cash index forecasts

Table 5.4. Robustness evaluation by different training and testing sample sizes

Relative Ratio	Training data set size	Testing data set size	Testing MAAPE (%)	Testing RMSPE (%)
50%	154	153	3.8644	1.8425
60%	184	123	3.1396	0.8324
70%	215	92	2.3316	0.7204
80%	246	61	1.1151	0.5895

5.5 Robustness Evaluation of the Neural Networks Model

To evaluate the robustness of the neural network approach, the performance of the designed network was tested using different combinations of training and testing sample sizes. The testing plan is based on the relative ratio of the size of the training data to the complete data sets. For example, given the data up to October 5, 1999 for training, the ratio is 80% (=246/307). (Here 307 is the total number of data points and 246 is the number of data points used to train the network.) In this section, four relative ratios are considered. They are 50%, 60%, 70% and 80%, having the training data up to May 24, July 6, August 16, and October 5, 1999, respectively. The prediction results for the opening cash market price index are summarized in Table 5.4 in terms of two criteria, the maximum absolute percentage error (MAAPE) and RMSPE. It is noted that both the MAAPE and RMSPE decrease as the ratio increases. And as the result indicates, the prediction error made by the

Table 5.5. BPNN prediction results with the 09:00 futures grey forecast as input nodes

[t]

Number of nodes in the hidden layer	Learning rate	Training RMSE	Testing RMSE
3	0.01	0.016688	0.012896
	0.05	0.016459	0.012623
	0.08	0.016291	0.012590
	0.1	0.016108	0.012185
4	0.01	0.016431	0.012479
	0.05	0.016152	0.012224
	0.08	0.016144	0.012236
	0.1	0.016088	0.012155
5	0.01	0.016301	0.012310
	0.05	0.016059	0.012034
	0.08	0.016125	0.012200
	0.1	0.016361	0.012006
6	0.01	0.016416	0.012343
	0.05	0.016157	0.012251
	0.08	0.016147	0.012209
	0.1	0.016064	0.012145
7	0.01	0.016463	0.012336
	0.05	0.016025	0.012034
	0.08	0.015976	0.012010
	0.1	0.016052	0.012038

proposed approach is not seriously impacted by different training data sizes. In other words, even though the training sample size to the complete data set size is only 50%, it can still provide well forecasting results (RMSPE $< 2\%$). This implies that the proposed backpropagation neural network model can successfully forecast the opening cash market price index after the "learning" procedure.

5.6 Conclusions and Areas of Future Research

This study uses 1998/10/01 to 1999/12/31 intraday 5-minute data to analyze the information content of the MSCI Taiwan futures prices in the NCT period. Originally all the five 5-minute futures prices are used as the input variables of the BPNN forecasting models. However, the training process is slow and the network is unable to converge when the training iterations are not big enough. Therefore the five 5-minute futures prices are first used to build a GM(1,1) in forecasting the 09:00 futures price. The obtained 09:00 futures grey forecasts and previous day's cash closing index are then used as the input variables of the BPNN model to forecast the 09:00 opening cash price

Table 5.6. BPNN prediction results with previous day's closing index as input nodes

Number of nodes in the hidden layer	Learning rate	Training RMSE	Testing RMSE
3	0.01	0.020839	0.015960
	0.05	0.020120	0.015298
	0.08	0.020401	0.015567
	0.1	0.020125	0.015270
4	0.01	0.020487	0.015921
	0.05	0.020115	0.015222
	0.08	0.020351	0.015436
	0.1	0.020553	0.015692
5	0.01	0.020287	0.015877
	0.05	0.020055	0.015133
	0.08	0.019948	0.015157
	0.1	0.019934	0.015013
6	0.01	0.020292	0.016021
	0.05	0.020099	0.015188
	0.08	0.019880	0.015020
	0.1	0.019572	0.014935
7	0.01	0.020382	0.016154
	0.05	0.019947	0.015163
	0.08	0.019821	0.015105
	0.1	0.019948	0.015011

index. The testing result indicates that the optimal BPNN model has a structure of 2-6-1 with the learning rate of 0.10. It is concluded that the BPNN model with the futures prices during the NCT period and previous day's closing index in the cash market as input variables provides significantly better forecasting results than those forecasts only considering previous day's closing cash price index. This study indicates that the futures prices in the NCT period contains valuable information in forecasting the opening cash price index. Besides, grey forecasts provide a better initial solution that speeds up the learning procedure for the neural networks and give better forecasting results.

It is suggested to integrate other techniques, such as the GARCH family or fuzzy theory models, in building the forecasting model. Besides, since the stock markets around the world are highly interrelated, it is also recommended to consider another market's MSCI Taiwan futures prices and/or other area or country's futures prices as input nodes in building the BPNN forecasting model.

Table 5.7. Random walk, ARIMA and BPNN forecasting results

(1)	(2)	(3)	(4)	(5)	(6)	(7)
10/06/99	332.92	333.74	343.13	336.77	334.79	335.81
10/07/99	335.50	331.19	332.88	334.12	335.58	335.11*
10/08/99	335.47	336.88	335.46	339.98	334.38	336.97
10/11/99	335.88	335.16	335.43	338.23	335.36	336.77*
10/12/99	341.94	337.85	335.84	340.96	337.89	339.39*
10/13/99	346.35	348.42	341.90	351.21	344.27	347.52*
10/14/99	347.08	348.56	346.31	351.34	343.67	347.24*
10/15/99	354.38	350.78	347.04	353.38	351.79	352.81*
10/16/99	344.46	347.52	354.34	350.37	343.77	346.86*
10/18/99	347.94	348.30	344.42	351.10	348.20	349.73
10/19/99	347.33	345.15	347.90	348.13	343.86	345.93*
10/20/99	343.19	342.78	347.29	345.84	345.14	345.68*
10/21/99	345.90	340.96	343.15	344.06	348.41	346.85*
10/22/99	341.78	340.21	345.86	343.32	343.61	343.71*
10/25/99	338.45	336.56	341.74	339.66	340.38	340.26*
10/26/99	342.38	342.28	338.41	345.35	339.41	342.18
10/27/99	346.62	344.48	342.34	347.48	342.82	345.06*
10/29/99	348.60	344.36	346.58	347.37	352.37	350.61*
10/30/99	351.38	346.38	348.56	349.30	357.44	354.50*

(1) Date; (2) Opening cash index; (3) Random walk forecast; (4) ARIMA forecast; (5) BPNN with previous day's closing index as input; (6) BPNN with 09:00 grey forecast as input; (7) BPNN with previous day's closing index and 09:00 grey forecast as inputs. *BPNN with the 09:00 futures grey forecast and previous day's closing index as inputs accurately forecasts the opening cash price index is up or down.

Acknowledgements

The paper was originally written in the Word format. The author is grateful to Chung-Chih Liao, Tzu-Wen Kuo and Li-Cheng Sun for their kind supports to transform it into the LaTeX format.

Table 5.8. Random walk, ARIMA and BPNN forecasting results (cont.)

(1)	(2)	(3)	(4)	(5)	(6)	(7)
11/01/99	354.20	354.30	351.34	356.53	356.61	357.01
11/02/99	350.15	352.29	354.16	354.74	352.61	353.88
11/03/99	348.29	347.53	350.11	350.38	350.36	350.68*
11/04/99	342.71	341.21	348.25	344.30	348.59	347.06*
11/05/99	338.69	337.17	342.67	340.27	341.43	341.13*
11/06/99	341.44	339.11	338.65	342.22	344.54	343.79*
11/08/99	331.45	334.93	341.40	337.99	331.66	334.57*
11/09/99	336.42	335.99	331.41	339.08	335.98	337.49*
11/10/99	335.66	334.45	336.38	337.50	333.91	335.62*
11/11/99	340.89	336.40	335.62	339.49	341.14	340.63*
11/15/99	344.53	342.42	340.85	345.49	348.45	347.47*
11/16/99	345.85	343.59	344.49	346.63	347.97	347.67*
11/17/99	350.59	346.59	345.81	349.49	351.76	351.13*
11/18/99	346.72	347.03	350.55	349.91	348.20	349.22
11/19/99	355.68	350.68	346.68	353.29	355.04	354.70*
11/22/99	363.70	351.97	355.64	354.46	363.70	360.51*
11/23/99	369.21	364.93	363.66	365.37	366.84	366.91*
11/24/99	363.28	366.05	369.17	366.25	358.96	362.62*
11/25/99	360.00	359.32	363.24	360.84	360.13	360.95*
11/26/99	360.32	357.89	359.96	359.63	359.72	360.19*
11/29/99	348.85	343.35	360.28	346.39	351.80	349.86*
11/30/99	356.68	354.39	348.81	356.61	357.91	357.82*
12/01/99	350.08	349.66	356.64	352.36	350.04	351.35*
12/02/99	354.95	351.78	350.04	354.29	354.01	354.51*
12/03/99	359.09	354.14	354.91	356.39	358.12	357.86*
12/04/99	366.17	359.10	359.05	360.65	367.57	365.44*
12/06/99	361.03	361.36	366.13	362.52	363.69	363.80
12/07/99	356.52	357.96	360.99	359.69	360.42	360.64
12/08/99	353.77	353.53	356.48	355.85	354.63	355.55*
12/09/99	351.90	352.24	353.73	354.70	351.30	353.09
12/10/99	349.84	349.10	351.86	351.84	350.41	351.34*
12/14/99	356.96	349.17	349.80	351.90	357.30	355.49*
12/15/99	353.97	353.79	356.92	356.08	351.30	353.70
12/16/99	355.67	353.57	353.93	355.89	352.70	354.43*
12/18/99	349.06	346.86	355.63	349.75	351.24	350.94*
12/20/99	351.52	351.26	349.02	353.82	353.16	353.80*
12/21/99	351.13	349.17	351.48	351.90	352.07	352.34*
12/22/99	361.49	355.99	351.09	358.00	359.89	359.60*
12/23/99	362.51	361.07	361.45	362.28	360.67	361.89*
12/24/99	369.57	364.72	362.47	365.21	366.86	366.85*
12/27/99	375.40	372.35	369.53	370.92	369.44	370.83
12/28/99	385.92	383.17	375.36	378.09	383.59	382.59

References

5.1 Abhyankar, A. H. (1995): Return and Volatility Dynamics in the FT-SE 100 Stock Index and Stock Index Futures Markets. Journal of Futures Markets, Vol. 15, 457–488

5.2 Admati, A., Pfleiderer, P. (1988): A Theory of Intraday Patterns: Volume and Price Variability. Review of Financial Studies, Vol. 1, 3–40

5.3 Anderson, J. A., Rosenfeld, E. (1988): Neurocomputing: Foundations of Research. MIT Press

5.4 Brock, W. A., Kleidon, A. W. (1992): Periodic Market Closure and Trading: A Model of Intraday Bids and Asks. Journal of Economic Dynamics and Control, Vol. 16, 451–490

5.5 Chan, K. (1992): A Further Analysis of the Lead-lag Relationship Between The Cash Market and Stock Index Futures Market. The Review of Financial Studies, Vol. 5, 123–152

5.6 Chan, K., Chung, Y., Johnson, H. (1995): The Intraday Behavior of Bid-ask Spreads for NYSE Stocks and CBOE Options. Journal of Finance, Vol. 30, 329–346

5.7 Cheng, B., Titterington, D. M. (1994): Neural Network: A Review from A Statistical Perspective. Statistical Science, Vol. 9, 2–54

5.8 Chuang, G. -X. (1993): Intraday Return Spillover Effects Between Taiwan and International Stock Markets: An Application of ARCH model. Master's thesis, Graduate Institute of Finance, National Chung-Cheng University, Taiwan

5.9 Cybenko, G. (1989): Approximation by Superpositions of A Sigmoidal Function. Mathematical Control Signal Systems, Vol. 2, 303–314

5.10 Deng, J. L. (1989): Introduction to Grey System Theory. Journal of Grey System, Vol. 1, No. 1, 1–24

5.11 Ekman, P. (1992): Intraday Patterns in the S&P 500 Index Futures Market. Journal of Futures Markets, Vol. 12, 365–381

5.12 French, K., Roll, R. (1986): Stock Return Variances: the Arrival of Information and the Reaction of Traders. Journal of Financial Economics, Vol. 17, 5–26

5.13 Harris, L. (1986): A Transaction Data Study of Weekly and Intradaily Patterns in Stock Returns. Journal of Financial Economics, Vol. 1, 99–117

5.14 Haykin, S. S. (1994): Neural Networks: A Comprehensive Foundation. Macmillan

5.15 Hecht-Nielsen, R. (1990): Neurocomputing. Addison-Wesley

5.16 Hiraki, T., Maberly, E. D., Takezawa, N. (1995): The Information Content of End-of-the-day Index Futures Returns: International Evidence from the Osaka Nikkei 225 Futures Contract. Journal of Banking and Finance, Vol. 19, 921–936

5.17 Hornik, K., Stinchcombe, M., White, H. (1989): Multilayer Feedforward Networks Are Universal Approximations. Neural Networks, Vol. 2, 336–359

5.18 Iihara, Y., Kato, K., Tokunaga, T. (1996): Intraday Return Dynamics Between the Cash and the Futures Markets in Japan. Journal of Futures Markets, Vol. 16, 147–162

5.19 Kang, S. (1991): An Investigation of the Use of Feedforward Neural Networks for Forecasting. Ph.D. Thesis, Kent State University

5.20 Lee, T. S., Chen, N. J. (2000): The Information Content of Futures Prices in Non-cash-trading Periods: Evidence from the SIMEX Nikkei 225 Futures Contracts. 2000 NTU International Conference on Finance-Finance Markets in Transaction, January

5.21 Lippmann, R. P. (1987): An Introduction to Computing with Neural Nets. IEEE ASSP Magazine, April, 4–22
5.22 Martikainen, T., Puttonen, V. (1994): A Note on the Predictability of Finnish Stock Market Returns: Evidence from Stock Index Futures Markets. European Journal of Operational Research, Vol. 73, 27–32
5.23 Mcinish, T. H., Wood, R. A. (1990): An Analysis of Transactions Data for the Toronto Stock Exchange: Journal of Banking and Finance, Vol. 14, 441–458
5.24 Min, J. H., Najand, M. (1999): A Further Investigation of the Lead-lag Relationship Between the Spot Market and Stock Index Futures: Early Evidence from Korea. Journal of Futures Markets, Vol. 19, 217–232
5.25 Pizzi, M. A., Economopoulous, A., O'Neill, H. M. (1998): An Examination of the Relationship Between Stock Index Cash and Futures Market: A Cointergration Approach. Journal of Futures Markets, Vol. 18, 297–305
5.26 Qnet 97 (1998): Neural Network Modeling for Windows 95/98/NT. Vesta Services
5.27 Repley, B. (1994): Neural Networks and Related Methods for Classification (with discussion). Journal of the Royal Statistical Society, Series B 56, 409–456
5.28 Rumelhart, E., Hinton, G. E., Williams, R. J. (1986): Learning Internal Representations by Error Propagation in Parallel Distributed Processing, Vol. 1. MIT Press, 318–362
5.29 Stern, H. S. (1996): Neural Networks in Applied Statistics. Technometrics, Vol. 38, No. 3, 205–216
5.30 Stoll, H. R., Whaley, R. E. (1990): The Dynamics of Stock Index and Stock Index Futures Returns. Journal of Financial and Quantitative Analysis, Vol. 25, 441–468
5.31 Tang, Z., Fishwick, P. A. (1993): Feedforward Neural Nets as Models for Time Series Forecasting. ORSA Journal on Computing, Vol. 5, No. 4, 374–385
5.32 Wong, F. S. (1991): Time Series Forecasting Using Backpropagation Neural Networks. Neurocomputing, Vol. 2, 147–159
5.33 Zhang, G., Patuwo, B. E., Hu, MY. (1998): Forecasting with Artificial Neural Networks: the State of the Art. International Journal of Forecasting, Vol. 14, 35–62

6. A Support Vector Machine Model for Currency Crises Discrimination

Claudio M. Rocco and José Alí Moreno

Universidad Central de Venezuela, Facultad de Ingeniería, Caracas 1041-A, Venezuela
email: {rocco, jose}@neurona.ciens.ucv.ve

This paper discusses the feasibility of using the support vector machine (SVM) to build empirical models of currency crises. The main idea is to develop an estimation algorithm, by training a model on a data set, which provide reasonably accurate model outputs. The proposed approach is illustrated to model currency crises in Venezuela.

6.1 Introduction

The object of this paper is to develop a model that helps policy makers to identify situations in which currency crises may happen. Two alternative methodologies have been used for early currency crisis detection. The first approach is based on multivariate logit or probit model [6.1, 6.2, 6.8, 6.10] and the second approach used, the "signals" approach, is a warning system. Essentially it considers the evolution of a number of economic indicators that tend to systematically behave differently prior to a crisis. If an indicator exceeds a specific threshold value, it is interpreted as a warning signal: a crisis may take place in the next months [6.8, 6.10].

In this work, empirical models built by training a Support Vector Machine (SVM) are proposed. SVM provides a new approach to the two-category classification problem (crisis or non- crisis) with clear connections to the underlying statistical learning theory [6.3]. Nowadays, SVM methods have reached a high level of maturity, with algorithms that are simple, easy to implement, faster and with good performance. To our best knowledge, the SVM approach has not been yet used in currency crisis evaluation.

The chapter is organized as follows: Section 6.2 contains an overview of Crisis Definition. The SVM approach is presented in Section 6.3, and Section 6.4 presents the proposed approach and the results of an example.

6.2 Definition of a Currency Crisis

A currency crisis is defined as a situation in which an attack on the currency leads to a sharp depreciation of the currency, a large decline in international reserves or a combination of the two [6.10]. According to most of the literature on this topic, a currency crisis is identified by the behavior of an index of "Speculative Pressure" (ISP), defined as [6.8]:

$$ISP_t = \Delta_{\%} Exchange_{rate} + \Delta_{\%} Interest_{rates} - \Delta_{\%} Intl_{reserve} \tag{6.1}$$

where all the variables (expressed in monthly percentage changes) are standardized to have mean zero and unit variance. An increase in the index due to variation on these variables, for example a loss of international reserves, reflects stronger selling pressure on the domestic currency [6.10].

A crisis is defined as a period with an unusual "pressure:" [6.8] $ISP_t > \mu + k\sigma$, where μ is the sample mean, σ the standard deviation of the ISP series and $k \geq 1$. As suggested in [6.2], $k = 1.5$ is selected to detect a crisis event, while $k = 1.0$ is used to detect a financially fragile event. Thus, the binary variable $Crisis_t$ is defined as:

$$Crisis_t = \left\{ \begin{matrix} 1, & \text{if } ISP_t > \mu + k\sigma \\ 0, & \text{otherwise} \end{matrix} \right\} \tag{6.2}$$

There is a wide set of variables that can be used to build a model to explain a crisis. In general, the choice of variables is dictated by theoretical considerations and by the availability of information on a monthly basis [6.10]. In the example to be presented below the following variables were used:

- Real Domestic Credit
- M2/International Reserves
- Inflation
- Oil Prices (Brent)
- An index of industrial equity prices
- Exchange rate
- Exchange rate overvaluation using the Hodrick-Prescott descomposition approach [6.7]

6.3 Support Vector Machines (SVMs)

Support Vector Machines provide a novel approach to the two-category classification problem (crisis or a non-crisis) [6.3]. The method has been successfully applied to a number of applications ranging from particle identification, face identification and text categorization to engine detection, bioinformatics, and data base marketing. The approach is motivated by statistical learning theory [6.6].

SVM is an estimation algorithm ("learning machine") based on [6.3], which has the following three major steps:

1. Parameter estimation procedure ("Training") from a data set

2. Computation of the function value ("Testing")
3. Generalization accuracy ("Performance")

Training involves optimization of a convex cost function, hence there are no local minima to complicate the learning process. Testing is based on the model evaluation using the most informative patterns in the data (the support vectors). Performance is based on error rate determination as test set size goes to infinity [6.4].

Suppose $\mathbf{X}_t$ is a set of variables and y_t is the corresponding crisis evaluation. In order to apply the SVM, we will change the notation to:

$$Crisis_t = \begin{Bmatrix} -1 & \text{if } ISP_t > \mu + k\sigma \\ 1 & \text{otherwise} \end{Bmatrix} \tag{6.3}$$

6.3.1 Linear SVM

Suppose a set of N training data points $\{(\mathbf{X}_1, y_1), \dots, (\mathbf{X}_N, y_N)\}$. Consider the hyperplane:

$$H : y = \langle \mathbf{w} \cdot \mathbf{X} \rangle - b = 0 \tag{6.4}$$

where $\mathbf{w}$ is normal to the H, $b/\|\mathbf{w}\|$ is the perpendicular distance to the origin and $\|\mathbf{w}\|$ is the Euclidean norm of $\mathbf{w}$. Consider also the two hyperplanes parallel to it:

$$H_1 : y = \langle \mathbf{w} \cdot \mathbf{X} \rangle - b = +1 \; , \tag{6.5a}$$

$$H_2 : y = \langle \mathbf{w} \cdot \mathbf{X} \rangle - b = -1 \; , \tag{6.5b}$$

with the condition, that there are no data points between H_1 and H_2. Fig. 6.1 shows such a case [6.9].

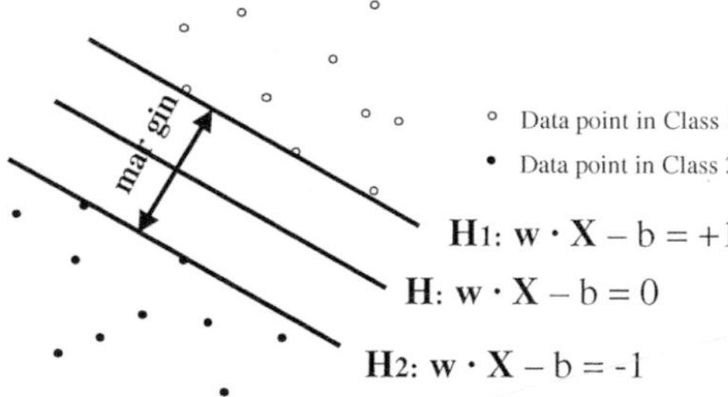

Fig. 6.1. Decision hyperplanes generated by a linear SVM

Let d_+ (d_-) be the shortest distance from the separating hyperplane to the closest positive (negative) example. The distance between H_1 and H_2

(the "margin") is: $(d_+ + d_-)$. It is easy to show that since $d_+ = d_- = 1/\|\mathbf{w}\|$, then the margin is simply $2/\|\mathbf{w}\|$ [6.3].

Therefore, the problem to find a separating hyperplane with the maximum margin is achieved by solving the following optimization problem.

$$\begin{aligned} &\underset{\mathbf{w},b}{Min}\ \frac{1}{2}\mathbf{w}^T\mathbf{w} \\ &\text{subject to: } y_i(\langle \mathbf{w}\cdot\mathbf{X}_i\rangle - b)\ \geq\ 1 \end{aligned} \tag{6.6}$$

The quantities $\mathbf{w}$ and b are the parameters that control the function and are referred as the weight vector and bias [6.6]. The problem (6.6) can be stated as a convex quadratic programming problem in $(\mathbf{w}, b)$. Using the Lagrangian formulation, the constraints will be replaced by constraints on the Lagrange multipliers themselves. Additionally, a consequence of this reformulation is that the training data will only appear in the form of inner product between data vectors [6.3].

By introducing Lagrange multipliers $\alpha_1, \ldots, \alpha_N \geq 0$, a Lagrangian function for the optimization problem can be defined:

$$L_p(\mathbf{w}, b, \alpha) = \frac{1}{2}\mathbf{w}^T\mathbf{w} - \sum_{i=1}^{N}(\alpha_i\, y_i\,(\langle \mathbf{w}\cdot\mathbf{X}_i\rangle - b) - 1) \tag{6.7}$$

To solve the optimization problem above, one has to find the saddle point of (6.7), i.e., to minimize (6.7) with respect to $\mathbf{w}$ and b and to maximize it over the nonnegative Lagrange multipliers $\alpha_i \geq 0$ [6.3].

$$\frac{\partial L_p}{\partial w_i} = 0,\ \ i = 1, \ldots, N \tag{6.8}$$

$$\frac{\partial L_p}{\partial b} = 0 \tag{6.9}$$

The gradients give the conditions [6.3]:

$$\mathbf{w} = \sum_{i=1}^{N} \alpha_i\, y_i\, \mathbf{X}_i \tag{6.10}$$

$$\sum_{i=1}^{N} \alpha_i\, y_i = 0 \tag{6.11}$$

By substituting (6.10) and (6.11) into (6.7), one obtains the Wolfe dual formulation [6.3, 6.5, 6.6]:

$$L_D(\alpha) = \sum_{i=1}^{N} \alpha_i - \frac{1}{2} \sum_{i,j=1}^{N} \alpha_i \alpha_j y_i y_j \langle \mathbf{X}_i \cdot \mathbf{X}_j \rangle. \tag{6.12}$$

The notation has been changed from $L_p(\mathbf{w}, b, \alpha)$ to $L_D(\alpha)$. Now to construct the optimal hyperplane one has to find the coefficients α_i that maximize the function (6.12) in the nonnegative quadrant, $\alpha_i \geq 0$ under the constraint (6.11). Solving for α_i and computing b gives $\mathbf{w} = \sum_i \alpha_i y_i \mathbf{X}_i$.

Once a SVM has been trained it is simple to determine on which side of the decision boundary a given test pattern $\mathbf{X}^*$ lies and assign the corresponding class label, using $sgn(\langle \mathbf{w} \cdot \mathbf{X}^* \rangle - b)$.

For the primal problem (6.7), the Karush-Kuhn-Tucker (KKT) conditions are:

$$\alpha_i \left(y_i \left(\langle \mathbf{w} \cdot \mathbf{x}_i \rangle + b \right) - 1 \right) = 0, \quad \forall i \tag{6.13}$$

From conditions (6.13) it follows that nonzero values α_i correspond only to vector $\mathbf{X}_i$ that satisfy the equality

$$y_i \left(\langle \mathbf{w} \cdot \mathbf{X}_i \rangle - b \right) - 1 = 0. \tag{6.14}$$

When the maximal margin hyperplane is found, only those points which lie closest to the hyperplane have $\alpha_i > 0$ and these points are the *support vectors*, that is, the critical elements of the training set. All other points have $\alpha_i = 0$. This means that if all other training points are removed and training is repeated, the same separating hyperplane would be found [6.3]. In Fig. 6.2, the points a, b, c, d and e are examples of support vectors [6.9].

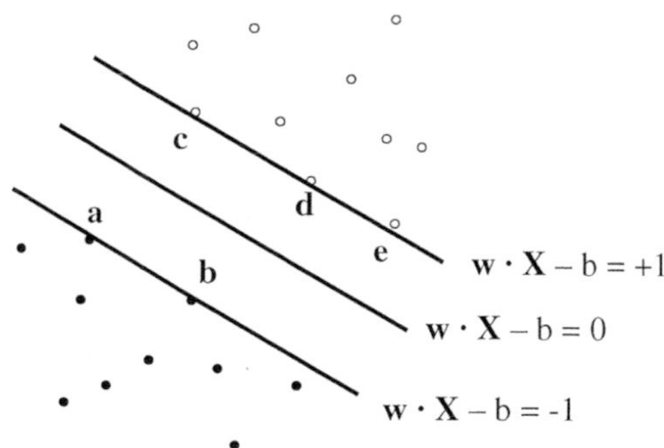

Fig. 6.2. Example of support vectors

Small problems can be solved by any general-purpose optimization package that solves linearly constrained convex quadratic programs. For larger problems, a range of existing techniques can be used [6.6]. The basic steps are [6.3]:

1. Note the optimality KKT conditions which the solutions must satisfy,
2. Define a strategy for approaching optimality and

3. Decide on a decomposition algorithm so that only a portion of the training data need be handled at a given time.

6.3.2 Nonlinear SVM

If the surface separating the two classes is not linear, the data points can be transformed to a high dimensional feature space where the problem is linearly separable. Fig. 6.3 denotes such transformation [6.3].

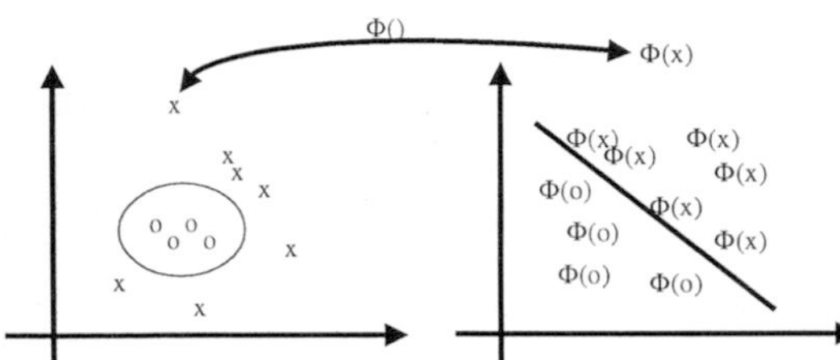

Fig. 6.3. A non-linear separating region transformed into a linear one

Let the transformation be $\Phi(\cdot)$. The Lagrangian function in the high dimensional feature space is:

$$L_D(\alpha) = \sum_{i=1}^{N} \alpha_i - \frac{1}{2} \sum_{i,j=1}^{N} \alpha_i \alpha_j y_i y_j \langle \Phi(\mathbf{X}_i) \cdot \Phi(\mathbf{X}_j) \rangle \tag{6.15}$$

Suppose, in addition, $\langle \Phi(\mathbf{X}_i) \cdot \Phi(\mathbf{X}_j) \rangle = k(\mathbf{X}_i, \mathbf{X}_j)$. That is, the dot product in that high dimensional feature space defines a kernel function of the input space. It is, therefore, not necessary to be explicit about the transformation $\Phi(\cdot)$ as long as it is known that the kernel function $k(\mathbf{X}_i, \mathbf{X}_j)$ corresponds to a dot product in some high dimensional feature space [6.3, 6.5, 6.6, 6.9, 6.12].

There are many kernel functions that can be used, for example [6.3, 6.6] Gaussian radial basis function kernel:

$$k(\mathbf{X}_i, \mathbf{X}_j) = e^{\left(\frac{-\|\mathbf{X}_i - \mathbf{X}_j\|^2}{2\sigma^2} \right)} \tag{6.16}$$

Polynomial kernel:

$$k(\mathbf{X}_i, \mathbf{X}_j) = (\langle \mathbf{X}_i \cdot \mathbf{X}_j \rangle + m)^p \tag{6.17}$$

The characterization of a kernel function $k(\mathbf{X}, \mathbf{Y})$ is done by means of Mercer's theorem [6.3, 6.6]. It establishes a positivity condition that in the case of a finite subset of data points means that the corresponding matrix of the function $k(\mathbf{X}, \mathbf{Y})$ has to be positive semi-definite.

With a suitable kernel, SVM can separate in the feature space the data that in the original input space was non-separable. This property means that we can obtain non-linear algorithms by using proven methods to handle linearly separable data sets [6.5].

6.3.3 Imperfect Separation

SVM can be extended to allow for imperfect separation [6.3, 6.6, 6.9]. That is, data between H_1 and H_2 can be penalized. The penalty C will be finite. Introducing non-negative slack variables $\xi_i \geq 0$, so that:

$$((\mathbf{w} \cdot \mathbf{X}_i) - b) \geq 1 - \xi_i \quad \text{for}\, y_i = +1 \tag{6.18}$$

$$((\mathbf{w} \cdot \mathbf{X}_i) - b) \leq -1 + \xi_i \quad \text{for}\, y_i = -1 \tag{6.19}$$

and adding to the objective function a penalizing term, the problem is now formulated as:

$$\underset{\mathbf{w},b,\xi}{Min} \; \frac{1}{2}\mathbf{w}^T\mathbf{w} + C\sum_{i=1}^{N}\xi_i \tag{6.20}$$

$$\text{subject to: } y_i((\mathbf{w} \cdot \mathbf{X}_i) - b) + \xi_i \geq 0$$
$$\xi_i \geq 0$$

Using the Lagrange multipliers and the Wolfe dual formulation, the problem is transformed to:

$$\underset{\alpha}{Max} \; L_D = \sum_{i=1}^{N}\alpha_i - \frac{1}{2}\sum_{i,j=1}^{N}\alpha_i\alpha_j y_i y_j \langle \Phi(\mathbf{X}_i) \cdot \Phi(\mathbf{X}_j)\rangle \tag{6.21}$$

$$\text{subject to: } 0 \leq \alpha_i \leq C$$
$$\sum_i \alpha_i y_i = 0$$

The only difference from the perfectly separating case is that the α_i are now bounded above by C.

6.3.4 SVM Properties

SVMs have the following properties [6.3]:

1. Both training and test functions depend on the data and the kernel function. Even if we need to evaluate a dot product, the complexity of computing the kernel can be smaller. Thus, SVM circumvent both forms of the "curse of dimensionality:" the proliferation of parameters causing intractable complexity and the proliferation of parameters causing overfitting.
2. Training algorithms may take advantage of parallel processing in several ways. For example, the evaluation of the kernel and the sum are highly parallelizable procedures.

3. SVMs usually exhibit good generalization performance.
4. The choice of the kernel is a limitation of the support vector approach. Some work has been done on limiting kernels using prior knowledge [6.5]. However, it has been noticed that when different kernel functions are used in SVM, they empirically lead to very similar classification accuracy [6.5]. In these cases, the SVM with lower complexity should be selected.
5. Although the size of the quadratic programming problem scales with the number of support vectors, decomposition algorithms have been proposed to avoid problems with large data sets. For example the Sequential Minimal Optimization technique [6.11] explores an extreme case, where the decomposition is made into sub-problems that are so small that an analytically solution can be found.
6. The performance of a binary classifier is measured during the test phase using its sensitivity, specificity and accuracy [6.13]:

$$sensitivity = \frac{TP}{TP+FN} \tag{6.22}$$

$$specificity = \frac{TN}{TN+FP} \tag{6.23}$$

$$accuracy = \frac{TP+TN}{TP+TN+FP+FN} \tag{6.24}$$

   where:
   - TP: Number of True Positive classified cases (SVM correctly classifies)
   - TN: Number of True Negative classified cases (SVM correctly classifies)
   - FP: Number of False Positive classified cases (SVM labels a case as positive while it is a negative)
   - FN: Number of False Negative classified cases (SVM labels a case as negative while it is a positive)

   For crises discrimination, *sensitivity* gives the percentage of correctly classified non-crisis and the *specificity* the percentage of correctly classified crisis events.

These indexes can be easily converted to Type I and II errors, as defined by *Kaminsky et al* [6.10].

6.4 Example

The SVM approach is applied to evaluate currency crises in Venezuela, during the period January 1980 - May 1999. A monthly database with 232 observations [6.1] for each variable mentioned in Section 6.2 is used. During this period 15 crises were detected, using the ISP and $k = 1.5$. The database used is available upon request.

Two SVM models were developed:

1. Classify any vector $\mathbf{X}^*$ as representing a crisis or a non-crisis event
2. Predict from a vector $\mathbf{X}^*$, corresponding to a specific month, if the next month can be classified as a crisis or a non-crisis event.

In the first case, the first 216 observation were used as the training set: 203 correspond to non-crisis event and 13 to crisis event. In Case 2, the SVM was trained with 215 observations (202 non-crisis event and 13 crisis event): for each vector $\mathbf{X}_t$ the corresponding ISP_{t+1} was used.

As suggested in [6.10] and [6.8], the information about sensitivity and specificity can be combined into a measure of the "noisiness" of the indexes, the Noise/Signal Ratio (NSR):

$$NSR = \frac{1 - Sensitivity}{Specificity} \tag{6.25}$$

This index measures the false signals as a ratio of the good signals issued. The selection rule is to choose the model that minimizes the NSR [6.8].

Different kernels were tried and the less complex kernel that completely separates the training data set with the best NSR was chosen. In Case 1, a third order polynomial was selected, while a fourth order polynomial was the best kernel for Case 2. In both cases the training NSRs were zero. During the testing phase, the NSRs were 0.143 and 0.153 for Case 1 and Case 2 respectively. All learning runs were performed on a 500 MHz Pentium III, using a C++ program based on the Sequential Minimum Optimization algorithm [6.11].

In order to compare the SVM, results previously reported in similar studies based on the "signals" approach [6.8] and on multivariate probit model [6.2] were used. In the first study, the data set corresponds to the period January 1980 – April 1998, while the data set of the second study was the same data set used in this paper. Both studies considered the whole data set and did not use training/testing subsets. Additionally, they present results only for experiment 1.

Table 6.1 presents SVMs performance indexes evaluated using all the observations.

The performance indexes achieved using these models are better than those previously reported. For example, the best NSR reported in [6.8] is 0.19, while the best NSR reported in [6.2] is 0.0458.

6.5 Conclusions

This paper has presented a novel approach to evaluate currency crises based on SVM. The excellent results obtained in the example show that currency crises are properly emulated using only a small fraction of the database and could be used as an evaluation tool as well as an early warning system.

Table 6.1. SVM results

	Case 1	Case 2
Polynomial Order	3	4
Number of SVs	28	39
Sensitivity(%)	96.31	97.68
Specificity(%)	93.33	86.67
Accuracy(%)	96.12	96.97
NSR	0.0394	0.0257

Acknowledgements

The paper was refereed by Bin-Tzong Chie and Shin-Ming Tsai. The authors are grateful to the two referees for their substantial comments on the earlier version of the paper.

References

6.1 Alvarez, F., Aponte, P., Zambrano, P. (2000): Logit y Probit: modelos para el tratamiento de variables polítomas. Universidad Central de Venezuela, Facultad de Ingeniería, Internal Report
6.2 Arreaza, A., Fernández, M. A., Mirabal, M. J., Alvarez, F. (2001): Fragilidad Financiera en Venezuela: Determinantes e Indicadores. Revista BCV, Vol. XV, No. 1
6.3 Burges, C. (1998): A Tutorial on Support Vector Machines for Pattern Recognition. http://www.kernel-machines.org/
6.4 Campbell, C. : Kernel Methods: A survey of Current Techniques. http://www.kernel-machines.org/
6.5 Campbell, C. (2000): An Introduction to Kernel Methods. Radial Basis Function Networks: Design and Applications, Howlett, R. J., Jain, L. C. (Eds.), 23. Springer Verlag
6.6 Cristianini, N, Shawe-Taylor, J. (2000): An introduction to Support Vector Machines. Cambridge University Press
6.7 Goldfajn, I., Valdéz, R. (1999): The Aftermath of Appreciations. Quaterly Journal of Economics, 114(1)
6.8 Herrera, S., García, C. (1999): A User's Guide to an Early Warning System of Macroeconomic Vulnerability for LAC Countries. XVII LAtin American Meeting of Economics Society
6.9 http://www.ics.uci.edu/~xge/svm
6.10 Kaminsky, G., Lizondo, S., Reinhart, C. (1998): Leading Indicators of Currency Crisis. IMF Staff Paper No. 45
6.11 Platt, J. (1999): Fast Trainning of Support Vector Machines using Sequential Minimal Optimization. http://www.research.microsoft.com/~jplatt
6.12 Schölkopf, B. (2000): Statistical Learning and Kernel Methods. Microsoft Research, MSR-TR-2000-23. http://www.research.microsoft.com/~bsc

6.13 Veropoulos, K., Campbell, C., Cristianini, N. (1999): Controlling the Sensitivity of Support Vector Machines. Proceedings of the International Joint Conference on Artificial Intelligence Stockholm, Sweden (IJCAI99), Workshop ML3, 55–60

7. Saliency Analysis of Support Vector Machines for Feature Selection in Financial Time Series Forecasting

Lijuan Cao[1] and Francis E. H. Tay[2]

[1] Institute of High Performance Computing, 89C Science Park Drive, 118261 Singapore
email: caolj@ihpc.nus.edu.sg

[2] Department of Mechanical Engineering, National University of Singapore, 10 Kent Ridge Crescent, 119260, Singapore
email: mpetayeh@nus.edu.sg

This chapter deals with the application of saliency analysis to Support Vector Machines (SVMs) for feature selection. The importance of feature is ranked by evaluating the sensitivity of the network output to the feature input in terms of the partial derivative. A systematic approach to remove irrelevant features based on the sensitivity is developed. Two simulated non-linear time series and five real financial time series are examined in the experiment. The simulation results show that that saliency analysis is effective in SVMs for identifying important features.

7.1 Introduction

Over the recent past years, support vector machines (SVMs) have been receiving increasing attention in the regression estimation area due to their remarkable characteristics such as good generalization performance, the absence of local minima and sparse representation of the solution [7.1, 7.2, 7.3, 7.4]. When SVMs are used to model time series, there are usually a vast number of features that can be used as inputs. For example, in financial markets, to predict the next day's closing price, any previous day's closing price can be used as predictor. In addition, other fundamental or technical indicators related to the predicted time series can also be included. Among all available features, some are really relevant to the prediction while others are not. To distinguish significant features from the irrelevant is crucial to the overall prediction performance of the network. Irrelevant features can prolong the training time of the network and increase the required training samples.

However, within the SVMs framework, there are very few established approaches for identifying important features. Selecting significant features from all candidate features is the first step in regression estimation, and this procedure can improve the network performance, reduce the network complexity , and speed up the training of the network. The issue of feature selection for SVMs is recently discussed in [16.1]. There it has been stated that feature selection is better to be performed in SVMs if many features exist, as this procedure can improve the network performance, speed up the training and reduce the complexity of the network.

The chapter deals with the application of saliency analysis in SVMs for identifying irrelevant features. Saliency Analysis (SA) measures the importance of features by evaluating the sensitivity of the network output with respect to the weights (weight-based SA) or the feature inputs (derivative-based SA) [7.6]. Based on the idea that important features usually have large absolute values of connected weights and unimportant features have small absolute values of connected weights, the weight-based SA detects irrelevant weights by evaluating the magnitude of weights, and then removes the features emanating these irrelevant weights. This method is also extended to other types of weight-pruning to remove irrelevant features by adding a penalty term in the cost function [7.7]. The derivative-based SA measures the importance of features by evaluating the sensitivity of the network output with respect to the feature inputs [7.8]. To irrelevant features which provide little information to the regressor, the output produces a small value of saliency metric which indicates that the network output is insensitive to those features. On the contrary, to significant features which contribute significant information to the regressor, the output will produce a large value of saliency metric. In SVMs, the weights lie in a high dimensional feature space rather than the original feature space, thus the magnitude of weights is actually a reflection of the importance of the high dimensional feature inputs instead of the original feature inputs. This will be explained in detail in Section 3. In this paper only the derivative-based SA is developed for SVMs.

SA is the feature selection method in the context of artificial intelligence, and it has been proved to be effective in the multi-layer perceptron [7.9, 7.10, 7.11]. However, unlike the multi-layer feedforward neural network which implements the *Empirical Risk Minimization Principle*, SVMs are based on the *Structural Risk Minimization Principle* which seeks to minimize an upper bound of the generalization error rather than minimize the empirical error. This results in the training algorithm and the form of decision in SVMs to be very different from those of the multi-layer feedforward neural network. Specifically, the decision function of SVMs is linearly constructed in a high dimensional feature space which is mapped from the original feature space. Training SVMs is equivalent to solving a linearly constrained quadratic programming, unlike the multi-layer perceptron training which requires non-linear optimization with the danger of getting stuck into local minima. The solution of SVM is unique global, and it is only dependent on a small subset of training data points which are referred to as support vectors. The support vectors are automatically determined by the unique training algorithm of SVMs. Considering these characteristics of SVMs, it is therefore very meaningful to check whether saliency analysis is effective in SVMs for detecting insignificant features. The saliency analysis considered in this paper is in the context of regression estimation.

This chapter is organized as follows. In section 7.2, we briefly introduce the theory of SVMs in the regression estimation. In section 7.3, the method of saliency analysis is described. The reason why the weight-based SA is not applicable to SVMs is also presented. Section 7.4 gives the experimental results for both simulated and real data, followed by the conclusions in the last section.

7.2 Theory of Support Vector Machines (SVMs) for Regression Estimation

Compared to other neural network regressors, there are three distinct characteristics when SVMs are used to estimate the regression function. First of all, SVMs estimate the regression by a set of linear functions which are defined in a high dimensional space. Secondly, SVMs define the regression estimation as the problem of risk minimization where the risk is measured using Vapnik's ε-insensitive loss function. Thirdly, SVMs use the risk function consisting of the empirical error and a regularization term, which is derived from the Structural Risk Minimization Principle.

Given a set of data points $G = \{(X_i, d_i)\}_i^N$ (X_i is the input vector, d_i is the desired value, and N is the total number of data patterns), SVMs approximate the function by

$$y = f(X_i) = W \cdot \phi(X_i) + b, \tag{7.1}$$

where $\phi(X_i)$ is the high dimensional feature space which is nonlinearly mapped from the input space X_i. The coefficient W and b are estimated by minimizing

$$R_{SVMs}(C) = C\frac{1}{N}\sum_{i=1}^{N} L_\varepsilon(d_i, y_i) + \frac{1}{2}\|W\|^2, \tag{7.2}$$

$$L_\varepsilon(d_i, y_i) = \begin{cases} |d_i - y_i| - \varepsilon & |d_i - y_i| \geq \varepsilon \\ 0 & \text{otherwise} \end{cases}. \tag{7.3}$$

In the regularized risk function (7.2), the first term $C\frac{1}{N}\sum_{i=1}^{N} L_\varepsilon(d_i, y_i)$ is the empirical error (risk). They are measured by the ε-insensitive loss function (7.3) because this loss function provides the advantage of using sparse data points to represent the decision function (7.1). The second term $\frac{1}{2}\|W\|^2$, on the other hand, is the regularization term. C and ε are referred to as the regularized constant and tube size. They are both user-defined parameters.

To get the estimations of W and b , Eq. (7.2) is transformed to the primal function (7.4) by introducing the positive slack variables ξ_i and ξ_i^*.

$$R_{SVMs}(W, \xi^{(*)}) = \frac{1}{2}\|W\|^2 + C\sum_{i=1}^{N}(\xi_i + \xi_i^*) \tag{7.4}$$

$$\begin{aligned} \text{subjected to: } & d_i - W \cdot \phi(X_i) - b_i \le \varepsilon + \xi_i, \\ & W \cdot \phi(X_i) + b_i - d_i \le \varepsilon + \xi_i^*, \\ & \xi_i^* \ge 0, \quad \xi_i \ge 0. \end{aligned}$$

Finally, by introducing Lagrange multipliers and exploiting the optimality constraints, the decision function (7.1) has the following explicit form [7.12]:

$$f(X, \alpha_i, \alpha_i^*) = \sum_{i=1}^{N}(\alpha_i - \alpha_i^*)K(X, X_i) + b. \tag{7.5}$$

Lagrange Multipliers and Support Vectors

In (7.5), α_i and α_i^* are the so-called Lagrange multipliers. They satisfy the equalities $\alpha_i \times \alpha_i^* = 0, \quad \alpha_i \ge 0$ and $\alpha_i^* \ge 0$ where $i = 1, \ldots, N$, and are obtained by maximizing the dual function of (7.4), which has the following form:

$$\begin{aligned} R(\alpha_i, \alpha_i^*) = & \sum_{i=1}^{N} d_i(\alpha_i - \alpha_i^*) - \varepsilon \sum_{i=1}^{N}(\alpha_i + \alpha_i^*) \\ & - \frac{1}{2}\sum_{i=1}^{N}\sum_{j=1}^{N}(\alpha_i - \alpha_i^*)(\alpha_j - \alpha_j^*)K(X_i, X_j), \end{aligned} \tag{7.6}$$

with the following constraints:

$$\begin{aligned} & \sum_{i=1}^{N}(\alpha_i - \alpha_i^*) = 0, \\ & 0 \le \alpha_i \le C, \ i = 1, 2, \ldots, N, \\ & 0 \le \alpha_i^* \le C, i = 1, 2, \ldots, N. \end{aligned}$$

Based on the Karush-Kuhn-Tucker (KKT) conditions of quadratic programming, only a number of coefficients $(\alpha_i - \alpha_i^*)$ in (7.5) will assume non-zero. The data points associated with them have approximation errors equal to or larger than ε and are referred to as support vectors.

Kernel Function

$K(X_i, X_j)$ is defined as the kernel function. The value of kernel is equal to the inner product of two vectors X_i and X_j in the feature space $\phi(X_i)$ and $\phi(X_j)$, that is, $K(X_i, X_j) = \phi(X_i) \cdot \phi(X_j)$. The elegance of using the kernel function is that one can deal with feature space of arbitrary dimensionality without having to compute the map $\phi(X)$ explicitly. Any function that satisfies Mercer's condition [7.12] can be used as the kernel function. Common examples of the kernel function are the polynomial kernel $K(X_i, X_j) = (X_i \cdot X_j + 1)^d$, the Gaussian kernel $K(X_i, X_j) = \exp(-\frac{1}{\delta^2}(X_i - X_j)^2)$, and the 2-lay tangent kernel $K(X_i, X_j) = \tanh(\beta_0 X_i \cdot X_j + \beta_1)$, where d, δ^2, β_0 and β_1 are all kernel parameters.

From the implementation point of view, training SVMs is equivalent to solving a linearly constrained Quadratic Programming (QP) with the number of variables equal to the number of training data points. The Sequential Minimal Optimization (SMO) algorithm extended by Scholkopf and Smola [7.13, 7.14] is very effective in training SVMs for solving the regression problem.

7.3 Saliency Analysis of SVMs

There are two types of SA. One is the weight-based SA in which the sensitivity of the network output is evaluated based on the weights, and the other is the derivative-based SA where the sensitivity of output is evaluated based on the feature inputs. The weight-based SA is not applicable to SVMs, and the reason is given as below.

As shown in (7.1), the output of X_i is equal to the sum of a constant b and the inner product of W and $\phi(X_i)$. This implies that W lies in the same dimensional space as $\phi(X_i)$. Suppose $\phi(X_i) = (\phi(X_i)_1, \phi(X_i)_2, \cdots, \phi(X_i)_d)$, a d-dimensional feature space, is mapped from $X_i = (X_{i1}, X_{i2}, \ldots, X_{iK})$, the original K-dimensional feature space, from (7.1) and (7.5), W is expressed as:

$$\begin{aligned} W &= (w_1, w_2, \ldots, w_d) \\ &= (\sum_{i=1}^{N_s}(\alpha_i - \alpha_i^*)\phi(X_i)_1, \sum_{i=1}^{N_s}(\alpha_i - \alpha_i^*)\phi(X_i)_2, \\ &\quad \ldots, \sum_{i=1}^{N_s}(\alpha_i - \alpha_i^*)\phi(X_i)_d). \end{aligned} \tag{7.7}$$

where N_s is the number of support vectors. The value of W is dependent on the converged Lagrange multipliers $(\alpha_j - \alpha_j^*)$, the support vectors X_i as well as the explicit form of $\phi(X_i)$. Thus $\{w_i\}_{i=1}^d$ denotes a set of linear weights connecting $\{\phi(X_i)_j\}_{j=1}^d$ and y_i (Fig. 7.1). In this sense, the magnitude of $\{w_i\}_{i=1}^d$ is a reflection of the importance of high dimensional feature input $\{\phi(X_i)_j\}_{j=1}^d$, rather than the original feature input $\{x_{ij}\}_{j=1}^K$. For example, if $X = (x_1, x_2)$, and the polynomial kernel $K(X_i, X_j) = (X_i \cdot X_j + 1)^2$ is used, then $\phi(X) = (x_1^2, \sqrt{2}x_1x_2, x_2^2, \sqrt{2}x_1, \sqrt{2}x_2)$, and the weights are a measure of the importance of 3-dimensional inputs x_1^2, $\sqrt{2}x_1x_2$, x_2^2, $\sqrt{2}x_1$ and $\sqrt{2}x_2$ not the 2-dimensional original inputs x_1 and x_2.

Furthermore, for most kernel functions such as the Gaussian kernel and the 2-layer tangent kernel, the explicit form of $\phi(X_i)$ and its dimension d are actually unknown, and it would not be easy to calculate the magnitude of $\{w_i\}_{i=1}^d$.

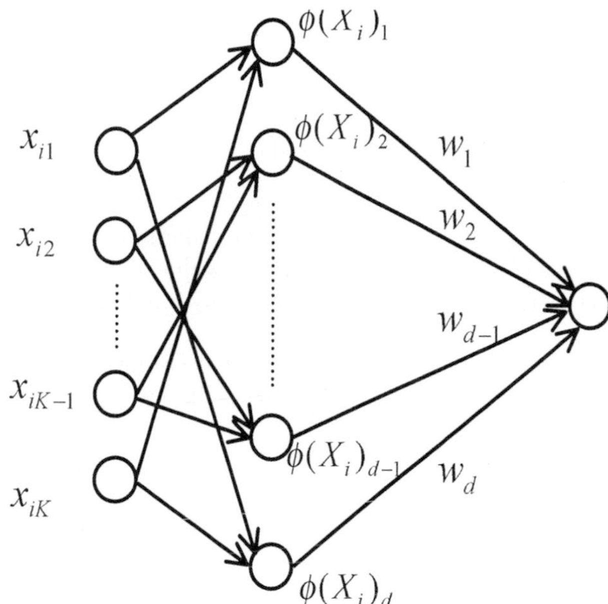

Fig. 7.1. A pictorial illustration of the output, input and weight in SVMs

7.3.1 Deriving the Sensitivity

In the derivative-based SA, the sensitivity of the network output to the input is approximated by the derivative:

$$\begin{aligned}\frac{\partial y_i}{\partial x_{ik}} &= \frac{\partial(\sum_{j=1}^{N_s}(\alpha_j - \alpha_j^*)K(X_j, X_i) + b)}{\partial x_{ik}} \\ &= \frac{\partial \sum_{j=1}^{N_s}(\alpha_j - \alpha_j^*)K(X_j, X_i)}{\partial x_{ik}} \\ &\quad + \frac{\partial b}{\partial x_{ik}} \\ &= \sum_{j=1}^{N_s}(\alpha_j - \alpha_j^*)\frac{\partial K(X_j, X_i)}{\partial x_{ik}}. \end{aligned} \tag{7.8}$$

- To the Gaussian kernel $K(X_i, X_j) = \exp(-\frac{\sum_{l=1}^{K}(x_{il}-x_{jl})^2}{\delta^2})$, then $\frac{\partial K(X_j,X_i)}{\partial x_{ik}} = -\frac{2}{\delta^2}(x_{ik} - x_{jk}) \exp(-\frac{\sum_{l=1}^{K}(x_{il}-x_{jl})^2}{\delta^2})$. In this case, $\frac{\partial y_i}{\partial x_{ik}} = -\frac{2}{\delta^2}\sum_{j=1}^{N_s}(\alpha_j - \alpha_j^*)(x_{ik} - x_{jk}) \exp(-\frac{\sum_{l=1}^{K}(x_{il}-x_{jl})^2}{\delta^2})$.
- To the polynomial kernel $K(X_i, X_j) = (\sum_{l=1}^{K} x_{il} \cdot x_{jl} + 1)^d$, then $\frac{\partial K(X_j,X_i)}{\partial x_{ik}} = d(\sum_{l=1}^{K} x_{il} \cdot x_{jl} + 1)^{d-1} x_{jk}$. In this case, $\frac{\partial y_i}{\partial x_{ik}} = d\sum_{j=1}^{N_s}(\alpha_j - \alpha_j^*)x_{jk}(\sum_{l=1}^{K} x_{il} \cdot x_{jl} + 1)^{d-1}$.
- To the 2-layer tangent kernel $K(X_i, X_j) = \tanh(\beta_0 \sum_{l=1}^{K} x_{il} \cdot x_{jl} + \beta_1)$, then $\frac{\partial K(X_j,X_i)}{\partial x_{ik}} = \frac{4\beta_0 x_{jk}}{(\exp(\beta_0 \sum_{l=1}^{K} x_{il} \cdot x_{jl} + \beta_1) + \exp(-(\beta_0 \sum_{l=1}^{K} x_{il} \cdot x_{jl} + \beta_1)))^2}$. In this case, $\frac{\partial y_i}{\partial x_{ik}} = \sum_{j=1}^{N_s} \frac{4(\alpha_j - \alpha_j^*)\beta_0 x_{jk}}{(exp(\beta_0 \sum_{l=1}^{K} x_{il} \cdot x_{jl} + \beta_1) + exp(-(\beta_0 \sum_{l=1}^{K} x_{il} \cdot x_{jl} + \beta_1)))^2}$.

The derivative of the network output to the input can be calculated for any type of the kernel function according to (7.8), and the value depends on the input x_{ik}, the support vectors X_j as well as the converged Lagrange multipliers $(\alpha_j - \alpha_j^*)$.

Then, the saliency metric for features is calculated as the absolute average of the derivative of the network output to the input over the entire training sample, which is:

$$s_k = \frac{\sum_{i=1}^{N} \left|\frac{\partial y_i}{\partial x_{ik}}\right|}{N}. \tag{7.9}$$

Other calculations could be $s_k = \sqrt{\frac{\sum_{i=1}^{N}(\frac{\partial y_i}{\partial x_{ik}})^2}{N}}$ and $s_k = \max\{\frac{\partial y_i}{\partial x_{ik}}\}$, $k = 1, \ldots, K$ [7.11].

7.3.2 Removing Irrelevant Features

After calculating the saliency value, a criterion needs to be set up to determine how many features could be removed from the whole feature set. Several methods are proposed in [7.9, 7.11]. Belue and Bauer suggest a method which adds a noise feature to the original feature set and classifies the features whose saliency values fall into an upper-one-sided confidence interval about that of the noise as irrelevant features. Zurada and Malinowski use the largest gap method to detect irrelevant features. But these two methods are not very effective for our problems. As shown in the experiment, a simple θ-threshold method works best to remove irrelevant features. That is, the features with saliency values lower than a predetermined threshold are the irrelevant features which could be deleted, and the features with saliency values larger than the threshold are the significant features which should be retained. A systematic procedure for eliminating insignificant features is outlined as follows.

1. Train SVMs using the full feature set.

2. Calculate s_i by (7.8) and (7.9) for each candidate feature.
3. Rank s_i in a descending order as $s'_1 > s'_2 > \ldots > s'_K$, where $s'_1 = \arg\max\{\{s_i\}_{i=1}^K\}, s'_2 = \arg\max\{\{s_i\}_{i=1}^K - s'_1\}, \ldots,$ and $s'_K = \arg\max\{\{s_i\}_{i=1}^K - \{s'_i\}_{i=1}^{K-1}\}$.
4. Choose a proper threshold θ.
5. If $s'_i > \theta$ and $s'_{i+1} < \theta$, delete the features corresponding to the saliency values $s'_{i+1}, \ldots, s'_K$.

In SVMs, one training run is sufficient as the solution is guaranteed to be optimal global. This is in contrast with the multi-layer perceptron in which multiple training runs are needed to avoid the results based on the local minima which may be generated by the neural network.

7.4 Experiments

7.4.1 Simulated Data

Two simulated non-linear time series taken from [7.15, 7.16] are examined in the first series of experiments. They are defined in 7.10 and 7.11 and referred to as data-1 and data-2.

$$data - 1 : x_t = -4x_{t-1}^2 + 4x_{t-1} \tag{7.10}$$

$$data - 2 : x_t = 0.3x_{t-6} - 0.6x_{t-4} + 0.5x_{t-1} + \mu_t, \quad u_t \text{ is a white noise} \tag{7.11}$$

Each data set generates a total of 900 data patterns. The whole data patterns are partitioned into three parts randomly. The first part is for training, the second part for validating which is used to select optimal parameters for SVMs, and the last part for testing. There are a total of 500 data patterns in the training set, 200 data patterns in both the validation set and the test set in the data-1 and data-2.

The inputs of SVMs are arbitrarily chosen as six lagged output variables. That is, to predict $x(t)$, the inputs are $x(t-1)$, $x(t-2)$, $x(t-3)$, $x(t-4)$, $x(t-5)$ and $x(t-6)$. So in data-1, $x(t-2)$, $x(t-3)$, $x(t-4)$, $x(t-5)$ and $x(t-6)$ are all irrelevant variables, and to data-2, $x(t-2)$, $x(t-3)$ and $x(t-5)$ are irrelevant variables. For the algorithm to be effective, the saliency values for these irrelevant variables will be small.

In the SA, the polynomial kernel, Gaussian kernel and two-layer tangent kernel are all tested in the two data sets. This is to investigate whether the results of SA are sensitive to the selection of the kernel function. The choice of kernel parameters, C and ε are based on the smallest normalized mean squared error (NMSE) on the validation set. The NMSE is calculated as follows.

$$NMSE = \frac{1}{N\delta^2} \sum_{i=1}^{N} (d_i - y_i)^2 \tag{7.12}$$

$$\delta^2 = \frac{1}{N-1} \sum_{i=1}^{N} (d_i - \overline{d})^2 \tag{7.13}$$

where $\overline{d}$ is the mean of actual value d_i. The SMO for solving the regression problem is implemented in SVMs and the program is developed using VC+ language [7.17].

The selected features for the two data sets are listed in Table 7.1. The detailed calculations for the saliency are found in Table 7.2 and Table 7.3. In both Table 7.2 and Table 7.3, θ could be choosen as any value from interval which is determined by the two continuous saliency value with the largest difference. Table 7.1 shows SA can detect the true features ($x(t-1)$ in data-1 and $x(t-1)$, $x(t-4)$, $x(t-6)$ in data-2) from the redundant features ($x(t-2)$, $x(t-3)$, $x(t-4)$, $x(t-5)$, $x(t-6)$ in data-1 and $x(t-2)$, $x(t-3)$, $x(t-5)$ in data-2). Furthermore, the SA is insensitive to the selection of the kernel function, as there are the same top ranks for the true features and slightly different ranks for the irrelevant features in the three kernel functions.

Table 7.1. Selected features in data-1 and data-2

Methods	data-1	data-2
SA-polynomial	$x(t-1)$	$x(t-1)$, $x(t-4)$, $x(t-6)$
SA-Gaussian	$x(t-1)$	$x(t-1)$, $x(t-4)$, $x(t-6)$
SA-two-layer tangent	$x(t-1)$	$x(t-1)$, $x(t-4)$, $x(t-6)$

Table 7.2. Saliency and rank in data-1. Note: A ranking of "1" indicates the best feature and a ranking of "6" indicates the worst feature

Features	Polynomial			Gaussian			2-layer tangent		
	s_i	Rank	s_i'	s_i	Rank	s_i'	s_i	Rank	s_i'
$x(t-1)$	2.4084	1	2.4084	2.0744	1	2.0744	0.8417	1	0.8417
$x(t-2)$	0.2869	2	0.2869	0.6321	2	0.6321	0.1212	2	0.1212
$x(t-3)$	0.0587	3	0.0587	0.5005	3	0.5005	0.0463	6	0.0729
$x(t-4)$	0.0137	4	0.0137	0.4830	4	0.4830	0.0642	5	0.0651
$x(t-5)$	0.0058	5	0.0058	0.4795	5	0.4795	0.0651	4	0.0642
$x(t-6)$	0.0057	6	0.0057	0.4751	6	0.4751	0.0729	3	0.0463
θ	(0.2869, 2.4084)			(0.6321, 2.0744)			(0.1212, 0.8417)		
selected features	x(t-1)			x(t-1)			x(t-1)		

Table 7.3. Saliency and rank in data-2

Features	Polynomial			Gaussian			2-layer tangent		
	s_i	Rank	s_i'	s_i	Rank	s_i'	s_i	Rank	s_i'
$x(t-1)$	0.2589	2	0.4801	0.4323	2	0.5420	0.3572	2	0.4919
$x(t-2)$	0.0148	6	0.2589	0.0591	6	0.4323	0.0289	6	0.3572
$x(t-3)$	0.0681	5	0.1702	0.0730	5	0.2103	0.0847	5	0.1874
$x(t-4)$	0.4801	1	0.0793	0.5420	1	0.0737	0.4919	1	0.0871
$x(t-5)$	0.0793	4	0.0681	0.0737	4	0.0730	0.0871	4	0.0847
$x(t-6)$	0.1702	3	0.0148	0.2103	3	0.0591	0.1874	3	0.0289
θ	(0.0793,0.1702)			(0.0737,0.2103)			(0.0871,0.1874)		
selected features	x(t-1), x(t-4), x(t-6)			x(t-1), x(t-4), x(t-6)			x(t-1), x(t-4), x(t-6)		

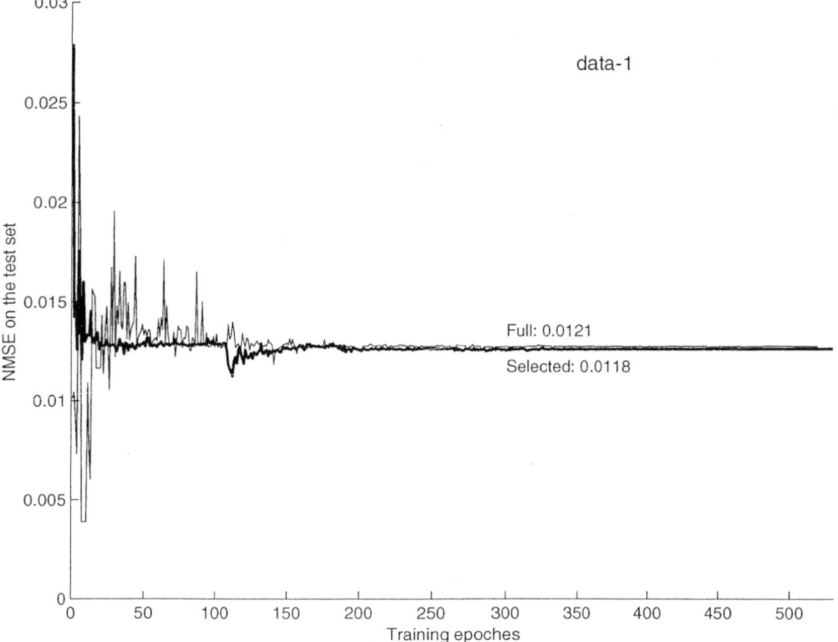

Fig. 7.2. The NMSE of the test set for the full feature set and the selected feature set in the data-1. There is a smaller converged NMSE in the selected feature set (0.0118) than that of the full feature set (0.0121)

Fig. 7.2 and Fig. 7.3 respectively give the behaviors of NMSE on the test set for the full feature set and the selected feature set in data-1 and data-2. All the settings of SVMs are kept the same in both feature subsets for comparison. From these figures, it can be observed that in both data-1 and data-2 there is a smaller converged NMSE in the selected feature set (0.0118 in data-1 and 2.3833 in data-2) than that of the full feature set (0.0121 in data-1 and 2.6060 in data-2). This indicates that using the selected feature set can enhance the generalization performance of SVMs.

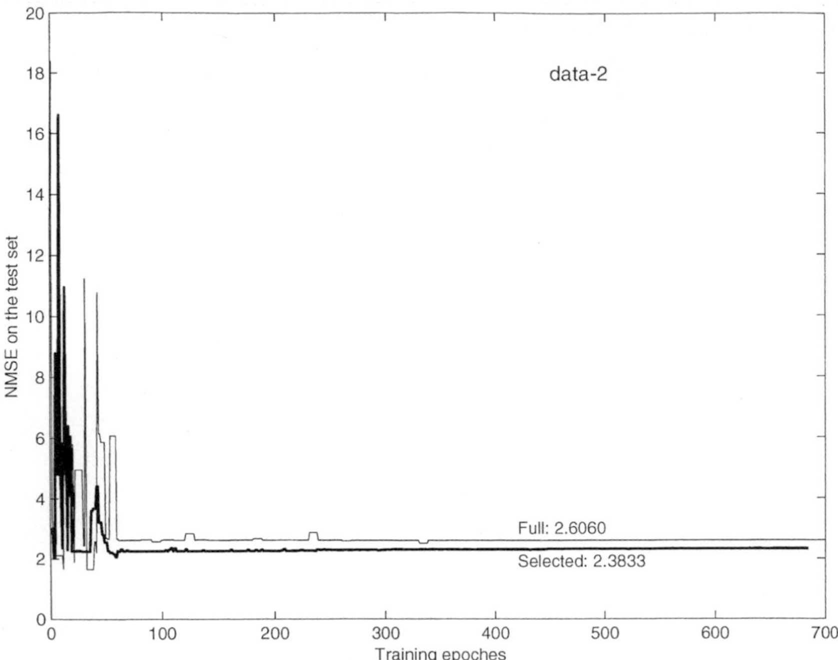

Fig. 7.3. The NMSE of the test set for the full feature set and the selected feature set in the data-2. There is a smaller converged NMSE in the selected feature set (2.3833) than that of the full feature set (2.6060)

7.4.2 Real Financial Data

Five real futures contracts collated from the Chicago Mercantile Market are examined in the second series of experiments. They are the Standard & Poor 500 stock index futures (CME-SP), United Sates 30-year government bond (CBOT-US), Unite States 10-year government bond (CBOT-BO), German 10-year government bond (EUREX-BUND) and French stock index futures (MATIF-CAC40). The time periods used in the experiment are listed in Table 7.4. The daily closing prices are used as the data set. And the original closing price is transformed into a five-day relative difference in percentage of price (RDP). As mentioned by Thomason [7.18], there are four advantages in applying this transformation. The most prominent advantage is that the distribution of the transformed data will become more symmetrical and follows more closely to a normal distribution as illustrated in Fig. 7.4 and Fig. 7.5. The modification to the trend of the data distribution will improve the predictive power of the neural network.

The input variables are constructed from 3 lagged transformed closing prices which is obtained by subtracting a 15-day exponential moving average from the closing price (x_1, x_2, x_3) and 14 lagged RDP values based on 5, 10, 15, 20-day periods (x_4, ... x_{17}). The subtraction in the transformaed close

Table 7.4. Five futures contracts and their corresponding time periods

Futures	Time period
CME-SP	30/12/1992 — 30/07/1996
CBOT-US	01/01/1993 — 01/08/1996
CBOT-BO	01/01/1993 — 01/08/1996
EUREX-BUND	01/01/1993 — 01/08/1996
MATIF-CAC40	01/06/1995 — 01/02/1999

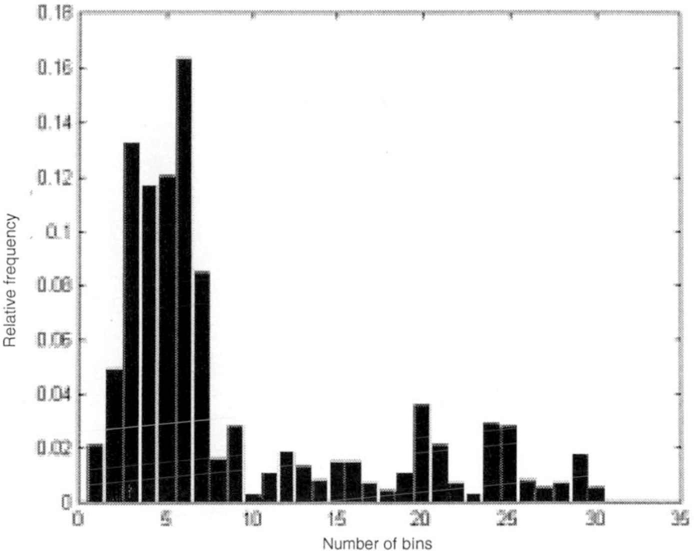

Fig. 7.4. Histograms of CME-SP daily closing price

prices is performed to eliminate the trend in price as there is a ratio of the maximum value to the minimum value about 2:1 in all of the five data sets. The optimal length of the moving average is not critical, but it should be longer than the forecasting horizon of 5 days [7.18]. The use of the transformed closing price is to maintain the information contained in the original closing price as much as possible, as the application of the RDP transform to the original closing price may remove some useful information. The lagged RDP values and transformed closing prices are recommended by Thomason [7.19]. However, a greater numbers of indicators are used here for feature selection. The output variable RDP+5 is obtained by first smoothing the closing price with a 3-day exponential moving average, because the application of a smoothing transform to the dependent variable generally enhances the prediction performance of neural network [7.19, 7.20]. The calculations for all the indicators are found in Table 7.5.

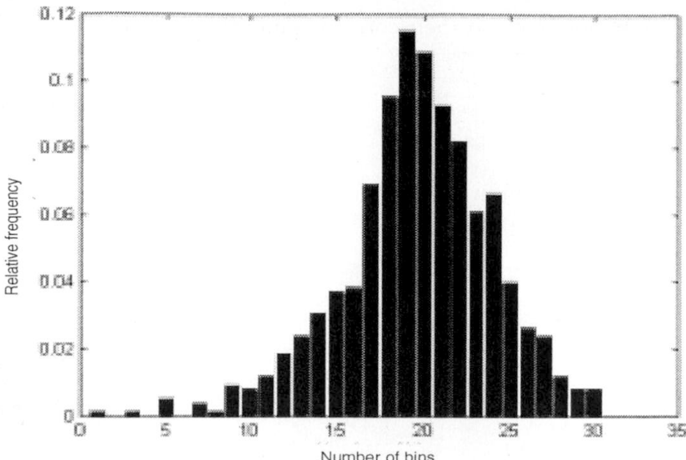

Fig. 7.5. Histograms of RDP+5. RDP+5 values are more symmetrical and have a normal distribution

The long left tail in Fig. 7.5 indicates that there are outliers in the data set. Since outliers may make it difficult or time-consuming to arrive at an effective solution for the SVMs, RDP values beyond the limits of ± 2 standard deviations are selected as outliers. They are replaced with the closest marginal values. The other preprocessing technique is data scaling. All the data points are scaled into the range of $[-0.9, 0.9]$ as the data points include both positive values and negative values. Each data set is partitioned into three parts according to the time sequence. The first part is for training, the second part for validating which is used to select optimal kernel parameter for SVMs, and the last part for testing. There are a total of 907 data patterns in the training set, 200 data patterns in both the validation set and the test set in all the data sets.

The Gaussian kernel is used for SVMs as it produces a smaller NMSE on the validation set than that of the polynomial kernel and two-layer tangent kernel based on the experiment. The kernel parameter, C and ε that produce the smallest NMSE on the validation set are used in the experiment. The values of δ^2, C and ε vary in futures due to different characteristics of the futures.

The results of SA are given in Table 7.6, with the detailed calculations of the saliency shown in Table 7.7. As it is unknown whether the irrelevant features have been correctly deleted, the selected features are used as the inputs of SVMs to retrain the network. Figs. 7.6 — 7.10 give the NMSE of the test set for the full feature set and the selected feature set in each futures contract. It is evident that the SVMs using the selected feature set converged to smaller NMSE of the test set than that of using full feature set. The result is consistent in all of the five contracts. This indicates that saliency analysis is effective in selecting important features for SVMs.

Table 7.5. Input and output variables

Indicator	Calculation
x_1	$p(i) - \overline{EMA_{15}(i)}$
x_2	$p(i-1) - \overline{EMA_{15}(i-1)}$
x_3	$p(i-2) - \overline{EMA_{15}(i-2)}$
x_4	$(p(i) - p(i-5))/p(i-5) \times 100$
x_5	$(p(i-1) - p(i-6))/p(i-6) \times 100$
x_6	$(p(i-2) - p(i-7))/p(i-7) \times 100$
x_7	$(p(i-3) - p(i-8))/p(i-8) \times 100$
x_8	$(p(i-4) - p(i-9))/p(i-9) \times 100$
x_9	$(p(i) - p(i-10))/p(i-10) \times 100$
x_{10}	$(p(i-1) - p(i-11))/p(i-11) \times 100$
x_{11}	$(p(i-2) - p(i-12))/p(i-12) \times 100$
x_{12}	$(p(i) - p(i-15))/p(i-15) \times 100$
x_{13}	$(p(i-1) - p(i-16))/p(i-16) \times 100$
x_{14}	$(p(i-2) - p(i-17))/p(i-17) \times 100$
x_{15}	$(p(i) - p(i-20))/p(i-20) \times 100$
x_{16}	$(p(i-1) - p(i-21))/p(i-21) \times 100$
x_{17}	$(p(i-2) - p(i-22))/p(i-22) \times 100$
$RDP+5$	$(\overline{p(i+5)} - \overline{p(i)})/\overline{p(i)} \times 100$
	$\overline{p(i)} = \overline{EMA_3(i)}$

$EMA_n(i)$ is the n-day exponential moving average of the i-th day.
$p(i)$ is the closing price of the i-th day.

Table 7.6. Selected features in the five futures contracts

Futures	SA
CME-SP	x_1, x_2, x_3, x_7, x_{12}
CBOT-US	x_1, x_2, x_3, x_{10}, x_{12}, x_{16}
CBOT-BO	x_1, x_2, x_3, x_8, x_9, x_{10}, x_{12}
EUREX-BUND	x_1, x_2, x_3, x_6, x_{12}, x_{17}
MATIF-CAC40	x_1, x_2, x_3, x_9, x_{12}, x_{17}

7.5 Conclusions

The saliency analysis for ranking the importance of features has been developed for SVMs by evaluating the partial derivative of the output to the feature input over the entire training data samples. The threshold method is applied to delete irrelevant features from the whole feature set. According to the simulation results on both the simulated and the real data, it can be concluded that saliency analysis is effective in SVMs for selecting important features. By deleting the irrelevant features, the generalization performance of SVMs is greatly enhanced.

There are still some aspects that require further investigation. Although using a simple threshold method for determining how many unimportant features are deleted works well in this study, more formal methods need to be

Table 7.7. Saliency ands rank in the five real futures contracts

Features	CME-SP		CBOT-US		CBOT-BO		EUREX-BUND		MATIF-CAC40	
	s_i	Rank	s_i	Rank	s_i	Rank	s_i	Rank	s_i	Rank
x_1	0.5350	3	0.3143	2	0.5726	2	0.4547	3	0.2768	5
x_2	0.5408	2	0.2504	4	0.5042	3	0.5638	2	0.2162	6
x_3	0.9326	1	0.5608	1	0.9002	1	0.8006	1	0.4696	1
x_4	0.1010	7	0.0116	13	0.0681	9	0.0726	9	0.0917	10
x_5	0.0228	14	0.1375	7	0.0505	10	0.0469	13	0.0043	17
x_6	0.0369	10	0.0211	12	0.1057	8	0.2424	4	0.0344	12
x_7	0.2307	5	0.0089	16	0.0488	12	0.1223	8	0.0395	11
x_8	0.0118	16	0.0800	11	0.2240	7	0.0139	17	0.0061	15
x_9	0.0199	15	0.1150	9	0.2411	5	0.0591	12	0.3736	2
x_{10}	0.1076	6	0.2473	5	0.2254	6	0.0605	11	0.1500	7
x_{11}	0.0789	8	0.0809	10	0.0120	14	0.0643	10	0.0071	14
x_{12}	0.2392	4	0.3138	3	0.2999	4	0.2113	6	0.3096	3
x_{13}	0.0094	17	0.0069	17	0.0097	15	0.1568	7	0.0045	16
x_{14}	0.0258	12	0.0095	14	0.0040	17	0.0180	16	0.0090	13
x_{15}	0.0352	11	0.0095	15	0.0377	13	0.0222	14	0.1282	8
x_{16}	0.0231	13	0.2233	6	0.0089	16	0.0180	15	0.1216	9
x_{17}	0.0458	9	0.1365	8	0.0491	11	0.2326	5	0.2805	4
θ	(0.1076, 0.2307)		(0.1375, 0.2233)		(0.1057, 0.2240)		(0.1568, 0.2113)		(0.1500, 0.2162)	
selected features	x_1, x_2, x_3 x_7, x_{12}		x_1, x_2, x_3 x_{10}, x_{12}, x_{16}		x_1, x_2, x_3, x_8 x_9, x_{10}, x_{12}		x_1, x_2, x_3 x_6, x_{12}, x_{17}		x_1, x_2, x_3 x_9, x_{12}, x_{17}	

explored for complex problems. It is also worthy to explore the weight-based

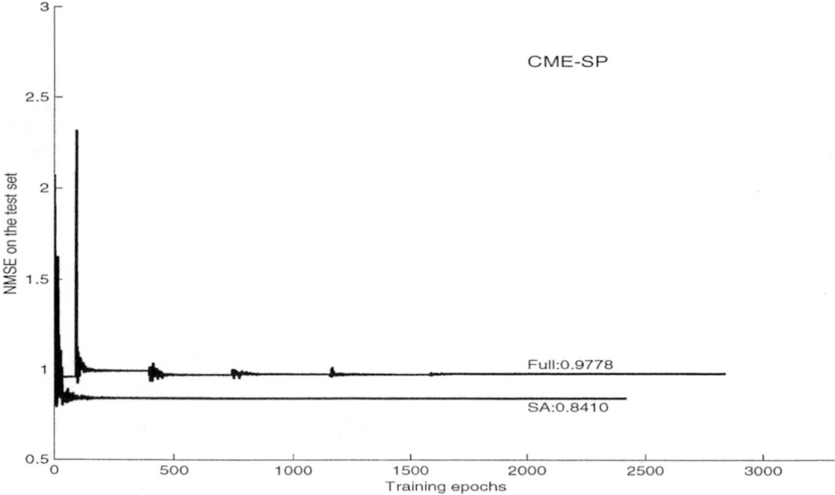

Fig. 7.6. The NMSE of the test set for the full feature set and the selected feature set in the CME-SP

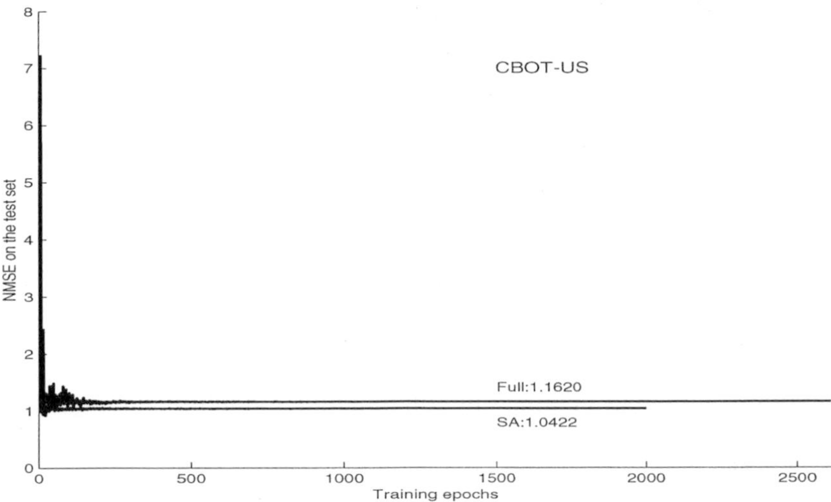

Fig. 7.7. The NMSE of the test set for the full feature set and the selected feature set in the CBOT-US

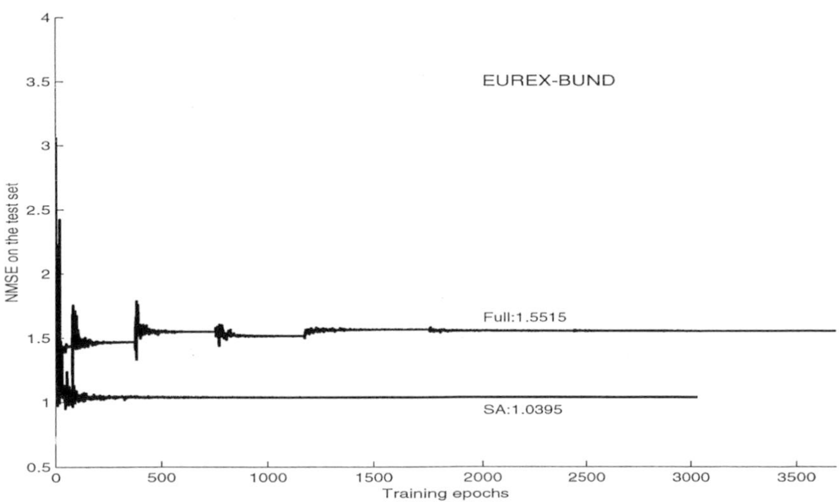

Fig. 7.8. The NMSE of the test set for the full feature set and the selected feature set in the EUREX-BUND

SA to SVMs with the kernel function constructed in explicit high dimensional feature space for further improving the performance of SVMs.

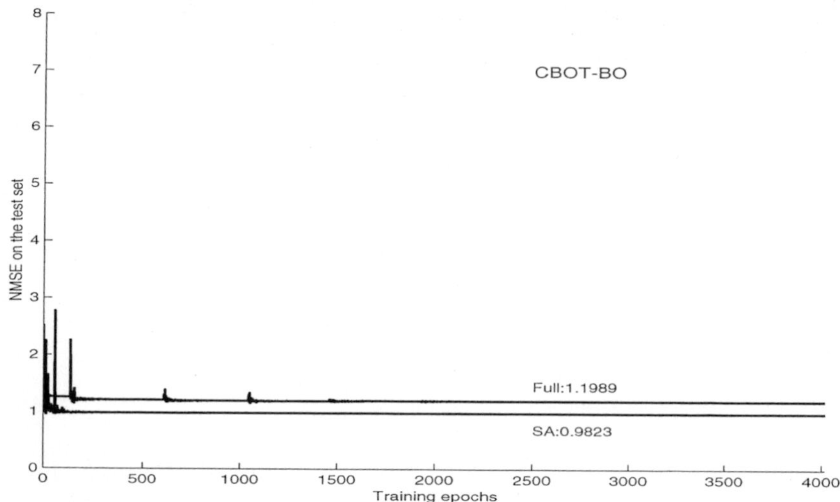

Fig. 7.9. The NMSE of the test set for the full feature set and the selected feature set in the CBOT-BO

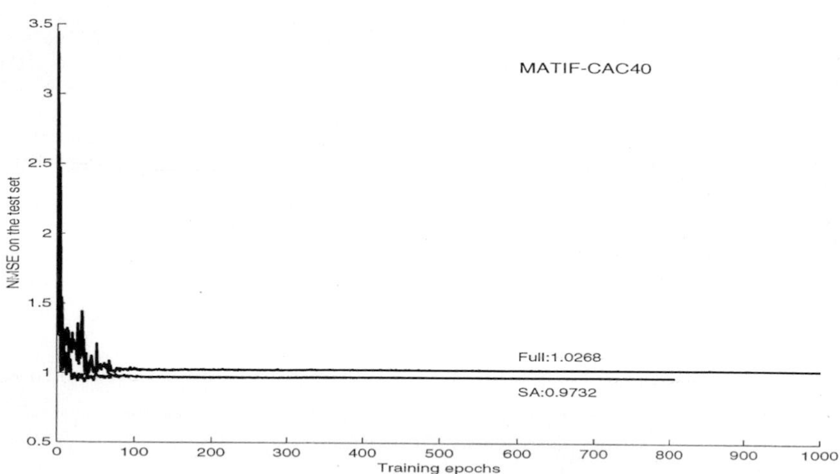

Fig. 7.10. The NMSE of the test set for the full feature set and the selected feature set in the MATIF-CAC40

References

7.1 Muller, R., Smola, J. A., Scholkopf, B. (1997): Prediction Time Series with Support Vector Machines. In: Proceedings of International Conference on Artificial Neural Networks, 999

7.2 Mukherjee, S., Osuna, E., Girosi, F. (1997): Nonlinear Prediction of Chaotic Time Series Using Support Vector Machines. Proc. Of IEEE NNSP'97, Amelia Isl, FL

7.3 Vapnik, V. N., Golowich, S. E., Smola, A. J. (1996): Support Vector Method for Function Approximation, Rregression Estimation and Signal Processing. Advances in Neural Information Processing Systems, Vol. 9, 281–287

7.4 Muller, K. R., Smola, J. A., Ratsch, G., Scholkopf, B., Kohlmorgen, J. (1999): Prediction Time Series with Support Vector Machines. Advances in Kernel Methods, The MIT Press

7.5 Barzilay, O., Brailovsky, V. L. (1999): On Domain Knowledge and Feature Selection Using a Support Vector Machine. Pattern Recognition Letters, Vol. 20, 475–484

7.6 Steppe, J. M., Bauer, Jr. K. W. (1997): Feature Saliency Measures. Computers Math. Application, Vol. 33, 109–126

7.7 Reed, R. (1993): Pruning Algorithms–a Survey. IEEE Transactions on Neural Networks, Vol. 4, 740–747

7.8 Ruck, D. W., Rogers, S. K., Kabrisky, M. (1990): Feature Selection Using a Multilayer Perceptron. Journal of Neural Network Computing, Vol. 2, 40–48

7.9 Belue, L. M., Bauer, Jr. K. W. (1995): Determining Input Features for Multilayer Perceptrons. Neurocomputing, Vol. 7, 111–121

7.10 Stepp, J. M., Bauer, Jr. K. W. (1996): Improved Feature Screening in Feedforward Neural Networks. Neurocomputing, 1347–58

7.11 Zurada, M. J., Malinowski, A., Usui, S. (1997): Perturbation Method for Deleting Redundant Inputs of Perceptron Networks. Neurocomputing, Vol. 14, 177–193

7.12 Vapnik, V. N. (1995): The Nature of Statistical Learning Theory. Springer-Verlag

7.13 Smola, A. J., Scholkopf, B. (1998): A Tutorial on Support Vector Regression. NeuroCOLT Technical Report TR, Royal Holloway College

7.14 Smola, A. J. (1998): Learning With Kernels. Ph.D. Thesis GMD, Birlinghoven

7.15 Cibas, T., Soulie, F. F., Gallinari, P., Raudys, S. (1996): Variable Selection with Neural Networks. Neurocomputing, Vol. 12, 223–248

7.16 Sexton, R. S. (1998): Identifying Irrelevant Input Variables in Chaotic Time Series Problems: Using a Genetic Algorithm for Training Neural Networks. Journal of Computational Intelligence in Finance, 34–41

7.17 Murray, W. H. (1992): Microsoft C/C++7: The Complete Reference. Osborne McGraw-Hill

7.18 Thomason, M. (1999): The Practitioner Methods and Tool. Journal of Computational Intelligence in Finance, Vol. 7, No. 3, 36–45

7.19 Thomason, M. (1999): The Practitioner Methods and Tool. Journal of Computational Intelligence in Finance, Vol. 7, No. 4, 35–45

7.20 Thomason, M. (1999) : The Practitioner Methods and Tool. Journal of Computational Intelligence in Finance, Vol. 7, No. 6, 35–45

Part IV

Self-organizing Maps and Wavelets

8. Searching Financial Patterns with Self-organizing Maps

Shu-Heng Chen[1] and Hongxing He[2]

[1] AI-ECON Research Group, Department of Economics, National Chengchi University, Taipei, Taiwan 11623
email: chchen@nccu.edu.tw

[2] CSIRO Mathematical and Information Sciences, Commonwealth Scientific & Industrial Research Organization, GPO Box 664, Canberra ACT 2601, Australia
email: Hongxing.He@csiro.au

Using *Self-organizing maps* (**SOM**), this paper formalizes chartists' behavior in searching of patterns (charts). By applying a 6 by 6 two-dimensional SOM to a time series data of TAIEX (Taiwan Stock Index), 36 patterns are established. To see whether these 36 patterns transmit profitable signals, a "normalized" equity curve is drawn for each pattern up to 20 days after observing the pattern. Many of these equity curves are either monotonically increasing or decreasing, and none of them exhibits random fluctuation. Therefore, it is concluded that the patterns established by SOM can help us foresee the movement of stock index in the near future. We further test profitability performance by trading on these SOM-induced financial patterns. The equity-curve results show that SOM-induced trading strategy is able to outperform the buy-and-hold strategy in quite a significant period of time.

8.1 Motivation and Introduction

It has been a long time that technical analysts detect trading signals with charts,[1] and for this, they are frequently called *chartists*. Since chartists has been a well-established profession in the financial industry, there is little doubt that charts, to some extent, did transmit signals. Nonetheless, from a scientific viewpoint, charts are somewhat subjective objects. In general, there is no well-articulated definition for charts. Alternatively speaking, analysts have to rely on their *experience* to identify charts. In this research, we attempt to provide a systematic study of charts. The ideas is simple: *if charts do transmit information, then the time series of assets' prices should have many unique patterns.* What we need is a mechanical tool to discover these patterns, and Self-organizing maps (**SOM**) seem to be a very natural tool to serve this purpose. In this paper, we use **SOM** to search for charts, and test whether these charts are profitable.

This paper is organized as follows. Section 2 is a brief introduction to SOM, and gives the experimental design given in this paper. Section 3

[1] Some of them even consider what they are doing as a science. See [8.2].

presents the 36 patterns established by SOM. Section 4 investigates the information contents of these 36 charts. Trading performance of these 36 patterns are analyzed in Section 5 followed by a few concluding remarks in Section 6.

8.2 Self-organizing Maps

In contrast to the artificial neural networks which are used for *supervised learning*, there is another special class of artificial neural networks known as *Self-organizing feature map* (**SOFM** or, simply, **SOM**). The SOM is used for *unsupervised learning* to achieve *auto classification*, *data segmentation* or *vector quantisation*. The SOM adopts a so-called *competitive learning* among all neurons. The output neurons that win the competition are called *winner-takes-all* neurons. In SOM, the neurons are placed on the sites of a n-dimensional lattice. The value of n is usually 1 or 2. Through competitive learning, the neurons are tuned to represent a group of input vectors in an organized manner. The mapping from continuous space to a discrete one or two-dimensional space achieved by SOM reserves the spatial order.

Among a number of training algorithms for SOM, the Kohonen's learning algorithm is the most popular one ([8.5],[8.3]). Kohonen's learning algorithm adopts a heuristic approach. Each neuron on the lattice has a weight vector of m components attached. The m is the number of input variables of the input data sets. The winning neuron and its close neighbors in the lattice have their weight vectors adjusted towards the input pattern presented on each iteration. Unlike other clustering methods such as k-means clustering([8.4]), the Kohonen's SOM has an advantage that the final training outcome is insensitive to the initial settings of weights. Therefore, Kohonen's SOM has found a wide variety of applications.

In this paper we present the result of the application of the SOM to financial time series data. The data set to be segmented is the daily closing price of Taiwan stock indices (TAIEX). The original data set covers the daily price from 1/5/71 to 3/26/97, which has 7435 observations. A sliding window whose width covers 125 trading days moves from the first day to the 7309th day. This gives us 7309 subsamples (windows) each with 125 observations. Each subsample represents a time series pattern. The SOM is then used to automatically divide all patterns to groups or clusters in a way such that members of the same group are *similar* (*close*) in the Euclidean metric space. The 125 observations of each subsample is normalized between 0 to 1. A two-dimensional 6×6 SOM is used to map the 7309 records into 36 clusters.[2] The 6×6 lattice of SOM is presented in Fig. 8.1.

[2] Regarding the question why the two-dimension lattice model is used for SOM, we do not have a theoretical answer here. However, the following observations may lend support for the usage of a two-dimensional map. First, in practice, only low-dimension lattice (dimension no more than three) has been used. Second, in

0	1	2	3	4	5
6	7	8	9	10	11
12	13	14	15	16	17
18	19	20	21	22	23
24	25	26	27	28	29
30	31	32	33	34	35

Fig. 8.1. Pattern indices and their corresponding positions in 6×6 lattice of SOM

In the training process, the weights of the winning neuron and its close neighbors are updated according to Eq. (8.1),

$$w_j(n+1) = w_j(n) + \eta(n)\pi_{j,i(x)}(n)[x - w_j(n)], \tag{8.1}$$

where $w_j(n)$ is the weight vector of the jth neuron at the nth iteration, $\pi_{j,i(x)}(n)$ is the *neighborhood function* (to be defined below) of node indices j and $i(x)$,

$$i(x) = arg \min_j \| x - w_j \|, j = 1, 2, ..., N, \tag{8.2}$$

and $\eta(n)$ is the *learning rate* at iteration n.

We take for the neighborhood function the *Gaussian* form,

$$\pi_{j,i(x)} = \exp^{-\frac{d^2_{j,i(x)}}{2\sigma^2(n)}}, \tag{8.3}$$

where $\sigma(n)$ is some suitably chosen, monotonically decreasing function of iteration times n. Here, the effective width σ decays with n according to Eq. 8.4.

$$\sigma(n) = \sigma_0 \exp^{-\frac{n}{\tau_1}}, \tag{8.4}$$

where σ_0 and τ_1 are constants. The learning rate decays in a similar manner:

$$\eta(n) = \eta_0 \exp^{-\frac{n}{\tau_2}}, \tag{8.5}$$

where η_0 and τ_2 are constants.

The training takes a long time with initially almost all neurons have their weights updated. This training phase is called the *ordering phase*. The

---

the commercial package, only low-dimension maps are implemented, and almost all use the two-dimensional lattice as a default setting. Third, in his book ([8.6]),

Table 8.1. Parameter setup for the implementation of the 2-dimensional 6 × 6 SOM

Number of total trading days	7435
Number of time series pattern	7309
Number of trading day in each time series	125
Dimensionality of SOM	2
Number of neuron	36
Ordering phase initial learning rate	0.900
Ordering phase learning rate decay rate	455.120
Ordering epoch	1000
Ordering phase initial radius	8.485
Ordering phase radius decay rate	467.654
Convergence phase initial learning rate	0.100
Convergence learning rate decay rate	333.808
Convergence phase initial radius	1.000
Convergence phase radius decay rate	434.294
Convergence phase epoch	1000

weights then are settle down gradually at the second phase of learning named *convergence phase* where only the weights of winning neuron and perhaps its nearest neighbors are updated according to the case presented. The control parameters chosen to conduct this experiment is summarized in Table 8.1.

8.3 Charts Constructed by SOM

After the training, each of the 7309 time series subamples (charts) is assigned a cluster index which is represented by its winning neuron. The charts assigned to the same cluster index are *closer* than the charts which belong to difference clusters in terms of Euclidean distance. Then the average behavior of each cluster can serve as a *representative chart* for that cluster. Since we are using 6 × 6 SOM, in the end, we have 36 representative charts (See Figs. 8.2 and 8.3). These 36 charts can then be considered as something "*equivalent*" to what chartists are looking for.

How can these SOM-induced charts be compared to those charts used by chartists? Well, we have to admit that there is no direct comparison, knowing that each of our charts is *only* a *representative* (*average*) of many similar charts, and hence is not referred to any specific chart, while the charts used and defined by chartists are based on a single price line. Therefore, not

Kohonen mentioned that the original idea of SOM is enlighten by the structure of the human brain. The brain has a surface area of 2400 cm^2. The cerebral cortex is like a two-dimension map. It is divided into a few areas like motor cortex, somatosensory cortex, visual cortex, auditory cortex, etc. Each area is responsible for one kind of memory or control. Therefore, the two-dimension lattice model is closest to the structure of the human brain.

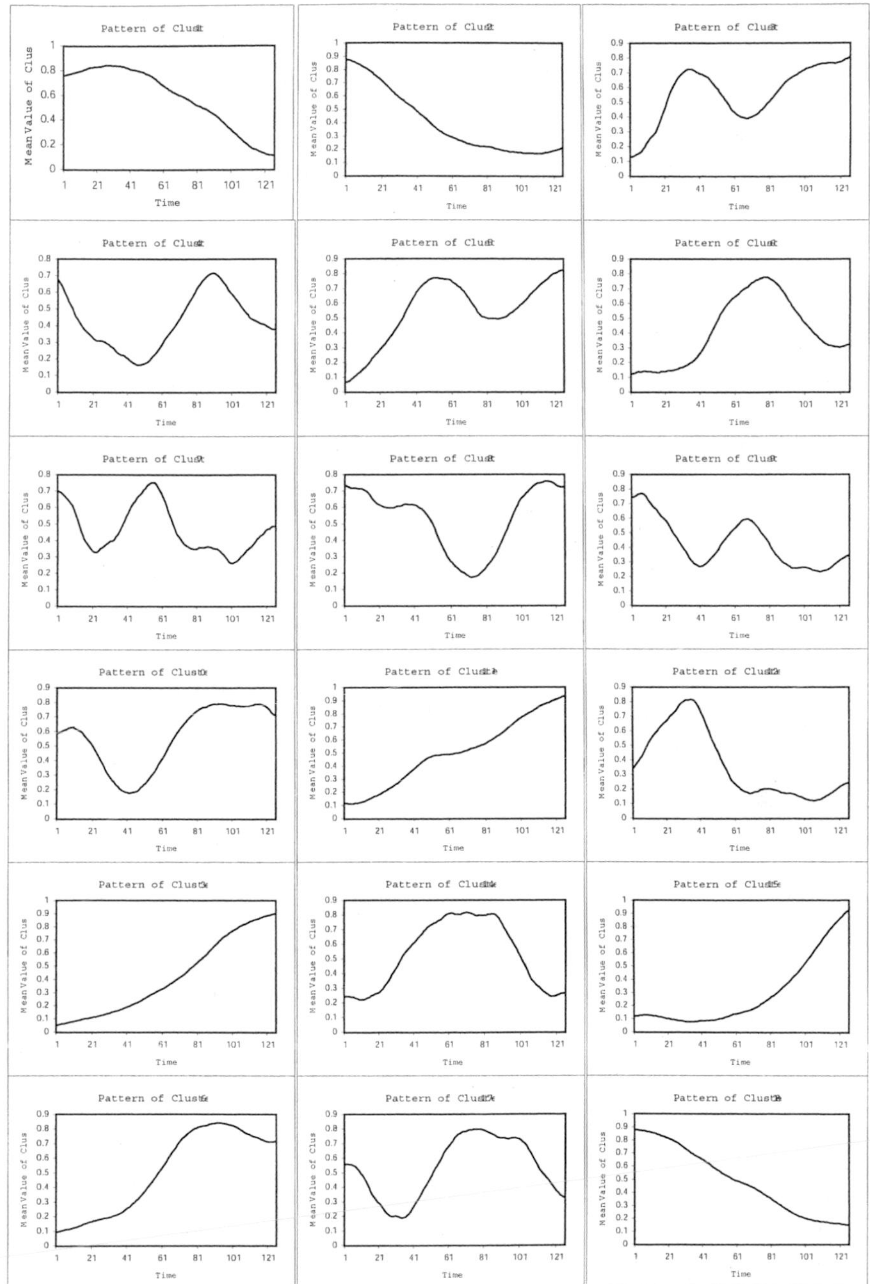

Fig. 8.2. Charts discovered by the 2-dimensional SOM: harts 1-18

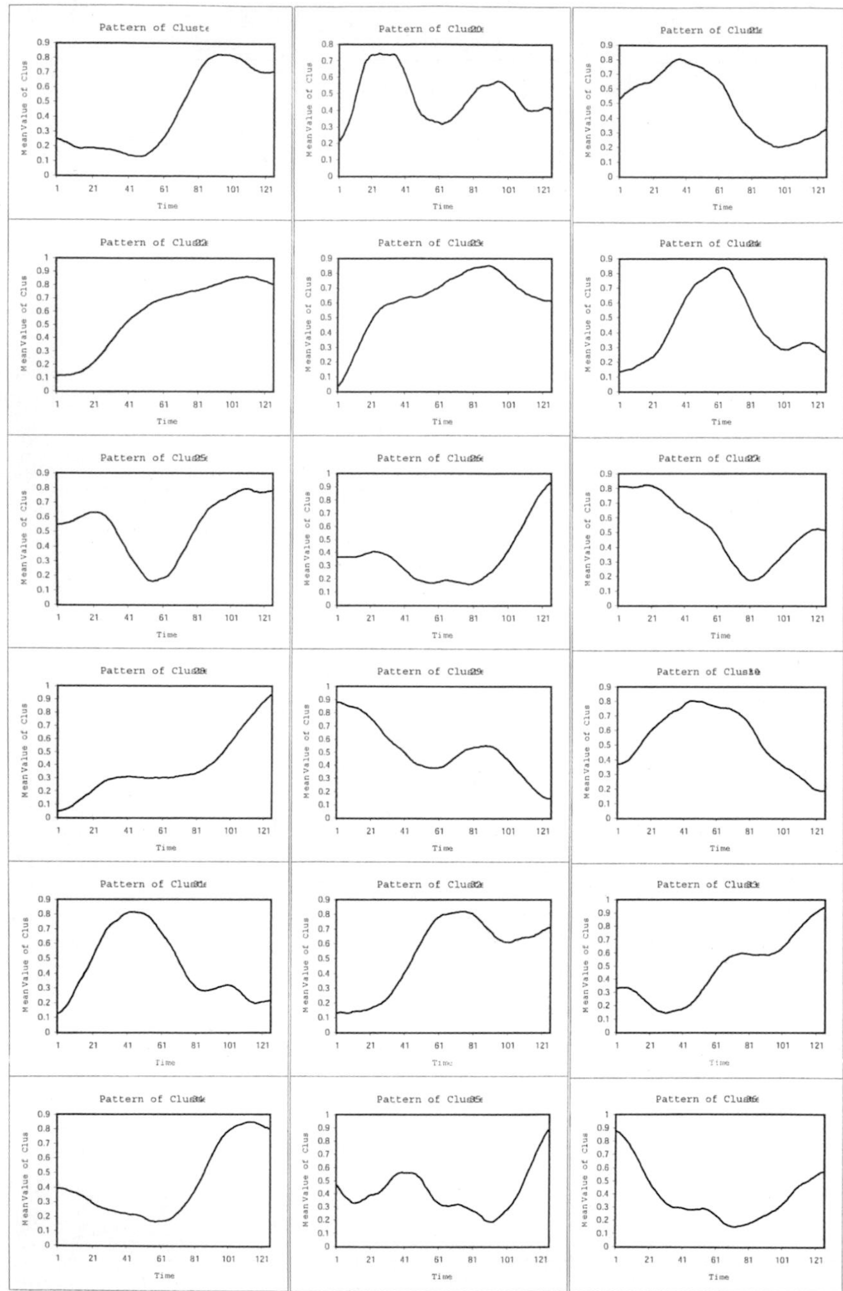

Fig. 8.3. Charts discovered by the 2-dimensional SOM: charts 19-36

surprisingly, all our charts behave quite *smoothly* as opposed to many of those used by chartists. This fundamental difference makes a direct comparison between these two types of charts not only infeasible but also meaningless. Nevertheless, some of our charts can roughly be given a name in chartists' eyes, such as *uptrends* (Charts 11, 13, 15), *downtrends* (Charts 1, 2, 18), *V-formations* (Charts 8, 24 and 25), *rounding bottom* (Chart 36), *rounding top* (Chart 30), *double tops* (Chart 20), *island reversal* (Chart 17).

8.4 Do SOM-Induced Charts Reveal Trading Signals?

So far, we only demonstrate how to use SOM to built charts. But, how do we know that these charts can indeed reveals trading signals rather than *just clusters*? To answer this question, we proposed the following test based on a *normalized equity curve*. First, we ask a simple question: *once a specific chart is observed, what are the stock returns in the following k days*? In other words, we are investigating the time series plot R_h,

$$R_{h,i} = \ln P_{t+h} - \ln P_t, \quad h = 1, 2, ..., k, \tag{8.6}$$

where t is the day which pattern (chart) i is observed ($i = 1, 2, ..., 36$). Call $R_{h,i}$ the *normalized equity curve* of the chart i. Then, we can make the following observation. Suppose that the patterns constructed by SOM are able to transmit trading signals, then what we expect from these equity curves are *systematic* movements, such as *monotone increasing* or *decreasing*, as the most ideal cases. They can be other types, but the bottom line is that they cannot be *erratic* or *random*. Finally, since each pattern appears many times in the whole series, instead of drawing a single equity curve, we are presenting an average of them, i.e., what we actually draw is the time series plot of $\bar{R}_{h,i}$:

$$\bar{R}_{h,i} = \frac{\sum_{j=1}^{m_i} R_{h,i:j}}{m_i}, \tag{8.7}$$

where j refers to the jth occurrence of chart i, and m_i is the total number of the occurrence of chart i.

Figs. 8.4 and 8.5 is the time series plot of these *representative* equity curves of the 36 charts. From these curves, most charts do seems to feature *strong* buy and sell signals. In 14 out of the 36 charts, the equity curve is monotonically increasing (Charts 2, 3, 4, 5,11, 12, 13, 16, 23, 26, 28, 32, 33, 35), featuring buying signals, and in 3 out of the 36, the equity curve is monotonically decreasing (Charts 1, 10, 18), featuring selling signals. For many others, even though the equity curve is not monotone, a trend to grow or decline is discernible. Therefore, we may roughly conclude from this initial analysis that the charts constructed by SOM are able to transmit buying and selling signals.

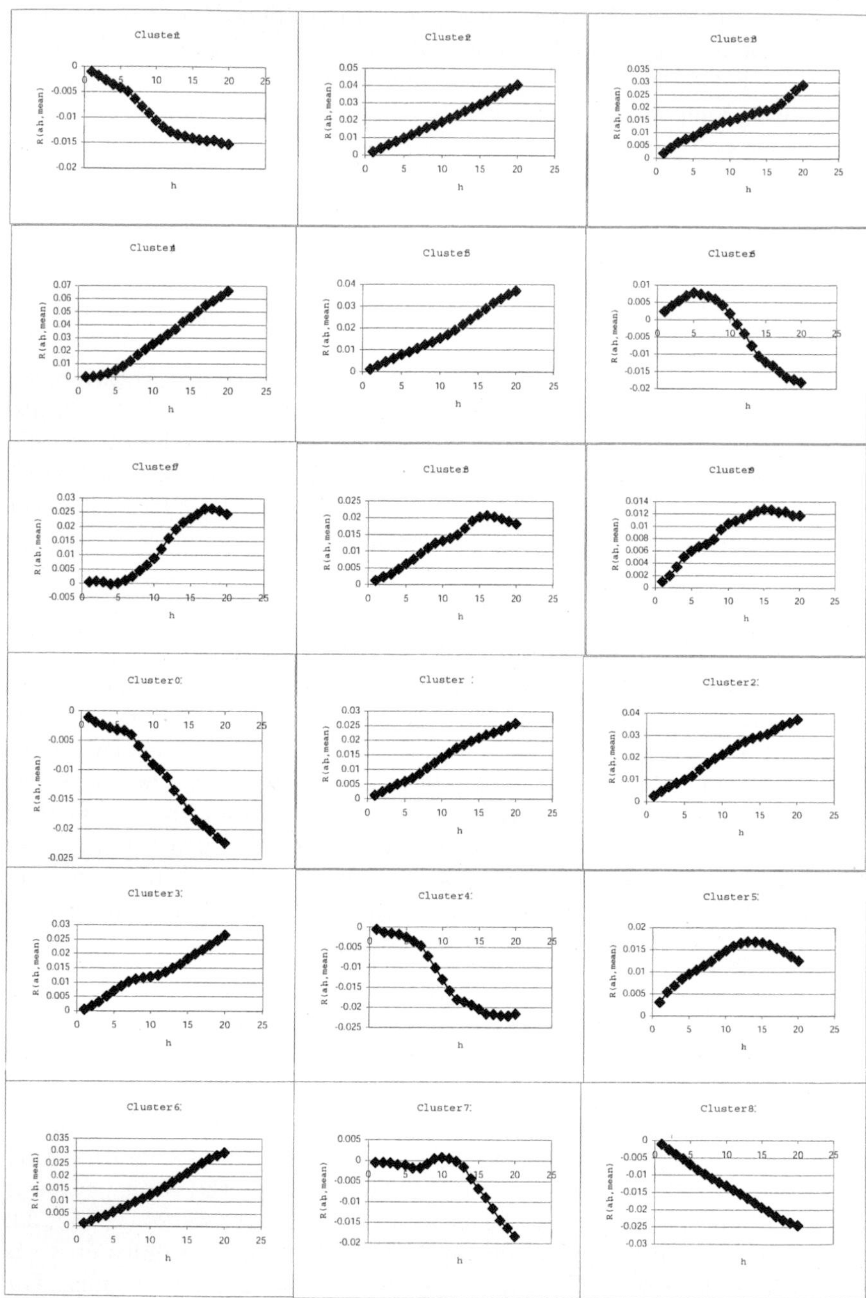

Fig. 8.4. The normalized equity curves of charts 1-18

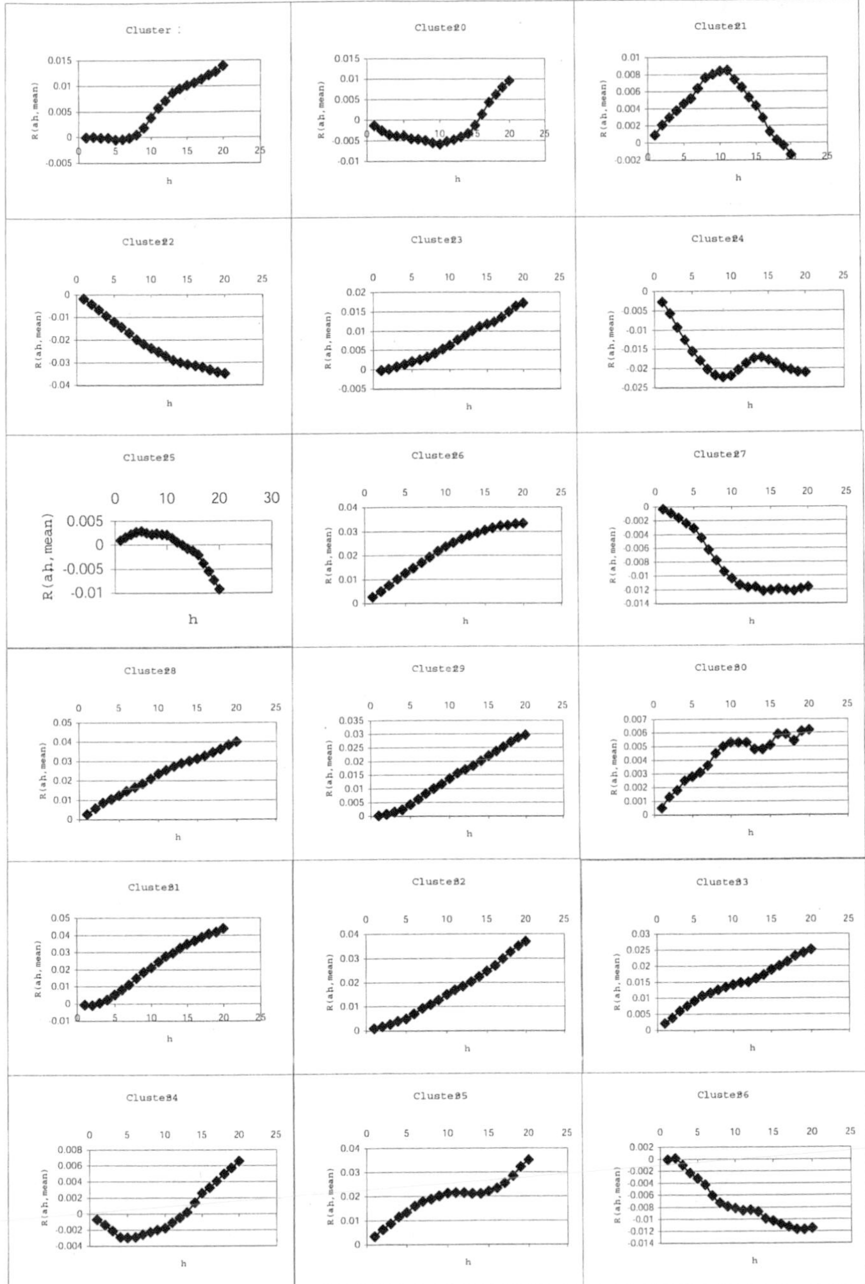

Fig. 8.5. The normalized equity curves of charts 19-36

8.5 SOM-Induced Trading Strategies

The previous analysis based on equity curves shows that the SOM-induced charts are not just revealing a bunch of similar patterns. The analysis indicates that many of these charts can be further used for helping us forecast the movement of stock price, at least, in the short term. In this section, we would like to go one step further, and ask: *if we actually use these charts to develop trading strategies, how well will they perform?* [3] Will they beat other well-known benchmarks, such as the *buy-and-hold* (**B&H** strategy)? But, how to trade with these charts? There are many possible ways to do so, the one considered in this paper is one of the simplest ones.

The first stage of this development is to associate with each chart a *buy* or *sell* signal. To do so, let us suppose that traders are interested in *day trading* only, and they are responsive to some patterns, say chart i ($i = 1, 2, ..., 36$) constructed by SOM. Usually, they will do two things when chart i appears, either *buy then sell* or *sell first then buy back later* (so-called *short selling*). Denote the first action, called the *buy* action, by **B**, and the second action, the *sell action*, by **S**. We can then calculate the rate of return for *one-time* appearance of Chart i. Now, let $R_{B,i}$ be the rate of return for Chart i under action B in the *whole* sample, and $R_{S,i}$ be that under action S. Then

$$R_{B,i} = \sum_{t=1}^{T} R_t I_{i,t} \tag{8.8}$$

where $I_{i,t}$ is an indicator function :

$$I_{i,t} = \begin{cases} 1, & \text{if Chart } i \text{ appears at time } t-1 \\ 0, & \text{otherwise.} \end{cases} \tag{8.9}$$

and $R_t = \ln P_t - \ln P_{t-1}$. The *mean* rate of return of trading on Chart i with the *buy* action can then be derived by taking the *simple average* as follows.

$$\bar{R}_{B,i} = \frac{R_{B,i}}{\sum_{t=1}^{T} I_{i,t}} \tag{8.10}$$

Similarly, we can have

$$R_{S,i} = -\sum_{t=1}^{T} R_t I_{1,t} \tag{8.11}$$

and

---

[3] To some extent, this inquiry is in line with the 2-stage process proposed by [8.8], namely, first, *visual segmentation* of the curve into primitive shapes of some portions, and second, a *knowledge-based correspondence* established between the time sequence of such primitive shapes found in the curve and specific investment decisions, including forecasts or trend assessments.

$$\bar{R}_{s,i} = \frac{R_{s,1}}{\sum_{t=1}^{T} I_{1,t}} \tag{8.12}$$

Table 8.2. Profitability of trading on the SOM-induced patterns: buy action

i	m_i	$\bar{R}_{B,i}$	A	i	m_i	$\bar{R}_{B,i}$	A
1	335	-0.101	S	19	195	-0.006	S
2	370	0.185	B	20	78	-0.132	S
3	123	0.203	B	21	181	0.089	B
4	88	0.003	B	22	207	-0.187	S
5	155	0.116	B	23	155	-0.016	S
6	150	0.236	B	24	154	-0.267	S
7	102	0.049	B	25	174	0.097	B
8	146	0.124	B	26	202	0.265	B
9	170	0.103	B	27	213	-0.032	S
10	136	-0.114	S	28	242	0.282	B
11	364	0.12	B	29	203	0.007	B
12	67	0.262	B	30	224	0.046	B
13	415	0.052	B	31	114	-0.073	S
14	164	-0.061	S	32	206	0.101	B
15	249	0.272	B	33	199	0.211	B
16	372	0.113	B	34	207	-0.072	S
17	108	-0.047	S	35	141	0.338	B
18	534	-0.112	S	36	166	-0.008	S

"i" denoted the ith chart, "m_i" indicates the frequencies of observing Chart i, and $\bar{R}_{B,i}$ is the mean rate of return for the buy action.

Table 8.2 summarizes these daily return statistics for each charts. We shall then use these return statistics described above to identify whether the appearance of a chart feature as a buy or a sell signal. For example, from Table 2, the first-day return for Chart 1 is negative (-0.00101), so it is recognized as a *sell* signal.[4] And it is 0.00185 for Chart 2, so Chart 2 features a *buy* signal. By this manner all charts are assigned to a **B** or **S** action, as shown in the fourth column of Table 8.2, and we can map the sequence of appearing charts to a sequence of "**B**" and "**S**" signals.

For example, in our case, the chart sequence starts with "34, 34, 34, 34, 34, 34, 34, 25, 25, 25, 25, 25, 25, 25, 25, 25, 25, 25, 25, 25, 25, 25, 25, 25, 25, 25, 25, 25, 25, 25, 33, 33, 33, 33, 33, 33, 33, 33, 33, 33, 33, 33, 33, 33, 18, 18, 18, 18, 18, 18, 18, 18, 18, 18, 18, 18, 18, 18, 18, 18, 5, 5, 5, 5, 5, 5, 5, 5, 5, 5, 5, 5, 5, 5, 5, 5, 5, 5, 5, 5, 5, ..." Then based on the description above, it can be mapped to the buy-or-sell sequence as below: "B, B, B, B, B, B, B, B, B, B, B, B, B, B, B, B, B, B, B, B, B, B, B, B, B, B, B, B, B, B, S, S, S, S, S, S, S, S, S, S, S, S, S, S, S, S, S, S, S, S, S, S, S, S, S, S, S, S, S, S, B, B, B, B,

[4] As seen above, $\bar{R}_{S,i}$ is simply just the negative of $\bar{R}_{B,i}$. So, if the buy action earn a negative profit, then the sell action must earn a positive profit.

B, B, B, B, B, B, B, B, B, B, B, B, B, B, B, B, B,..." So, at the beginning, we start with the long position, i.e., to buy, and the trading strategy since then is constructed as follows.

Suppose at time $t-1$, the trader is in a long position (i.e., holding one unit of the stock on hand, **Holding** $=$ **1**). If at time t, there is a **B** signal, and the trader simply continue to **Hold** (i.e., do nothing, **Holding** $=$ **1**). However, if at time t, it is a **SELL** signal, then the trader with long position shall **SELL** her one unit of the stock kept in the previous period, and her holding is now 0 (**Holding** =**0**). In addition, she shall further **SELL SHORT** one more unit of stock and wait to buy it back later (i.e. waiting for the next coming buy signal). For the time being, she has a short position (**Holding** $=$ **-1**). Now, consider the other possibility. Suppose that at time $t-1$, the trader are in a short position (**Holding** $=$ **-1**), and at time t, there is a **B** signal. Then the trader will buy back the stock sold short earlier, and her position is therefore back to 0. But, in addition, she shall go further to buy one more unit of stock, and she end up with a long position (**Holding** $=$ **1**). On the other hand, if at t, it is a sell signal, then the trader will do nothing (**Holding** $=$ **-1**).

This automatic trading rule can be represented as a *mapping* G given below. Function G is a mapping from a Cartesian product space, which is a product space of the signal and position, to the position space,

$$G(x, y) = z, \tag{8.13}$$

where $x \in X = (B, S)$, and $y, z \in Y = (-1, 0, 1)$, with the following mapping: $G(B, 0) = 1, G(S, 0) = -1, G(B, 1) = 1, G(S, 1) = -1, G(B, -1) = 1, G(S, -1) = -1$. With this trading rule, we can then transform a **B-S** sequence into a sequence of 1, 0 and -1. Now, let the position at t as H_t (Clearly, $H_t = 1, 0,$ or -1). Then (*accumulated*) trading profits at any point in time, π_t, can be calculated as follows,

$$\pi_1 = 0, \tag{8.14}$$

and

$$\pi_t = \begin{cases} \pi_{t-1}, \text{if } \ H_t = H_{t-1}, \\ \pi_{t-1} + P_t(1 - c_1 - c_2) - P_{s(t)}(1 + c_1), \\ \text{if } \ H_t = -1, H_{t-1} = 1, \\ \pi_{t-1} + P_{s(t)}(1 - c_1 - c_2) - P_t(1 + c_1), \\ \text{if } \ H_t = 1, H_{t-1} = -1, \end{cases} \tag{8.15}$$

where

$$s(t) = \max_j \{ j | H_j \neq H_t, H_{j-1} = H_t, 1 < j < t \}. \tag{8.16}$$

c_1 is the tax rate of each transaction, and c_2 the tax rate of securities exchange income. In the case of Taiwan, $c_1 = 0.001425$, and $c_2 = 0.003$.[5]

[5] This way to evaluate the accumulated profits is very standard and has been extensively used by practitioners. Also see [8.1].

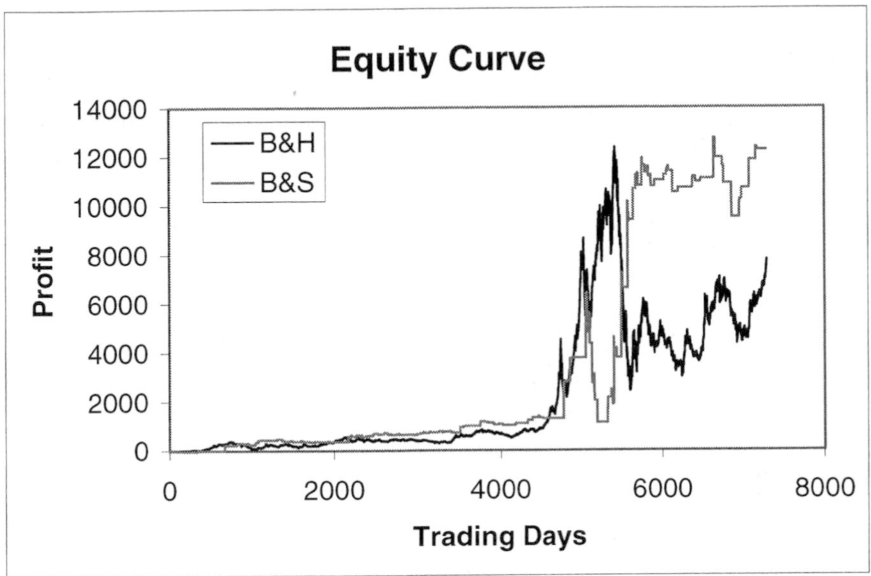

Fig. 8.6. The equity curves of the buy-and-hold strategy and the SOM-based trading strategy

Time series plot for π_t of the SOM-based trading strategy has been overlaid on that of the benchmark **B&H** strategy in Fig. 8.6. From these two plots, there are *roughly* two noticeable crossing points, one appears around the 5000th trading day, and the other appears near the 5500th day. These two crossing points suggest that neither trading strategy can dominate the other. However, the period that **B&H** outperforms **SOM** is much shorter than the other, and it occurs only between the first crossing point and the second one, which is about 500 days. For the rest of time, in particular, after the second crossing point, **SOM** dominates **B&H**. And SOM keeps the lead till end with the accumulated profits NT$ 12,258, which is 60% higher than **B&H** at that day.

8.6 Concluding Remarks

The ability to recognize patterns is an essential aspect of human intelligence. Herbert Simon, who won a Nobel Prize in economics in 1978, consider pattern recognition critical and advocates the need to pay much more explicit attention to teaching pattern recognition. Chartists appear to be good at doing pattern recognition for many decades, yet little academic research has been devoted to a systematic study of this kind of activities. [6]Lo, Mamaysky

[6] Apart from this paper, to our best knowledge, the only other attempt is [8.7].

and Wang (1999). On the contrary, sometimes it was treated as nothing more than the astrology, and hardly considered as a science. Using Self-organizing maps, this paper proposes a systematic and automatic approach to *charting*. To some extent, what SOM does is to *simulate* human intelligence in finding or creating patterns that summarize and store useful aspects of our perceptions.

In addition to using SOM simulate this cognitive process, our initial analysis also evidences why charting can be essential in technical analysis. Applying SOM to a 25-year time series data of Taiwan stock index, we have two interesting results. First, charts constructed by SOM can reveals profitable signals. Second, a simple trading strategy built upon charts can beat the buy-and-hold strategy. Moreover, the efforts to get these results is quite limited; we do not tried lots of variants of running SOM, which is certainly an interesting direction for the further study.

Acknowledgements

Research support from NSC grant No. 90-2415-H-004-018 is gratefully acknowledged. The chapter was refereed by Chueh-Yung Tsao. The authors are grateful to his corrections of some mistakes made in the earlier version of the paper.

References

8.1 Chen, S. -H., Chen, C. -F. (1998): Can We Believe that Genetic Algorithms Would Help without Actually Seeing Them Work in Financial Data Mining? Part 2, Empirical Tests. In: Xu, L., Chan, L. W., King, I., Fu, A. (Eds.), Intelligent Data Engineering and Learning: Perspectives on Financial Engineering and Data Mining, Springer
8.2 DeMark, T. R. (1994): The New Science of Technical Analysis. Wiley
8.3 Haykin, S. (1994): Neural Networks: A Comprehensive Foundation. MacMillan
8.4 Huang, Z. (1997): A Fast Clustering Algorithm to Cluster Very Large Categorical Data Sets in Data Mining. First Asia Pacific Conference on Knowledge Discovery and Data Mining. World Scientific
8.5 Kohonen, T. (1982): Self-organized Foundation of Topologically Correct Feature Maps. Biological Cybernetics, Vol. 43, 59–69
8.6 Kohonen, T. (1994): Self-Organizing Maps. Springer
8.7 Lo, A., Mamaysky H., Wang, J. (2000): Foundations of Technical Analysis: Computational Algorithms, Statistical Inference, and Empirical Implementation. Journal of Finance, Vol. 55, No. 4, 1705–1770
8.8 Pau, L. F. (1991): Technical Analysis for Portfolio Trading by Syntactic Pattern Recognition. Journal of Economic Dynamics and Control, 715–730

9. Effective Position of European Firms in the Face of Monetary Integration Using Kohonen's SOFM

Raquel Flórez López

Faculty of Economics and Business Administration, University of Leon, Campus de Vegazana s/n, 24071 Leon, Spain
email: dderfl@unileon.es

The Economic and Monetary Union (EMU) is the culmination of the European integration process from a financial perspective, whose main aim is the implantation of a one and only one currency for all the member states included in this integration project taking place in early 2002. To decide the relation of countries included in this phase it was established a group of economical regulations (macroeconomics rules) known as Convergence Criteria or Maastricht Criteria which must be fulfilled to guarantee the economic convergence among countries sharing the same currency. Nevertheless, these criteria are not enough to assure the effective convergence among states as far as enterprises is concerned, due to internal national differences in microeconomic structures, which affect competitiveness among firms and could distort a lot the effect of the union in favour of some countries and against others. The use of unsupervised artificial neural networks, specifically the employment of self-adaptive models based on Kohonen's proposal, makes easier to analyze the microeconomic differences by getting a visual image of a concrete dispersion of the differences through two-dimensional topological maps.

9.1 The Economic and Monetary Union Process

The Economic and Monetary Union (EMU) has been a goal to pursue by the European countries for a long time. Although the Rome Treat (1957) did not have a formal reference to this goal, it did anticipate the co-ordination of the economic policies of the member states via establishing a collaboration among the administration and Central Banks of European countries. In June 1989, the Delors Plan, approved in Madrid European Council, structured this process in three phases with two objectives:

1. the Economic Union: to facilitate the co-ordination of the economics policies among member states and the openness and the integration of domestic markets.
2. the Monetary Union: to come to the fixation of the irrevocable exchange rates among member states' currencies, and to establish of a common monetary policy to enhance the stability of prices.

In 1992, the Maastricht Treaty made a further close step to establish the EMU. It was decided to carry out the Delors Plan to facilitate the integration process by three stages.

1. Preparation stage (1-7-1990 to 1-12-1993), characterised by the total realisation of the interior market in 1993, the ratification of the EU Treaty by all countries and the increase of monetary funds to limit regional differences among nations and to free capital movement. In addition, it was established the Convergence Program, including five Convergence Criteria:
   a) Inflation Rate: Less than or equal to 1.5 points plus the average inflation rate of the three most stable nations.
   b) Exchange rate stability: within the fluctuation bands proposed by the European Monetary System, which was 6% for Spanish peseta, and 2.5% for all other currencies.
   c) National Deficit: Not greater than 3% of the Gross Domestic Product (GDP).
   d) National Debt: Not greater than 60% of the GDP.
   e) Interest Rate: Less than or equal to 2 points plus the median interest rate of the three countries with the lowest inflation rates.
2. Consolidation stage (1-1-1994 to 1-12-1998). It was a transition phase whose aim was to increase the convergence speed of economic and monetary policies among member countries and to make the Central Bank of each member countries be more independent from their own governments, which implied the heavy restriction to monetarise national deficits and the prohibition to provide privileged financing to the public sector.
3. Launch stage (1-1-1999 to 2002), whose aim was the fixation of the final exchange rates, the progressive implantation of the Euro as the Only European Currency and the definition of a common Monetary Policy for the European Central Bank, created in this phase together with the European System of Central Banks (ESCB), organism that assumes the responsibility of the common monetary policy.

While the two earlier phases were obligatory for all the member states, the access to the third stage was conditioned on the accomplishment of the Convergence Criteria. In addition, United Kingdom and Denmark were authorized to voluntarily exclude themselves of this third phase (which they did), and Sweden could decide whether to participate or not in the Launch stage (which she declined too). In the end, there were eleven countries entering into the third phase as shown by Table 9.1 (Greece was admitted in 2001).

Table 9.1. Fixed exchange rates for European currencies

Country	Exchange rate (per euro)
Germany	1.95583 DEM
France	6.55957 FRF
Italy	1936.27 ITL
Spain	166.386 ESP
Netherlands	2.20371 NLG
Belgium	40.3399 BEF
Portugal	200.482 PTE
Austria	13.7603 ATS
Finland	5.94573 FIM
Ireland	0.787564 IEP
Luxembourg	40.3399 LUF

9.2 ANNs and SOFMs

9.2.1 Artificial Neural Networks

Artificial Neural Nets (ANN) are models that appeared as a result of an attempt to build the mathematical forms of the human brain [9.7]. By imitating the behaviour of the nerve system of the human brain, these models are able to learn and to extract knowledge from experience. Basic elements in biological brains are those known as "neurons," which are grouped in a layer structure with its own functionality. An ANN presents a similar hierarchical structure, so it is defined as a collection of elemental processors or cells connected to others cells or to input data; these artificial neurons receive inputs or signals from different layers, process these signals, and produce either final outputs or signals which become inputs to other layers. As a whole, these layered structures are able to learn from mistakes in an inductive style, and are also able to generalize what they learn from specific example [9.8].

Based on the direction of information flow, there are three types of ANNs [9.10]. Firstly, in the *feedforward neural networks*, information flows in only one direction, namely, a forward direction from the input layer to hidden layers, and further to the output layer, with neither backward nor lateral connections in between. The desired input-output relation is usually determined by adjusting the connecting parameters (weights) among neurons based on the desired outputs given externally. Secondly, in *feedback neural networks*, information can flow in either direction between layers so the input information defines the initial state of the system and, after few runs based on a resonance mechanism, the system may reach a state asymptotically, which is taken as the output of the network. Finally, the *competitive, unsupervised* or *Self-organizing networks* are built based on competition among neighbouring neurons in the net by means of their mutual lateral adaptive interactions, which in the end develop neurons to specific detectors of different input patterns. The Self-organizing feature map (SOFM) used in this paper belongs to this category [9.16].

9.2.2 Self-organizing Maps

In human brain there are biological neurons organized into areas, so all data coming from the environment through sensory organs are represented internally by multi-dimensional maps, whose origin is thought to be predetermined genetically in a part, while the other part is affected by learning. Consequently, it is thought that human brain could have the inherent property of forming virtual topological maps of data coming from the environment [9.2]. With this inspiration, in 1982 professor Teuvo Kohonen presented ANNs which are able to demonstrate the self-formation of topological maps. These ANNs are knows as "Self-organizing maps" or "Self-organizing feature maps." [9.10] The SOFM algorithm allows us to construct multi-dimensional feature maps (usually two or three dimensions) by organizing the output layer, such as a plane or a hyper-plane, so that it can improve the geometric interpretation of the relation among patterns [9.11]. Fig. 9.1 shows the basic architecture of a Self-organizing feature map, model used in the empirical study of this chapter.

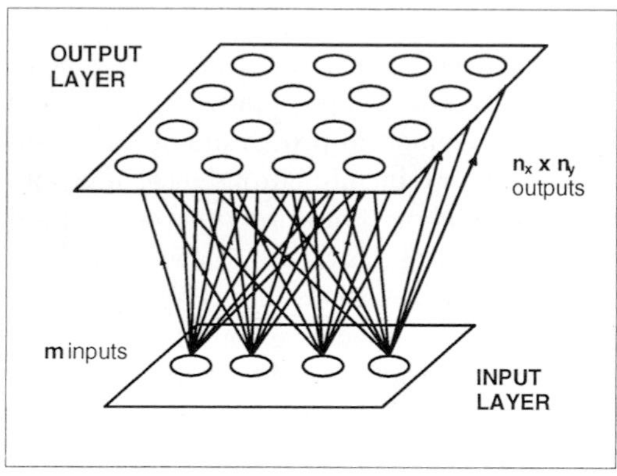

Fig. 9.1. The self organising feature map (SOFM)

The model is basically composed of two layers; the first one, or the input layer, is formed by m cells, one for each system input variable, whose task is to pick up the data from the environment; the second one, or the output layer, usually presents a rectangular feature and is mainly used for information processing. The construction of a SOFM model undergoes two stages.

1. *Learning or Training Stage*: In the learning stage the network establishes the categories which will be used in the operation stage to classify new data. The training process is characterised by the successive presentation of different patterns to the system. Each time when an individual $x(t)$ is

presented to the system, a "winner cell" $m^*_{ij}(t)$ (the nearest neuron to the pattern), denoted by c, is singled out, which will modify its weight vector together with other neurons situated in its neighbourhood (determined by a "neighbourhood radio" function $h(|\underline{ij} - c|, t))$ in the following way: $\forall k$,

$$m_{ijk}(t+1) = m_{ijk}(t) + \alpha(t) \cdot h(|\underline{ij} - c|, t) \cdot (x_k(t) - w_{ijk}(t)), \quad (9.1)$$

where $0 < \alpha < 1$, and k is the index for both input variables and weight vector components ($1 \leq k \leq m$). To avoid the case that the net get slanted, patterns are usually presented in a random manner, and we call it an *epoch* when all patterns have been presented to the network. The training process finishes after a pre-fixed number of epochs (e.g., 100,000 epochs).
In addition, it is possible to perform a second training stage, known as "fine tuning" to get a better fit of the weight vectors. In this stage the learning rate employed shall be very small (e.g., near 0.01) and the neighbourhood radio is set to 1 for the winning neuron, and 0 for others.
2. *Operation or Working Stage*: After the learning stage the working stage begins, where the weight vectors remain fixed. The operation phase is relatively simple. It generates one output from each input vector. This output is obtained through the parallel calculus for each neuron in the output layer the distance between input vector and the weight vector. The neuron with the maximum similarity is established as the winner, and the input vector is assigned (classified) to this cell. In that way, each cell acts like a specific feature detector, where the feature or the pattern of the winning cell represent that of the corresponding input vector.

In this chapter the SOFM will be used to perform a topological map of the relative competitive advantages of firms of countries in the EMU.

9.3 Empirical Analysis

9.3.1 The BACH Database

The database used in the study was obtained from the "BACH Project Database" (Bank for the Accounts of Companies Harmonized) [9.5], which collects in a harmonised way the accounting states of non-financial companies in thirteen countries: Germany, Finland, France, Italy, Spain, Belgium, the Netherlands, Portugal, Austria, Denmark, Sweden, the United Stated and Japan, broken down by major activity sector and by size, for a period different for each country, while with a consistent coverage of years from 1993 to 1998. Not all countries provide complete information on the items requested by the BACH project, in particular, the data on the Balance-sheet and Income and Loss State, which makes some specific analysis difficult to proceed.

The BACH database provides information on six different industrial groups (energy and water, manufacture, building and civil engineering, trade, transport and communication and other services) and three sizes of enterprises (small, medium and large). It is possible to get access to variable samples (data from different years with distinct samples' composition per year) and two-years sliding balanced samples (covering the period under review in overlapping sections to reduce the sample composition bias).

In the current analysis, samples covering financial data were selected from 1993 to 1998 (the period of the Consolidation Stage plus a year before the stage) for nine countries: Germany, France, Italy, Spain, Belgium, Netherlands, Portugal, Austria and Finland; the BACH database does not include data from the other three countries in the EMU (United Kingdom, Ireland and Greece), so it was not possible to analyze them. The study only focuses on the manufacturing sector. The initial data were converted into ratios to net out the effect of firms' sizes.

To evaluate the competitive position of European enterprises, the study is concentrated in the economic analysis based on cost and profitability measures. Although it would be also interesting to analyse the financial and patrimonial position of these enterprises too, we shall leave it to the next stage of this study. Since our analysis is much closer to the study of enterprises' efficiency or competitiveness, it would be useful to remind the reader a definition given by [9.4], which suggested that the measure of efficiency or competitiveness of an economic subject is the gap between its real situation and the best it can possibly achieve to meet the same aim.

9.3.2 Economic Analysis (I): Cost Analysis

The objective of the economic analysis is to study the profitability of firms (capacity to generate profits), and the information required for this analysis basically comes from the "Income and Loss State" as well as the "Balance-sheet" in the BACH database. The analysis was further divided into two parts:

1. analysis of the cost structure, and
2. analysis of the effective profitability.

Regarding the cost structure, an important variable on which many other ratios depend is the Adjusted Added Value (AAV) [9.1], which was calculated from the BACH database as:

Net turnover[1]
± Charge in stocks, finished goods and work in programme

[1] Net turnover = Net sales (VAT excluded)

+ Capitalised production
+ Operating income
- Cost of materials and consumables
- Other operating charges and taxes
= ADJUSTED ADDED VALUE (AAV)

It can be observed that the AAV represents the value generated for the enterprise from its productive operations. If the AAV is high, it means that the firm is able to generate an important amount of resources from its productive operations, that is to say, the technical cycle of exploitation behaves in a positive way (productive incomes are much higher that productive expenses). Under this circumstance, the enterprise stands in a good position to face competitive challenge (even when the way of financing its assets may influence its final capability).

Fig. 9.2 shows the AAV of the European firms, as percentage of net turnover, obtained from 1993 to 1998.

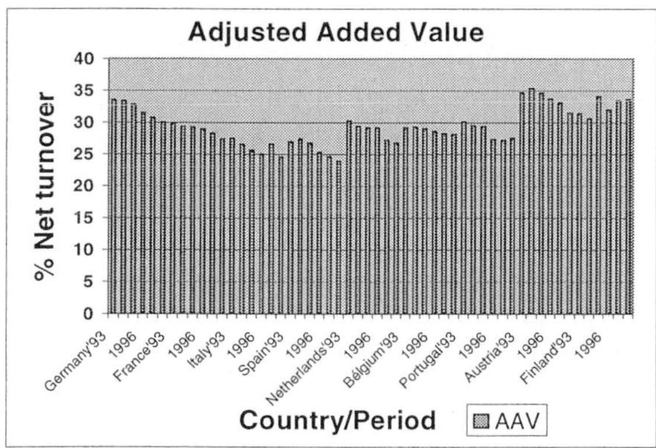

Fig. 9.2. Adjusted added value for European firms

It can be observed that most countries have rates situated in the interval 25%-35%, which could be considered as a medium-high position. In addition, all countries with the exception of Finland experienced a downward trend in the AAV. The AAV roughly dropped down to 3%. Emphasis must be placed on the case of Spain, which presents the lowest rates in this period, next to that of Italy. The situation is even worse than the period 1986-1991. In fact, the AAV in Spain came to its peak in the period from 1986 to 1991. At that moment, it was even higher than the AAV in Germany, France and Italy. In addition to Spain, other countries which are worth noticing are Finland, Austria and Germany, which have the highest AAV over the period 1993-1998.

To measure the efficiency by which enterprises employ their resources to get outputs, three variables are described as follows.

1. Staff costs/AAV (Labour Cost Rate): This variable, which could be interpreted as the labour unit cost, is one of the frequently cited indicators on measuring the competitiveness of firms.
2. Cost of materials and consumables/Net turnover (Material Cost Rate): This magnitude represents the percentage of turnover employed in tangible consumption needed to develop the production cycle.
3. Interest and similar charges/AAV (Financial Cost Rate): This ratio informs about the portion of AAV used to reward external creditors of the enterprise. In that way, the financial policy of the firm will determine in a very important way the evolution of this indicator.

From the BACH database, it was found that the third ratio (the financial ratio) could not be calculated for the Netherlands (period 1993-1994) and Austria (period 1993-1998), so it was not used in Kohonen's map and was evaluated separately. This indicator showed that the financial cost decreased in all countries and slipped to an interval of 7-9% for most of countries considered in this analysis. The highest decreases were evidenced by Finland, Spain and Portugal, whose financial ratio dropped from 20% to 5-10%.

Finally, Kohonen's map was derived by taking only the first two cost ratios and is shown in Fig. 9.3.

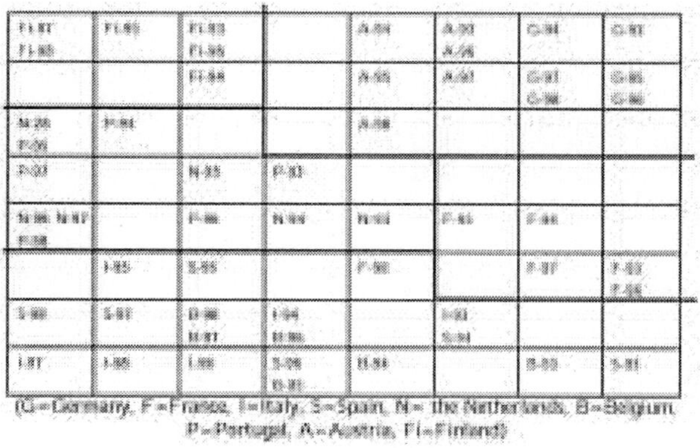

(G=Germany, F=France, I=Italy, S=Spain, N= the Netherlands, B=Belgium, P=Portugal, A=Austria, FI=Finland)

Fig. 9.3. Relative situation on European firms considering cost data

The analysis of weights of Kohonen's map makes us see some discernible trends with five separated topological regions.

1. Finland constitutes an isolate region, characterised by low labour cost rates (near 55-60%, see Fig. 9.4) and low material cost rates (about 55-60%, Fig. 9.5) which confers it a very good position for the Euro stage.
2. Germany and Austria form another separated region, characterised by very high labour cost rates (about 70-80%, Fig. 9.4) but quite low material cost rates (near 55%, Fig. 9.5). As a result, in spite of their good cost of materials position, these countries could have some difficulties to compete with nations with lower labour costs.

3. France forms the third separated region, with medium labour cost rates (65-70%, Fig. 9.4), which were decreasing over the sample period, and medium-high material ratio (above 70%, Fig. 9.5). Hence, this country is ranked as "medium."
4. Portugal and the Netherlands constitute the fourth region, with low labour cost rates (about 55-65%, Fig. 9.4) but high material cost rates (about 70-75%, Fig. 9.5). Since these countries are near to the region of France, they can be a good match as far as these two cost rates are concerned.
5. Spain, Italy and Belgium constitute the last region, characterised by a general downward movement in the labour cost rates from 70-80% to 60% (Fig. 9.4), and a general upward movement in the material cost rates from near 70% to 75% and higher (Fig. 9.5). These countries present the lowest labour costs, which can be its main competitive advantage. However, this advantage may be offset by its high material costs.

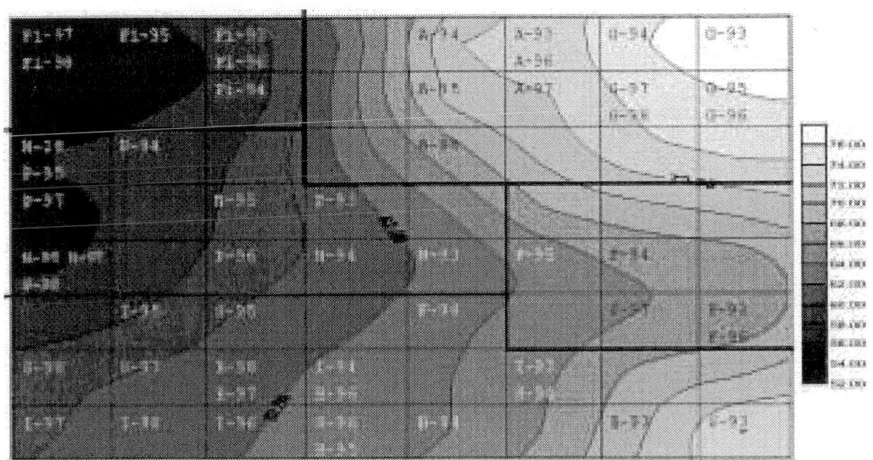

(G=Germany, F=France, I=Italy, S=Spain, N= the Netherlands, B=Belgium, P=Portugal, A=Austria, FI=Finland)

Fig. 9.4. Weights' map for ratio "Staff costs / AAV" (in percentage); first component of weights' vector for cost analysis Kohonen's map (Fig. 9.3)

In sum, these five regions shows sharp differences in cost ratios among European countries. The contrast is particular significant between countries on the north and countries on the south.[2]

[2] Belgium is an exception whose topological features of competitiveness are closer to the south. Portugal is another exception, whose features are similar to the Netherlands, a country on the north.

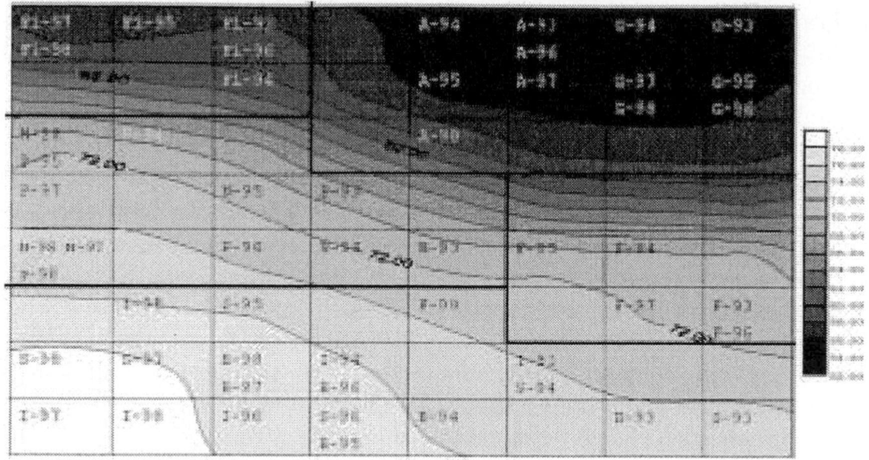

(G=Germany, F=France, I=Italy, S=Spain, N= the Netherlands, B=Belgium, P=Portugal, A=Austria, Fi=Finland)

Fig. 9.5. Weights' map for ratio "cost of material and consumables / net turnover" (in percentage); second component of weights' vector for cost analysis Kohonen's map (Fig. 9.3)

9.3.3 Economic Analysis (II): Profitability Analysis

The profitability, defined as the firm's capacity to generate profits, indicates firms' competence to face the increasing competitive pressure. The main variable used in this analysis is the *economic profitability*, calculated as "(Profit before financial taxes + interest expenses) / Net turnover." This item, which shows the efficiency of the business in the use of its resources, can be divided in two components, namely, the *margin effect* and the *rotation effect.* Four variables described below are used to measure economic profitability from different perspectives.[3]

1. Economic profitability 1 (Margin Effect): It is defined as "(Profit before financial taxes + interest expenses) / Net turnover." A high ratio implies a high profit per each unit of product sold.
2. Economic profitability 2 (Rotation Effect): It is "Net turnover/Net Assets." This measure indicates the frequencies of sale transaction per unit of investment.
3. Financial profitability: It is "Profit before taxes/Own resources (data available from "capital and reserves" in the BACH database)." By taking

[3] It should be noticed that some measures are not available for some countries in specific years, such as Netherlands (1993-1994), Finland (1993) and Austria (1993-1998). These countries with those specific years will not be included into the Kohonen's SOFM.

into account the liabilities of firms, this ratio shows the profit contributed by the proprietors of enterprises.

4. Cost of payable liabilities: It is the cost of the non-owned liabilities. The cost depends both on the interest rate and the financial policies of the enterprise. Since the interest rate is taken as one of the Convergence Criteria in this period of analysis, the cost mainly reflects financial policies of firms.

In addition, there is another important measure, called the "Financial Leverage" (FL), which is the difference between Financial profitability and Economic profitability 1:

Financial Leverage = Financial profitability - Economic profitability 1

Fig. 9.6 shows the FL of the countries studied in this chapter. From Fig. 9.6, one may see some differences among these countries. Germany, Finland, the Netherlands and France may be categorized as "aggressive ones," whereas Belgium, Spain, Italy and Portugal are less aggressive. In 1993, the last three countries even took negative leverage. After 1993, partially thanks to the decreasing interest rate, the FL of all countries is positive. Portugal and Spain seems to be the most two conservative countries in this period. The FL of Portugal in 1998 is low to 4,17%, and, for Spain, it is 7.29%.

By taking the four measures of profitability as the input variables, the Kohonen's map is drawn in Fig. 9.7.

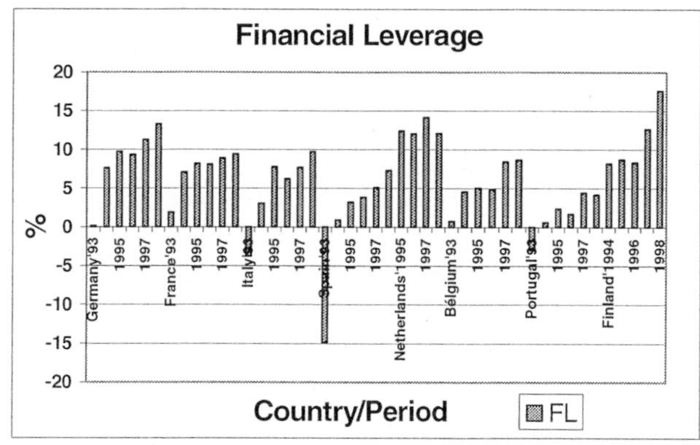

Fig. 9.6. Financial leverage for European firms

Unlike Fig. 9.3, the second SOFM does not have a noticeable boundaries among different countries. Roughly, six regions stand out.

1. Finland constitutes again a specific region, characterized by the highest "margin effect" (higher than 10%, see Fig. 9.8) but a relative low "rotation effect" (about 0.7, Fig. 9.9). The economic profitability obtained can be categorized as medium or medium-high. Its financial profitability, however, is the highest (greater than 20%, Fig. 9.10), and its financial

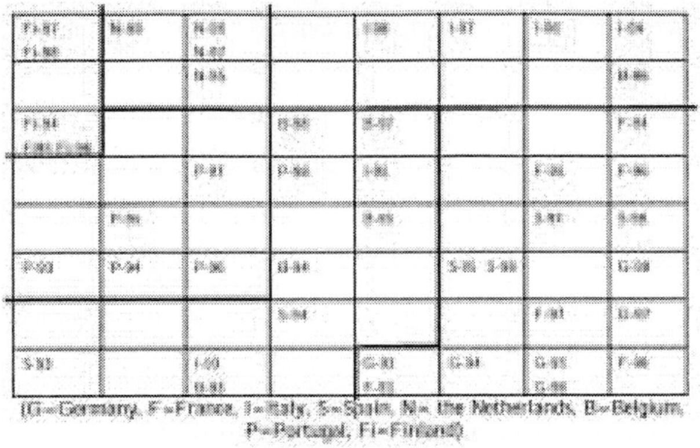

(G=Germany, F=France, I=Italy, S=Spain, N= the Netherlands, B=Belgium, P=Portugal, FI=Finland)

Fig. 9.7. Relative situation on European firms considering profitability data

cost is medium-low (between 4.2% to 5,7%, Fig. 9.11). So, overall, its profitability situation performs very well over the entire period.

2. Right to Finland comes The Netherlands. The main differences are a lower "margin effect" (7-8,5%, Fig. 9.8) and a higher "rotation effect" (approximately 0.75 times, Fig. 9.9). As a result, its general economic profitability may be lower that Finland's. Moreover, its financial profitability is lower too (about 14%, Fig. 9.10), but its financial cost ratio is also lower (near to 3.5%, Fig. 9.11). Hence, its profitability performance is steadily sound over time.
3. Even though its 1993 is a little far away from the rest of the years, Portugal defines the third region, characterised by a medium-high "margin effect" (5-7%, Fig. 9.8) and "rotation effect" (near 1, Fig. 9.9). Despite the different weights on the two effects, its economic profitability is quite similar to Finland. The financial profitability is a little low (7-10%, Fig. 9.10), and its financial cost ratio is the highest of all countries, although it decreased for 12% to 6% (Fig. 9.10). So, behind the generally good profitability performance, what should be taken care for Portugal is its financial policies toward reducing the financial costs.
4. The fourth region is mainly taken by Italy.[4] This region is characterised by a medium low, but progressively increasing "margin effect" (Fig. 9.8) with a medium-high "rotation effect" (a little above than 1, Fig. 9.9). Its financial profitability increased from 10,47% in 1996 to 14,58% in 1998 (Fig. 9.10). Also, its financial costs decreased to 3,46% in 1996 (Fig. 9.11), which also is a plus side. As a result, Italy seems to make a good progress when the Euro stage is approaching.
5. France, Germany and the last year of Spain give rise to the fifth region, characterized by a quite low "margin effect" (2-4%, Fig. 9.8) but the highest rotation effect (1.2-1.3, Fig. 9.9). Together, its economic profitability is relatively lower than that of other countries. Nonetheless, its financial

[4] Belgium (1996) shares a small part with it.

profitability is medium-high (9-17%, Fig. 9.10), and was increasing in the last few years. The financial cost is also low in this region (3.5-5%, Fig. 9.11), while it was increasing in France and decreasing in Germany and Spain. (with medium values in all ratios)

6. This last region is a mix of different countries and years, including the last few years of Belgium and some early years of Italy and Spain. All measures for Belgium in this region is about medium. For Italy and Spain, they are all low except with a medium "rotation effect" and high financial costs (Figs. 9.8-9.11).[5]

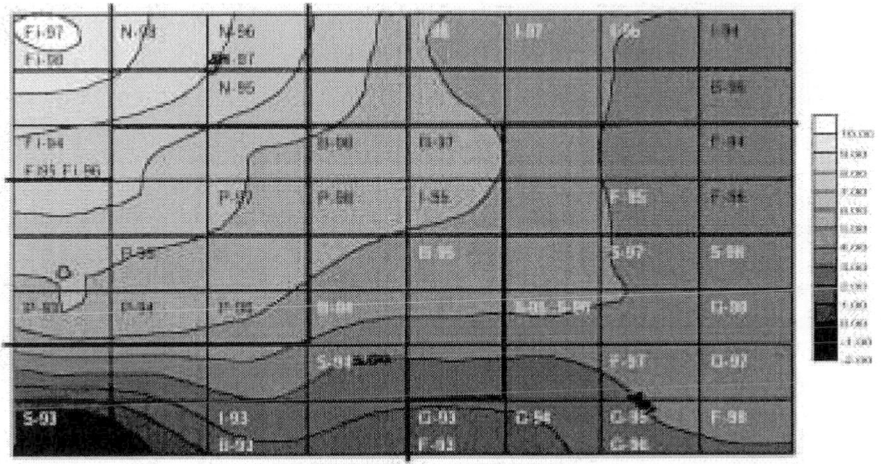

(G=Germany, F=France, I=Italy, S=Spain, N= the Netherlands, B=Belgium, P=Portugal, FI=Finland)

Fig. 9.8. Weights' map for ratio "(profit before financial taxes + interest expenses) / net turnover" (in percentage); first component of weights' vector for profitability analysis Kohonen's map (Fig. 9.7)

9.4 Conclusions

With the help of SOFM, the analysis based on the data from the BACH database in the period 1993-1998 enables us to have maps of competitive advantages for some Euro member countries. Below is a brief summary of their positions.

[5] Special care must be taken for the case of Spain. The year 1993 was a bad year year for Spain. Its margin effect and financial profitability was even negative, and the financial cost is high up to 9%. Nonetheless, since then Spain had made a substantial progress by reducing the financial cost. In 1998, this cost ratio is only 3.28%.

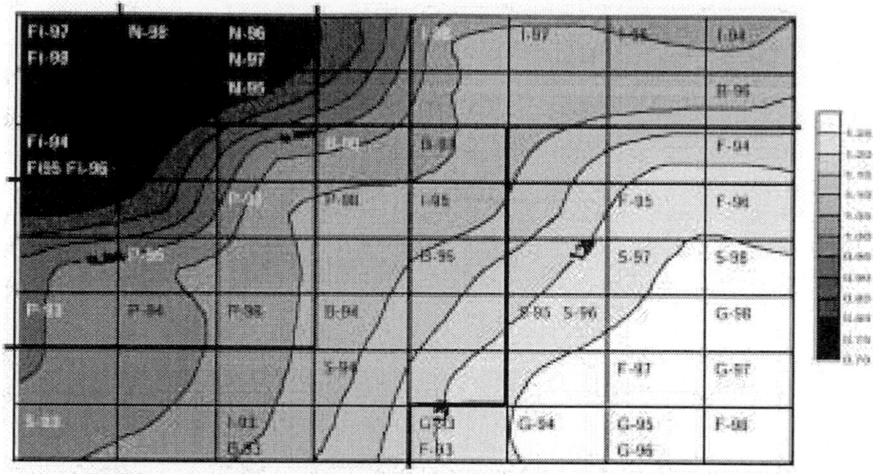

Fig. 9.9. Weights' map for ratio "net turnover / net assets;" second component of weights' vector for profitability analysis Kohonen's map (Fig. 9.7)

1. It has been shown that enterprises in European countries did not converge to a similar state as far as their competitiveness is concerned.
2. Finland had demonstrated to most competitive country and was well-prepared for the age of Euro.
3. Germany's competitiveness was also very strong. Its profitability performance was particular striking. However, the high labor cost would be its potential concern.
4. The profitability data of Austria is not available from the BACH database. But, its cost advantage is similar to Germany.
5. France was also situated in a good position, but the increasing financial cost should be watched. Also, the material cost should be kept under control.
6. The Netherlands presented a similar cost advantage to that of France and Portugal: the material cost could be its potential concern. Its profitability performance, however, was just a little worse of Finland's.
7. Portugal behaved quite differently from its geographical neighbors: Spain, France and Italy. It was more akin to the Netherlands, especially its cost advantages. Its main concerns came from the financial cost rates.
8. Spanish competitive advantage was improving, in particular, its profitability performance. However, it suffered from its high material cost.
9. Italy's profitability performance was good, but its general competitive advantage was adversely affected by a medium-low financial leverage and cost disadvantages.

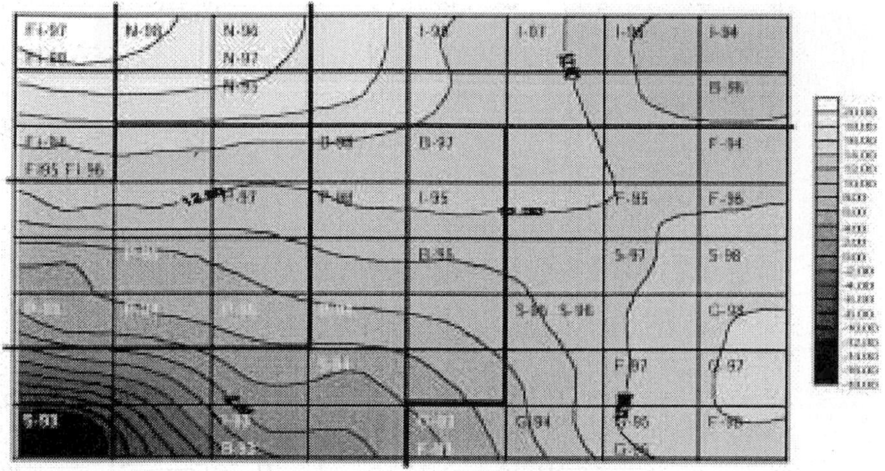

(G=Germany, F=France, I=Italy, S=Spain, N= the Netherlands, B=Belgium, P=Portugal, Fi=Finland)

Fig. 9.10. Weights' map for ratio "profit before taxes / own resources" (in percentage); third component of weights' vector for profitability analysis Kohonen's map (Fig. 9.7)

10. Belgium's profitability performance was in the middle. Its cost advantages was similar to Spain's, but its situation was deteriorating.

To wrap it up, even though the situation of European enterprises must have changed in these recent three years, our analysis based on the data from 1993 to 1998 may be still quite useful to indicate the potential challenges of the age of Euro, at which these enterprises are now situating. The analysis also suggests the direction to move for each country's enterprises, if they want to survive the age of Euro.

Acknowledgements

The paper was originally written in the Word format. The author is grateful to Tzu-Wen Kuo for her kind support to transform it into the Latex format. The English of this paper has been significantly rewritten by the volume editor, Shu-Heng Chen. His painstaking efforts to make this paper readable is greatly acknowledged. The paper was refereed by Chueh-Yung Tsao and Chia-Ling Chang, and the author also thanks for their valuable suggestions.

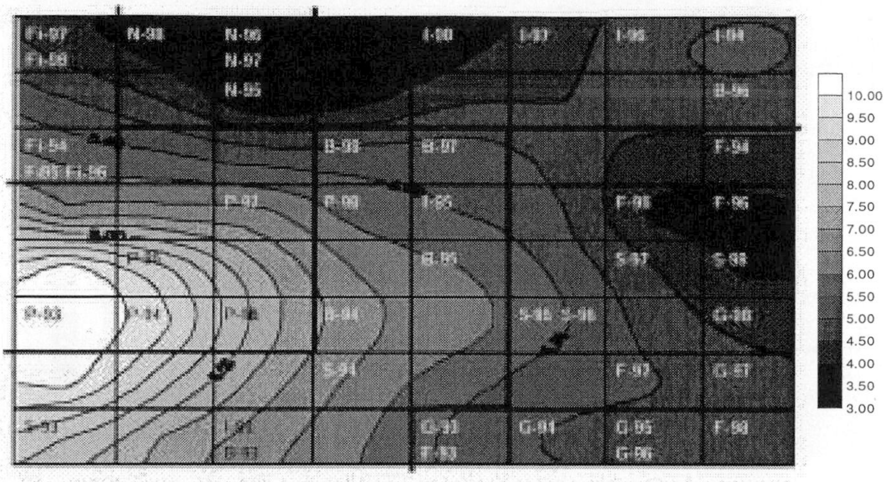

Fig. 9.11. Weights' map for ratio "cost of payable liabilities" (in percentage); fourth component of weights' vector for profitability analysis Kohonen's map (Fig. 9.7)

References

9.1 Amat, O. (1994): Analisis de estados financieros. Fundamentos y aplicaciones (Spanish). Gestion 2000
9.2 Azoff, E. M. (1994): Neural network. Time series forecasting of financial markets. Wiley Internacional
9.3 Barea, J. (1997): La Union Europea, una zona de estabilidad (Spanish). Boletin AECA Especial IX Congreso AECA, 6–7
9.4 Bueno, E. (1989): La competitividad de la empresa espanola (Spanish). AECA
9.5 European Commision (1991): BACH (Bank for the Accounts of Companies Harmonized). DOC II/383/91-EN, DG II
9.6 Cuervo, A. (1993): El papel de la empresa en la competitividad (Spanish). Papeles de Economia Espanola 56, 363–378
9.7 Hilera, J. R., Martinez, V. J. (1995): Redes neuronales artificiales: fundamentos, modelos y aplicaciones (Spanish). Ra-Ma
9.8 Hrycej, T. (1992): Modular learning in Neural Networks. Wiley Internacional
9.9 Jordan, J. (1994): Economia de la Union Europea (Spanish). Biblioteca Civitas de Economia y Empresa, Coleccion Economia, Civitas
9.10 Kohonen, T. (1990): The Self Organizing Map. Proceedings of the IEEE 78:9, september, 1464–1480
9.11 Kohonen, T. (1997): Self-Organizing Maps. Springer Verlag
9.12 Maroto, J. A. (1993): La financiacion empresarial y el sistema financiero (Spanish). Economia Industrial 293, 77–88
9.13 Prado, J. M. (1997): El euro, una nueva cultura para la Vieja Europa (Spanish). Boletin AECA Especial IX Congreso AECA, 8–9

9.14 Rivera, O, Olarte, F. J., Navarro, M. (1993): La situacion economico-financiera de la empresa espanola frente a la comunitaria (Spanish). Economia Industrial 293, 59–75
9.15 Velarde, J. (1997): Repercusiones del euro en el mundo empresarial espanol (Spanish). Boletin AECA Especial IX Congreso AECA, 10–11
9.16 Wasserman, P. D. (1993): Advanced methods in neural networks. Van Nostrand Reinhold

10. Financial Applications of Wavelets and Self-organizing Maps

Dimitrios Moshou and Herman Ramon

Katholieke Universiteit Leuven, Leuven B-3001, Belgium
email: {dimitrios.moshou, herman.ramon}@agr.kuleuven.ac.be

A methodology on how to combine wavelets with Self-organizing Maps (SOM) for financial time-series visualisation and interpretation is presented. For the denoising of the stock time-series wavelet packets are used because of their optimal signal compression and denoising capabilities. The visualisation of transient shocks like crashes, in higher order wavelet coefficients is presented. The Self-organising Map Neural Network is introduced to aid the visualisation of the behaviour of the daily closing value of S&P 500 and the daily closing value of two example stocks. The features that are used for the visualisation are based on the wavelet coefficients of 32-day trading periods with daily sampling of the closing value. The trajectory formed on the U-matrix of SOM shows the evolution of the individual stock and indicator data and aids the detection of abrupt changes in the behaviour of the time-series.

10.1 Introduction

Existing approaches to analysing financial data include the important concept of volatility. The volatility is described as a measure of the stock's tendency to move up or down based on daily price history over a fixed period of time. Volatility uses the quantitative analysis of price by concentrating on pure increments of price change. However, pure volatility has little value if it can not help with price prediction. An important characteristic of volatility is that volatility moves cyclically between active and inactive states. Many techniques now focus on the pattern of wide and narrow range price bars. A technique that uses such a concept is that of Bollinger Bands. These lines represent two standard on-going deviation occurrences away from a simple moving average. The volatility can be measured as an adjusting percentage away from a defined moving average. Bollinger Bands can widen during a volatile training period, and contract during a less volatile trading period. What the Bollinger Bands provide is information as to where the stock price stands with respect to its moving average. However the concept of Bollinger bands uses a simple moving average. More advanced denoising methods by using the discrete wavelet transform have several advantages over conventional moving average filtering of the noise. Most market analyses sample indicators at a low resolution (e.g. monthly), or use a 30-, 60-, or 90-day moving average in order to remove the noise. However, high frequency financial data are irregularly spaced. Methods, which address the irregularity of ultra-high frequency financial data are needed and wavelet analysis is one possible tool. By design the wavelet's usefulness is its ability to localize data in time-scale space. If the wavelet coefficients do go to zero then the series is smooth at a certain point. The smoothness of the wavelet coefficients at a certain sampling point can be associated with the presence of crashes. Given the recording errors that occur during short, intense trading periods, and transient shocks that are caused by news reports, wavelet-based denoising is important in processing financial data. The wavelet transform can reveal hidden short-term market patterns or clarify long-term patterns. Also, wavelet coefficients are used in combination with Self-organizing maps to visualise abnormal behaviour in the stock transaction profile. Such a visualisation tool is the formation of trajectories of the wavelet coefficients on the Self-organizing Map.

For the algorithm implementation in the presented work the software Matlab™ provided from Mathworks (Natick, Massachussets) has been used.

10.2 Wavelets

Wavelets are building blocks for general functions. That means that any general function can be expressed as an infinite series of wavelets. The basic idea underlying wavelet analysis consists of expressing a signal as a linear combination of a particular set of functions (wavelet transform, WT), obtained by shifting and dilating one single function called a mother wavelet. Several different mother wavelets have been studied in the literature [10.2] and [10.8]. The decomposition of the signal into the basis of wavelet functions implies the computation of the inner products between the signal and the basis functions, leading to a set of coefficients called wavelet coefficients. The signal can consequently be reconstructed as a linear combination of the basis functions weighted by the wavelet coefficients. In order to obtain an accurate reconstruction of the signal, a sufficient number of coefficients has to be computed. The procedure followed for calculating the wavelet coefficients is shown graphically in Fig. 10.1.

The main characteristic of wavelets is the time-frequency localisation. In effect that means that most of the energy of the wavelet is restricted to a finite time interval. Frequency localisation means that the Fourier transform is bandlimited. The advantage of time-frequency localisation is that contrary to the short-time Fourier transforms, a wavelet analysis varies the time-frequency aspect ratio, producing good frequency localisation at low frequencies (long time windows), and good time localisation at high frequencies (short time windows). This produces segmentation, or tiling of the time-frequency plane that is appropriate for most physical signals, especially those of a transient nature. The difference between the Short Time Fourier Transform and the Wavelet transform is illustrated in Fig. 10.2.

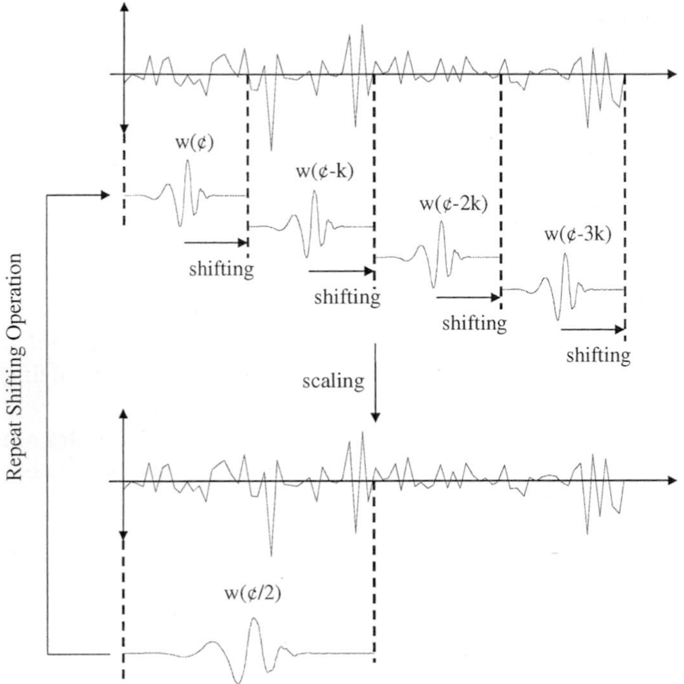

Fig. 10.1. The operations of shifting and dilation of the mother wavelet to calculate the wavelet coefficients

10.3 Wavelet Packets

The DWT is a subset of a far more versatile transform, the Wavelet Packet Transform (WPT). The WPT has been developed by professor Coifman of Yale University [10.1]. The wavelet packet transform generalizes the time-frequency analysis of the DWT. It yields a family of orthonormal transform bases of which the wavelet transform basis is but one member. The set of detail and approximation coefficients at each level of the transform form a pair of subspaces of the approximation coefficients of the next higher level of scale and, ultimately, the original data set. The subspaces created by the wavelet transform roughly correspond to the frequency subbands partitioning the frequency bandwidth of the data set. The subspaces have no elements in common and the union of the frequency subbands span the frequency span of the original data set. Any set of subspaces which are a disjoint cover of the original data set is an orthonormal basis. The wavelet transform basis is then but one of a family of orthonormal bases with different subband intervals. As with the wavelet transform basis, each disjoint cover roughly corresponds to a covering of the frequency space of the original signal. This family of transforms is called a wavelet packet library. The variety of orthonormal bases which can be formed by the Wavelet Packet Transform algorithm, coupled

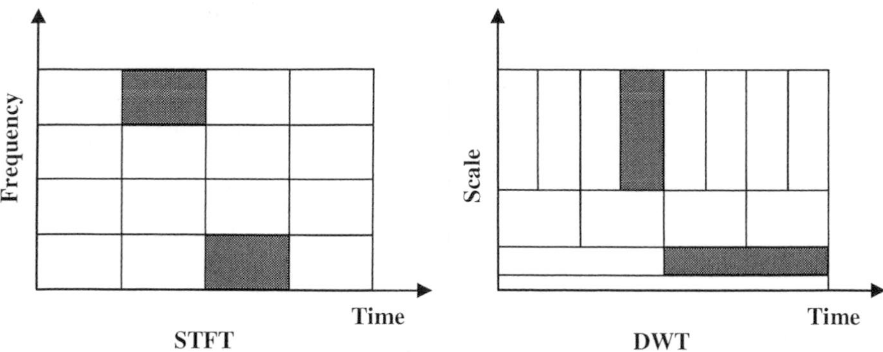

Fig. 10.2. The tiling of the time-frequency plane in the case of Short-Time Fourier Transform (STFT) and Discrete Wavelet Transform (DWT) respectively

with the infinite number of wavelet and scaling functions which can be created, yields a very flexible analysis tool. The WPT allows tailoring of the wavelet analysis to selectively localize spectral bands in the input data as well as correlate the signal to the wavelet. Not only can the best wavelet be chosen to analyze a particular signal but the best orthonormal basis as well. In signal processing terminology, the various bases of the WPT can be used as arbitrary adaptive tree-structured filter banks. A single wavelet packet decomposition gives a lot of bases from which you can look for the best representation with respect to a design objective. This can be done by finding the best tree based on an entropy criterion.

10.4 Wavelet-Based Denoising

Wavelets are a powerful tool for signal processing. Wavelets are used for the processing of signals that are non-stationary and are time varying in nature. The noisy nature of financial indicator data is a serious impeding factor that prohibits further processing in order to identify useful features of the signal. The wavelets can tackle the denoising problem optimally from the point of view that the wavelet-based denoising attempts to remove whatever noise is present and retain whatever signal is present regardless to the frequency content of the signal. This is a far more efficient way of handling noisy signals as opposed to filtering or smoothing where high frequencies are removed and low frequencies are retained. An illustrating example is that of music corrupted by noise where we would like to preserve both the treble and the bass.

The wavelet-based denoising of a signal, called (s), follows three steps:

- Choose a wavelet and level N. Compute the wavelet decomposition of the signal (s) at level N.

- For each level select a threshold and apply thresholding to the detail coefficients.
- Compute the wavelet reconstruction based on the original approximation coefficients of level N and the already thresholded detail coefficients of levels 1 to N.

There are two kinds of thresholding, hard and soft. Both set coefficients less than the threshold to zero. However soft thresholding shrinks nonzero coefficients while hard thresholding leaves them intact.

The wavelet transform maps white noise in the signal domain to white noise in the transform domain. In effect, that means that while signal energy becomes more concentrated into fewer coefficients in the transform domain, noise energy does not. It is this important principle that enables the separation of signal from noise.

Wavelet-based denoising has been theoretically proven to be nearly optimal from the following perspectives: spatial adaptation, estimation when local smoothness is unknown, and estimation when global smoothness is unknown. There is no alternative procedure that can perform better without knowing a priori the smoothness class of the signal.

Denoising and compression are interesting applications of wavelet packet analysis. The wavelet packet de-noising or compression procedure involves four steps:

- Decomposition: For a given wavelet, compute the wavelet packet decomposition of signal x at level N.
- Computation of the best tree: For a given entropy, compute the optimal wavelet packet tree.
- Thresholding of wavelet packet coefficients: For each packet (except for the approximation), select a threshold and apply thresholding to coefficients.
- Reconstruction: Compute wavelet packet reconstruction based on the original approximation coefficients at level N and the modified coefficients.

10.5 Self-organizing Map

The Self-organizing Map (SOM) [10.4] is a neural network (NN) that maps signals (x) from a high-dimensional space to a one- or two-dimensional discrete lattice of neuron units (s). Each neuron stores a weight (ws). The map preserves topological relationships between inputs in a way that neighbouring inputs in the input space are mapped to neighbouring neurons in the map space. A graphical representation illustrating the functioning of the SOM is shown in Fig. 10.3.

After training, each node acts as a representation of a particular class of the data vectors that were input into the network. Nodes which are adjacent to one another represent similar, but not identical vector classes.

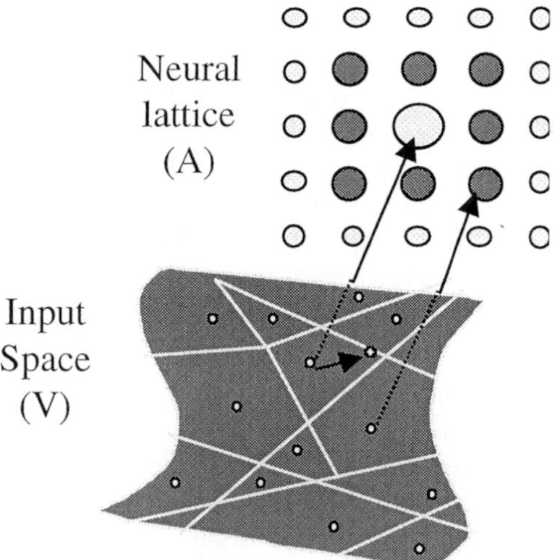

Fig. 10.3. Graphical representation of the SOM

The SOM algorithm is based on unsupervised, competitive learning. It provides a topology preserving mapping from the high dimensional space a two-dimensional grid. The main features of SOM are:

- SOM is a non-parametric method;
- No a priori assumptions are required about the distribution of the data; and
- SOM is a method that can detect unexpected structures or patterns by learning without supervision.

The following is a list of financial applications of Self-organizing maps. Several of these applications are documented in the literature [10.6], [10.7], [10.10], [10.11].

- initial financial analysis
- analysis of financial statements
- financial forecasting
- prediction of failures
- rating of financial instruments
- analysis of investment opportunities
- selection of investment managers
- commercial credit risk analysis
- country credit risk analysis
- construction of benchmarks
- strategic portfolio allocations
- monitoring of financial performance

10.6 Wavelet-Based Denoising of Financial Data

High frequency financial data are irregularly spaced. Methods, which address the irregularity of ultra-high frequency financial data are needed and wavelet analysis is one possible tool. By design the wavelet's usefulness is its ability to localize data in time-scale space. At low scales, the wavelet characterizes the data's coarse structure. If both the low and high-scaled wavelet coefficients are non-zero then something structural is occurring in the data. A wavelet coefficient that does not go to zero as the scale increases indicates a jump has occurred in the data. By downsampling the DJIA time-series the first level of detail can be reconstructed using 9420 coefficients of the first detail level as shown in Fig. 10.4.[1] If the wavelet coefficients do go to zero then the series is smooth at this point. The non-decay of the time series reconstruction based on the first detail level of wavelet coefficients (Haar analysed at 12 levels) in Fig. 10.4 of the Dow industrial clearly points out and identifies the stock market crashes of October 1987 and October 1929.

Given the recording errors that occur during short, intense trading periods, and transient shocks that are caused by news reports, wavelet packet based de-noising (Fig. 10.5) is important to financial data. The wavelet transform can reveal hidden short-term market patterns or clarify long-term patterns. The examples presented here indicate clearly that the real trend can be extracted from the noise. Given the high level of noise the real trend of the stock values is calculated with a much higher sensitivity based on the wavelet packet denoising than based on monthly or over -60 and -90 days moving averages (which are usually used to average out the noise). The mother wavelet that was used was Daubechies wavelet db10 and the decomposition consisted of 10 levels. The SURE (Stein Unbiased Risk Estimate) entropy criterion was used for selecting the best decomposition. There are two kinds of thresholding, hard and soft. Both set coefficients less than the threshold to zero. However soft thresholding shrinks nonzero coefficients while hard thresholding leaves them intact. A criterion that minimises the thresholding risk of cutting part of the signal during shrinkage denoising is the Stein Unbiased

[1] By using the pyramidal algorithm introduced by Mallat ([10.5]) averages and differences are calculated iteratively. The averages are used further for the calculation of the detail coefficients. The detail coefficients thus represent the variation of the average time signal on two consecutive time intervals. Starting from detail coefficients at level 1, which are 9420, we end up with 5 detail coefficients at level 12. Each step in the forward Haar transform calculates a set of wavelet coefficients and a set of averages. If a data set $s_0, s_1, ...s_{N-1}$ contains N elements, there will be $N/2$ averages and $N/2$ coefficient values. The averages are stored in the lower half of the N element array and the coefficients are stored in the upper half. The averages become the input for the next step in the wavelet calculation, where for iteration $i+1, N_i+1 = Ni/2$. The recursive iterations continue until a single average and a single coefficient are calculated. This replaces the original data set of N elements with an average, followed by a set of coefficients whose size is an increasing power of two up to $N/2$. In the case of DJIA we took N=18840 daily closing values hence $N/2$ is equal to 9420.

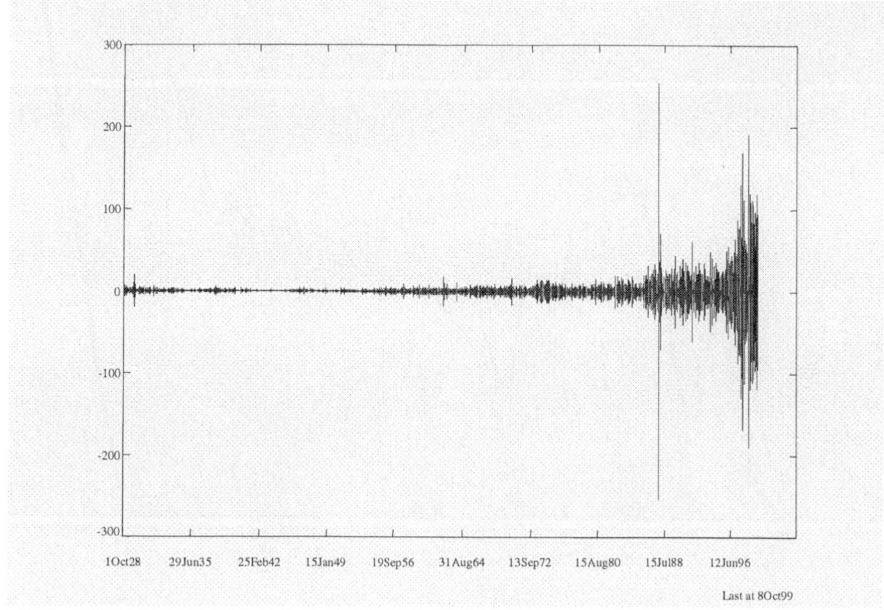

Fig. 10.4. Crash indication in the reconstruction based on the finest level detailed coefficients of Haar wavelet decomposed at 12 levels. The total amount of daily closing values is equal to 18840 between 1 October 1928 and 8 October 1999

Risk Estimate (SURE) criterion [10.3]. The SURE criterion has been used for thresholding the noisy wavelet coefficients.

10.7 SOM and Wavelet Combination for Abnormality Detection

Several SOMs have been trained with the wavelet-based features to investigate the hypothesis that certain statistical changes in the trading profile of a specific stock will be reflected in a certain level of wavelet coefficients. Subsequently the wavelet coefficients can be used to train a SOM. The features associated with segments of the trading data form a trajectory on the SOM since different periods will activate different SOM units.

Certain codebook vectors of the SOM have a much larger distance from the rest of the codebook vectors. This distortion of the SOM appears because certain of the input vectors have exhibited abnormally large values and have therefore created a codebook vector that can detect changes in the input signal when these occur. Such behaviour can be actually visualised through the utilisation of a U-matrix. The U-matrix [10.9] is a matrix of distances between neighboring map units. Distance matrices essentially show the density of prototype vectors in different parts of the map. Color-coding can be used

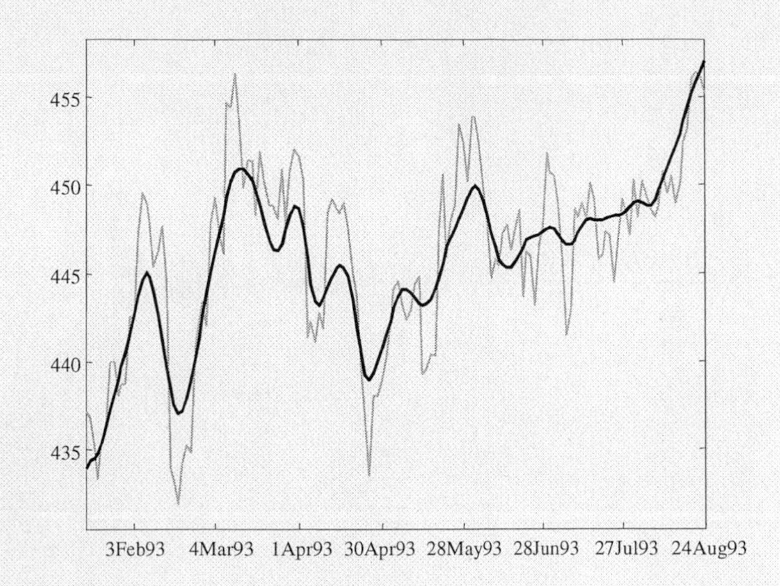

Fig. 10.5. Wavelet packet denoising of S&P 500 over 150 trading days. The smooth line represents the wavelet packet based reconstruction using denoising and soft thresholding

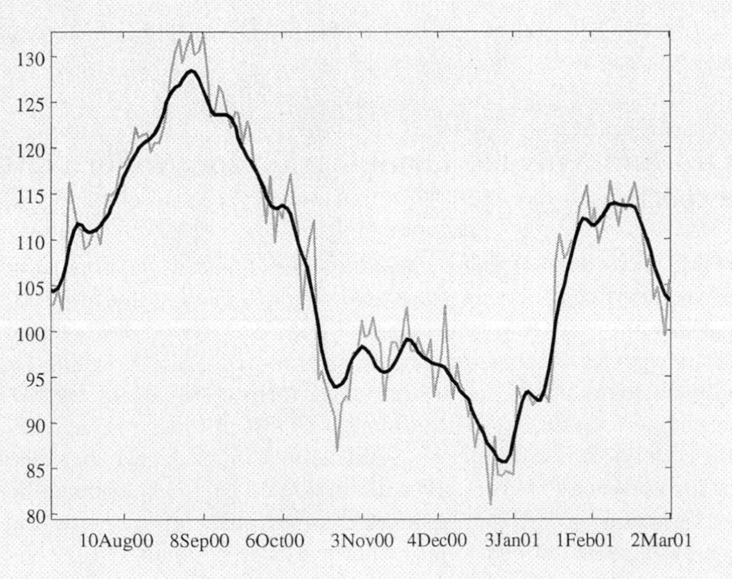

Fig. 10.6. Wavelet packet denoising of the stock of IBM over 160 trading days. The smooth line represents the wavelet packet based reconstruction using denoising and soft thresholding

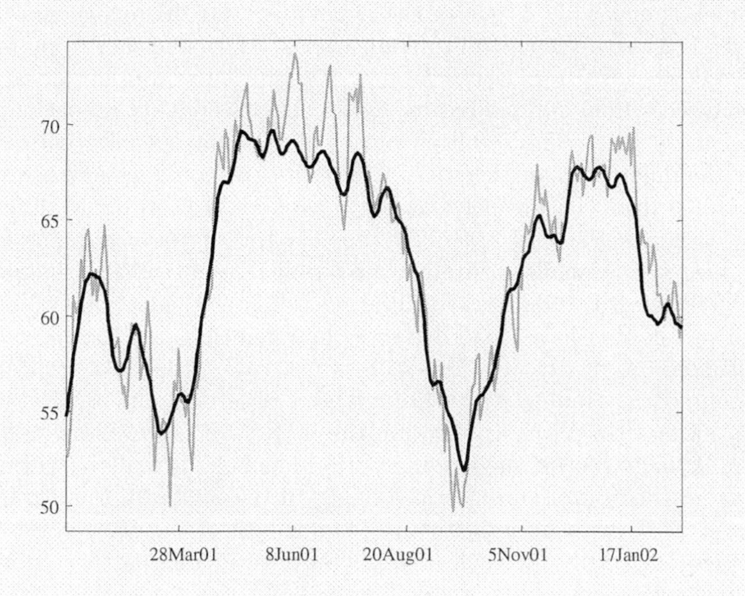

Fig. 10.7. Wavelet packet denoising of the stock of Microsoft over 270 trading days. The smooth line represents the wavelet packet based reconstruction using denoising and soft thresholding

to indicate distortion in case of prototype vectors that have large distance from their neighbors thus indicating abnormal events occurring with respect to the normal periods that have been used for training the SOM.

As example stocks have been selected the daily closing value of S&P 500 and the daily closing values of IBM and Microsoft corporation. A decomposition in 5 levels of the Daubechies db5 wavelet was used in a series of 32-day trading periods with daily sampling of the closing value for each day. For constructing feature vectors to train the SOM, the Window Wavelet Transform (WWT) has been used. The window size has been selected so as to correspond to wavelet dyadic decomposition values (i.e. powers of 2). Local time information is already present in the wavelet transform. The choice of WWT has the drawback that its use might do harm to the low-frequency data. However, the necessity of using the WWT has to do with the fact that the SOM has to be trained with fixed dimension vectors and by selecting a fixed window the resulting window wavelet transform gives always the same number of coefficients. The reason for selecting windows that are not overlapping was based on the idea that the SOM can be trained sequentially and then can be left to classify the behaviour of future trading periods of 32-days. This way previously trained SOMs can reveal tendencies in data that was not included in the original training set. The wavelet features that were 73 for each 32-day period have been concatenated in one vector. The dimension of the input vector which is equal to 73 results from concatenating 20 coefficients from detail level 1, 14 coefficients from detail level 2, 11 coefficients from detail level 3, 10 coefficients from detail level 4, 9 coefficients from detail level 5 and 9 coefficients from approximation level 5. Dimensionality reduction was performed using PCA. The resulting principal components were between 2 and 5 for the different example stocks. The principal components that have been extracted from the wavelet features have been used to train different SOMs for each example stock. After training the prototype vectors and their distances have been visualized through the use of a U-matrix. The units that have the greatest distortion are indicated with lighter shade while all other prototype vectors that have small distances between them are shown in with a darker shade.

By running the trading data through and observing the BMUs (best matching units) a trajectory connecting the BMUs for each 32-day trading period can actually show if the trading period has been actually assigned to BMU with low distortion and thus belonging to the pool of normal trading periods or a BMU with high distortion. The length of the trajectories has been selected arbitrarily for each stock. The critical issue is the tendency of the trajectory to enter high distortion areas on the sequentially trained SOM and the correlation of this event to subsequent trend behaviour in the following 32-day period. In the case of high distortion the trading period that has activated the SOM unit that has high distortion must be structurally different than the preceding periods. The SOM is activated based on the minimal Euclidean distance between the presented features for the tested trading period of 32-days and the prototype vectors of the SOM. By observing the closing values it can be seen that *the units with high distortion are activated exactly before periods that show an abrupt change in the trend.* Such a change is reflected in the wavelet coefficients, which are then mapped to principal components and subsequently activate SOM units with a high distortion indicating a structural change in the trading time-series.

More specifically, in the presented examples, Figs. 10.8, 10.10 and 10.12 show respectively the U-matrices for S&P 500 indicator, the IBM stock closing value and the Microsoft closing value. The corresponding trading periods are shown in Figs. 10.9, 10.11 and 10.13. The trajectory of the wavelet coefficients activations on the SOM runs through unit pools of small distortion, which indicates that these units are quite close in the weight space. At a certain trading period a high distortion unit is activated indicating a possible change in the wavelet structure, which can be due to an abrupt change in trading behaviour. In all examples the shown trajectories have appeared before dramatic falls in the values of the indicated stocks, which leads to the possibility to use the presented scheme of SOM based on wavelet features as an intelligent alarm that indicates a forthcoming change in the trend of the stock value.

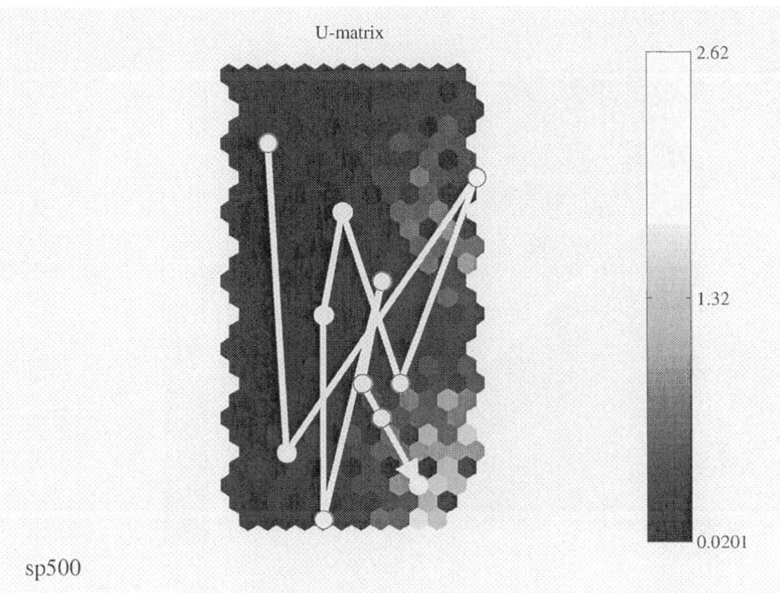

Fig. 10.8. Trajectory of the S&P 500

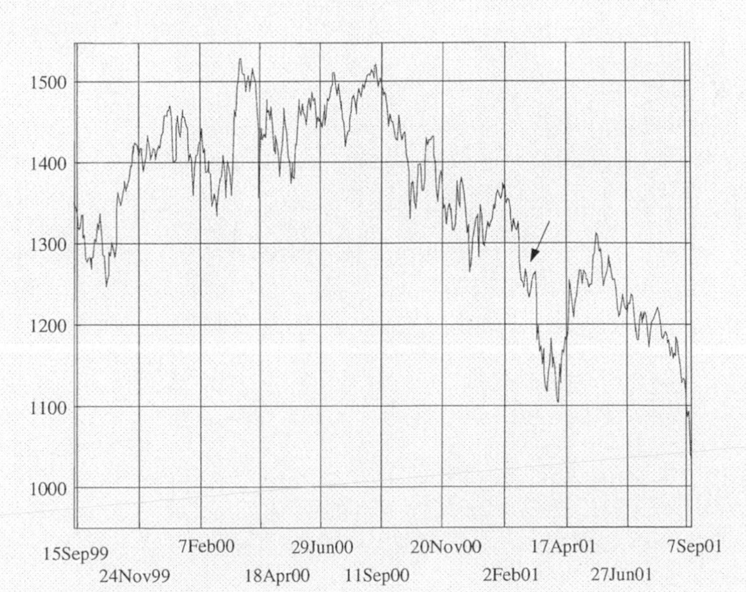

Fig. 10.9. Sample of the trading history of S&P 500 that produced the trajectory entering high distortion area in the SOM. Especially the trading days causing the highest distortion are shown by an arrow

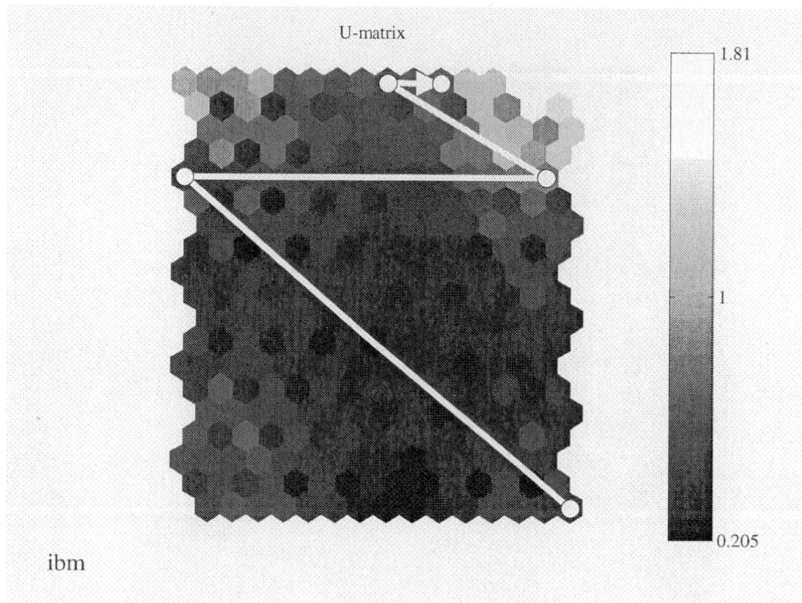

Fig. 10.10. Trajectory of the IBM stock

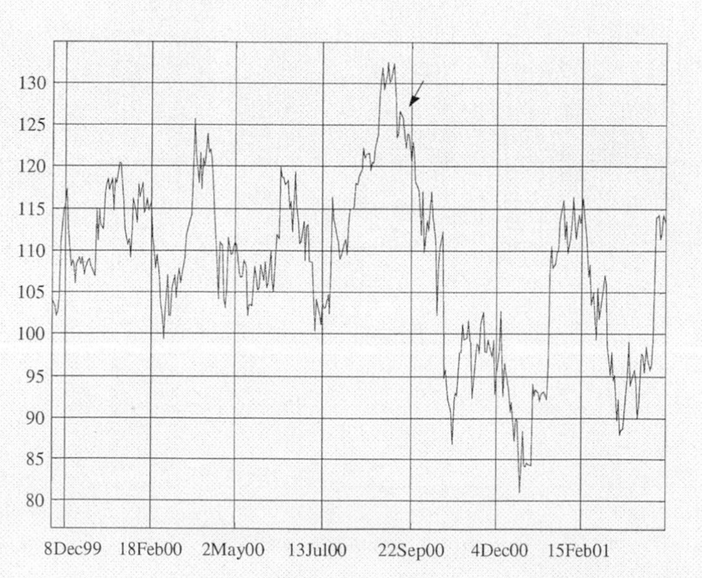

Fig. 10.11. Sample of the trading history of the IBM stock that produced the trajectory entering high distortion area in the SOM. Especially the trading days causing the highest distortion are shown by an arrow

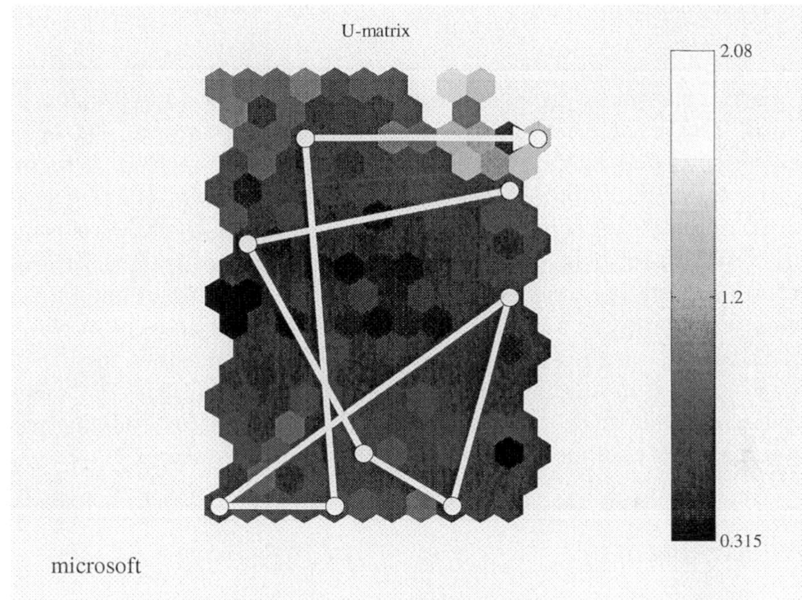

Fig. 10.12. Trajectory of the Microsoft stock

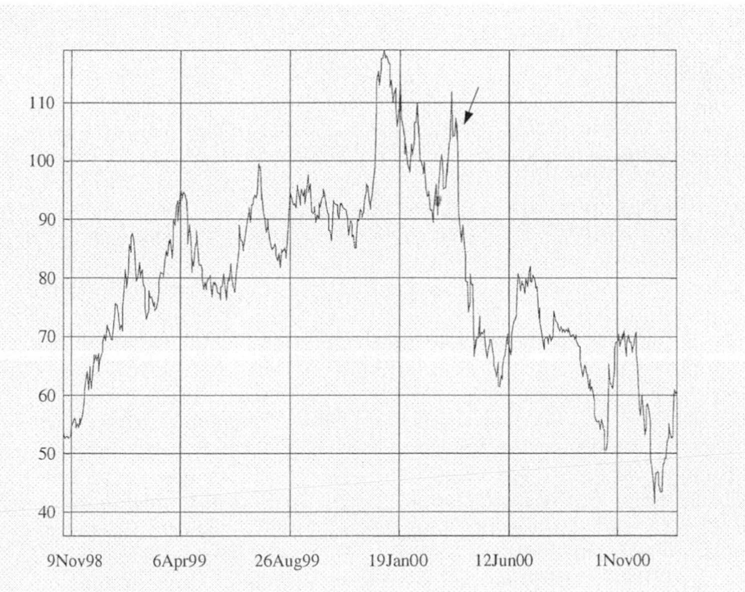

Fig. 10.13. Sample of the trading history of the Microsoft stock that produced the trajectory entering high distortion area in the SOM. Especially the trading days causing the highest distortion are shown by an arrow

10.8 Conclusions

Time-scale analysis afforded by the Discrete Wavelet transform and the Wavelet packet transform provides a new approach to studying financial market data. The capabilities of the wavelet packet transform in denoising stock time-series are demonstrated. Clear indication of the real trend is found through noise for some example stocks. The Self-organizing Map is introduced and financial applications of the SOM are presented. Then the SOM is combined with DWT in order to visualize and detect abnormal trading periods based on the properties of the prototype vectors and their distortion. Such a combination is proposed as an intelligent alarm for abrupt changes of trend in the stock values, which can be indicated in advance. Further testing of the proposed method in a larger number of stocks from different companies and also with high frequency data based on hourly sampling or sampling per transaction may lead to a more sensitive and fast reacting alarm.

Acknowledgements

The chapter was referred by Chueh-Yung Tsao. The authors are grateful for his painstaking review of the chapter. This chapter has been revised by incorporating many of his thoughtful suggestions.

References

10.1 Coifman, R. R., Meyer, Y., Wickerhauser, M. V. (1992): Wavelet analysis and signal processing. (Wavelets and their applications), Jones and Bartlett

10.2 Daubechies, I. (1988): Orthonormal bases of compactly supported wavelets. Commun. Pure Appl. Math. 41, 909–996

10.3 Donoho, D., Johnstone, I. (1994): Ideal spatial adaptation via wavelet shrinkage. Biometrika 81, 425–455

10.4 Kohonen, T. (1982): Self-Organized formation of topographically correct feature maps. Biol. Cyb., 43, 59–69

10.5 Mallat, S. (1989): A theory for multiresolution signal decomposition: the wavelet representation. IEEE Transactions on Pattern Analysis and Machine Intelligence, Vol. 11, 674–693

10.6 Martín del Brío, B., Serrano Cinca, C. (1993): Self- Organizing Neural Networks for the Analysis and Representation of Data: Some Financial Cases. Neur. Comput. & Appl., 1, 193–206

10.7 Martín del Brío, B., Serrano Cinca, C. (1995): Self- Organizing Neural Networks: The Financial State of Spanish Companies. (Neural Networks in the Capita Markets) John Wiley & Sons

10.8 Meyer, Y. (1989): Orthonormal Wavelets, Wavelets, time- frequency methods and phase-space. (Wavelets: Time-Frequency Methods and Phase Space) Springer-Verlag

10.9 Ultsch, A., Siemon, H. P. (1990): Kohonen Self-Organizing Feature Maps for Exploratory Data Analysis. (Proceedings of International Neural Network Conference - INNC90) Kluwer

10.10 Varfis, A., Versino, C. (1991): Clustering of European Regions on the Basis of Socio-Economic Data. A Kohonen Feature Map Approach. (Proceedings of the PASE international Workshop "Parallel Problem Solving From Nature. Applications in Statistics and Economics")
10.11 Varfis, A., Versino, C. (1992): Clustering of socio-economic data with Kohonen Maps. Neural Network World 2, 6, 813–834

Part V

Sequence Matching and Feature-Based Time Series Models

11. Pattern Matching in Multidimensional Time Series

Arnold Polanski

Facultad de Económicas, Universidad de Alicante, E-03071 Alicante, Spain
email: arnold@merlin.fae.ua.es

Based on a algorithm for pattern matching in character strings, a pattern description language (PDL) is developed. The compilation of a regular expression, that conforms to the PDL, creates a non deterministic pattern matching machine (PMM) that can be used as a searching device for detecting sequential patterns or functional (statistical) relationships in multidimensional data. As an example, a chart pattern of ex ante unknown length is encoded and its occurrences are searched for in financial data.

11.1 Introduction

"The explanation and the prediction of natural phenomena, be it to forecast the future or to elicit the hidden laws that underlie unknown dynamics, are the ultimate aims of science" ([11.10], p. 267). A relatively new research area concerned with revealing hidden relationships and regularities in observed variables is data mining, a host of methods that aim at extracting previously unknown and potentially useful information from large sets of data. Of particular importance is the class of data mining problems concerned with the discovery of frequently occurring patterns in sequential data (time series). Many systems (e.g. financial markets), that generate sequential data, can be interpreted as feed-back driven, i.e. their past output can be seen as one of their important inputs. Standard methods of complex system forecasting, like kernel regression or neural networks, take it into account when trying to estimate future output values as "functions" of a fixed number of past data. In contrast to these methods, the present approach does not restrict the "relevant past" or the reaction time of the system to time windows of fixed lengths. It is only important that the past state of the system and the system´s response to that state frequently generate sequences of data with some (possibly ex ante unknown) characteristics. Those characteristics, encoded in a suitable language, could be searched for in a time series. A concise and flexible pattern description language can be therefore a powerful tool for data mining and would serve two purposes: on the one hand as a language in which theories, concerned with the underlying data generating process, can be formulated and tested, and, on the other, as a forecasting instrument for practical applications.

There is some literature on mining sequential patterns in a database of customer transactions (e.g. [11.1],[11.2]). The approach pursued there can be briefly stated as follows: given a set of data sequences, all frequent subsequences should be found. By frequent is meant that the percentage of such subsequences in the data must exceed an user-specified minimum support. This line of research concentrate on finding efficient search algorithms.

Perhaps the most closely related work to the present one is Packard [11.6]. Packard develops a GA to address the problem of predicting complex systems. In the simplest form his approach can be stated as follows (cp. [11.5]):

- There are two time series: a vector $\mathbf{x}$ with T observations of an independent variable and a vector $\mathbf{y}$ with the same number of observations of a dependent variable. For example, in a stock market prediction task, the observations, $x_1...x_T$, might be the prices of a particular stock at successive days and the values, $y_1...y_T$, might be representing the prices of the same stock at some time in the future, $y_t = x_{t+k}$.
- There is a population of conditions on the independent variable that are expected to give good predictions for the dependent variable. Consider for example the following condition,

  $C = [(20 \leq x_t) \wedge (30 \leq x_{t+1} \leq 40) \wedge (x_{t+2} \leq 30)]$

  It can be used to extract from $\mathbf{x}$ all sets of three successive days in which C was met.

Packard uses GA to search for conditions that are good predictors of *something*. In particular, he looks for conditions that define sequences of data points with dependent-variable values close to being uniform. The fitness of a condition C is calculated by running all data points in $\mathbf{x}$ through C and collecting corresponding values of $\mathbf{y}$ for each sequence in $\mathbf{x}$ that satisfies this condition. If these values are all close to a particular value $\bar{y}$, then C is a candidate for a good predictor. We refer the interested reader to [11.6] and [11.5] for details of the GA used for the selection of the fittest conditions and for the discussion of the results.[1] In this paper we will extend Packard´s ideas and develop a language for succinct encoding of quite general, non deterministic conditions or patterns that describe fragments of a multidimensional time series.

11.2 String Search in a Time Series

As a first approximation, the problem of finding sequences of data points in a time series, that satisfy certain class of conditions, can be recast as a problem of pattern matching in a character string. To start with, suppose there is an one-dimensional time series $\mathbf{x} = (x_1...x_T)$, that will play the role of the string, where patterns are searched for. We will represent the data points in $\mathbf{x}$ as a

[1] In 1991 Packard left the University of Illinois to help form a company to predict financial markets.

sequence of letters written in an alphabet. One obvious possibility to define an alphabet, in that a class of conditions can be encoded, is to normalize the values in $\mathbf{x}$, such that all observations lie, for instance, in the interval $[0, 1]$. This interval is then divided into subintervals (of constant or variable length) and to each subinterval a letter is assigned.

Example 11.2.1. In a normalized time series $\mathbf{x}$ ten subintervals, [0.0,0.1), [0.1,0.2),..,[0.9,1.0], are created and the letters A..J are assigned to them. The condition,

$$C = [(0.2 \leq x_t < 0.3) \wedge (0.3 \leq x_{t+1} < 0.4)]$$

can now be expressed as the pattern P = "CD" with the meaning, that a match exists, whenever D follows C in the string $\mathbf{x}$. Searching for the condition C in time series $\mathbf{x}$ is equivalent to looking for the occurrences of the corresponding pattern P in the string $\mathbf{x}$. This is obviously an instance of a simple pattern matching problem as well known from any standard text editor.

It is often desirable to do searching with somewhat more general description of a set of characteristics. We will consider, in what follows, pattern descriptions made up of symbols (letters) tied together with the following three fundamental operators:

-*concatenation* as used for instance in the pattern P="CD." If two characters are adjacent in the pattern description, then there is a match iff the same two characters are adjacent in the string.

-*or* (+) between two letters. There is a match iff either of the letters occurs in the string searched.

-*closure operator*(*). If we have the closure of a symbol, then there is a match iff the symbol occurs any number of times in the string.

Example 11.2.2. With the time series (string) $\mathbf{x}$ from Example 11.2.1 and the operators defined above we are able to do searching for more general, non deterministic patterns. A pattern definition $P = (A*) + (J*)$ looks e.g. for sequences (of any length) of letters A or sequences (of any length) of symbols J in the string $\mathbf{x}$, i.e. P searches for consecutive successions of very low or very high values in the time series $\mathbf{x}$.

11.3 The Pattern Description Language (PDL)

In this section a definition of a language will be given that allows for a quite general specification of patterns in multivariate time series.

Be $\mathbf{x}$ a $N \times T$ matrix composed of N row vectors, $x_1..x_N$, with T data points each vector. $\mathbf{x}$ will be the space where patterns will be searched for. The notation x_{it}, $1 \leq i \leq N, 1 \leq t \leq T$, refers to the tth observation of the time series x_i. For notational convenience we will sometimes write x_t instead of x_{it}. A pattern description will consist of a sequence of *letters*, operators and parentheses combined in an arbitrarily complicated way. A letter is an expression composed of arithmetical and logical operators, elements of the vectors $x_1..x_N$, indexed by functions of the time index t, and (user defined) variables and constants. An expression encodes a condition and is evaluated as a function of t (and possibly of any other variable it contains) yielding as result the Boolean value true or false. The evaluation of a sequence of expressions takes place in the order in which the letters, that represent them, are found in the pattern description. After an expression is evaluated, and before passing to the next one, the time index is increased by 1. This means, in particular, that two letters adjacent in the pattern description and both containing the term x_t, will refer to adjacent observations in $\mathbf{x}$. To delimit letters in a pattern description, they will be enclosed in square brackets. A pattern encoded with these rules will be called a regular expression (RE).

Example 11.3.1. One of the most useful letters is the expression that encodes the rise or the fall of a time series. It is simply stated as $[x_t < x_{t+1}]$ or $[x_t > x_{t+1}]$, respectively. If we look for longest sequences with an indefinite number of observations with rising values, we can use a regular expression of the form,

$$[x_t < x_{t+1}] * [NOT(x_t < x_{t+1})]$$

(the condition $[NOT(x_t < x_{t+1})]$ marks the end of a rising series).

As an another example suppose that in a two-dimensional time series, $\mathbf{x} = (x_1, x_2)$, x_1 is the vector of prices of a stock A and x_2 is the vector of values of an indicator (for instance, the moving average calculated for A). To find all pairs of adjacent data points, where the plot of prices of stock A crosses from below the plot of the indicator, we use the simple RE,

$$[x_{1t} < x_{2t}][x_{1t} > x_{2t}].$$

In practical applications some enhancements of the pattern description language proved useful. First, to measure the number of occurrences of a letter, followed by the closure operator "*," the variables $l_1..l_n$ have been defined. During the search for the pattern P in $\mathbf{x}$, l_i contains the number of evaluations of the letter, followed by the ith asterisk, that yielded true, i.e. l_i measures the length of the fragment of data in which the condition, encoded by that letter, was met. l_i's are set to zero every time a complete pattern match is found or the search is started from the beginning of the pattern (after possible match failure at some position).

An another extension makes possible the suppression of the time index increase after an expression has been evaluated. To this end the letter containing this expression has to be enclosed in braces. Finally, to keep track of the length of a match, the variable t_0, which points to the beginning of the match, is defined.

Example 11.3.2. The regular expression,

$$[x_t < x_{t+1}] * \{l_1 = 7\}[x_t > x_{t+1}] * \{l_2 = 3\}$$

searches for a week-long increase followed by a 3 days fall of values in an one-dimensional time series $\mathbf{x}$. The braces are necessary, because the use of $[l_1 = 7]$ and $[l_2 = 3]$ would return a match of 12 data points instead of the correct length of 10.

The pattern description language represents a tool for feature extraction in time series. For example the classical technical trading methods, based on indicators like moving average, RSI or momentum indicator, can easily be described by means of regular expressions. Chartist methods aimed at finding graphical patterns in price plots (head & shoulders, spikes, flags etc.) require more sophisticated pattern descriptions but, their stylized shapes can also be expressed as REs (a toy example of a simple pattern description of this type is given in Section 5). An interesting characteristic of the PDL relates to the fact that the length of a pattern has not to be specified ex ante. Perhaps the most commonly used tools for extracting information from time series work with data vectors of fixed lengths. Kernel regression, nearest-neighbor estimators and neural networks are examples of procedures that "earn" output values from input data composed of a fixed number of observations. In contrast, a regular expression can identify patterns without knowing at the designing time the number of observations that compose it. In this manner, the same regular expression can capture qualitatively identical phenomena, which unfold on different time scales (fractal patterns) or stretch over time windows of variable lengths. Furthermore, regular expressions can be applied to test functional and/or statistical relationships in multidimensional data. Consider e.g. the simple condition $[x_{3t} = \sin(x_{2t}) * x_{1t}]$. It recognizes fragments of a multivariate time series, in which an exact, functional relation between x_1, x_2 and x_3 holds. As a further example, a RE $[\mid x_t - a_1 x_{t-1} - a_2 x_{t-2} \mid < \varepsilon]$ will match the data from an AR(2) process, whenever no realization of the innovations exceeds ε.

11.4 The Pattern Matching Machine (PMM)

This section describes briefly, how the algorithm for the pattern recognition works. This algorithm is based on a pattern matching machine for text search, as described for example in [11.9]. First a regular expression, that encodes a pattern, is compiled to a non deterministic pattern matching machine (PMM). The latter is implemented as a finite state automaton (FSA). Essentially, every state of the FSA represents a letter (condition) or an operator ("+" or "*") from the pattern description. The machine has an unique initial state and an unique final state. When started out in the initial state, the machine should be able to recognize any fragment of the matrix $\mathbf{x}$ that satisfies the pattern description. Intuitively, the matching algorithm works as follows: the PMM, constructed by the compiler, moves forward along the time axis in discrete steps, verifying at each step the condition encoded in its current internal state and changing this state according to the result (true or false) of the verification. This result depends on the position in the matrix $\mathbf{x}$, indicated by the time index t, where the verification takes place. The PMM can travel from a state A to a state B, pointed to by state A, whenever the evaluation of the state B yields the value true. If no such state exists, the machine returns to the initial state. If a state that points to the final state is eventually reached, a match has been found. What makes the machine non deterministic are some states which can point to different successor states (for example, the state which corresponds to the "or"-operator points to 2 successor states that check the expressions in the left and right operand, respectively). The non deterministic character of the PMM allows for alternative interpretations of the data and, if possible, for choosing an interpretation compatible with the encoded condition.

Example 11.4.1. Consider a two-dimensional vector $\mathbf{x} = (x_1, x_2)$ and a RE with only two letters, $P = [x_{1t} < x_{2t}][x_{1t} > x_{2t}]$. The PMM, that represents P, has four states (numbered 0..3): the initial state (0), the final state (3) and the states corresponding to the first and the second letter (states 1 and 2, respectively). These states are linked by pointers: state i points to the state $i + 1$, $i = 0..2$. The search for P in $\mathbf{x}$ starts at $t = 1$ and with the machine staying in the initial state. If state´s 1 condition, $[x_{11} < x_{21}]$, is true the PMM passes to that state. It stays in the initial state else. After the evaluation of state´s 1 condition the time index t is advanced by 1 and the condition in the state pointed by the current machine´s state is checked: $[x_{12} < x_{22}]$ if the machine has not moved from the initial state or $[x_{12} > x_{22}]$ if it has passed to state 1. If in the latter case the evaluation of the second state's condition for $t = 2$ yields true, the PMM passes to that state. The state 2 has an unique successor, the final state, and therefore a match has been found.

11.5 An Application

The PDL developed in the previous sections can be applied to a variety of areas. Here an example of an application to technical analysis of price data will be given. A wide variety of theoretical and empirical models have been proposed to explain why technical trading is widespread in financial markets (see for instance references in [11.11] and [11.4]). Without entering the discussion between fundamentalists and chartists, we will select one of the patterns, applied by technical analysts to predict time series, and search for it in the data from New York Stock Exchange. Consider the technique of price targeting,[2] based on the "golden ratio" [11.7], as illustrated in the Fig. 11.1. The price target P3 is given there by $P3 = P2 - (P2 - P1) * 2.618$ (the constant 2.618 is closely related to the "golden ratio" $\frac{\sqrt{5}+1}{2} \approx 1.618$).

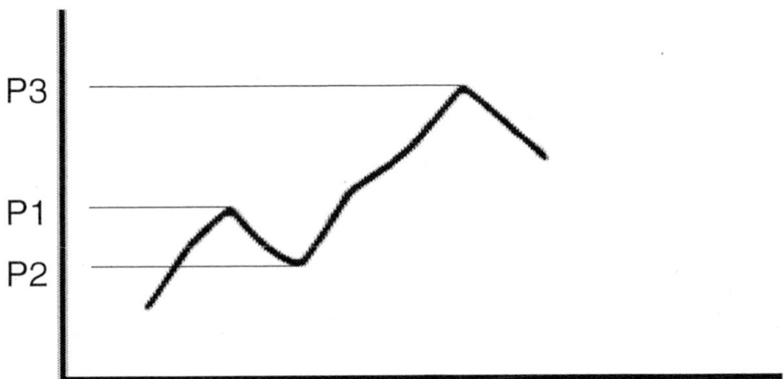

Fig. 11.1. The "golden ratio rule"

With the pattern description P, defined below, we will look for sequences (of indefinite length) of observations from NYSE that verify this stylized version of price targeting.

$$P = [x_t < x_{t+1}] * [x_t > x_{t+1}] * [x_t < x_{t+1}] * \{x_t \geq x_{t+1}\}$$
$$\{((x_{t_0+l_1+l_2} + (x_{t_0+l_1} - x_{t_0+l_1+l_2}) * 2.618) \approx x_t) \wedge l_1 > l_2 \wedge l_3 > l_2\}$$

(note that an asterisk has a different meaning inside and outside of brackets). The first part of the pattern definition,

$$[x_t < x_{t+1}] * [x_t > x_{t+1}] * [x_t < x_{t+1}] * \{x_t \geq x_{t+1}\}$$

stands simply for "a rising followed by a falling followed by the longest, rising sequence of data points" ($\{x_t > x_{t+1}\}$ marks the end of the last rising sequence). The second part (composed of the last expression in braces) is decisive in selecting only those fragments of data that satisfy the "golden ratio rule." Here, t_0 is the time index of the first, t, of the last observation in a fragment of data matched, "$\approx$" is a user defined operator meaning "approximately equal" and l_1, l_2, l_3 contain the lengths of the three subsequences in the rising-falling-rising sequence defined in the first part. Consequently, $x_{t_0+l_1}$ stands for P1, $x_{t_0+l_1+l_2}$ for P2 and x_t ($= x_{t_0+l_1+l_2+l_3}$) for P3. $(x_{t_0+l_1+l_2} + (x_{t_0+l_1} - x_{t_0+l_1+l_2}) * 2.618) \approx x_t$ encodes therefore the condition $P2 + (P1 - P2) * 2.618 \approx P3$ that checks, whether the price target has been (approximately) reached. The inequalities $l_1 > l_2 \wedge l_3 > l_2$ require that the two rising sequences have to be longer than the falling sequence between them and capture the typical shape of a rising market.

[2] Price targets are forecasted maximum (minimum) values of a time series after a certain constellation in the past data has been observed.

When running the search for the pattern P in the data from NYSE in the period from the 2nd of January 1990 to the 30th of December 1998 we have found matches with lengths ranging between 6 and 19 observations. As an example, the next figure shows the matches for the pattern P that have been found in the data between the 15.10.1997 and the 10.02.1998.

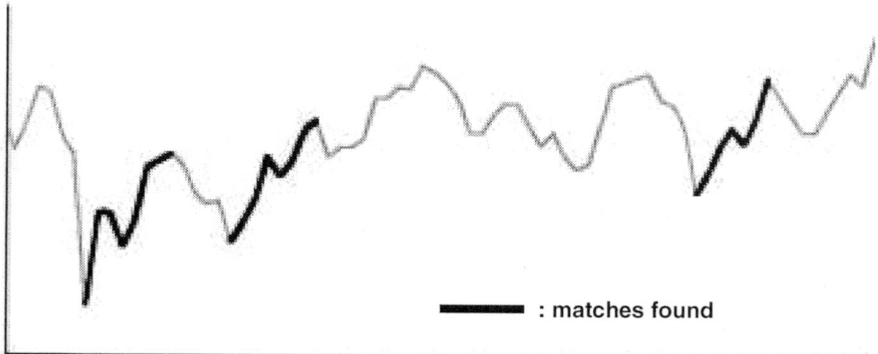

Fig. 11.2. Matches for the pattern P found in the data from NYSE between the 15.10.1997 and the 10.02.1998 (80 data points). The two first matches are 8 and the last match 7 observations long

11.6 Conclusions and Further Research

Based on the algorithm for pattern matching in a character string, a pattern description language has been developed. The compilation of a regular expression, that conforms to the PDL, creates a pattern matching machine that can be used as a searching device for detecting patterns or functional (statistical) relationships in multidimensional data.

This is just the first step on the way to isolate "pockets of predictability" in data streams. It remains the non trivial task of finding patterns that, at the same time, can be formulated as regular expressions and are good predictors. The future work will concentrate therefore on two issues: firstly, on encoding of empirically observed patterns (e.g. the chart patterns commonly used in the technical analysis) and verifying their predictive power. Secondly, following a strand of papers on genetic evolution of expressions (see e.g. [11.12], [11.3], [11.8], [11.10]), it is planned to use GA to build up regular expressions and to run evolutionary selection on them in order to determine the fittest ones. Regular expressions are just strings of symbols that conform to a simple grammar. A GA, like the one used in [11.10], can build up such expressions from building blocks and then break up the strings again. The natural selection process causes suitable blocks to survive and combine among themselves while useless parts disappear. The "usefulness" or fitness of a pattern can be assessed, for instance, by the variance of the observations, calculated in a suitable chosen time window after the occurrence of that pattern in the time series.

Acknowledgements

I profited immensely from seminars and discussions, I participated in, at the 6th International Conference on Computing in Economics and Finance in Barcelona 2000. Particularly helpful were the talks with M.A. Kaboudan. Special thanks to J. Sikora who shared with me his rich experience as (technical) trader. The chapter was refereed by Chueh-Yung Tsao, I am also grateful for his corrections made to an early draft.

References

11.1 Agraval, R., Srikant, R. (1995): Mining Sequential Patterns, In: Proceedings of the 11th International Conference on Data Engineering, ICDE
11.2 Garafalakis, M., Rastogi, R., Shim, K., SPIRIT (1999): Sequential Pattern Mining with Regular Expression Constraints, In Proceedings of the 25th VLDB Conference, Edinburgh, Scotland
11.3 Iba, H., Nikolaev, N. (2000): Genetic programming polynomial models of financial data series, In: Proceedings of the 2000 Congress on Evolutionary Computation CEC00, La Jolla, USA. IEEE Press
11.4 LeBaron, B., Arthur, W. B., Palmer, R. (1999): Time series properties of an artificial stock market, Journal of Ec. Dynamics & Control, 23, 1487–1516
11.5 Mitchell, M. (1998): An Introduction to Genetic Algorithms, MIT Press
11.6 Packard, N. (1990): A Genetic Learning Algorithm for the Analysis of Complex Data, Complex Systems 4, No. 5, 543–572
11.7 Plummer, T. (1998): Forecasting Financial Markets, Kogan Page
11.8 Santini, M., Tettamanzi, A. (2001): Genetic Programming for Financial Time Series Prediction, In: Proceedings of the 4th European Conference, EuroGP 2001, Lake Como, Italy. Springer
11.9 Sedgewick, R. (1988): Algorithms, Addison Wesley
11.10 Szpiro, G. (1997): A Search for Hidden Relationships: Data Mining with Genetic Algorithms, Computational Economics, 10, 267–277
11.11 Wolberg, J. R. (2000): Expert Trading Systems, John Wiley & Sons
11.12 Yoshihara, I., Aoyama, T., Yasunaga, M. (2000): GP-based modeling method for time series prediction with parameter optimization and node alternation, In Proc. of the 2000 Congress of Evolutionary Computation CEC00, La Jolla, USA. IEEE Press

12. Structural Pattern Discovery in Time Series Databases

Weiqiang Lin[1], Mehmet A. Orgun[1], and Graham J. Williams[2]

[1] Department of Computing, Macquarie University
Sydney, NSW 2109, Australia,
email: {wlin, mehmet}@comp.mq.edu.au

[2] CSIRO Mathematical and Information Sciences, GPO Box 664,
Canberra ACT 2601, Australia
email: Graham.Williams@cmis.csiro.au

This study proposes a temporal data mining method to discover qualitative and quantitative patterns in time series databases. The method performs discrete-valued time series (DTS) analysis on time series databases to search for any similarity and periodicity of patterns that are used for knowledge discovery. In our method there are three levels for mining patterns. At the first level, a structural search based on distance measure models is employed to find pattern structures; the second level performs a value-based search on the discovered patterns using a local polynomial analysis; the third level, based on hidden Markov models (HMMs), finds global patterns from a DTS set. As a result, similar and periodic patterns are successfully extracted. We demonstrate our method on the analysis of "Exchange Rate Patterns" between the U.S. dollar and Australian dollar.

12.1 Introduction

Temporal data mining is concerned with discovering qualitative and quantitative temporal patterns in a temporal databases or in a discrete-valued time series (DTS) dataset. DTS commonly occur in temporal databasess (e.g., the weekly salary of an employee or the daily rainfall at a particular location). Recently, there are two kinds of major problems that have been studied in temporal data mining:

1. The similarity problem: finding fully or partially similar patterns in a DTS, and
2. The periodicity problem: finding full or partial periodic patterns in a DTS.

Although there are various results to date on discovering periodic patterns and similarity patterns in discrete-valued time series (DTS) datasets (e.g. [12.22], [12.21], [12.2], [12.1], [12.6]), a general theory of discovering patterns for DTS data analysis is not yet well known [12.16]. Since there are different types of discovery problems that have been addressed in the literature, it is important to characterise those problems using some formal framework. Most studies in temporal data mining pay little or no attention to basic statistical methods and models which are fundamental in DTS data analysis. Although there are mainly two popular models, the similarity model and periodicity model, a general method of analysis is still lacking even when using different measures on a given time series dataset or its samples.

The proposed framework is based on a new model of DTS that addresses the problem of using a small subset of a real DTS to represent the whole DTS and to find important characteristics in the DTS that may lead to the discovery of similar or periodic patterns. The framework focuses on a special problem of discovering patterns using hidden Markov chain regression modeling via functional data analysis. The first step of the framework consists of a distance measure function for discovering structural patterns (shapes). In this step, only the rough shapes of patterns are decided from the DTS and a distance measure is employed to compute the nearest neighbours (NN) to, or the closest candidates of, given patterns among the similar ones selected. In the second step the degree of similarity and periodicity between the extracted patterns are measured based on local polynomial models. The third step of the framework consists of a hidden Markov model for discovering all level patterns by combining the results of the first two steps.

The paper is organised as follows. Section 12.2 briefly reviews some examples in discrete-valued time series and surveys the models that have been proposed. The section 12.3 presents the basic methods and results of local polynomial and hidden Markov modeling. Section 12.4 presents our new method for pattern discovery in more detail. Section 12.5 applies it to "Daily Foreign Exchange Rates" data. In section 12.6 we discuss related work on temporal data mining. The last section concludes the paper with a brief discussion and future work.

12.2 Discrete-Valued Time Series Examples and Their Models

One of the basic tasks of temporal data mining and temporal data analysis is to automatically discover the qualitative and quantitative patterns in a temporal databases. According to temporal characteristics, objects in temporal databases can be classified into three categories [12.9]:

- Time-invariant objects;
- Time-varying objects; and
- Time-series objects (e.g., such as DTS).

In the rest of the section we focus only on temporal databasess for DTS objects, which are often called time series databases.

12.2.1 Some Examples in Discrete-Valued Time Series

Temporal databases naturally arise in different business organizations as well as scientific decision-support applications where events and data evolve over a time domain. Many time series which occur in practice are by their very nature discrete-valued, although it is often quite adequate to represent them by means of models based on their distributions. For example, most DTS datasets can be represented by means of models based on normal distribution. Some examples of discrete-valued time series are:

- stock exchange daily records;
- the sequence of wet and dry days at some place;
- road accidents or traffic counts; and
- multiple spectral astronomical data.

Although sometimes models based on a well-defined distribution are suffice (e.g., Poisson distribution), this is not always the case. There are a number of challenges in mining DTS databases. For example, when a time series dataset is given, discovering the inherent patterns and regularities is a challenging task. Another challenge is when no time-domain knowledge is available or its knowledge is too weak or incomplete; this often is called the time-dimension problem [12.24].

12.2.2 Discrete-Valued Series Models

There are many discrete-valued data models proposed in the applied statistics literature, but these models have not found their way into major data-mining research areas. It may be that there is no well-known family of models that are structurally simple, sufficiently versatile to cater for a useful variety of data types, and readily accessible to the practitioner. We briefly outline some discrete-valued time series models which we consider to have very good potential to be applied in temporal data mining:

1. Markov models using Markov chains on a finite space of m states as possible models for time series taking values in the space, which have $m^2 - m$ independently-determined transition probabilities.
2. Hidden Markov models using an unobserved stationary Markov chain and either a Poisson or a binomial distribution.
3. Geometric models using dependence marginal distributions and correlation structure to compare with individual distribution.
4. Markov regression models extending the ideas of generalized linear models and quasi-likelihood to the context of time series.
5. State-space models using a distribution of observations with the mean given by a distribution process.

6. Local regression analysis models based on an application of local modeling techniques to explore the association between dependent and independent variables, to assess the contribution of the independent variables, and to identify their impact on the dependent variables.

In the rest of the paper, we focus on hidden periodicity analysis to discover patterns in a temporal databases only related to the problem of similarity and periodicity via multiple models.

12.3 Local Polynomial and Hidden Markov Modeling

In this section we introduce the data analysis of a DTS by a combination of distance measures and local polynomial models with Markov models to form local polynomial hidden Markov models (LPHMM). The first level, built by the distance functions, performs a structural pattern search via a Markov chain model, and then local polynomials are used for a value-based search for the classification of discovered structural patterns. At the end, the overall result is obtained by combing the results of the first two levels of discovery.

12.3.1 Definitions and Basic Models

In this section we first define the basic concepts of DTS as they relate to our work.

The term pattern is a word in everyday vocabulary and means something exhibiting certain regularities, something able to serve as a model,or something representing a concept of what was observed. For our purposes a behaviour or a pattern is a pair of variables:

$$\mathsf{Pattern/Behaviour} = \{\Theta, \Psi\},$$

where Θ is a collection of observations that can, at least in principle, be executed by some technical apparatus and Ψ is the concept behind the observations.

Suppose that we have a random experiment with sample space Γ. The probability of an event A is a measure of how likely the event will occur when the experiment is run. Mathematically, a probability measure (or distribution) Σ for a random experiment is a real-valued function defined on a collection of events Ω that satisfies the following axioms:

1. $0 \leq \Sigma(A) \leq 1$ for any event A in Ω.
2. $\Sigma(\Omega) = 1$.
3. $\Sigma[\bigcup_{i \in N} A] = \sum_{i \in N} \Sigma(A_i)$, where $\{A_i \mid i \in N\}$ is a countable, pairwise disjointed collection of events.

We now have defined the three essential ingredients that model a random experiment:

1. The collection of events Ω.
2. The sample space Γ.
3. The probability measure Σ.

Together these define a probability space $\mathcal{P} = \{\Omega, \Gamma, \Sigma\}$:[1]

Definition 1 *A measure space $\mathcal{P} = \{\Omega, \Gamma, \Sigma\}$ is called a probability space if $\Sigma(\Omega) = 1$. Term Ω is the sample space or sure event, the measurable sets are called events, and the measurable functions are called random variables. If ξ_1, ξ_2, ..., ξ_n are random variables, then $\xi = (\xi_1, \xi_2, \ldots, \xi_n)$ is a vector-valued random variable.*

Definition 2 *Suppose that $\mathcal{P} = \{\Omega, \Gamma, \Sigma\}$ is a probability space, and $\mathcal{T}$ is a linear time index set. If for any $t \in \mathcal{T}$ there exists a random variable ξ_t defined on $\{\Omega, \Gamma, \Sigma\}$, then the family of random variables $\{\xi_t : t \in \mathcal{T}\}$ is called* a time series. *If $\mathcal{T}$ is a discrete-valued time index set, then the family of random variables $\{\xi_t \mid t \in \mathcal{T}\}$ is called* a discrete-valued time series(DTS).

The random variables ξ_t $(t \in \mathcal{T})$ in the above definition should be understood as complex-valued variables in general. We assume that for every successive pair of time points in DTS, say t_{i+1} and t_i, the time-gap between the time points t_{i+1} - t_i = $f(t)$ is a function (in most cases, $f(t)$ is constant for all pairs of time points). For every successive triple time points (t_{i-1}, t_i), and t_{i+1} (belonging to $\mathcal{T}$), any triple value of $(X_{t_{i-1}}, X_{t_i}, X_{t_{i+1}})$ of a given time series X has only nine distinct states (or called nine local features, e.g., in Fig. 12.1), depending on whether the values increase, decrease, or stay the same.

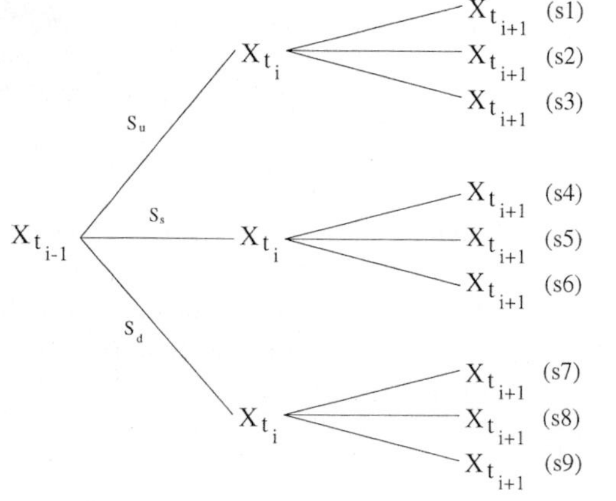

Fig. 12.1. $\mathcal{S}$ = {s1, s2, s3, s4, s5, s6, s7, s8, s9}

[1] See any standard textbook of time series analysis or its applications, i.e.,[12.18, 12.19].

Definition 3 *Let S_s be the same state as the prior one, S_u is the go-up (or stronger one) state compared with the prior one, and S_d is the go-down (or weaker one) state compared with the prior one. Let $\mathcal{S}$ = {s1, s2, s3, s4, s5, s6, s7, s8, s9} = {$(X_{t_{i-1}}, S_u, S_u)$, $(X_{t_{i-1}}, S_u, S_s)$, $(X_{t_{i-1}}, S_u, S_d)$, $(X_{t_{i-1}}, S_s, S_u)$, $(X_{t_{i-1}}, S_s, S_s)$, $(X_{t_{i-1}}, S_s, S_d)$, $(X_{t_{i-1}}, S_d, S_u)$, $(X_{t_{i-1}}, S_d, S_s)$, $(X_{t_{i-1}}, S_d, S_d)$ }. Hence, $\mathcal{S}$ is called a state-space.*

The meaning of state $s1 = (X_{t_{i-1}}, S_u, S_u)$ is that $X_{t_{i-1}} < X_{t_i} < X_{t_{i+1}}$, in other words, the value of the time series has increased at two consecutive time points. For any given time series, we can obtain a structural sequence based on these nine states.

A sequence is called a *full periodic sequence* if its every point in time contributes (precisely or approximately) to the cyclical behavior of the overall time series (that is, there are cyclic patterns with the same or different periods of repetition).

A sequence is called a *partial periodic sequence* if the behavior of the sequence is periodic at some, but not all, points in the time series.

Definition 4 *Let $y = \{y_1, y_2, \ldots\}$ be a real value sequence. If y_i(mod 1) with $0 \leq y_j < 1$ for all $j = 1, 2, \ldots$, then we say that y is uniformly distributed if every subinterval of the interval [0, 1] gets its fair share of the terms of the sequence in the long run. More precisely,*

$$\lim_{n \to \infty} \frac{number\ of\ \{j \leq n : y_j \in J\}}{n} = length\ of\ J, \tag{12.1}$$

for all subintervals J of [0, 1).

A uniform distribution of a DTS is a rather trivial random pattern. However, M independent uniformly distributed datasets can be superimposed to form a new dataset pattern. In the more general case, an independent-statistical distribution $\mathcal{Y}$ of a dataset construction with the base distribution B of the dataset can be superimposed to form every different pattern in the dataset.

Definition 5 *Let $y = \{y_1, y_2, \ldots\}$ be a sequence of real numbers with $I - \delta < y_k < I + \delta$ for all k ($k \geq 1$, I is a constant, and $\delta \geq 0$). We say that y has an approximate constant sequence distribution of $y = \{I, I, \ldots\}$. More generally, if $h(t) - \delta < y_k < h(t) + \delta$ for all k, then we say that y has an approximate distribution function $h(t)$.*

We have the following results:[2]

Lemma 12.3.1. *A discrete-valued dataset contains periodic patterns if and only if there exist periodic patterns both in state space S and probability space $\mathcal{P}$ with or without an independently identical distribution(i.i.d.).*

[2] The proofs are straightforward from the above definitions.

Lemma 12.3.2. *A discrete-valued dataset contains similarity patterns if and only if there exist periodic patterns in state space S and there exist similarity patterns in probability space $\mathcal{P}$ with or without an independently identical distribution.*

12.3.2 Hidden Markov Models

In a hidden Markov model (HMM), an underlying and unobserved sequence of states follow a Markov chain with a finite state space, and the probability distribution of the observation at any time is determined only by the current state of that Markov chain. In this subsection we briefly introduce the hidden Markov time series models which are based on an unobserved stationary Markov chain.

Let $\{S_t : t \in \mathsf{N}\}$, where N is the set of natural numbers, be an irreducible homogeneous Markov chain on the state space $\mathcal{S} = \{s1, s2, \ldots, sm\}$, with transition probability matrix Δ. Here, $\Delta = (\eta_{ij})$, where for all states i and j, and times t:

$$\eta_{ij} = \mathsf{P}(S_t = sj \mid S_{t-1} = si). \tag{12.2}$$

For $\{S_t\}$, there exists a unique, strictly positive, stationary distribution $\gamma = (\gamma_1, \ldots, \gamma_m)$, where we suppose $\{S_t\}$ is stationary, so that γ is, for all t, the distribution of S_t.

Suppose there exists a non-negative random process $\{\xi_t;\ t \in \mathsf{N}\}$ such that, conditional on $S^{(T)} = \{S_t : t = 1, \ldots, T\}$, the random variables $\{\xi_t : t = 1, \ldots, T\}$ are mutually independent and, if $S_t = si$, then ξ_t takes the value v with probability π^t_{vi}. That is, for $t = 1, \ldots, T$, the distribution of ξ_t conditional on $S^{(T)}$ is given by the formula

$$\mathsf{P}(\xi_t = v | S_t = si) = \pi^t_{vi}, \tag{12.3}$$

where the probabilities π^t_{vi} are called the "state-dependent probabilities." If the probabilities π^t_{vi} do not depend on t, then the subscript t will be omitted.

A useful device for depicting the dependent structure of such a model is the conditional independence graph. Fig. 12.2 displays the independence of the observations $\{C_t\}$ given the states $\{S_t\}$ occupied by the Markov chain, as well as the conditional independence of C_{t-1} and C_{t+1} given C_t.

12.3.3 Local Polynomial Modeling

The key idea of local modeling is explained in the context of least squares regression models. Regression models are generalized linear data analysis models. Thus, they also include many commonly-used distributions such as the Gaussian, Binomial, Poisson, and Gamma distributions and can be used to analyze both discrete and continuous types of data.

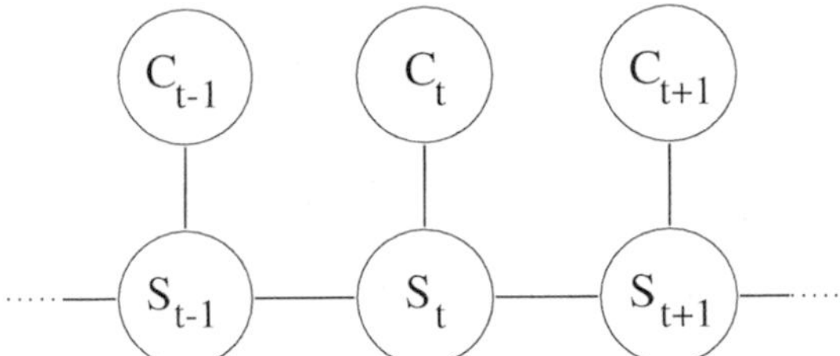

Fig. 12.2. Conditional independence graph of Hidden Markov Model

Without loss of generality, we consider the bivariate data (X_1, Y_1), ..., (X_n, Y_n), which form an independent and identically-distributed sample from a population (X, Y). For given pairs of data (X_i, Y_i), $i = 1, 2, \ldots, \mathsf{N}$, we can regard the data as being generated from the model

$$\mathbf{Y} = m(\mathbf{X}) + \sigma(\mathbf{X})\varepsilon, \tag{12.4}$$

where $E(\varepsilon) = 0$, $Var(\varepsilon) = 1$, and X and ε are independent. We always denote the conditional variance of Y given $X = x_0$ by $\sigma^2(x_0)$ and the density of X by $f(\cdot)$. We then approximate the unknown regression function $m(x)$ locally by a polynomial of order p in a neighbourhood of x_0:

$$m(x) \approx m(x_0) + m'(x_0)(x - x_0) + \ldots, + \frac{m^{(p)}(x_0)}{p!}(x - x_0)^p. \tag{12.5}$$

This polynomial is fitted locally by a weighted least squares regression problem:

$$\mathsf{minimize}\{\sum_{i=1}^{n}\{Y_i - \sum_{j=0}^{p}\beta_j(X_i - x_0)^j\}^2 K_\delta(X_i - x_0)\}, \tag{12.6}$$

where δ is a constant greater than zero, and $K_\delta(\cdot)$ has K with a kernel function assigning weights to each datum point.

From the generalized limit theorem point of view, all real-world datasets are functional observations, where the data observed are, or can be considered to be, functions observed continuously. The analysis of such a dataset is called *Function Data Analysis.* In our new method we always think of a functional observation as a single datum rather than as a real dataset on its own. In this paper, for example, we think of the local polynomial functional observation as "Exchange rate data" and then apply our new method to two kinds of groups by using statistical techniques for finding patterns.

We should mention some important issues on parameter δ, p, and kernel function K. First of all, if δ is too large, then it causes a large modeling bias; if it is too small, then it results in noisy estimates. Second, since the modeling bias is primarily controlled by δ and we also consider computational cost, according to the basic idea of our new method, the order of p of the polynomial function is small. Finally, for kernel function K, since no negative weight K should be used, we choose $K(z) = \frac{3}{4}(1 - z^2)$ as a kernel function (Epanechnikov kernel), which minimizes the asymptotic MSE (i.e., mean squared error) of the resulting local polynomial estimators.

12.4 Local Polynomial Hidden Markov Models

A real-world temporal dataset may contain different kinds of patterns such as complete or partial similarity patterns and periodicity patterns, and complete or partially different order patterns. There are many different techniques for efficient sequence or subsequence matching to find patterns in a DTS (e.g, [12.2]). A limitation on those techniques is that they do not provide a coherent language for expressing prior knowledge and handling uncertainty in the matching process. The existence of different patterns also does not guarantee the existence of an explicit model. In this section we present our new method of pattern discovery on a DTS which uses HMM as an explicit model, combining Markov chain models and local polynomial models to deal with certain kinds of uncertainty.

12.4.1 Modeling DTS

One of the aims in time series analysis is to find a statistical model for a given dataset. We stipulate that for any successive pair of time points over which elements of a DTS are observed, we have t_{i+1} - $t_i = c$ (a unit constant). We then use the *time gap* as a time variable between events instead of natural time (e.g., day).

According to our new method, the data-generated model (12.4) for the structural base sequence and value-point process data model has become:

$$\mathbf{U} = m(\mathbf{V}) + \sigma(\mathbf{V})\varepsilon, \tag{12.7}$$

where $\mathbf{U}$ is a given sample sequence under a sub-structural base or given structural base of the sample sequence.

We first apply pattern searching on state-space $\mathcal{S}$ and the structural base sequence as a set of vector sequences $\{\mathbf{S_1}, \mathbf{S_2}, \cdots, \mathbf{S_m}\}$, where $\mathbf{m}$ is the size of the set and each $\mathbf{S_i} = (s1, s2, \ldots s8, s9)^T$ denotes the 9-dimensional observation on an object that is to be assigned to a prespecified group. Here, each si is just a structural sequence defined over $\mathcal{S}$, and indexed by elements of T. The problem of structural pattern discovery for the sequence and each subsequence $\mathbf{S}_{ij} = \{s_{i1}, s_{i2}, \cdots, s_{ij} : 1 \leq i \leq 9, 1 \leq j \leq m\}$ of $\mathbf{S}$ on finite-state space can then be formulated as an irreducible homogeneous Markov chain with a distance function on the state space $\mathcal{S}$.

We may also view the value-point process data model as a local polynomial model:

$$y(x) = \beta_0 + \beta_1 x + \ldots + \beta_p x^p + \varepsilon. \tag{12.8}$$

The problem of the value-point pattern discovery can then be formulated as the local polynomial analysis of discrete-valued time series. In fact, many practical problems in temporal data mining related to local modeling can be explained in the context of regression models.

12.4.2 Structural Pattern Discovery

We now introduce an approach to discovering patterns in structural sequences which uses a distance measure within Markov chain models.

From the viewpoint of our method in temporal data analysis, we use squared distance functions which are provided by a class of positive semidefinite quadratic forms. Specifically, if $\mathbf{u} = (u_1, u_2, \cdots, u_p)$ denotes the p-dimensional observation of each different distance of patterns in a state on an object that is to be assigned to one of the g prespecified groups,[3] then for measuring the squared distance between $\mathbf{u}$ and the centroid of the ith group, we can consider the function

$$D^2(i) = (\mathbf{u} - \bar{\mathbf{y}})' \mathbf{M}(\mathbf{u} - \bar{\mathbf{y}}), \tag{12.9}$$

where $\mathbf{M}$ is a positive semi-definite matrix to ensure that $D^2(i) \geq 0$. Different choices of the matrix $\mathbf{M}$ lead to different metrics, and the class of squared distance functions represented by the above equation is not unduly narrow.[4]

In our new models, we use the matrix $\mathbf{M}$ as a generalized distance of the "unknown" from the centroid of the ith group in a p-dimensional space of the original variables. Specifically, let $\mathbf{M} = \mathbf{C}_\mathrm{i}^{-1}$ and $\mathbf{C}_\mathrm{i}^{-1}$ be the covariance matrix derived from the sample observations. An important feature for using the matrix $\mathbf{M} = \mathbf{C}_\mathrm{i}^{-1}$ is that we can use the distribution of the time-gap for either a linear or non-linear relationship among the dataset. For example, given two conditional distributions with about the same covariance matrices and means that are far away from each other, we use the *Mahalanobis distance* instead of the Euclidean distance function (e.g., distribution time-gap analysis).

12.4.3 Value-Point Pattern Discovery

We introduce here an enhancement to the local polynomial modeling approach through functional data analysis.

[3] If there exists a large number of groups g each of which has p-dimensional space, then we apply an ad hoc procedure based on using the first few *discriminant coordinates*. In recent years, there have been a lot of work in the study of a small number of groups, but only a few for a large number of groups.

[4] For instance, in our experiment, if we let $\mathbf{M} = \mathbf{I}$, then we obtain the familiar Euclidean squared distance between the "unknown" and the centroid of the ith group in a p-dimensional space of the responses (this is akin to cluster analysis).

On the value-point pattern discovery, given the bivariate data (X_1, Y_1), $\cdots$, (X_n, Y_n), one can replace the L_2-loss function (e.g., Eq. 12.6) in section 4.3

$$\sum_{i=1}^{n}\{Y_i - \sum_{j=0}^{p}\beta_j(X_i - x_0)^j\}^2 K_\delta(X_i - x_0)$$

by

$$\sum_{i=1}^{n}\ell\{Y_i - \sum_{j=0}^{p}\beta_j(X_i - x_0)^j\}K_h(X_i - x_0), \tag{12.10}$$

where $\ell(\cdot)$ is a loss function. For the purpose of predicting future values, we use a special case of the above function with $\ell_\alpha(t) = |t| + (2\alpha - 1)t$. This is often called *quantile regression* (or *local quantile regression*).

One of the important applications of quantile regression is to construct *predictive intervals*. A predictive interval is an interval that forecasts, with certain coverage probability, the future value of the response variable Y for a given covariate $X = x$. More precisely, the (1 - α)100% predictive interval is the interval $[\xi_{\alpha/2}(x),\ \xi_{1-\alpha/2}(x)]$. This kind of an interval can easily be constructed using quantile regression.

Suppose that the observations can be modeled as follows:

$$Y_i = m(X_i) + \sigma(X_i)\varepsilon_i. \tag{12.11}$$

The quantile function can then be written as

$$\xi_\alpha(x) = m(x) + H_0^{-1}(\alpha)\sigma(x), \tag{12.12}$$

where $H_0(\cdot)$ denotes the distribution function of ε.[5]

12.4.4 Local Polynomial Hidden Markov Model for Pattern Discovery (LPHMM)

To use LPHMM in pattern discovery, we combine the two kinds of pattern discovery discussed above. In structural pattern searching, let structural sequence $\{S_t : t \in \mathsf{N}\}$ be an irreducible homogeneous Markov chain on the state space $\mathcal{S} = \{s1, s2, \ldots, sn\}$, with the transition probability matrix Δ.

In value-point pattern searching, suppose the local polynomial value sequence is a non-negative random process $\{V_t;\ t \in \mathsf{N}\}$ such that, conditional on $S^{(T)} = \{S_t : t = 1, \ldots, T\}$, the local polynomial value sequences $\{V_t : t = 1, \ldots, T\}$ are mutually independent and, if $S_t = si$, then

[5] When $\alpha = 0.5$, then $H_1^{-1}(\alpha) = 0$, where $H_1^{-1}(\alpha)$ is the αth sample quantile of the residuals and we obtain a polynomial fit for the $\xi_\alpha(x)$, such as $\xi_\alpha(x) = \alpha_0 + \alpha_1 + \ldots + \alpha_{p+3}x^{p+3}$.

V_t takes the value v with probability π^t_{vi}. That is, for $t = 1, \ldots, T$, the distribution of V_t conditional on $S^{(T)}$ is given by

$$\mathsf{P}(V_t = v | S_t = si) = \pi^t_{vi}. \tag{12.13}$$

The above models $\{V_t\}$ are defined as *Local Polynomial-Hidden Markov Models*. We now give one example:[6] "Binomial-hidden Markov model (BHMM)."

Example

Suppose that, if $S_t = si$, V_t has a Binomial distribution with parameters n_t (a known positive integer) and p_i. That is, the conditional Binomial distribution of V_t has parameters n_t and $p(t)$, where

$$p(t) = \sum_{i=1}^{m} p_i W_i(t),$$

and $W_i(t)$ is, as before, the indicator of the event $\{S_t = si\}$. For $v = 0, 1, \ldots, n_t$:

$$\pi^t_{vi} = \binom{n_t}{v} p_i^v (1 - p_i)^{n_t - v}.$$

The model $\{V_t\}$ is defined as a Binomial-hidden Markov model. In this case there are m^2 parameters: m parameters p_i, and $m^2 - m$ transition probabilities η_{ij}, e.g. the off-diagonal elements of Δ, to specify the "hidden Markov chain" $\{S_t\}$.

12.5 Experimental Results

We now illustrate our method by applying it to the analysis of daily exchange rates between the U.S. dollar and the Australian dollar.[7] The data consist of daily exchange rates for each business day between 3 January 1994 and 9 August 1999. All experiments were performed on a Unix system and a Windows NT 4.0 system; the prototype was written in Gawk language. The time series is plotted in Fig. 12.3.

There are three steps of experiments on "Daily Foreign Exchange Rates" between the U. S. dollar and the Australian dollar in order to investigate "Exchange Rates Patterns":[8]

---

[6] In fact, all hidden distributional models are based on conditional distributions.

[7] Due to space limitations, we only present a small subset of our experimental results.

[8] The Federal Reserve Bank of New York for the trade-weighted value of the dollar = index of the weighted average exchange value of the U.S. dollar against the Australian dollar: `http://www.frbchi.org/econinfo/finance/finance.html`.

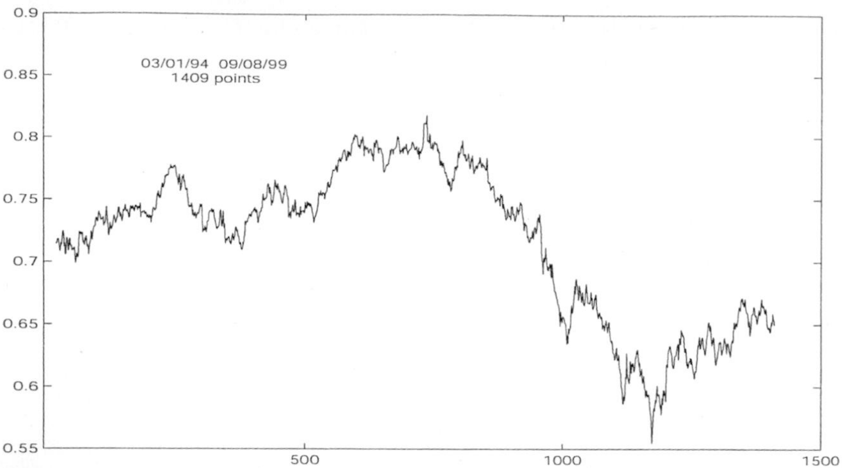

Fig. 12.3. 1409 business exchange rates between the U.S. dollar and the Australian dollar, since 1994

- structural pattern searching
- exchange rate values pattern searching
- a combination of structural and value-point pattern searching based on HMM.

12.5.1 Structural Pattern Searching

We investigate the sample of the structural base to test the naturalness of the similarity and periodicity on a Structural Base distribution. The sample dataset consists of daily exchange rates for each business day between 3 January 1994 and 3 January 1999. The length of this discrete-valued time series is about 1409 points.

In this experiment we only consider eight different states that are = {s1, s2, s3, s4, s6, s7, s8, s9} = $\{(X_j, S_u, S_u), (X_j, S_u, S_s), (X_j, S_u, S_d), (X_j, S_s, S_u), (X_j, S_s, S_d), (X_j, S_d, S_u), (X_j, S_d, S_s), (X_j, S_d, S_d)\ \}$. The reason for omitting state $s5$ is that the daily exchange rate has been changing for three successive business days and state 5 (i.e, s5 = (X_j, S_s, S_s)) means that there are no changes for three successive days; hence, state s5 is an empty state.

For example, if the original data sequence is

.6845, .686, .684, .6865, .6905, .6902, .6908, .69, .696, .701, .7034, .7025, .703, .708, .7085, ..., .6175, .6204, .6155, .6165, .6195, .6208, .6235, .6187, .6231, .6211, .617, .6145, .6114, .611, .607, .6114, .6129, .6123,

then the structural sequence will be

s1, s3, s7, s1, s3, s7, s3, s7, s1, s1, s3, s7, s1, ..., s7, s1, s1, s3, s7, s3, s7, s3, s9, s9, s9, s9, s9, s7, s1, s3.

The first step of structural pattern searching is to find patterns between the states by using distance functions. For example, we look for patterns of timegaps for individual states. We then look for patterns of timegaps between pairs of states, for example, between states $s1$ and $s3$.

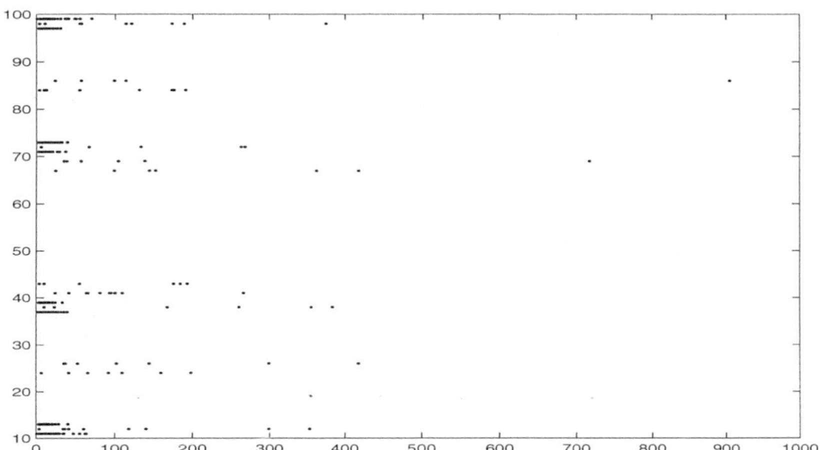

Fig. 12.4. Plot of time distances between states for all 8 states in 1409 business days

In Fig. 12.4, the x-axis represents the distances of time between two events.[9] The y-axis represents patterns of moves among different states, for example, the fact that a pattern moves from state one (e.g., s1) to state two (e.g., s2) is represented at point 12 on the y-axis. For example, given a point P(x, y) = (300, 12) in Fig. 12.4, we can interpret the point as a pattern (moving from state 1 to state 2) that has occurred at least twice. The timegap between the occurrences of the pattern is 300 days long.

In summary, the plot in Fig. 12.4 explains two facts:

1. There exists a hidden periodic distribution which corresponds to patterns on the same line (e.g., pattern $s_{i,j}$ between state s_i to state s_j) with different distances.
2. There exist partial periodic patterns on and between the same lines and different lines (e.g., at different states, $s_{i,j}$ and $s_{k,h}$.). This means that different pairs of states do not always exhibit the same pattern structures.

[9] The time distance between two events is often called time gap $\mathcal{G}_t$. Sometimes timegap $\mathcal{G}_t$ is used as a variable in statistical analysis methods, for example, in this experiment.

For example, in Fig. 12.5, at $y = 24$ and $y = 26$ and $0 < x < 10$, the pattern of the two points of $s_{2,4}$ and $s_{2,6}$ appeared a few times with other patterns (e.g., points) like $s_{3,7}$, and then it disappears even though the pattern for $s_{3,7}$ appears again in a time distance longer than 10 days.

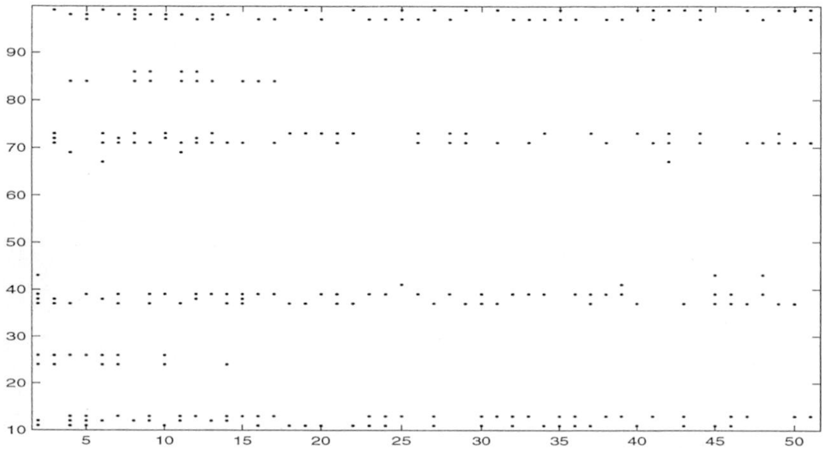

Fig. 12.5. Plot of time distances not exceeding 50 business days between the patterns for all 8 states

To explain this further, we can look at the plot of distances between patterns at a finer granularity over a selected portion of the daily exchange rates. For example, in Fig. 12.5 the dataset consists of daily exchange rates for time distances not exceeding 50 business days. In the same state of Fig. 12.5, each point represents the distance between two occurrences of the same pattern. And between some combined pattern classes there exist some similar patterns; moreover, for example, between patterns $s_{3,7}$ and $s_{7,3}$.

In Fig. 12.6, the y-axis represents how many times the same distance is found between repeating patterns, and the x-axis represents the distance between the two occurrences of each repeating pattern. In other words, we classify repeating patterns based on a distance classification technique. Again, we can look at the plot over a selected portion to observe the distribution of distances in more detail.

In Fig. 12.7, the dataset consists of daily exchange rates for the time distance not exceeding 50 business days. It can be observed from Fig. 12.6 and 12.7 that the distribution of time distances is a normal distribution. The significance of the time distance normal distribution is that between each time distance interval (for example, an interval between 0 and 25 in Fig. 12.7), the frequency of patterns with regards to their time distances is a normal distribution. This also explains why partial periodic patterns exist in Fig. 12.4.

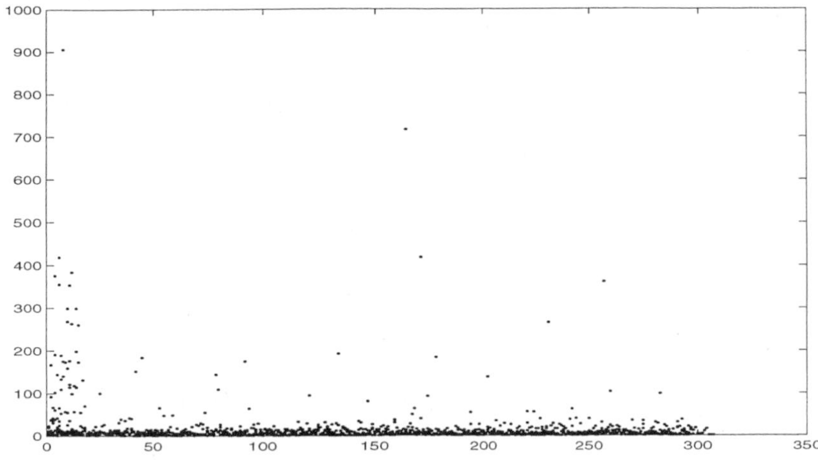

Fig. 12.6. Plot of all time distance repeating patterns between the same state for all 8 states in 1409 business days

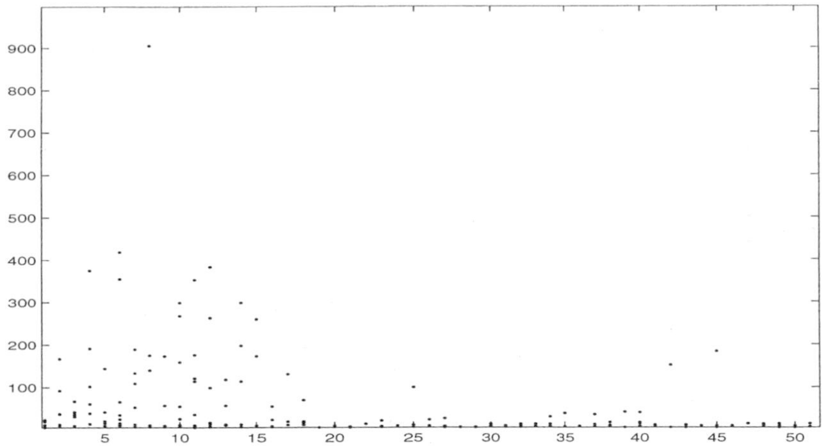

Fig. 12.7. Plot of different patterns appearing in different distances for the first 50 business days

In summary, some results for the structural base experiments are as follows:

- Structural distribution is a hidden periodic distribution with a periodic length function $f(t)$ (at this stage we are not concerned with the exact form of this function, however, there are techniques available to approximate the form of this function such as higher-order polynomial functions).
- Some partial periodic patterns exist based on a distance shifting.
- Some similarity patterns exist with a small distance shifting.

12.5.2 Value-Point Pattern Searching

We now illustrate our technique to construct predictive intervals on the value-point sequence to search for periodic or similar patterns. We consider the sample size of 1409 data points (the data consist of daily exchange rates for every business day since 03/01/1994). The linear regression of the valuepoint of X_t against X_{t-1} explains about 99% of the variability of the data sequence, but it does not help us much in the analysis and prediction of future exchange rates.

In light of our structural base experiments, we find that the series $Y_t = X_t - X_{t-2}$ has non-trivial autocorrelation. The correlation between Y_t and Y_{t-1} is 0.4898. In this experiment, linear regression modeling has been used as a special case of local polynomial modeling (e.g., the order p of the local polynomial model takes value 1 in Eq. 12.8):

$$y(x) = \beta_0 + \beta_1(x - x_0) + \varepsilon.$$

It is more convenient to work with the matrix notation for the solution to the above least squares problem. The first stage of value-point pattern searching when using a local polynomial modeling is to detect the characteristics of those records by implementing the local linear regression analysis. We may assume that the local linear model is:

$$\mathbf{Y} = \mathbf{X}\beta + \varepsilon. \tag{12.14}$$

Let $\mathbf{Y} = (Y_1, \cdots, Y_n)^T$, $\hat{\beta} = (\hat{\beta_0}, \cdots, \hat{\beta_p})^T$ and the linear model based upon least square estimation LSE be[10]

$$\mathbf{Y}' = \mathbf{X}'\beta + \varepsilon'$$

$$\hat{\beta} = (\mathbf{X}^T\mathbf{X})^{-1}\mathbf{X}^T\mathbf{Y}.$$

We then have:

$$\hat{\beta} \sim N(\beta, Cov(\hat{\beta})).$$

Particularly, for $\hat{\beta}_i$ we have

$$\hat{\beta}_i \sim N(\beta_i, {\sigma_i}^2),$$

[10] An alternative result on the solution vector is provided by weighted least squares theory: let $\mathbf{W}$ be the $n \times n$ diagonal matrix of the weights:

$$\mathbf{W} = diag\{K_\delta(X_i - x_0)\}.$$

and is given by

$$\hat{\beta} = (\mathbf{X}^T\mathbf{W}\mathbf{X})^{-1}\mathbf{X}^T\mathbf{W}\mathbf{Y}.$$

where ${\sigma_i}^2 = \sigma^2 a^{ii}$, and a^{ii} is the ith diagonal element of $(\mathbf{X}^T\mathbf{X})^{-1}$.

For each valuepoint, we may fit a local linear model as above and the parameters can be estimated under LSE. Since $\hat{\beta}_i$ follows a normal distribution, we may have the t-test for the means of each given sample set. If we cannot reject the hypothesis, then we have the a possibility that the features remain uncovered in the fluctuation periods of observation curves. Therefore, we first remove the trend effect of each curve from the original record by subtracting the above regression function at x from the corresponding value to obtain a comparatively stationary series.

The observations can then be modeled as a special case of the local polynomial regression function, say

$$Y_t = X_t - X_{t-2} + \sigma(X_t)\varepsilon_t, \qquad t = 1, 2, \ldots, N \tag{12.15}$$

and then the following new series

$$y(t) = Y_t + Y_{t-1} + \varepsilon_{t'} = \mathbf{X}\beta + \varepsilon_{t'} \qquad t = 1, 2, \ldots, N \tag{12.16}$$

may be obtained.

We also consider the $\varepsilon(t)$ as an auto-regression $AR(2)$ model

$$\varepsilon_{t'} = a\varepsilon_{t'-1} + b\varepsilon_{t'-2} + e_{t'}, \tag{12.17}$$

where a, b are constants dependent on the sample dataset, and $e_{t'}$ is a small variance constant which can be used to improve the predictive equation. Our analysis focuses on the series Y_t, which is presented in Fig. 12.8 . It is a scatter plot of lag 2 differences: Y_t against Y_{t-1}. Fig. 12.8 tells us that Y_t and Y_{t-1} have a *strong positive relationship*.

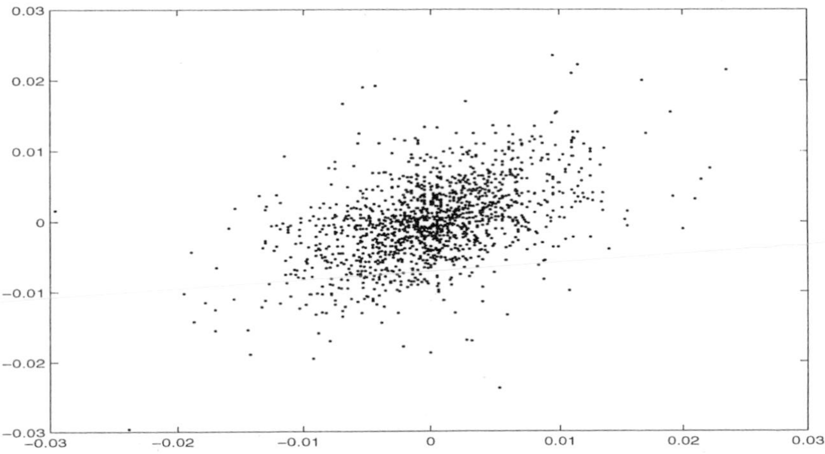

Fig. 12.8. Scatter plot of lag 2 differences: Y_t against Y_{t-1}

We obtain the exchange rates model according to the nonparametric quantile regression theory (from Eq. 12.15):

$$Y_t = 0.4732Y_{t-1} + \varepsilon_t. \tag{12.18}$$

From the distribution of ε_t, the $\varepsilon(t)$ can be modeled as an $AR(2)$

$$\varepsilon_t = 0.2752\varepsilon_{t-1} - 0.4131\varepsilon_{t-2} + e_t, \tag{12.19}$$

with a small $Var(e_t)$ (about 0.00041) to improve the predictive equation.

According to Eq. (12.18) and Eq. (12.19), there is a important result for the valuepoint of the experiments:

- The value-point distribution is a hidden and approximate Poisson distribution.[11] It tells us that the number of increasing (decreasing) *exchange rates* is similar for a certain period of time.

12.5.3 LPHMM for Pattern Searching

Using the *Local Polynomial-Hidden Markov Model for Pattern Discovery*, we combine the two kinds of pattern discovery discussed above. For structural pattern searching, let structural sequence $\{S_t : t \in \mathsf{N}\}$ be an irreducible homogeneous Markov chain on the state space $\mathcal{S} = \{s1, s2, \ldots, sn\}$, with the transition probability matrix Δ. The steps in the pattern discovery are as follows:

Using structural sequence and hidden Markov model properties

The first step uses the structural sequence $\{S_t : S_t \in \mathcal{S}\}$ and finds its transition probability matrix (`TPM`) (e.g., probability of transition from state si to state sj) Δ.[12] According to the properties of Markov models, we then know that $\{S_t : S_t \in \mathcal{S}\}$ is an irreducible homogeneous Markov chain on the state space $\{s1, \ldots, s4, s6, \ldots, s9\}$. For example, the transition probability matrix (`TPM`) Δ of this experiment is as follows:

$$\Delta = \begin{array}{c|cccccccc|} & s1 & s2 & s3 & s4 & s6 & s7 & s8 & s9 \\ s1 & 0.4765 & 0.0302 & 0.4933 & 0 & 0 & 0 & 0 & 0 \\ s2 & 0 & 0 & 0 & 0.5 & 0.5 & 0 & 0 & 0 \\ s3 & 0 & 0 & 0 & 0 & 0 & 0.4815 & 0.0202 & 0.4983 \\ s4 & 0.5882 & 0 & 0.4118 & 0 & 0 & 0 & 0 & 0 \\ s6 & 0 & 0 & 0 & 0 & 0 & 0.5 & 0 & 0.5 \\ s7 & 0.4932 & 0.0170 & 0.4898 & 0 & 0 & 0 & 0 & 0 \\ s8 & 0 & 0 & 0 & 0.6667 & 0.3333 & 0 & 0 & 0 \\ s9 & 0 & 0 & 0 & 0 & 0 & 0.4723 & 0.0293 & 0.4984 \end{array} .$$

[11] Distributions are identified by statistical techniques.

[12] Probability values in matrix Δ are calculated from the structural sequence for each state in state-space $\mathcal{S}$.

We are always interested in the future distribution of TPM, $f(t) = \Delta^t$. In light of the structural patterns found in Section 12.5.1 and their TPM, we also find the TPM:[13]

$$\lim_{t\to\infty} \Delta^t = \begin{pmatrix} 1 \\ 1 \\ 1 \\ 1 \\ 1 \\ 1 \\ 1 \\ 1 \end{pmatrix} \times \begin{pmatrix} 0.2365 \\ 0.0111 \\ 0.2373 \\ 0.0136 \\ 0.0096 \\ 0.2348 \\ 0.01198 \\ 0.2452 \end{pmatrix}^T$$

According to the definition of a Markov chain, this is a stationary distribution of a Markov chain. From this results of the powers of the TPM , we cannot use present exchange rates to predict distant future exchange rates if t is large (e.g., the largest element in the matrix is less than 0.3), but we are able to predict near future exchange rates if $t = t_n$ and $n < N$, where N is constant and $<< \infty$.

Using Local Polynomial-Hidden Markov Model for Pattern Discovery

According to the results of value-point pattern searching, the value-point distribution is a hidden and approximate Poisson distribution.

In state space $\mathcal{S}$, the structural sequence $\mathsf{S} = \{si_1, si_2, \cdots, si_j : 1 \le i \le 9, 1 \le j \le m\}$ on a finite-state space can be formulated as an irreducible homogeneous Markov chain by a distance function. In probability space $\mathcal{P}$, the value-point sequence can be created by the following quantile regression function:

$$Y_t = 0.4732 Y_{t-1} + \varepsilon_t,$$

and $\varepsilon(t)$ can be modeled as an $AR(2)$

$$\varepsilon_t = 0.2752\varepsilon_{t-1} - 0.4131\varepsilon_{t-2} + e_t.$$

If $S_t = si$, then Y_t has a Poisson distribution with mean λ_i. Let $E(Y_t \mid S_t)$, the conditional mean of Y_t, be

$$\mu(t) = \sum_{i=1}^{m} \lambda_i W_i(t),$$

where random variable $W_i(t)$ is the indicator of the event $\{S_t = si\}$. The state-dependent probabilities are then given for all non-negative v by the formula:

[13] The Ergodic theorem in stationary Markov chain, e.g.,[12.17]

$$\pi_{vi}^t = \frac{e^{-\lambda_i}\lambda_i^v}{v!}.$$

Model Y_t is defined as a local polynomial-hidden Markov model. In fact, the model can also be called a Poisson-hidden Markov model.

We now use a quantile regression function to build a statistical model. In Section 12.4.3, the predictive interval is the interval $[\xi_{\alpha/2}(x), \xi_{1-\alpha/2}(x)]$. Suppose our prediction of the future exchange rate of the value-point sequence is a non-negative random process $\{Y_t;\ t \in \mathsf{N}\}$ and satisfies the equation $Y_t = \alpha Y_{t-1} + \theta_t$. The quantile regression function of dataset Y_t is nearly linear when $\alpha = 0.25, 0.5$, and 0.75, but the fan shape for these values is not good enough. To deal with this problem, we use the transition probability matrix of the structural sequence to control the structural sequence of the future exchange rates.

Suppose the distributions of the sequence of the transition probability matrix under time order $\Delta_1, \Delta_2, \cdots, \Delta_t (t \in \mathsf{N})$ corresponds to the prediction value-point sequence in the predictive interval.[14] In other words, Δ_1 corresponds to all the original data (1409 points); Δ_2 corresponds to all the original data plus the first predictive interval (for example, seven days); Δ_3 corresponds to all the original data plus the first two predictive intervals and so on. So, Δ_i always corresponds to all the historical data before the computation of the next predictive interval.

For example, under the prediction of future exchange rates for the next 150 business days, we combine the simple equation $Y_t = 0.4732 Y_{t-1}$ with an average error of -0.000057 (where the average error is calculated from Eq. 12.19) and structural distribution $\{S_t\}$ to build up a new model which is called the local polynomial-hidden Markov model so as to search for global temporal patterns.

The main method is described as follows. The following steps are repeated (for $i = 1, \ldots, \mathsf{N}$) until the predicted values are out of the range of the i.i.d. error term e_t.

1. Given the historical data and the structural sequence for Δ_i, we use Eq. (12.18) and (12.19) to form a non-negative value sequence for the future values for the next predictive interval.[15] For example, from Eq. (12.18) and (12.19) we have
$$X_t = 0.4732(X_{t-1} - X_{t-3}) + X_{t-2} + 0.2752\varepsilon_{t-1} - 0.4131\varepsilon_{t-2} + e_t$$
where the X_{t-1}, X_{t-2} and X_{t-3} are historical data for predicting future data X_t; ε_{t-1}. ε_{t-2} are estimated from historical data, and e_t is an i.i.d. variable.
2. Now we can construct the corresponding structural sequence incrementally from the historical data (including the computed future values).

[14] Z_t is used to decide the length of a local prediction interval.

[15] In fact, the non-negative value sequence is a Poisson distribution.

3. We compute Δ_{i+1} from the historical data and the new structural sequence as a stationary Markov chain sequence.
4. Combine both sequences by conditional distribution so that we can use the distribution function for predicting future values for the next predictive interval.

According to the method of conditional distribution function from statistics, it is not surprising that results from structural pattern searching have been used to decide the structural behaviour of future exchange rates. The results from value-point pattern searching have been used to decide the numerical behaviour of future exchange rates. The combined results from our experiments with exchange rates are as follows:

- We are only able to predict near future exchange rates by using all present information.
- There does not exist any full periodic pattern, but there exist some partial periodic patterns do exist.
- Some similar patterns with a small distance shifting do exist.

Fig. 12.9 plots the actually-observed series and predicted series from the local polynomial-hidden Markov model.

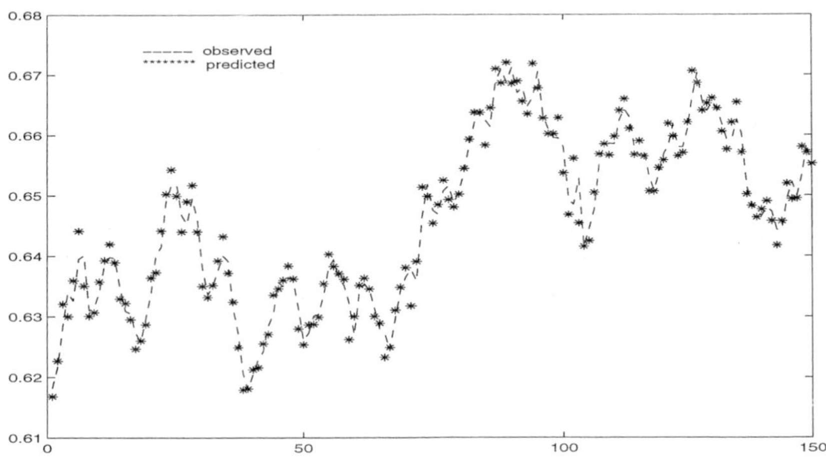

Fig. 12.9. Plot of future exchange rates only for 150 business days by using the local polynomial-hidden Markov model

12.6 Related Work

According to general pattern theory, objectives in pattern searching can be classified into three categories:

- Create representations in terms of algebraic systems with probabilistic superstructures intended for the representation and understanding of patterns in nature and science.
- Analyze the regular structures from the perspectives of mathematical theory.
- Apply regular structures in particular applications, and implement the structures by algorithms and code.

In recent years, various studies have been conducted in temporal databasess to search out different kinds of and/or different levels of patterns, but these studies have only covered one or two of the above categories. For example, most researchers use statistical techniques such as the Metric-distance based technique, the Model-based technique, or a combination of techniques (e.g, [12.8], [12.25]) to solve various pattern discovery problems such as those in periodic pattern searching, e.g, [12.7, 12.10] and for similar pattern searching.

Some studies cover all the above three categories for searching patterns in data mining. For instance, Agrawal *et al* [12.3] present a "shape definition language," called $\mathcal{SDL}$, for retrieving objects based on shapes contained in the histories associated with these objects. Das *et al* [12.5] describe adaptive methods which are based on similar methods for finding rules and discovering local patterns, and Baxter *et al* [12.4] consider three alternative feature vectors for representing variable-length patient health records.

Our work is different from their studies. First, we use a statistical language to perform all the search work. The significance of the use of a statistical language is that it allows us to build up understandable models for prediction. Second, we divide the data sequence, or data vector sequence, into two groups: one group is structural-based and the other is pure value-based.

In the structural-based group our techniques are a combination of the works of Agrawal[12.3] and Baxter [12.4]. In this group we use distance measuring functions on structural sequences based on each state in state-space $\mathcal{S}$. This approach is similar to that of G. Berger's work in [12.23]. For example, we can use a model-based clustering program, which is called Snob, on state-space $\mathcal{S}$ (such as [12.4] did) to find clusters, but it does not help us very much to understand pattern distributions within the dataset.

In the value-based group we apply statistical techniques such as frequency distribution functions to deal with values under conditions of structural distribution. This is similar to the Das work [12.5], but it is stronger. The main reason is that we use local polynomial techniques (e.g., [12.20]) based on the results of structural pattern searching. In [12.20], there is an example for using quantile regression to construct predictive intervals, but not pure local polynomial techniques. We use each value sequence according to each state sequence which is separated by a time distance. Our work combines significant information from two groups (e.g., structural distribution functions and values distribution function and prediction intervals) to gather information which is behind the dataset.

12.7 Concluding Remarks

This paper has presented a new method combining hidden Markov models and local polynomial analysis. The rough decision for pattern discovery comes from the structural level, which is a collection of certain predefined similarity patterns. The clusters of similarity patterns are computed at this level by the choice of certain distance measures. The value-point patterns are decided in the second level and the similarity and periodicity of the DTS are extracted. In the final level we combine structural and value-point pattern searching into the HMM model to obtain a global pattern picture and to better understand the patterns in the dataset.

Other approaches to finding similar and periodic patterns have been reported in [12.11, 12.14, 12.12, 12.15, 12.13]; there, the model used is based on hidden periodicity analysis. However, we find that using different models at different levels produces better results. "Daily Foreign Exchanges Rates" data are used to find the similar patterns and periodicities. The existence of similar and partial periodic patterns are observed even though there is no clear full periodicity observed in this analysis.

The method guarantees finding different patterns with a structural and valued probability distribution of a real dataset. The results of preliminary experiments have been shown to be promising. The work will be extended by new DTSB, such as temporal health records and scientific time series databases.

Acknowledgements

This research has been supported in part by an Australian Research Council (ARC) grant and a Macquarie University Research Grant (MURG). The English of this paper has been significantly polished by the volume editor, Shu-Heng Chen. His corrections of some technical mistakes made in an early draft and painstaking efforts to make this paper readable is highly appreciated.

References

12.1 Agrawal, R., Srikant, R. (1994): Fast algorithms for mining association rules in large databases. In: Bocca, J., Jarke, M., Zaniolo, C. (Eds.), 20th International Conference on Very Large Data Bases, September 12–15, 1994, Santiago, Chile Proceedings, 487–499, Morgan Kaufmann Publishers

12.2 Agrawal, R., Lin, K. -I.., Sawhney, H. S., Shim, K. (1995): Fast similarity search in the presence of noise, scaling, and translation in time-series databases. In: Proceedings of the 21st VLDB Conference, 490–501, Morgan Kaufmann
12.3 Agrawal, R., Psaila, G., Wimmers, E. L., Zait, M. (1995): Querying shapes of histories. In: Proceedings of the 21st VLDB Conference, 490–501, Morgan Kaufmann
12.4 Baxter, R. A., Williams, G. J., He, H. (2001): Feature selection for temporal health records. available from Graham.Williams@cbr.dit.csiro.au, APR
12.5 Das, G., Lin, K., Mannila, H., Renganathan, G., Smyth, P. (1998): Rule discovery from time series. In: Proceedings of the international conference on KDD and Data Mining (KDD-98)
12.6 Han, J. W., Yin, Y., Dong, G. (1999): Efficient mining of partial periodic patterns in time series database. In: Proceedings of the 15th international conference on Data Engineering, Sydney, Australia, 106–115, IEEE CS Press, March
12.7 Elder IV, J., Pregibon, D. (1995): A statistical perspective on knowledge discovery in databases. In: Fayyad, U., Piatetsky-Shapiro, G., Smyth, P., Uthurusamy, R. (Eds.), Advances in Knowledge Discovery and Data Mining, 83–115. The MIT Press
12.8 MacQueen, J. B. (1967): Some methods for classification and analysis of multivariate observations. In: 5th Berkeley Symposium on Mathematical Statistics and Probability, 281–297
12.9 Lee, J. Y., Elmasri, R., Won, J. (1997): An integrated temporal data model incorporating time series concept
12.10 Li, C., Biswas, G. (1999): Temporal pattern generation using hidden markov model based unsuperised classifcation. In: Proceedings of the 3rd International Symposium, Springer-Verlag, LNCS 1642, 245–256
12.11 Lin, W. Q., Orgun, M. A. (1999): Applied hidden periodicity analysis for mining discrete-valued time series. In: Proceedings of ISLIP-99, 56–68, Demokritos Institute, Athens, Greece
12.12 Lin, W. Q., Orgun, M. A. (2000): Mining patterns in discrete time series databases. In: Intensional programming II, World Scientific, 182–196
12.13 Lin, W. Q., Orgun, M. A., Williams, G. (2000): Temporal data mining using multilevel-local polynomial models. In: Proceedings of IDEAL-2000, The Chinese University of Hong Kong, Hong Kong, Springer-Verlag, 180–186, LNCS 1983, December
12.14 Lin, W. Q., Orgun, M. A.,Williams, G. (2001): Multilevels hidden markov models for temporal data mining. In: Temporal data mining workshop Proceedings of KDD-2001, San Franciso, California, USA
12.15 Lin, W. Q., Orgun, M. A.,Williams, G. (2001): Temporal data mining using local polynomial-hidden markov models. In: Proceedings of PAKDD-2001. The University of Hong Kong, Hong Kong, Springer-Verlag, 324–335, LNCS 2045, April
12.16 MacDonald, I. L., Zucchini, W. (1997): Hidden Markov and other models for discrete-valued time series. Chapman and Hall
12.17 Keilson, J. (1979): Markov Chain models–Rarity and Exponentiality. In: Applied Mathematical Sciences Volumn 28, Springer-Verlag, 15–19
12.18 Zhongjie, X. (1993): Case Studies in Time Series Analysis. World Scientific
12.19 Hatfield, C. (1992): The Analysis of Time Series. Chapman & Hall
12.20 Fan, J., Gijbels, I. (1992): Local Polynomial Modelling and Its Applications. Chapman & Hall

12.21 Mannila, H., Bollobas, B. (1997): Time-series similarity problem and well-separated geometric sets. In: Proc. 13th Annual ACM symposium on computational geometry, 454–476, Nice, France

12.22 Ng, M. K., Huang, Z. (1997): Temporal data mining with a case study of astronomical data analysis. In G. Golub, S. H. Lui, F. Luk, and R. Plemmons, editors, Proceedings of the Workshop on Scientific Computing 97, 258–264. Springer-Verlag, Hong Kong, March

12.23 Jajodia, S., Sripada, S., Etzion, O. (1998): Temporal Databases: Research and Practice. 281–309, LNCS1399, Springer-Verlag

12.24 Wong, A. K. C., Wang, Y. (1997): High-order pattern discovery from discrete-valued data. IEEE Transaction on Knowledge and Data Engineering, 9(6): 887–892, Nov., Dec

12.25 Huang, Z. (1997): Clustering large data set with mixed numeric and categorical values. In: 1st Pacific-Asia Conference on Knowledge Discovery and Data Mining, Singapore, World Scientific, 21–34, February

13. Are Efficient Markets Really Efficient?: Can Financial Econometric Tests Convince Machine-Learning People?

Shu-Heng Chen and Tzu-Wen Kuo

AI-ECON Research Center, Department of Economics National Chengchi University, Taipei, Taiwan 11623
email: chchen@nccu.edu.tw, kuo@aiecon.org

Using Quinlan's Cubist, this paper examines whether there is a consistent interpretation of the efficient market hypothesis between financial econometrics and machine learning. In particular, we ask whether machine learning can be useful only in the case when the market is not efficient. Based on the forecasting performance of Cubist in our artificial returns, some evidences seems to support this consistent interpretation. However, there are a few cases whereby Cubist can beat the random walk even though the series is independent. As a result, we do not consider that the evidence is strong enough to convince one to give up his reliance on machine learning even though the efficient market hypothesis is sustained.

13.1 Introduction

A series of applications of *genetic programming* (**GP**) to model agent-based markets has been conducted in [13.1], [13.3], [13.4] and [13.6]. One of the issues typically examined in these studies is whether the agent-based markets are satisfied with the *efficient market hypothesis* (**EMH**). The conjecture that agent-based markets can generate time series data which are satisfied with the EMH was first established in [13.1].

[13.1] simulated *a cobweb model* with GP-based producers and speculators. From their PSC([13.9]) testing of their artificial financial time series, among the 40 series examined, 38 have no linear processes at all, i.e., they are all identified as $ARMA(0,0)$. [1] This result indicates that *the GP-based artificial market is so efficient that there are hardly any linear signals left.* To some extent, this can be considered as a match for the classical version of the EMH.

[13.1]'s result is further strengthened by [13.3] and [13.6] in the agent-based modeling of stock markets. These two papers do not allow traders to imitate other traders directly. Instead, the only way that traders can learn is through a built-in *business school.* With this setup, they find that the stock

[1] One of the by-products of the PSC filter is to inform us of the linear AutoRegressive-MovingAverage process, i.e., the $ARMA(p,q)$ process, extracted from the original series.

return series are not only linearly uncorrelated, but also nonlinear uncorrelated or independent.[2] Therefore, in terms of statistical independence, these two markets are even more efficient than the real financial markets.

Thanks to [13.8], there is a *puzzle* about this result. If the return series is indeed so efficient (no signals left at all), then *what is the incentive for traders to search?* In fact, both [13.3] and [13.6] examine traders' profiles and produce two interesting findings. First, most traders are not believers of the EMH, or technically, they are not martingale believers. Second, during the dynamics of the stock markets, there are many traders that show a "successful" search for new ideas.

[13.2] attempt to solve this puzzle by assuming the existence of *brief signals* in an agent-based stock market. They employ the technique known as the *complexity function* to detect whether there are brief signals in the series, and find that brief signals are very weak as opposed to what we experience from real financial data. Therefore, the puzzle remains unsolved. However, there is another possibility, i.e., while the market is efficient in the *econometric* sense, it may not be efficient in the *machine-learning* sense. Those statistics of market efficiency, such as the BDS statistics, may not be able to tell us how hard one must forecast a series in question. In this paper we shall examine whether machine-learning people can share a similar interpretation as econometricans on the BDS testing results. We can do this by using a promising machine-learning tool. The machine-learning tool chosen in this paper is the well-known Quinlan's **Cubist**.

The rest of the paper is organized as follows. Section 13.2 gives a brief introduction to Quinlan's **Cubist**. Section 13.3 describes the data used in the paper. Experimental results with some discussions are given in Section 13.4. Section 13.5 then leaves the concluding remarks.

13.2 Cubist

The following is a very brief introduction to **Cubist**.

The "**Data Mining with Cubist**" website address is:
http://www.rulequest.com/cubist-info.html, which is also the company page of Professor Quinlan. Tutorials describing and illustrating the use of Cubist can be found on http://www.rulequest.com/cubist-win.html.
The following materials are adapted from the tutorial materials put up on the websites.

Cubist is a tool for generating *rule-based predictive models* from data. The *target value* for prediction in our case is the return for the next period, r_t, and the set of *attributes* are simply the historical returns $\{r_{t-i}\}_{i=1}^{10}$. Cubist's job

[2] In both [13.3] and [13.6], the ARMA(p,q) processes extracted from the return series were all ARMA(0,0). Furthermore, in [13.6], the BDS test failed to reject the null of IIDness in all return series, while it was rejected once in [13.3].

Table 13.1. Data description

High-Frequency Financial Data		
Time Series (Type)	Period	# of Obs. (N_1, N_2)
HSIX (one-minute)	12/1/98-12/31/98	4585 (3475,1100)
HSIX (one-minute)	1/4/99-1/29/99	4695 (3584,1100)
$ECU/$US (tick-by-tick)	2/25/99 (8:00)- 2/26/99 (7:59)	11571 (8678, 2893)
$ECU/$US (tick-by-tick)	3/1/99 (1:00)- 3/2/99 (0:59)	11206 (8396, 2799)
Artificial Data		
MARKET	High BDS	Low BDS
A	10 (7.02)	1 (4.05)
B	5 (2.69)	1 (-0.18)
C	6 (-1.62)	2 (-0.68)
D	1 (4.92)	4 (3.95)

N_1 refers to the size of in-sample data, and N_2 refers to the size of hold-out sample.

is to find out how to predict r_t from $\{r_{t-i}\}_{i=1}^{10}$. Cubist does this by building a *model* consisting of *rules* (conjunctions of conditions) associated with a *linear regression*. Cubist thus constructs a *piecewise linear model* to predict r_t. See Fig. 13.1 and 13.2 for some illustrations.

Cubist allows those piecewise linear models extracted to be combined with the *nearest neighbor* (**NN**) *models*. For some applications, the predictive accuracy of a rule-based model can be improved by combining it with a nearest-neighbor model. However, Cubist uses an unusual method for combining rule-based and nearest neighbor models.

In this paper we experiment with both versions of Cubist, namely, the one using simple rule-based models (coded as "[Cubist]") and the one using the rule-based models and the NN models (coded as "[Cubist + NN]").

13.3 Data Description

The real financial data considered in this paper are described in Table 13.1. Two types of financial data are employed and they are both high-frequency data. One is the one-minute Hang Seng stock index and the other is the tick-by-tick $ECU/$US exchange rate. For both series, we transform the original data into the return series, and then the forecasting is made based on the return series. The time series plot of the return series is given in Fig. 13.3-13.6. The dotted lines in these figures show the cutoff point where the whole sample is divided into the training sample and the hold-out sample.

The artificial data employed in this paper is sampled from the artificial returns of Markets A, B, C, and D. (See [13.7] for a description of the gener-

```
Market D ( High BDS )
    Rule 1:   (1195 cases)
      If   Rt_1 <= -5e-006
      Then   return = 0.00023 + 0.967 Rt_1 - 0.018 Rt_4 + 0.016 Rt_9 - 0.001 Rt_3
                  - 0.001 Rt_7 + 0.001 Rt_10 - 0.001 Rt_6
    Rule 2:   (56 cases)
      If   Rt_1 > -5e-006   and   Rt_4 > -5e-006
      Then   return = 0.00036 - 0.077 Rt_4 - 0.037 Rt_5 + 0.031 Rt_9 + 0.028 Rt_10
                  - 0.027 Rt_8 - 0.025 Rt_3 - 0.024 Rt_6 - 0.017 Rt_2
                  + 0.014 Rt_1 - 0.006 Rt_7
    Rule 3:   (249 cases)
      If   Rt_1 > -5e-006   and   Rt_4 <= -5e-006
      Then   return = 0.01059 + 0.777 Rt_4 - 0.334 Rt_5 - 0.241 Rt_3 + 0.206 Rt_9
                  - 0.195 Rt_8 + 0.177 Rt_10 - 0.143 Rt_6 - 0.134 Rt_2
                  - 0.071 Rt_7 + 0.052 Rt_1
Market D ( Low BDS )
    Rule 1:   (1227 cases)
       If   Rt_1 <= -4.1e-005
       Then   return = 0.00043 + 1.002 Rt_1 - 0.001 Rt_4
    Rule 2:   (209 cases)
       If   Rt_1 > 0.003977   and   Rt_4 > -0.004665
       Then   return = 0.00668 + 0.33 Rt_9 - 0.193 Rt_6 - 0.177 Rt_2 - 0.164 Rt_4
                 - 0.117 Rt_1 + 0.073 Rt_10 - 0.039 Rt_5 - 0.025 Rt_3
                 - 0.001 Rt_7
    Rule 3: (42 cases)
       If   Rt_1 > -4.1e-005   and   Rt_1 <= 0.003977
       Then   return = 0.00192 - 0.032 Rt_4 - 0.03 Rt_5 - 0.025 Rt_2 - 0.017 Rt_3
                 - 0.016 Rt_6 + 0.009 Rt_9 - 0.002 Rt_10 - 0.002 Rt_8
                 - 0.001 Rt_7
    Rule 4:   (22 cases)
       If   Rt_1 > 0.003977   and   Rt_4 <= -0.004665
       Then   return = 0.01899 + 2.134 Rt_4 - 0.697 Rt_5 - 0.322 Rt_3 - 0.246 Rt_2
                 - 0.136 Rt_6 - 0.124 Rt_1 - 0.083 Rt_10 + 0.054 Rt_9
                 + 0.026 Rt_8 - 0.001 Rt_7
```

Fig. 13.1. Decision trees extracted from the artificial data by using Cubist

```
HSIX ( Dec, 98 )
    Rule :   return = 6e-006
HSIX ( Jan, 99 )
    Rule :   return = 4e-006 + 0.005 Rt_4
$EUC/$US ( 2/ 25/ 99 – 2/ 26/ 99 )
    Rule :   return = 0
$EUC/$US ( 3/ 1/ 99 – 3/ 2/ 99 )
    Rule :   return = -1e-006
```

Fig. 13.2. Decision trees extracted from the financial data by using Cubist

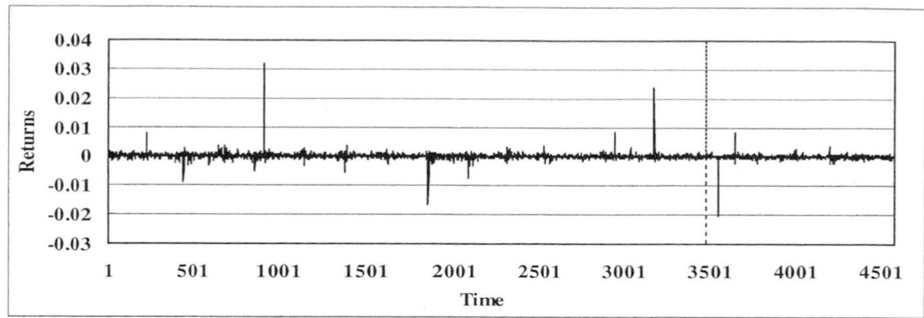

Fig. 13.3. Time series polt of stock returns: one-minute HSIX (12/1/98-12/31/98)

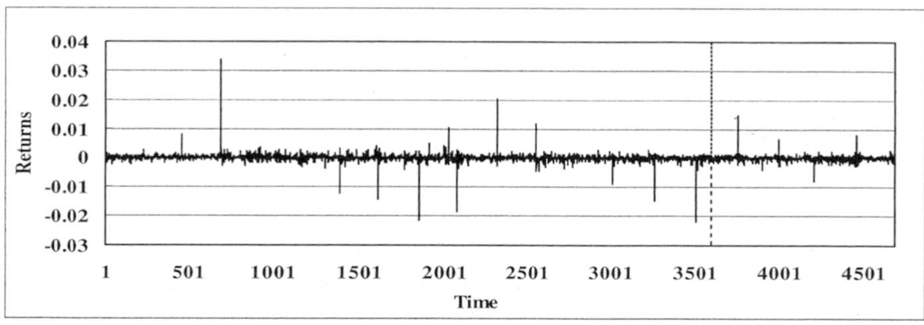

Fig. 13.4. Time series polt of stock returns: one-minute HSIX (1/4/99-1/29/99)

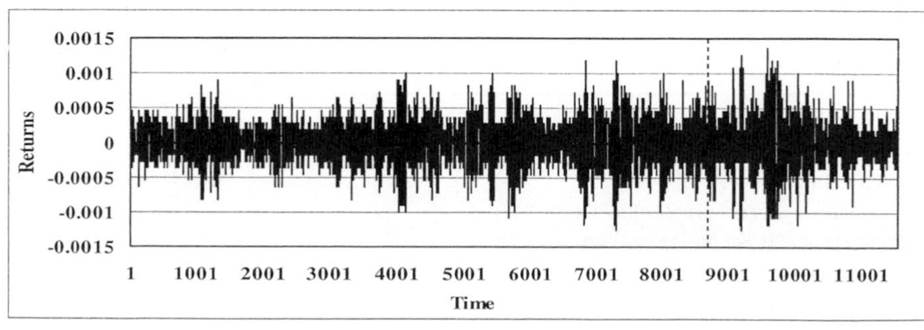

Fig. 13.5. Time series plot of *ECU*/US returns (tick-by-tick, 2/25/99-2/26/99)

Table 13.2. The forecasting performance of Cubist on high-frequency returns

	MSE	MAPE	MAE
×	10^{-4}	10^{0}	10^{-3}
Hang-Shen Index: Dec. 98			
Cubist	0.01	98	0.36
Cubist+NN	0.01	207	0.47
RW	0.01	100	0.36
Hang-Shen Index: Jan. 99			
Cubist	0.01	99	0.54
Cubist+NN	0.01	202	0.59
RW	0.01	100	0.54
\$ECU/\$US: 2/25/99 8:00 - 2/26/99 7:59			
Cubist	0.00	100	0.20
Cubist+NN	0.00	90	0.18
RW	0.00	100	0.20
\$ECU/\$US: 3/1/99 1:00 - 3/2/99 0:59			
Cubist	0.00	100	0.19
Cubist+NN	0.00	92	0.18
RW	0.00	100	0.19

"Cubist" refers to the *simple rule-based model*, whereas "Cubist+NN" refers to the composite of the *simple rule-based model* and the *nearest-neighbor model.*

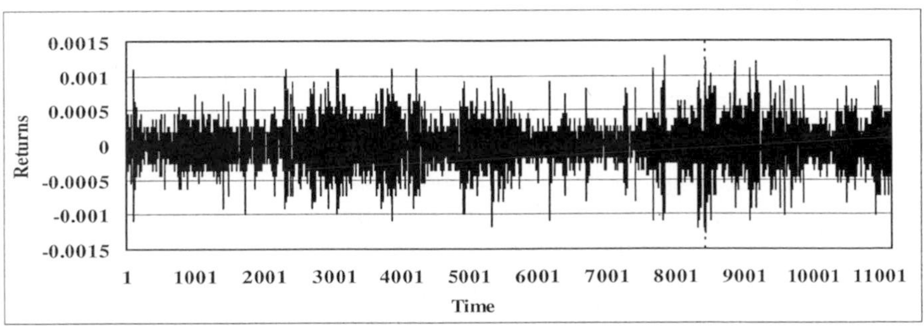

Fig. 13.6. Time series plot of *ECU*/US returns (tick-by-tick, 3/1/99-3/2/99)

ation process of these return series.) However, we do not use the whole series. Instead, based on the BDS value, we only select two subseries, namely, the one with the smallest absolute value and the one with the largest absolute values. (See Table 5 of [13.3].) Therefore, we have two series, each with 1999 observations, from these four markets. The series which are singled out by this criterion and their corresponding BDS values are given in Table 13.1. For example, for Market A the series with the highest absolute value of the BDS statistic is Series 10, which has a BDS value of 7.02. We then take the first 1,499 observations as the training sample and the others (500 observations) as the hold-out sample.

Table 13.3. Experimental results: markets A and B

	MSE	MAPE	MAE
×	10^{-4}	10^{0}	10^{-3}
Market A: High BDS			
Cubist	0.29	103	2.58
Cubist+NN	0.18	91	2.16
RW	1.21	100	7.01
Market A: Low BDS			
Cubist	0.46	115	2.99
Cubist+NN	0.33	136	2.43
RW	1.52	100	7.48
Market B: High BDS			
Cubist	0.75	140	4.36
Cubist+NN	0.74	255	4.87
RW	0.74	100	5.03
Market B: Low BDS			
Cubist	3.07	269	9.60
Cubist+NN	3.18	247	10.79
RW	2.86	100	9.93

"Cubist" refers to the *simple rule-based model*, whereas "Cubist+NN" refers to the composite of the *simple rule-based model* and the *nearest-neighbor model*.

13.4 Experimental Results

The performance criteria to evaluate Cubist are the *mean squared error* (**MSE**), *mean absolute percentage error* (**MAPE**) and *mean absolute error* (**MAE**). The benchmark is the *random walk* (**RW**). For each financial or artificial return series, the two versions of Cubist mentioned in Section 13.2 are applied and the results are given in Tables 13.2, 13.3, and 13.4.

Let us take a look at the financial data first. In all cases and all criteria, [Cubist] performs at least as good as RW. On the other hand, [Cubist+NN] is the best in the tick-by-tick return series of the exchange rate. However, the number of rules discovered by the Cubist is *only one*, and the contents of those rules are also very simple and naive (See Fig. 13.2). In a few cases, the rules simply took "0" as the prediction value. In other words, *what Cubist did was to rediscover the EMH.* Therefore, if someone argues that these series are efficient, then Cubist would not argue over that.

An important question remains: how about the artificial return series generated from the agent-based stock markets? Several interesting results deserve our attention. First, when we apply Cubist in a highly efficient market, it does not mean that it will simply give "0" as the prediction value. For example, consider Market C. This market is shown to be highly efficient in [13.6]. However, the Cubist suggests 17 rules for the series with a high BDS value, and 20 rules for the one with a low BDS value. Moreover, in terms of

Table 13.4. Experimental results: markets C and D

	MSE	MAPE	MAE
×	10^{-4}	10^{0}	10^{-3}
Market C: High BDS			
Cubist	4.65	209	10.55
Cubist+NN	4.58	335	11.87
RW	4.11	100	11.61
Market C: Low BDS			
Cubist	4.53	160	9.56
Cubist+NN	4.39	197	11.78
RW	4.24	100	11.45
Market D: High BDS			
Cubist	1.13	141	4.94
Cubist+NN	1.11	756	5.46
RW	1.58	100	7.21
Market D: Low BDS			
Cubist	1.36	125	4.51
Cubist+NN	1.18	416	4.93
RW	1.74	100	7.19

"Cubist" refers to the *simple rule-based model*, whereas "Cubist+NN" refers to the composite of the *simple rule-based model* and the *nearest-neighbor model.*

the criterion MAE, both series beat the random walk. Therefore, the efficient market hypothesis does not necessarily imply that machine-learning people can do nothing.

Let us consider Market D. In [13.3], this market is shown to be not efficient. Therefore, we anticipate that Cubist can do something on these series, and it in fact does. For the series with a high BDS value, 3 rules are discovered, and for the series with a low BDS value, 4 rules are discovered. However, does it work? The answer is "not quite". If we take MAPE as the criterion, then Cubist clearly is beaten by the random walk in both cases.

13.5 Conclusions

Examples like this are many and can be found easily from Tables 13.3 and 13.4. Due to these negative examples, machine-learning people may think that econometric test results are irrelevant. In agent-based stock markets, we may have many traders like this. Therefore, the puzzle proposed in Section 13.1 may be solved from this direction.

References

13.1 Chen, S. -H., Kuo, T. -W. (1999): Towards an Agent-Based Foundation of Financial Econometrics: An Approach Based on Genetic-Programming Artificial Markets. In: Banzhaf, W., Daida, J., Eiben, A. E., Garzon, M. H., Honavar, V., Jakiela, M., Smith, R. E. (Eds.), GECCO-99: Proceedings of the Genetic and Evolutionary Computation Conference, Vol. 2. Morgan Kaufmann, 966–973

13.2 Chen, S. -H., Tan, C. -W. (1999): Brief Signals in the Real and Artificial Stock Markets: An Approach Based on the Complexity Function. In: Arabnia, H. R. (Ed.), Proceedings of the International Conference on Artificial Intelligence, Vol. II. Computer Science Research, Education, and Application Press, 423–429

13.3 Chen, S. -H., Yeh, C. -H. (1999): On the Consequence of "Following the Herd": Evidence from the Artificial Stock Market. In: Arabnia, H. R. (Ed.), Proceedings of the International Conference on Artificial Intelligence, Vol. II. Computer Science Research, Education, and Application Press, 388–394

13.4 Chen, S. -H., Yeh, C. -H. (2001): Evolving Traders and the Business School with Genetic Programming: A New Architecture of the Agent-Based Artificial Stock Market. Journal of Economic Dynamics and Control, 25, 363–393

13.5 Chen, S. -H., Wang, H. -S., Zhang, B. -T. (1999): Forecasting High-Frequency Financial Time Series with Evolutionary Neural Trees: The Case of Hang-Seng Stock Index. In: Arabnia, H. R. (Ed.), Proceedings of the International Conference on Artificial Intelligence, Vol. II. Computer Science Research, Education, and Application Press, 437–443

13.6 Chen, S. -H., Yeh, C. -H. (2002): On the Emergent Properties of Artificial Stock Markets: The Efficient Market Hypothesis and the Rational Expectations Hypothesis. Journal of Economic Behavior and Organization, 49(1), 217–239

13.7 Chen, S. -H., Liao, C. -C. (2002): Testing for Granger Causality in the Stock-Price Volume Relation: A Perspective from the Agent-Based Model of Stock Markets. Information Sciences, forthcoming

13.8 Grossman, S. J., Stiglitz, J. (1980): On the Impossibility of Informationally Efficiency Markets. American Economic Review, 70, 393–408

13.9 Rissanen, J. (1986): Stochastic Complexity and Modeling. Annals of Statistics, 14(3), 1080–1100

14. Nearest-Neighbour Predictions in Foreign Exchange Markets

Fernando Fernández-Rodríguez[1], Simón Sosvilla-Rivero[2], and Julián Andrada-Félix[1]

[1] Universidad de Las Palmas de Gran Canaria, Islas Canarias, Spain
[2] Universidad Complutense and Fedea, Madrid, Spain
email: sosvilla@fedea.es

The purpose of this paper is to contribute to the debate on the relevance of non-linear predictors of high-frequency data in foreign exchange markets. To that end, we apply nearest-neighbour (NN) predictors, inspired by the literature on forecasting in non-linear dynamical systems, to exchange-rate series. The forecasting performance of univariate and multivariate versions of such NN predictors is first evaluated from the statistical point of view, using a battery of statistical tests. Secondly, we assess if NN predictors are capable of producing valuable economic signals in foreign exchange markets. The results show the potential usefulness of NN predictors not only as a helpful tool when forecast daily exchange data but also as a technical trading rules.

14.1 Introduction

Exchange rates have proven extremely difficult to forecast, despite a vast literature strewn with attempts. The structural exchange rate models have little success forecasting exchange rate compared with the forecast from a simple random walk model, as first noted by [14.69] and confirmed subsequently by many others (see, e. g., [14.42] and [14.41]). Therefore, beating the random walk still remains the standard metric by which to judge empirical exchange rate models.

Several explanations for the failure of structural models have been suggested, including misspecification of the models and poor modelling of expectations (see [14.44] for a survey). Recently, interest has been shown in the possibility that non-linearities account for the apparent unpredictability of exchange rates, and some papers have highlighted the importance of non-linear adjustment of the exchange rate to the value implied by fundamentals, including [14.92] and [14.57].

Dynamical systems theory suggests that non-linear behaviour may explain fluctuations in financial asset prices that appear to be random. One strand of the financial economics literature attempts to provide theoretical justification of the presence of non-linearities in assets prices. Another strand empirically tests for the presence of non-linearities in such prices. Regarding foreign exchange markets, economic theory highlights a number of potential sources for the presence of non-linearities in exchange rates. The main explanations include the concept of time deformation ([14.24]), diversity in agents' beliefs ([14.43]), fads or noise trading ([14.29]), technical traders ([14.10]), hysteresis model with transaction cost bands ([14.7]), bubbles with self-fulfilling expectations ([14.45]), target zone models ([14.60]), non-linear risk premium model ([14.76]) and heterogeneity in investors' objectives arising from varying investment horizons and risk profiles ([14.78]). On the other hand, many empirical studies have uncovered significant non-linearities in exchange rates (see, for example, [14.51], [14.58] and [14.23]).

Recent breakthroughs pertaining to non-linear dynamics have helped empirical investigation of non-linear models and have fuelled the research agenda in this area, dramatically increasing the number of approaches to forecasting exchange rates in an attempt to guard against the failure to benefit from non-linearities due to the incorrect choice of functional form: chaotic dynamics ([14.28]), turbulent motions ([14.47]), neural networks ([14.79]), GARCH models ([14.11]), self-exciting threshold autoregressive models ([14.59]), etc.

One of these approaches to forecasting is the non-parametric, nearest neighbour (NN hereafter) forecasting technique. There are several ways of computing such forecasts. The basic idea behind these predictors, inspired in the literature on forecasting in non-linear dynamical systems, is that pieces of time series sometime in the past might have a resemblance to pieces in the future. In order to generate predictions, similar patterns of behaviour are located in terms of nearest neighbours. The time evolution of these nearest neighbours is exploited to yield the desired prediction. Therefore, the procedure only uses information local to the points to be predicted and does not try to fit a function to the whole time series at once (see, e. g., [14.4], [14.5] and [14.36]).

The NN approach has been used in forecasting several financial time series, but the results are rather inconclusive. For exchange rates, while [14.30], [14.52], [14.70] or [14.71] concluded that there was little gain in predictive accuracy over a simple random walk, more promising results were reported in [14.4], [14.5], [14.66], [14.37], or [14.18]. Applications of NN to other financial variables include stock markets ([14.39], [14.40] and [14.3]) and interest rates ([14.6]).

In this chapter we will survey our contribution to this programme of research on predictability in financial markets, addressing the question of whether NN prediction methods can improve out-of-sample forecasting in foreign exchange markets, with special reference to the European Monetary System (EMS). The chapter is organised as follows. The NN predictors are presented in Section 14.2. In Section 14.3 we discuss the EMS. In Section 14.4 the forecast accuracy of the NN predictors is assessed from a statistical point of view, while in Section 14.5 we use the NN predictions in a simple trading strategy to assess the economic significance of predictable patterns in the foreign exchange market. Some concluding remarks are provided in Section 14.6.

14.2 The NN Approach to Nonlinear Forecasting

14.2.1 Literature Review

NN forecasting of time series has a dual and recent history. On one hand, in the statistical literature there were some attempts to develop a wide class of non-parametric regression estimates (see [14.85] for a review) that culminated with the work of [14.25] on robust locally weighed regression, perhaps the first form of a NN predictor. On the other hand, independently and about ten years later than the statistics community, in the dynamical system literature some authors began to use NN predictors for nonlinear time series, associated with studies on deterministic chaos. Seminal papers by [14.34], [14.20] and [14.86] gave an important impulse to NN predictions.

The non-parametric regression approach. Accordingly to the non-parametric regression approach, if y_t (for $t = 1, ..., T$) are observations of a dependent variable and $x_t = (x_{t1}, ..., x_{tp})$ (for $t = 1, ..., T$) is a vector of explanatory independent variables, the model of local regression is given by

$$y_t = g(x_t) + \varepsilon_t \tag{14.1}$$

where $g(x)$ is an unknown smooth function to be estimated from the data, and ε_t are i.i.d. normal random variables with mean 0 and variance σ^2. Note the sharp difference of the former assumption with respect to the standard paradigm of parametric regression: instead of a function from a specific parametric class, we just have an assumption about smoothness.

A smoother $\hat{g}(x_t)$ is an estimate of the conditional mean of the dependent variable, which is obtained by local averaging of the dependent variable at any given neighbourhood of the explanatory variables.

The estimator is constructed through the following steps:

1. Let W be a weighted function with the following properties:
   - $W(x) > 0$ for $|x| < 1$;
   - $W(-x) = W(x)$;
   - $W(x)$ is a non-increasing function for $x \geq 0$; and
   - $W(x) = 0$ for $|x| \geq 1$.

   An example of function verifying the properties is $W(x) = (1 - |x|^3)^3$, for $|x| < 1$, and, $W(x) = 0$, for $|x| \geq 1$.
2. Let x be a point where we are going to compute $\hat{g}(x)$. Let f be a smoothing constant such that $0 < f \leq 1$, and let be $q = int(fT)$, where $int(.)$ rounds down to the nearest integer. Then rank the x_t's by Euclidean distance from x. Call these $x_{t_1}, x_{t_2}, ..., x_{t_k}$, so that x_{t_1} is closest to x, x_{t_2} is second closest to x and so on.

3. Let $d(x, x_{t_k}) = \left[\sum_{j=1}^{p} (x_j - x_{t k_j})^2 \right]^{1/2}$ be the Euclidean distance from x to its k-th closest neighbour.
4. Now a set of weights for the points (x_t, y_t) is defined by

$$w_t(x) = W(\frac{d(x_t, x)}{d(x, x_k)}) \tag{14.2}$$

5. The value of the regression surface at is then computed using ordinary least squares as:

$$\hat{y} = \hat{g}(x) = \hat{\alpha}_0 + \hat{\alpha}_1 x_{t_1} + ... + \hat{\alpha}_p x_{t_p} \tag{14.3}$$

where $\hat{\alpha}_i$ minimise the expression

$$\sum_{t=1}^{T} w_t(x) \left(y_t - \alpha_0 - \alpha_1 x_{t_1} - ...... - \alpha_p x_{t_p} \right)^2 \tag{14.4}$$

The dynamical system approach. A second approach to NN predictors is related to the forecast in chaotic time series. A key result in this approach is given by Taken [14.89] theorem.

Following [14.89], we say that the time series of real numbers x_t (for $t = 0, ..., \infty$) has a smoothly deterministic explanation if there exists a system (h, F, a_0) such that $h : R^n \to R$ and $F : R^n \to R^n$ are smooth (i.e. twice differentiable almost everywhere) and

$$x_t = h(a_t), \quad a_t = F(a_{t-1}), \quad t = 1, 2, ...(a_0 \text{ } is \text{ } given) \tag{14.5}$$

Note that the relationship $a_t = F(a_{t-1})$ is an unknown law of motion involving the unobserved n vector of state variables. Nevertheless, Takens suggests that we can learn about the underlying state variable a_t from embeddings of the observed x_t's in R^m for m sufficiently large. To that end, we compute a *m-history* as a vector of m observations sampled from the original time series:[1]

$$x_t^m = (x_t, x_{t-1}, x_{t-2}, ..., x_{t-(m-1)}), \quad t = m, m+1, ..., T \tag{14.6}$$

with m referred to as the *embedding dimension* and the m-dimensional space R^m is referred to as the *reconstructed phase space* of time series. The Takens embedding theorem states that, if m is sufficiently large ($m \geq 2n + 1$) the dynamics of the unknown law $a_t = F(a_{t-1})$ is equivalent to a dynamics of x_t^m. This means that the sequence of m-histories in this case can mimic the true data generation process. Therefore, the dynamics of the m-histories x_t^m provide a correct topological picture of the unknown chaotic dynamics.

[1] Note that we are assuming daily data, and therefore the m observations are sampled from the original time series at intervals of one unit.

More formally, if the time series has a smoothly deterministic explanations Takens showed that, for a embedding dimension m sufficiently large, there exists a function

$$\widetilde{F} : R^m \rightarrow R^m \tag{14.7}$$

such that $x^m_{t+1} = \tilde{F}(x^m_t)$ and this map has the same dynamic behaviour as that of the original unknown system in the sense of topological equivalence.

The Takens embedding theorem provides a new geometrical interpretation for the NN predictors. In this case, the NN method can be seen as selecting some geometric segments in the past of the time series, similar to the last segment available before the observation we want to forecast (see [14.34]). So m-histories with a similar dynamic behaviour are detected in the series and employed afterward in the prediction of the next term at the end of the series. This term is computed as some adequate average of the actually observed terms next to the m-history involved.

14.2.2 The Approach in this Paper

In our empirical implementations, we have followed the dynamical system approach. In what follows we shall succinctly described this approach for both the univariate and multivariate cases follows (see [14.37] for a more detailed account).

Univariate NN predictors. A forecast of a variable using information from its own time series can be produced through the following steps:

1. We first transform the scalar series x_t $(t = 1, ..., T)$ into a series of m-dimensional vectors, $x^m_t, t = m, ..., T$:

   $$x^m_t = \left(x_t, x_{t-1}, x_{t-2}, ..., x_{t-(m-1)}\right)$$

   with m referred to as the *embedding dimension*. These m-dimensional vectors are often called *m-histories.*
2. As a second step, we select the k m-histories

   $$x^m_{t_1}, x^m_{t_2}, ..., x^m_{t_k} \tag{14.8}$$

   most similar to the last available vector

   $$x^m_T = (x_T, x_{T-1}, x_{T-2}, ..., x_{T-(m-1)})$$

   where $k = int(\lambda T)$ $(0 < \lambda < 1)$, with $int(\cdot)$ standing for the integer value of the argument in brackets, and where the subscript $t_r (r = 1, 2, ..., k)$ is used to denote each of the k chosen m-histories.
   To that end, we look for the closest k vectors in the phase space R^m, in the sense that they maximise the function:

   $$\rho\left(x^m_t, x^m_T\right) \tag{14.9}$$

(i.e. looking for the highest serial correlation of all m-histories, x_t^m, with the last one, x_T^m)[2]

3. Finally, to obtain a predictor for x_{T+1}, we consider the following local regression model:

$$\hat{x}_{T+1} = \hat{\alpha}_0 x_T + \hat{\alpha}_1 x_{T-1} + ... + \hat{\alpha}_{m-1} x_{T-(m-1)} + \hat{\alpha}_m \tag{14.10}$$

whose coefficients have been fitted by a linear regression of x_{t_r+1} on $x_{t_r}^m = \left(x_{t_r}, x_{t_r-1}, x_{t_r-2}, ..., x_{t_r-(m-1)}\right), \quad (r = 1, ..., k)$. Therefore, the $\hat{\alpha}_i$ are the values of α_i that minimise

$$\sum_{r=1}^{k} \left(x_{t_{r+1}} - (\alpha_0 x_{t_r} + \alpha_1 x_{t_r-1} + ... + \alpha_{m-1} x_{t_r-(m-1)} + \alpha_m)\right)^2 \tag{14.11}$$

Therefore, in order to obtain a predictor for x_{T+1}, we look for similar behavioural patterns (or occurring analogues) in the time series x_t by means of choosing k m-histories whose resemblance to the last available m-history could help us to infer the likely future evolution of x_t from instant time T onward.

Alternatively, one can consider the weighted least squares algorithm in order to estimate the regression coefficients,[3] that is minimising the expression

$$\sum_{r=1}^{k} w(x_{t_r})\left(x_{t_{r+1}} - (\alpha_0 x_{t_r} + \alpha_1 x_{t_r-1} + \cdots + \alpha_{m-1} x_{t_r-(m-1)} + \alpha_m)\right)^2 \tag{14.12}$$

where the weights $w\left(x_{t_r}\right)$ are assigned using the so-called tri-cube weight function:

$$w(x_{t_r}) = W\left(\frac{\left\|x_{t_r}^m - x_T^m\right\|}{\sum_{r=1}^{k} \left\|x_{t_r}^m - x_T^m\right\|}\right)$$

where $\|.\|$ is the Euclidean distance of the last m-history x_T^m to the r-th nearest neighbour $x_{t_r}^m$ and, for any u, $W(u) = (1 - u^3)^3$ for $0 \leq u \leq 1$ and $W(u) = 0$ otherwise.

[2] Alternatively, we could have established nearest neighbours to x_T^m by looking for the closest k points x_t^m that minimise the following functions $\|x_t^m - x_T^m\|$ or $1 - \cos(x_t^m, x_T^m)$ (i.e. looking, respectively for the minimum distance or the lowest angle) (see [14.35], for the relationship between the nearest neighbours obtained by using these three functions).

[3] The fit of Eq. (14.10) may be also done by using more sophisticated tools as singular-valued decomposition of [14.16] (see [14.66] for an application of this method predicting exchange rates).

Although this last weighting scheme has the theoretical attraction of allowing us to use an arbitrary amount of NN, weighted algorithms may be numerically less stable than unweighted algorithms. [14.94] showed that, in presence of noise, unweighted algorithms tend to yield superior results, because when calculating the parameters of the local regression, it is necessary to invert an $X'X$ matrix whose rows are highly collinear since they are formed by the selected nearest neighbours that are all very similar to x_T^m. When the $X'X$ matrix is close to singular, the parameter estimates are numerically highly unstable. Numerical experiments reveal that this problem is specially acute for the weighted $X'X$ matrix, because it is much more closer to singular than the unweighted matrix. We therefore focus our attention only on unweighted algorithms (see also [14.54]).

Multivariate NN predictors. Hitherto we have consider a univariate version of the NN predictor. However, when we have a set of simultaneous time series, the NN prediction can be extended to a multivariate case using the *simultaneous nearest neighbour* (SNN hereafter) predictors in an attempt to also consider the information content in other related time series. To simplify, let us consider a set of two time series:

$$x_t(t = 1, ..., T), \quad y_t(t = 1, ..., T)$$

We are interested in making predictions of an observation of one of these series (e.g., x_{T+1}), by simultaneously considering nearest neighbours in both series. To this end, we embed each of these series in the vectorial space R^{2m}, paying attention to the following vector:

$$(x_t^m, y_t^m) \in R^m \times R^m$$

which gives us the last available m-history for each time series.

In order to establish nearest neighbours to the last m-histories (x_t^m, y_t^m) we can look for the closest k points that maximise the function:

$$\rho\,(x_t^m, x_T^m) + \rho\,(y_t^m, y_T^m) \tag{14.13}$$

The predictor for x_{T+1} and y_{T+1} can be obtained from a linear autoregressive predictor with varying coefficients estimated by ordinary least squares:

$$\hat{x}_{T+1} = \hat{\alpha}_0 x_T + \hat{\alpha}_1 x_{T-1} + ... + \hat{\alpha}_{m-1} x_{T-(m-1)} + \hat{\alpha}_m \tag{14.14}$$

$$\hat{y}_{T+1} = \hat{\beta}_0 y_T + \hat{\beta}_1 y_{T-1} + ... + \hat{\beta}_{m-1} y_{T-(m-1)} + \hat{\beta}_m$$

The difference between this SNN predictor (14.14) and the NN predictor (14.10) is that now the nearest neighbours are established using criteria in which information on both series is used. Therefore, to obtain a predictor for x_{T+1}, we look now for similar behavioural patterns not only in the time series x_t , but also in the related time series y_t, whose resemblance to the last available m-history of x_t could be used to help us when projecting the likely future evolution of x_t from instant time T onward.

As in the NN case, the weighted least squares algorithm could also be used to estimate the regression coefficients, but the same arguments on numerical instability apply.

14.2.3 Selecting the Key Parameters in (S)NN Predictions

Note that the NN predictors (both the univariate and the multivariate cases) depend on the values of two key parameters: the embedding dimension m and the number of closest k points in the phase space R^m.

The Takens embedding theorem provides no information upon these key parameters. The only limitation is to take $m \geq 2n+1$ here n is the dimension of compact manifold that have to be embedding in the R^m Euclidean space ([14.89]).

Even though these parameters are crucial in the determination of predicting map $\tilde{F}(.)$ in (14.7) since they are essential for the accuracy of the predictions, in empirical applications they are usually selected "based on a combination of judgement and of trial and error" ([14.26], p. 96). Nevertheless, some methods have been proposed in the literature to choose these key parameters.

Selection of m. A capital issue in (S)NN prediction is the appropriated choice of the embedding dimension for the time series. Among the basic methods proposed in the literature, we can name the following:

- Methods based on dynamical properties of the strange attractor, which rely upon computing some invariant of the chaotic motion as correlation dimension or Lyapunov exponents (e.g. [14.48]). However, one could argued that these methods are not relevant in financial time series where the chaotic nature of the series is not so clear.
- Singular value decomposition methods ([14.16]).
- False neighbourhoods methods ([14.56]) and averaged false neighbours methods ([14.17]) and
- The zero order approximation method was developed by [14.19]. This method consists of finding the embedding dimension by minimising the average absolute one-step ahead prediction errors, using a zero-order approximation predictive model.

Selection of k. In chaotic time series, the ideal number of nearest neighbours depends on the complexity of the attractor and the number of observations in the time series. For experimental and financial data, there are a great diversity of approaches for selecting the number of nearest neighbours.

To fit the Eq. (14.10), k must be at least as big as $m+1$. When $k = m+1$, this method is equivalent to linear interpolation and the least-square problem has a unique solution. In practice, in order to ensure stability of the solution and to reduce the forecasting error, it would be an advantage to take $k > m+1$. [14.22] recommend to take $k = 2(m+1)$.

In general there is not a uniform guideline for selecting the number of nearest neighbours when forecasting a time series. It is usually claimed that standard forecasting errors diminish with the number of neighbours (see [14.22], [14.54]).

[14.22] have proposed a forecasting algorithm based on study the behaviour of the normalised root mean square error (RMSE) of forecasts when m and k vary. Although [14.22] uses this algorithm in order to distinguish low dimensional chaotic behaviour from linear stochastic behaviour by comparing the accuracy of short-term forecasts, this procedure may be used for selecting the embedding dimension and number of nearest neighbours. The procedure implies the selection of a "fitting set" $F_t = \{x_t : 1 < t \leq T\}$ and a "prediction set" $P_t = \{x_t : T < t \leq N - 1\}$, for some $T < N$. We choose the parameters k and m in such a way that they minimise the normalised RMSE of forecasts in the testing set:

$$RMSE_m(k) = \frac{\sqrt{\sum_{t=1}^{T} \left(\hat{x}_t(k) - x_t\right)^2}}{\sigma} \tag{14.15}$$

where σ is the standard deviation of the time series in the fitting set.[4] In order to prevent the over-fitting problem, we can assign a cost to the introduction of each additional unity of embedding dimension using the well-known Akaike information criterion [14.1].

In the empirical applications reviewed in Sections 14.4 and 14.5, m and k are chosen according to [14.21], [14.22] algorithm.

14.3 The European Monetary System

The European Monetary System (EMS) was initially planned as an agreement to reduce exchange rate volatility for a Europe in transition to a closer economic integration. Following its inception in March 1979, a group of European countries linked their exchange rates through formal participation in the Exchange Rate Mechanism (ERM). The ERM was an adjustable peg system in which each currency had a central rate expressed in the European Currency Unit (ECU), predecessor of the Euro. These central rates determined a grid of bilateral central rates *vis-à-vis* all other participating currencies, and defined a band around these central rates within which the exchange rates could fluctuate freely. In order to keep these bilateral rates within the margins, the participating countries were obliged to intervene in the foreign exchange market if a currency approached the limits of its band, for which some special credit facilities were established.[5] If the participating countries

[4] See [14.50] for a discussion of this "cross-validation" method.

[5] Following Basel/Nyborg agreement of September 1987, central banks were also empowered to intervene within margins before limits were reached.

decided by mutual agreement that a particular parity could not be defended, realignments of the central rates were permitted.

The ERM was the most prominent example of a target zone exchange-rate system. In the 1990s an extensive literature appeared, building on the seminal paper by [14.60], which studied the behaviour of exchange rates in target zones. The main result of the simple target zone model was that, with perfect credibility, the zone would exert a stabilising effect (the so-called honeymoon effect), reducing the exchange rate sensitivity to a given change in undamentals. However, in a target zone with credibility problems, expectations of future interventions would tend to destabilise the exchange rate, making it less stable than the underlying fundamentals ([14.9]). Therefore, credibility (i.e. the degree of confidence that the economic agents assign to the announcements made by policymakers) becomes a key variable. In the context of an exchange-rate target zone, credibility refers to the perception of economic agents with respect to the commitment to maintain the exchange rate around a central parity. Therefore, the possibility for the official authorities to change the central parity could be anticipated by the economic agents, triggering expectations of future changes in the exchange rate that can act as a destabilising element of the system. Nevertheless, in both cases (with perfect credibility or with credibility problems), the target zone model implies that the exchange rate is a non-linear function of the fundamentals, therefore providing a further theoretical foundation for the exchange rate to exhibit a non-linear data generating process.

The fact that the currencies participating in the ERM are institutionally related makes them a natural candidate among financial time series for the application of the SNN methodology in order to examine if forecast accuracy can be obtained by considering the information content of other related exchange rates. In other words, the use of SNN predictions in this context can be seen as an attempt to incorporate structural information into the non-parametric analysis.

14.4 Assessing the Forecast Accuracy of the NN Predictors

A first contribution applying the NN methods is [14.4], where daily data for the Spanish peseta-US dollar, spot and one- and three-month forward exchange rates, during the period January 1985-May 1991, were used in the empirical application; a deeper discussion of the methods used in the paper can be found in [14.5]. In that paper, several predictors based on the NN methodology were computed, and their performance was compared with that of a simple random walk, by calculating their respective forecasting errors, as measured by the root mean square error (RMSE). In general, the non-linear predictors outperformed the random walk in all cases for the forward rates, whereas for the spot rate this only occurred in four over nine cases.

The objective of [14.83] was to compute an indicator of volatility, defined as the (absolute value of the) forecast error, derived from the NN predictors, weighted by the standard deviation of the original series. This indicator was applied to six EMS currencies experiencing different evolutions after the crisis that affected the system after the summer of 1992, that lead to the broadening of the fluctuation bands in August 1993: two of them temporarily leaving the ERM (Italian lira and Pound sterling), two others forced to devalue (Spanish peseta and Portuguese escudo), and the remaining two not devaluing (French franc and Dutch guilder), with the sample period running from January 1974 to April 1995. The volatility indicators showed an initial low degree of exchange rate volatility, with a sudden increase from September 1992 onwards. Then, volatility remained high for the currencies that abandoned the ERM; however, for the rest of the currencies, the broadening of the bands after August 1993 would have led to a decrease in volatility to levels comparable with those prevailing before the crisis. These results were interpreted, rather than in terms of an unexpected loss of credibility, as being a consequence of the fragility of the EMS in a world of very high international capital mobility, which became evident with the problems associated with German reunification at the end of 1989 and the effects of self-fulfilling speculative attacks ([14.33]).

The above papers compute predictions for each variable using information from the own series [i.e. in a univariate (NN) context]. In a later contribution, [14.37] applied the SNN predictors, therefore using the information content of a wider set of time series (nine currencies participating in the ERM) in an attempt to incorporate structural information into the non-parametric analysis. The data set includes daily observations of nine exchange rates (Belgian franc, Danish crown, Portuguese escudo, French franc, Dutch guilder, Irish pound, Italian lira, Spanish peseta, and Pound sterling) covering the period January 1978-December 1994. Given the central role of Germany in the European Union (see [14.6]) they are expressed *vis-à-vis* the Deutschemark.

As mentioned above, the predictors depend on the values of the embedding dimension m and the number of closest k points in the phase space R^m. [14.37] chose them according to [14.21] algorithm, obtaining an embedding dimension $m = 6$ and a number of nearest neighbourhood points k around a 2% of the sample. On the other hand, it is necessary to chose related exchange-rate series in order to establish occurring analogues in the SNN predictor. These authors consider three groups of currencies according to the credibility with respect to the commitment to maintain the exchange rate around the central parity (see [14.63]). It is interesting to note that these groups roughly correspond to those found in [14.53], applying principal component and cluster analyses to a wide set of structural and macroeconomic indicators, to form a homogeneous group of countries. Moreover, these groups are basically the same that those found in [14.84] when examining the duration of the central parities in the ERM.

After founding evidence of non-linear dependence in the series using the well-known BDS test statistic (see [14.14]), hence supporting their approach to forecasting, [14.37] evaluated the forecasting performance for the whole sample using Theil's U statistic, a summary statistic that is based on standard symmetric loss function:

$$U = \frac{\sqrt{\sum_{t=T+1}^{T+N} (x_t - \hat{x}_t)^2}}{\sqrt{\sum_{t=T+1}^{T+N} (x_t - x_{t-1})^2}} \tag{14.16}$$

where x_t is the actual value and $\hat{x}_t$ the forecast value. Note that we have defined the U statistic is defined as the ratio of the RMSE of forecasts from a particular predictor to the RMSE of the naive random walk forecast. Therefore, a value of U less than one indicates better performance than the random walk specification.

Table 14.1 shows the forecasting performance, relative to the random walk, from both the SNN predictors and the traditional (linear) ARIMA(1,1,0) model. As can be seen, the U statistics were, for the SNN predictors, above one only in three of the nine cases, suggesting that the non-linear predictors marginally outperformed the random walk, despite the forecasting period being very long and heterogeneous, with the best SNN predictor presenting an improvement of 18.9%. We can also see that the predictors from an ARIMA(1,1,0) model always show U statistics below one, the best one showing an improvement of 12.2% out of sample. Nevertheless, in six out of nine cases, the SNN predictors show lower U statistics than the ARIMA(1,1,0) model.

As [14.12] observed, a relevant test of forecasting performance relative to a random walk is the accuracy in predicting the direction of exchange rate movements. This is because obtaining the sign right in the prediction matters in markets with low transaction costs, like foreign exchange markets. To explore this possibility, [14.37] also computed the percentage of correct predictions as a further test of forecasting accuracy. Table 14.2 reports the results, which are rather promising. In eight of nine cases, the SNN predictors show a value higher than 50%, clearly outperforming the random walk directional forecast.[6] Note also that in all the cases, the predictors from an ARIMA(1,1,0) model show a value greater than 50%. Finally, in seven out of the nine cases, the SNN predictors offer higher values than the ARIMA model.

To formally assess the forecast accuracy of the local predictors, [14.37] used the test proposed by Diebold and [14.31] on the (corrected) value of the

[6] The value of 50% is the ususl benchmark. However, the results must be treated with caution, since numbers of positive changes do not necessarily coincide with the number of negative changes.

Table 14.1. Forecast accuracy (1)

	SNN predictor	ARIMA (1,1,0) predictor
BFR (2)	0.984	0.995
DKR (2)	0.939	0.954
ESC (3)	1.016	0.997
FF (2)	0.908	0.952
HFL (2)	0.811	0.878
IRL (4)	1.014	0.997
LIT (3)	0.973	0.981
PTA (3)	0.995	0.999
UKL (4)	1.022	0.999

Notes:
(1) Theil's U statistic.
(2) Time series used in establishing occurring analogues in the SNN predictor: BFR, DKR, FF, and HFL.
(3) Time series used in establishing occurring analogues in the SNN predictor: ESC, LIT and PTA.
(4) Time series used in establishing occurring analogues in the SNN predictor: IRL and UKL.
Source: [14.37].

differential between two forecasting errors. Let $\hat{x}_{1_t}$ and $\hat{x}_{2_t}$ denote alternative forecasts of the variable x_t $t = T+1, ..., T+N$, where N is the number of forecast. Let e_{1_t} and e_{2_t} denote the corresponding forecast errors $(x_t - \hat{x}_{1_t})$ and $(x_t - \hat{x}_{2_t})$, respectively). The Diebold-Mariano test is given by:

$$DM = \frac{\bar{d}}{\sqrt{\frac{2\pi \hat{f}_d(0)}{N}}} \tag{14.17}$$

where $\bar{d}$ is an average (over N observations) of a general loss differential function and $\hat{f}_d(0)$ is a consistent estimate of the spectral density of the loss differential at frequency zero. Diebold and Mariano show that the DM statistic is asymptotically distributed as standard normal under the null of equal forecast accuracy. Therefore, a significant and positive (negative) value for DM would indicate a significant difference between the two forecasting errors, which would mean a better accuracy of the $\hat{x}_{2_t}$ $(\hat{x}_{1_t})$ predictor. In line with a large literature (see, e. g., [14.68] and [14.55]), the loss differential function considered was the difference between the absolute forecast error.

As can be seen in Table 14.3, the null hypothesis of no difference in the loss function can be rejected in eight out of the nine cases when assessing forecast accuracy of the random walk *versus* the SNN predictor. The results suggest that the SNN predictor outperforms the random walk at the 1% significance level for the Belgian franc, Danish crown, French franc and Dutch guilder, while the random walk presents superior performance relative to the NN local

Table 14.2. Directional forecast (1)

	SNN predictor	ARIMA (1,1,0) predictor
BFR (2)	63.23	52.19
DKR (2)	67.35	63.73
ESC (3)	55.95	53.86
FF (2)	67.43	62.02
HFL (2)	69.19	65.34
IRL (4)	57.73	51.77
LIT (3)	57.25	54.83
PTA (3)	50.64	55.33
UKL (4)	47.58	51.21

Notes:
(1) Percentage of correct forecast direction.
(2) Time series used in establishing occurring analogues in the SNN predictor: BFR, DKR, FF, and HFL.
(3) Time series used in establishing occurring analogues in the SNN predictor: ESC, LIT and PTA.
(4) Time series used in establishing occurring analogues in the SNN predictor: IRL and UKL.
Source: [14.37].

predictor for the Portuguese escudo and Pound sterling at the 1% significance level.

When comparing the predictors from an ARIMA model with the SNN predictors, the null hypothesis of equal absolute error is rejected in four of the nine cases. The results indicate that the SNN predictor outperforms the ARIMA model at the 1% significance level for the Belgian franc, French franc and Dutch guilder, while the ARIMA model presents superior performance relative to the SNN predictor for the Pound sterling at the 1% significance level.

Next, [14.37] computed the Pesaran-Timmerman non-parametric test proportion of correctly predicted signs. Let $z_{t+l} = 1$ if $(x_{t+l}\hat{x}_{t+l} > 0)$ and $z_{t+l} = 0$ otherwise. Let $P_x = \Pr(x_{t+l} > 0)$ $P_{\hat{x}} = \Pr(\hat{x}_{t+l} > 0)$ and

$$\hat{P} = \frac{1}{N} \sum_{t=T+1}^{T+N} z_{t+l} \tag{14.18}$$

be the percentage of correct sign predictions. Denoting the *ex ante* probability that the sign will correctly be predicted as P^* , then

$$\hat{P}^* = \Pr(z_{t+l} = 1) = \Pr(x_{t+l}.\hat{x}_{t+l} > 0) = P_x \cdot P_{\hat{x}} + (1 - P_x) \cdot (1 - P_{\hat{x}})$$

[14.77] show that

$$PT_l = \frac{\hat{P} - \hat{P}^*}{\sqrt{v\hat{a}r(\hat{P}) - v\hat{a}r(\hat{P}^*)}} \tag{14.19}$$

Table 14.3. The Diebold Mariano test statistic(1)

	SNN predictor	ARIMA (1,1,0) predictor
BFR (2)	5.11[a]	4.16[a]
DKR (2)	12.12[a]	0.95
ESC (3)	-2.17[b]	-0.54
FF (2)	14.35[a]	11.22[a]
HFL (2)	12.83[a]	7.72[a]
IRL (4)	2.97[a]	-0.21
LIT (3)	3.19[a]	1.53
PTA (3)	1.14	0.38
UKL (4)	-3.14[a]	-3.16[a]

Notes:
(1) [a] and [b]denote significance at the 1% and 5% levels, respectively.
(2) Time series used in establishing occurring analogues in the SNN predictor: BFR, DKR, FF, and HFL.
(3) Time series used in establishing occurring analogues in the SNN predictor: ESC, LIT and PTA.
(4) Time series used in establishing occurring analogues in the SNN predictor: IRL and UKL.
Source: [14.37].

is asymptotically distributed as a standard normal variable under the null hypothesis of independence between corresponding actual and forecast values, where $\hat{P}^*$ denotes estimated values,

$$v\hat{a}r(\hat{P}) = \frac{1}{T}\hat{P}^*(1-\hat{P}^*),$$

and

$$\begin{aligned} v\hat{a}r(\hat{P}^*) =& \frac{1}{T}(2\hat{P}_x - 1)^2\hat{P}_{\hat{x}}(1-\hat{P}_{\hat{x}}) + \frac{1}{T}(2\hat{P}_{\hat{x}} - 1)^2\hat{P}_x(1-\hat{P}_x) \\ &+ \frac{4}{T^2}\hat{P}_x\hat{P}_{\hat{x}}(1-\hat{P}_x)(1-\hat{P}_{\hat{x}}) \end{aligned}$$

Notice that l denotes the number of periods ahead in the prediction, being $l = 1$ in this case of daily forecasts.

As shown in Table 14.4, the results indicated that the probability of correctly predicting the sign of change is higher for the SNN predictors than the ARIMA case.

Overall, the evidence presented in [14.37] suggested that, when predicting exchange-rate time series, some forecast accuracy can be gained by considering the information content of other related exchange rates through SNN predictors.

Table 14.4. The Pesaran Timmerman test statistic(1)

	SNN predictor	ARIMA (1,1,0) predictor
BFR (2)	11.33[a]	3.45[a]
DKR (2)	13.10[a]	10.83[a]
ESC (3)	2.80[a]	1.54
FF (2)	13.01[a]	9.08[a]
HFL (2)	14.97[a]	11.89[a]
IRL (4)	6.91[a]	3.01[a]
LIT (3)	6.31[a]	3.79[a]
PTA (3)	0.53	1.15
UKL (4)	0.18	-0.37

Notes:
8!) [a] denotes significance at the 1% .
(2) Time series used in establishing occurring analogues in the SNN predictor: BFR, DKR, FF, and HFL.
(3) Time series used in establishing occurring analogues in the SNN predictor: ESC, LIT and PTA.
(4) Time series used in establishing occurring analogues in the SNN predictor: IRL and UKL.
Source: [14.37].

14.5 Assessing the Economic Value of the NN Predictors

As pointed out by [14.80], standard forecasting criteria are not necessarily particularly suited for assessing the economic value of predictions in financial time series. To assess the economic significance of predictable patterns detected in the EMS series, it is necessary to explicitly consider how investors may exploit the computed (S)NN predictions as a trading rule.

In fact, an important line of research has evaluated the relevance of technical analysis in foreign exchange market. As is well known, technical analysis involves the use of charts of financial price movements to infer the likely course of future prices and therefore constructs forecasts and determine trading decisions. Technical analysis is used by the vast majority of active foreign exchange participants, who are mostly interested in the short term movements of the currencies (see, e. g., [14.2], [14.91], [14.67] and [14.74]). A considerable amount of work has provided support for the view that technical trading rules are capable of producing valuable economic signals in foreign exchange markets. [14.32] presented some of the earliest evidence suggesting that technical trading rules might be detecting changes in conditional mean returns in foreign exchange rate series, generating profits in excess of the buy-and-hold strategy. Later, [14.87] also found results supportive of the profitability of similar rules, whereas [14.90] documented similar evidence for even more extensive sets of rules and data series. Moreover, [14.61] and [14.65] followed the methodology of [14.15] and used bootstrap simulations to demonstrate the statistical significance of the technical trading rules against several parametric null models of exchange rates, while Lee and [14.64] showed that only in two of the six cases examined the trading rules were marginally profitable. More recently, [14.88], [14.62], [14.82] and [14.73] discovered that excess returns from extrapolative technical trading rules in foreign exchange markets are high during periods of central bank intervention. Finally, [14.72], using genetic programming methodology, found that *ex ante* trading rules generate significant excess returns in three of the four cases considered, whereas [14.46] using both feed-forward networks and NN regressions, found statistically significant forecast improvements for the current returns over the random walk model of foreign exchange returns.

[14.38] transformed the forecast from NN and SNN predictors into a simple technical trading strategy in which positive returns are executed as long positions and negative returns are executed as short positions. As shown in [14.27], the non-linear NN forecasting technique can be viewed as a generalisation of graphical methods ("heads and shoulders", "rounded tops and bottoms", "flags, pennants and wedges", etcetera) widely used by practitioners. The profitability of this NN strategy is evaluated against the traditional moving average trading rules. Furthermore, unlike previous empirical evidence, when evaluating trading performance, they considered both interest rates and transaction cost, as well as a wider set of profitability indicators than those usually examined.

Regarding the MA trading rules, if E_t is the daily exchange rate, the moving average $m_t(n)$ when it is defined as:

$$m_t\left(n\right)=\frac{1}{n}\sum_{i=0}^{n}E_{t-i}$$

where n is the length of the moving average. Very simple technical trading rules consider the signal $s_t(n_1, n_2)$ defined by

$$s_t\left(n_1,n_2\right)=m_t\left(n_1\right)-m_t\left(n_2\right)$$

where $n_1 < n_2$, and where n_1 and n_2 are the short and the long moving averages, respectively. When $s_t\left(n_1, n_2\right)$ exceeds zero, the short term moving average exceeds the long term moving average to a certain extent, and a "buy" signal is generated. Conversely, when $s_t\left(n_1, n_2\right)$ is negative, and a "sell" signal is given. As can be seen, the moving average rule is essentially a trend following system because when prices are rising (falling), the short-period average tends to have larger (lower) values than the long-period average, signalling a long (short) position. In particular, the following popular moving average rules were evaluated: [1,50], [1,150], [1, 200], [5, 50], [5,150] and [5,200], where the first number in each pair indicates the days in the short period (n_1) and the second number shows the days in the long period (n_2) (see, e. g., [14.15]).

In order to assess the economic significance of a trading strategy, we can consider the estimated total return of such a strategy:

$$R_t = \sum_{t=1}^{N} z_t \cdot r_t \tag{14.20}$$

where r_t is the return from a foreign currency position over the period $(t, t+1)$, z_t is a variable interpreted as the recommended position which takes either a value of -1 (for a short position) or +1 (for a long position), and N is the number of observations.

Given that trading in the spot foreign exchange market requires consideration of interest rates when evaluating trading performance, we can use overnight interest rates to compute r_t as follows:

$$r_t = \ln(E_{t+1}) - \ln(E_t) - (\ln(1 + i_t) - \ln(1 + i_t^*))$$

where E represents the spot exchange rate expressed vis--vis the Deutsche mark, i is the domestic daily interest rate and i^* is the German daily interest rate.

On the other hand, assuming that transaction costs of $c\%$ are paid each time a new position (i. e., from short to long or from long to short) is established, the net return of the technical trading strategy is given by:

$$R_n = \sum_{t=1}^{N} z_t \cdot r_t + nrt \cdot (\ln(1 - c) - \ln(1 + c)) \tag{14.21}$$

where nrt is the number of round-trip trades. The last term in the equation reflects the transaction costs that are assumed to be paid whenever a new position is established. Following [14.65] and [14.75], transaction costs of 0.05% were considered.

Tables 14.5 and 14.6 report the estimated mean annual total and net return, respectively, as well as its associated t-statistics (with corrections for serial correlation and heteroskedasticity, see [14.49]) as a measure of the statistical significance of the results. As can be seen in Table 14.5, only in 1 of the 9 cases considered (Pound sterling) the MA trading rules outperformed the trading strategy based on a non-linear (NN or SNN) predictor. Nevertheless, the t-statistic suggests that in this case the null hypothesis that the mean annual return is equal to zero cannot be rejected. From Table 14.5, we can also see that in 6 out of 9 cases (Belgian franc, Danish crown, Portuguese escudo, Dutch guilder, Italian lira and Spanish peseta), the mean annual total returns when using the SNN predictors are the highest, while for the cases of the French franc and Irish pound, it is the trading system based on the NN predictor that yields the highest mean annual total returns. It should be noted that for the later cases all t-statistics reject the null hypothesis that the mean annual total returns are equal to zero. On the other hand, and as can be seen from Table 14.6, the mean annual net return from the non-linear trading rule dominates that from the MA trading rules in all the cases, except for the Pound sterling case. In 6 such cases (Belgian franc, Danish crown, Portuguese escudo, Dutch guilder, Italian lira and Spanish peseta), the trading system based on the SNN predictor give the highest mean annual net returns, whereas in the cases of the French franc and Irish pound, the highest mean annual net return is obtained when using the NN predictors. Finally, note that in 7 of the 9 exchange rate examined (Belgian franc, Danish crown, French franc, Dutch guilder, Irish pound, Italian lira and Pound sterling) the results are statistically significant (at least at the 5% level) as indicated by the t-statistics.

Besides the total and net returns, [14.38] also took into account other two profitability measures: the ideal profit and the Sharpe ratio. We consider a version of the ideal profit that measures the net returns of the trading system against a perfect predictor and is calculated by:

$$R_I = \frac{\sum_{t=1}^{N} z_t \cdot r_t + nrt \cdot (\ln(1-c) - \ln(1+c))}{\sum_{t=1}^{N} |r_t| + nrt \cdot (\ln(1-c) - \ln(1+c))} \tag{14.22}$$

According to this equation, $R_I = 1$ if the indicator variable takes the correct trading position for all observations in the sample. If all trade positions are wrong, then the value of this measure is $R_I = -1$. An $R_I = 0$ value is considered as a benchmark to evaluate the performance of an investment strategy. Regarding the Sharpe ratio ([14.81]), it is simply the annual mean net return of the trading strategy divided by its standard deviation:

$$S_R = \frac{\mu_{R_n}}{\sigma_{R_n}} \tag{14.23}$$

As can be seen, the higher the Sharpe ratio, the higher the mean annual net return and the lower the volatility. The results for these additional profitability measures are reported in Tables 14.7 and 14.8.

As shown in Table 14.7, the MA trading rules always render negative values for the ideal profit ratio. In contrast, the trading strategy based on the non-linear (NN or SNN) predictors renders positive values in 4 out of the 9 cases considered (Danish crown, French franc, Dutch guilder and Italian lira). In all cases, except for the Pound sterling, the use of non-linear predictors to generate sell/buy signals produces higher values of this profitability measure than those from the MA trading rules. As for the Sharpe ratio, a similar pattern emerges from Table 14.8: the trading strategy based on the non-linear predictors yields the highest Sharpe ratios in 8 out of the 9 cases (Belgian franc, Danish crown, Portuguese escudo, French franc, Dutch guilder, Irish pound, Italian lira and Spanish peseta), while for the Pound sterling the highest (less negative) value is obtained from the MA trading rules.

Table 14.5. Mean annual total return (1) (2)

Exchange rates	Non-linear trading rules NN predictor	SNN predictor	Linear trading rules [1,50]	[1,150]	[1,200]	[5,50]	[5,150]	[5,200]
BFR (3)	0.0631	0.0670	-0.0437	-0.0282	-0.0298	-0.0035	-0.0071	-0.0057
	(5.9861[a])	(7.8334[a])	(−5.0054[a])	(−3.5077[a])	(−3.6200[a])	(-0.4854)	(-0.9706)	(-0.7981)
DKR (3)	0.2614	0.2822	-0.1734	-0.1312	-0.1367	-0.0516	-0.0323	-0.0406
	(11.7277[a])	(11.5596[a])	(−6.1999[a])	(−5.0620[a])	(−5.4208[a])	(−2.7768[a])	(−2.0516[b])	(−2.5656[b])
ESC (4)	0.0323	0.0891	0.0317	0.0385	0.0381	0.0673	0.0329	0.0391
	(0.6578)	(2.2230[a])	(0.7455)	(1.1145)	(1.0782)	(1.7961[c])	(0.9258)	(1.1333)
FF (3)	0.2448	0.2427	-0.1115	-0.0745	-0.0657	-0.0085	-0.0204	-0.0211
	(13.9101[a])	(13.6570[a])	(−5.5922[a])	(−4.2358[a])	(−4.4343[a])	(-0.6926)	(-1.8477)	(−1.8279[c])
HFL (3)	0.1403	0.1471	-0.1131	-0.0984	-0.0845	-0.0324	-0.0237	-0.0222
	(17.6453[a])	(18.1531[a])	(−11.6381[a])	(−8.8918[a])	(-8.3545)	(−5.9658[a])	(−4.7103[a])	(−4.6854[a])
IRL (5)	0.0527	0.0305	-0.0378	-0.0242	-0.0308	-0.0065	-0.0094	-0.0099
	(3.1079[a])	(1.8536[c])	(−2.3394[b])	(-1.6387)	(−1.9767[b])	(-0.4002)	(-0.6427)	(-0.7225)
LIT (4)	0.2681	0.3348	-0.1664	-0.0449	-0.0425	-0.0332	-0.0032	-0.0008
	(6.2970[a])	(9.1434[a])	(−3.5986[a])	(-1.2011)	(-1.1516)	(-0.8943)	(-0.0922)	(-0.0244)
PTA (4)	0.0405	0.0677	0.0626	0.0193	0.0473	0.0483	0.0368	0.0349
	(0.9921)	(2.0879[b])	(1.7485[c])	(0.4924)	(1.2748)	(1.2894)	(0.9768)	(0.9290)
UKL (5)	-0.0663	-0.0079	0.0561	0.0076	0.0139	0.0352	0.0111	0.0076
	(-1.4951)	(-0.2688)	(1.4503)	(0.1952)	(0.3612)	(0.9124)	(0.2851)	(0.1943)

Notes:
(1) Returns generated by each forecasting method over the forecast sample, before transaction fees are taken into account [see Eq. (14.20) in the text].
(2) t- statistics (corrected for serial correlation and heteroskedasticity) in parenthesis: [a],[b] and [c] denote significance at the 1%, 5% and 10% levels, respectively.
(3) Time series used in establishing occurring analogues in the SNN predictor: BFR, DKR, FF and HFL.
(4) Time series used in establishing occurring analogues in the SNN predictor: ESC, LIT and PTA.
(5) Time series used in establishing occurring analogues in the SNN predictor: IRL and UKL.
Source: [14.38].

Table 14.6. Mean annual net return (1) (2)

Exchange rates	Non-linear trading rules NN predictor	SNN predictor	Linear trading rules [1,50]	[1,150]	[1,200]	[5,50]	[5,150]	[5,200]
BFR (3)	-0.0619	-0.0580	-0.1687	-0.1532	-0.1548	-0.1285	-0.1321	-0.1307
	(−5.8806^{a})	(−6.7737^{a})	(-19.3104)	(−19.0678^{a})	(−18.8113^{a})	(−17.8063^{a})	(−17.9502^{a})	(−18.3446^{a})
DKR (3)	0.1364	0.1572	-0.2984	-0.2562	-0.2617	-0.1766	-0.1573	-0.1656
	(6.1189^{a})	(6.4393^{a})	(−10.6697^{a})	(−9.8857^{a})	(−10.3778^{a})	(−9.5070^{a})	(−9.9967^{a})	(−104596^{a})
ESC (4)	-0.0927	-0.0359	-0.0934	-0.0865	-0.0869	-0.0577	-0.0921	-0.0859
	(−1.8860^{c})	(-0.8958)	(−2.1990^{b})	(−2.5006^{b})	(−2.4568^{b})	(-1.5379)	(−2.5919^{a})	(−2.4909^{b})
FF (3)	0.1198	0.1177	-0.2365	-0.1995	-0.1907	-0.1335	-0.1454	-0.1461
	(6.8069^{a})	(6.6234^{a})	(−11.8608^{a})	(−11.3436^{a})	(−12.8718^{a})	(−10.8287^{a})	(−13.1503^{a})	(−12.6432^{a})
HFL (3)	0.0153	0.0221	-0.2381	-0.2234	-0.2095	-0.1574	-0.1487	-0.1472
	(1.9241^{c})	(2.7314^{a})	(−24.4976^{a})	(−20.1831^{a})	(−20.7122^{a})	(−28.9952^{a})	(−29.5279^{a})	(−31.0157^{a})
IRL (5)	-0.0723	-0.0945	-0.1628	-0.1492	-0.1558	-0.1315	-0.1344	-0.1349
	(−4.2672^{a})	(−5.7488^{a})	(−10.0772^{a})	(−10.0892^{a})	(−9.9999^{a})	(−8.1011^{a})	(−9.2135^{a})	(−9.8889^{a})
LIT (4)	0.1431	0.2098	-0.2914	-0.1699	-0.1675	-0.1582	-0.1282	-0.1258
	(3.3608^{a})	(5.7296^{a})	(−6.3014^{a})	(−4.5449^{a})	(−4.5404^{a})	(−4.2609^{a})	(−3.7280^{a})	(−3.6508^{a})
PTA (4)	-0.0845	-0.0573	-0.0624	-0.1057	-0.0777	-0.0767	-0.0883	-0.0901
	(−2.0689^{b})	(−1.7653^{c})	(−1.7425^{c})	(−2.6970^{a})	(−2.0937^{b})	(−2.0490^{b})	(−2.3459^{b})	(−2.3998^{b})
UKL (5)	-0.1913	-0.1329	-0.0690	-0.1174	-0.1111	-0.0898	-0.1139	-0.1174
	(−4.3125^{a})	(−4.4966^{a})	(−1.7841^{c})	(−3.0194^{a})	(−2.8958^{a})	(−2.3284^{b})	(−2.9346^{a})	(−2.9940^{a})

Notes:
(1) Returns generated by each forecasting method over the forecast sample, after transaction fees are taken into account [see Eq. (14.21) in the text].
(2) t- statistics (corrected for serial correlation and heteroskedasticity) in parenthesis: ^{a},^{b} and ^{c} denote significance at the 1%, 5% and 10% levels, respectively.
(3) Time series used in establishing occurring analogues in the SNN predictor: BFR, DKR, FF and HFL.
(4) Time series used in establishing occurring analogues in the SNN predictor: ESC, LIT and PTA.
(5) Time series used in establishing occurring analogues in the SNN predictor: IRL and UKL.
Source: [14.38].

Table 14.7. Ideal profit ratio (1)

Exchange rates	Non-linear trading rules NN predictor	SNN predictor	Linear trading rules [1,50]	[1,150]	[1,200]	[5,50]	[5,150]	[5,200]
BFR (2)	-0.5479	-0.5127	-1.4925	-1.3548	-1.3691	-1.1366	-1.1688	-1.1559
DKR (2)	0.2388	0.2753	-0.5225	-0.4486	-0.4582	-0.3092	-0.2754	-0.2900
ESC (3)	-0.1793	-0.0695	-0.1807	-0.1673	-0.1681	-0.1116	-0.1782	-0.1663
FF (2)	0.2599	0.2554	-0.5132	-0.4328	-0.4137	-0.2897	-0.3155	-0.3170
HFL (2)	0.1356	0.1963	-2.1112	-1.9809	-1.8574	-1.3953	-1.3186	-1.3054
IRL (4)	-0.2839	-0.3710	-0.6389	-0.5858	-0.6115	-0.5161	-0.5274	-0.5293
LIT (3)	0.1538	0.2255	-0.3133	-0.1827	-0.1801	-0.1701	-0.1378	-0.1353
PTA (3)	-0.1359	-0.0921	-0.1004	-0.1701	-0.125	-0.1234	-0.142	-0.1450
UKL (4)	-0.2932	-0.2037	-0.1057	-0.1799	-0.1703	-0.1376	-0.1746	-0.1799

Notes:
(1) The ideal profit measures the returns of the trading system against a perfect predictor [see Eq. (14.22) in the text].
(2) Time series used in establishing occurring analogues in the SNN predictor: BFR, DKR, FF and HFL.
(3) Time series used in establishing occurring analogues in the SNN predictor: ESC, LIT and PTA.
(4) Time series used in establishing occurring analogues in the SNN predictor: IRL and UKL.
Source: [14.38].

Table 14.8. Sharpe ratio (1)

Exchange	Non-linear trading rules		Linear trading rules					
rates	NN predictor	SNN predictor	[1,50]	[1,150]	[1,200]	[5,50]	[5,150]	[5,200]
BFR (2)	-0.1583	-0.1484	-0.4285	-0.3875	-0.3917	-0.3243	-0.3335	-0.3298
DKR (2)	0.1635	0.1900	-0.3483	-0.2964	-0.3031	-0.2023	-0.1800	-0.1897
ESC (3)	-0.0880	-0.0342	-0.0886	-0.0821	-0.0825	-0.0548	-0.0874	-0.0816
FF (2)	0.1835	0.1801	-0.3437	-0.2878	-0.2748	-0.1916	-0.2087	-0.2097
HFL (2)	0.0555	0.0814	-0.8277	-0.7685	-0.7046	-0.5119	-0.4825	-0.4775
IRL (4)	-0.1045	-0.1363	-0.2349	-0.2152	-0.2247	-0.1895	-0.1936	-0.1943
LIT (3)	0.0941	0.1392	-0.1898	-0.1101	-0.1085	-0.1025	-0.0830	-0.0815
PTA (3)	-0.0732	-0.0497	-0.0541	-0.0916	-0.0673	-0.0665	-0.0765	-0.0781
UKL (4)	-0.1646	-0.1142	-0.0593	-0.1008	-0.0955	-0.0772	-0.0979	-0.1008

Notes:
(1) The Sharpe ratio is obtained dividing the mean return of the trading system by its standard deviation [see Eq. (14.23) in the text].
(2) Time series used in establishing occurring analogues in the SNN predictor: BFR, DKR, FF and HFL.
(3) Time series used in establishing occurring analogues in the SNN predictor: ESC, LIT and PTA.
(4) Time series used in establishing occurring analogues in the SNN predictor: IRL and UKL.
Source: [14.38].

14.6 Concluding Remarks

There is a growing consensus in the literature on the fact that foreign exchange markets have become increasingly complex and therefore less amenable to forecasting over time when using standard linear models. As an alternative approach, recent advances in both analytic and computational methods have suggested new lines of research exploring the possible non-linear dynamics in foreign exchange rates. This approach has been supported by a burgeoning literature providing a theoretical justification for the presence of non-linearities in foreign exchange rates, while many empirically papers have found non-linearities in foreign exchange markets.

The purpose of this chapter has been to contribute to the debate on the relevance of these non-linear forecasting methods for high-frequency data presenting the results of applying the nearest neighbour (NN) approach to forecasting to nine currencies participating in the exchange rate mechanism of the EMS.

The NN forecasting approach relies on the premise that short-term predictions can be made based on past patterns of the time series, therefore circumventing the need to specify an explicit econometric model to represent the time series. Therefore, it is philosophically very different from the Box-Jenkins methodology. In contrast to the traditional Box-Jenkins (linear) models (see [14.13]), where extrapolation of past values into the immediate future is based on correlation among lagged observations and error terms, NN methods select relevant prior observations based on their levels and geometric trajectories, not their location in time.

The papers reviewed in the chapter illustrate the relevance of the approach, both from the point of view of the statistical forecasting accuracy and from the point of view of the economic value as derived from it use as a trading rule. The results have showed the potential usefulness of nearest neighbour predictors not only as useful tool when forecast daily exchange data but also as a technical trading rule capable of producing valuable economic signals in foreign exchange markets. Therefore, further consideration of NN predictors could be a fruitful enterprise.

There are a number of directions that extensions from the present research might take. Two avenues that seem worthy of further research are:

1. the combination of NN methods with other non-linear forecasting methods such as artificial neural networks.
2. the application of NN methods to other currencies, such as the Dollar/Euro or the Dollar/Yen exchange rates.

References

14.1 Akaike, H. (1973): Information Theory and an Extension of the Maximum Likelihood Principle. In: Petrov, N., Csake, F. (Eds.), Second International Symposium on Information Theory.

14.2 Allen, H., Taylor, M. P. (1990): Charts, Noise and Fundamentals in the London Foreign Exchange Market. Economic Journal, 100, 49–59

14.3 Andrada-Félix, J., Fernández-Rodríguez, F., García-Artiles, M. D., Sosvilla-Rivero, S. (2001): An Empirical Evaluation of Non-Linear Trading Rules. Working Paper 2001-16, FEDEA (available online at ftp://ftp.fedea.es/pub/Papers/2001/dt2001-16.pdf)

14.4 Bajo-Rubio, O., Fernández-Rodríguez, F., Sosvilla-Rivero, S. (1992a): Chaotic Behaviour in Exchange-Rate Series: First Results for the Peseta-U.S. Dollar Case. Economics Letters, 39, 207–211 (reprinted in Trippi (1995), 327–334)

14.5 Bajo-Rubio, O., Fernández-Rodríguez, F., Sosvilla-Rivero, S. (1992b): Volatilidad y predecibilidad en las series del tipo de cambio peseta-délar: Un enfoque basado en el caos determinista. Revista Española de Economía, Monographic issue on "Spanish Financial Markets", 91–109

14.6 Bajo-Rubio, O., Sosvilla-Rivero, S., Fernández-Rodríguez, F. (2001): Asymmetry in the EMS: New Evidence Based on Non-Linear Forecasts. European Economic Review, 45, 451–473

14.7 Baldwin, R. E. (1990): Re-Interpreting the Failure of Foreign Exchange Market Efficiency Tests: Small Transaction Costs, Big Hysteresis Bands. CEPR Discussion Paper No. 407

14.8 Barnett, W. A., Gallant, A. R., Hinich, M. J., Jungeilges, J. A., Kaplan, D. T., Jensen, M. J. (1998): A Single-Blind Controlled Competition Among Tests for Nonlinearity and Chaos. Journal of Econometrics, 82, 157–192

14.9 Bertola, G., Caballero, R. (1992): Target Zones and Realignments. American Economic Review, 82, 520–530

14.10 Bilson, J. (1990): 'Technical' Currency Trading. In: Thomas, L. (Ed.), The Currency-Hedging Debate

14.11 Bollerslev, T., Chou, R. Y., Kroner, K. F. (1992): ARCH Modeling in Finance: A Review of the Theory and Empirical Evidence. Journal of Econometrics, 52, 5–59

14.12 Boothe, P., Glassman, D. (1987): Comparing Exchange Rate Forecasting Models: Accuracy versus Profitability. International Journal of Forecasting, 3, 65–79

14.13 Box, G. E. P., Jenkins, G. M. (1976): Time Series Analysis Forecasting and Control.

14.14 Brock, W., Dechert, W., Scheinkman, J., LeBaron, B. (1996): A Test for Independence Based on the Correlation Dimension. Econometric Reviews, 15, 197–235

14.15 Brock, W., Lakonish, J., LeBaron, B. (1992): Simple Technical Rules and the Stochastic Properties of Stock Returns. Journal of Finance, 47. 1731–1764

14.16 Broomhead, D. S., y King, G. P. (1986): Extracting Qualitative Dynamics from Experimental Data. Physica D, 20, 217–231

14.17 Cao, L. (1997): Practical Method for Determining the Minimum Embedding Dimension of a Scalar Time Series. Physica D, 110, 43–50

14.18 Cao, L., Soofi, A. S. (1999): Nonlinear Deterministic Forecasting of Daily Dollar Exchange Rates. International Journal of Forecasting, 15, 421–430

14.19 Cao, L., Mees, A., Judd, K. (1998): Dynamics from Multivariate Time Series. Phisica D, 110, 43–50

14.20 Casdagli, M. (1989): Nonlinear Prediction of Chaotic Time Series. Physica D, 35, 335–356

14.21 Casdagli, M. (1991): Nonlinear Forecasting, Chaos and Statistics. Working Paper 91-05-022, Santa Fe Institute

14.22 Casdagli, M. (1992): Chaos and Deterministic versus Stochastic Nonlinear Modeling. Journal of the Royal Statistical Society B, 54, 303–328
14.23 Cecen, A. A., Erkal, C. (1996): Distinguishing between Stochastic and Deterministic Behaviour in Foreign Exchange Rate Returns: Futher Evidence. Economics Letters, 51, 323–329
14.24 Clark, P. K. (1973): A Subordinated Stochastic Process Model with Finite Variance for Speculative Prices. Econometrica, 41, 135–155
14.25 Cleveland, W. S. (1979): Robust Locally-Weighted Regression and Smoothing Scatterplots. Journal of the American Statistical Association, 74, 829–836
14.26 Cleveland, W. S. (1993): Visualizing Data (New Jersey: Hobart Press)
14.27 Clyde, W. C., Osler, C. L. (1997): Charting: Chaos Theory in Disguise?. Journal of Futures Markets, 17, 489–514
14.28 De Grauwe, P., Vansanten, K. (1990): Deterministic Chaos in the Foreign Exchange Market. CEPR Discussion Paper No. 370
14.29 De Long, J. B., Shleifer, A., Summers, L. H., Walfmann, R. J. (1990): Noise Trader Risk in Financial Markets. Journal of Political Economy, 98, 703–738
14.30 Diebold, F. X., Nason, J. (1990): Nonparametric Exchange Rate Prediction? Journal of International Economic, 28, 315–332
14.31 Diebold, F. X., Mariano, R. S. (1995): Comparing Predictive Accuracy. Journal of Business and Economic Statistics, 13, 253–263
14.32 Dooley, M. P., Shafer, J. (1983): Analysis of Short-Run Exchange Rate Behaviour: March 1973 to November 1981. In: Bigman, D., Taya, T. (Eds.), Exchange Rate and Trade Instability: Causes, Consequences and Remedies.
14.33 Eichengreen, B., Wyplosz, C. (1993): The Unstable EMS. Brookings Papers on Economic Activity, 1, 51–143
14.34 Farmer, D., Sidorowich, J. (1987): Predicting Chaotic Time Series. Physical Review Letters, 59, 845–848
14.35 Fernández-Rodríguez, F. (1992): El problema de la prediccién en series temporales: Aplicaciones del uso del caos determinista. Ph.D. Thesis, Universidad de Las Palmas de Gran Canaria
14.36 Fernández-Rodríguez, F., Sosvilla-Rivero, S., Andrada-Félix, J. (1997): Combining Information in Exchange Rate Forecasting: Evidence from the EMS. Applied Economics Letters, 4, 441–444
14.37 Fernández-Rodríguez, F., Sosvilla-Rivero, S., Andrada-Félix, J. (1999): Exchange-Rate Forecasts with Simultaneous Nearest-Neighbour Methods: Evidence from the EMS. International Journal of Forecasting, 15, 383–392
14.38 Fernández-Rodríguez, F., Sosvilla-Rivero, S., Andrada-Félix, J. (2002): Technical Analysis in Foreign Exchange Markets: Evidence from the EMS. Applied Financial Economics, forthcoming
14.39 Fernández-Rodríguez, F., Sosvilla-Rivero, S., García-Artiles, M. D. (1997): Using Nearest-Neighbour Predictors to Forecast the Spanish Stock Markets. Investigaciones Econémicas, 21, 75–91
14.40 Fernández-Rodríguez, F., Sosvilla-Rivero, S., García-Artiles, M. D. (1999): Dancing with Bulls and Bears: Nearest Neighbour Forecast for the Nikkei Index. Japan and the World Economy, 11, 395–413
14.41 Flood, R. P. and Rose, A. K. (1995): Fixing Exchange Rates: A Virtual Quest for Fundamentals. Journal of Monetary Economics, 36, 3–37
14.42 Frankel, J. A. (1984): Test of Monetary and Portfolio-Balance Models of Exchange Rate Determination. In: Bilson, J., Marston, R. (Eds.): Exchange Rate Theory and Practice. University of Chicago Press, Chicago, 239–260
14.43 Frankel, J. A., Froot, K. A. (1986): The Dollar as a Speculative Bubble: A Tale of Chartists and Fundamentalists. NBER Working paper 1854

14.44 Frankel, J. A., Rose, A. K. (1995): Empirical Research on Nominal Exchange Rates. In: Grossman, G. M., Rogoff, K. (Eds.), Handbook of International Economics, Vol. III. North-Holland, 1689–1729

14.45 Froot, K. A., Obstfeld, M. (1991): Intrinsic Bubbles: The Case of Stock Prices. American Economic Review, 81, 1189–1214

14.46 Gençay, R. (1999): Linear, Non-Linear and Essential Foreign Exchange Rate Prediction with Some Simple Technical Trading Rules. Journal of International Economics, 47, 91–107

14.47 Ghahghaie, S., Breynabb, W., Peinke, J., Talkner, P., Dodge, Y. (1996): Turbulent Cascades in Foreign Exchange Markets. Nature, 281, 767–770

14.48 Grassberger, P., Procaccia, I. (1983): Characterization of Strange Attractors. Physical Review Letters, 50, 346–349

14.49 Hamilton, J. D. (1994): Time Series Analysis. Princeton University Press

14.50 Härdle, W. (1990): Applied Nonparametric Regression. Cambridge University Press

14.51 Hsieh, D. A. (1989): Testing for Nonlinear Dependence in Daily Foreign Exchange Rates. Journal of Business, 62, 339–368

14.52 Hsieh, D. A. (1991): Chaos and Nonlinear Dynamics: Applications to Financial Markets. Journal of Finance, 46, 1839–1877

14.53 Jacquemin. A., Sapir, A. (1996): Is a European Hard Core Credible? A Statistical Analysis. Kyklos, 49, 105–117

14.54 Jaditz T., Sayers C. (1998): Out-of-Sample Forecast Performance as a Test for Nonlinearity in Time Series. Journal of Business and Economic Statistic, 16, 110–117

14.55 Kilian, L. (1999): Exchange Rates and Monetary Fundamentals: What do We Learn from Long-Horizon Regressions? Journal of Applied Econometrics, 14, 491–510

14.56 Kennel, M., Brown, R., Abarbanel, H. (1992): Determining Embedding Dimension for Phase-Space Reconstruction Using a Geometrical Construction. Physical Review A, 45, 3403–3411

14.57 Kilian, L., Taylor, M. P. (2001): Why is It So Difficult to Beat the Random Walk Forecast of Exchange Rates? CEPR Discussion Paper 3024

14.58 Kugler, P., Lenz, C. (1993): Chaos, ARCH and the Foreign Exchange Market: Empirical Results from Weekly Data. Rivista Internazionale di Scienze Echonomiche e Commerciali, 40, 127–140

14.59 Kräger, H., Kugler, P. (1993): Non-Linearities in Foreign Exchange Markets: A Different Persoective. Journal of International Money and Finance, 12, 195–208

14.60 Krugman, P. (1991): Target Zones and Exchange Rate Dynamics. Quarterly Journal of Economics, 106, 669–682

14.61 LeBaron, B. (1992): Do Moving Average Rule Results Imply Nonlinearities in Foreign Exchange Markets? SSRI Working Paper No. 9222, University of Wisconsin-Madison

14.62 LeBaron, B. (1999): Technical Trading Rule Profitability and Foreign Exchange Intervention. Journal of International Economics, 49, 125–143

14.63 Ledesma-Rodríguez, F., Navarro-Ibáñez, M., Pérez-Rodríguez, J., Sosvilla-Rivero, S. (2001): Assessing the Credibility of a Target Zone: Evidence from the EMS. Working Paper 2001-04, FEDEA (available online at ftp://ftp.fedea.es/pub/Papers/2001/dt2001-04.pdf)

14.64 Lee, C. I., Mathur, I. (1996): Trading Rule Profits in European Currency Spot Cross-Rates. Journal of Banking and Finance, 20, 949–962

14.65 Levich, R., Thomas, L. (1993): The Significance of Technical Trading Rule Profits in the Foreign Exchange Market: A Bootstrap Approach. Journal of International Money and Finance, 12, 451–474

14.66 Lisi, F., Medio, A. (1997): Is a Random Walk the Best Exchange Rate Predictor? International Journal of Forecasting, 13, 255–267

14.67 Lui, Y. H., Mole, D. (1998): The Use of Fundamental and Technical Analyses by Foreign Exchange Dealers: Hong Kong Evidence. Journal of International Money and Finance, 17, 535–545

14.68 Mark, N. C. (1995): Exchange Rates and Fundamentals: Evidence of Long-Horizon Predictability. American Economic Review, 85, 201–218

14.69 Meese, R. A., Rogoff, K. (1983): Empirical Exchange Rate Models of the Seventies: Do They Fit out of Sample? Journal of International Economics, 14, 3–24

14.70 Meese, R. A., Rose, A. K. (1991): An Empirical Assessment of Non-Linearities in Models of Exchange Rate Determination. Review of Economic Studies, 58, 603–619

14.71 Mizrach, B. (1992): Multivariate Nearest-Neighbour Forecasts of EMS Exchange Rates. Journal of Applied Econometrics, 7, S151-S163

14.72 Neely, C. J., Weller, P. (1999): Technical Trading Rules in the EMS. Journal of International Money and Finance, 18, 429–458

14.73 Neely, C. J., Weller, P. (2001): Technical Analysis and Central Bank Intervention. Journal of International Money and Finance, 20, 949–970

14.74 Oberkechner, T. (2001): Importance of Technical and Fundamental Analysis in the European Foreign Exchange Market. International Journal of Finance and Economics, 6, 81–93

14.75 Osler, C. L., Chang, P. H. K. (1995): Head and Shoulders: Not Just a Flaky Pattern. Federal Reserve Bank of New York Staff Paper No. 4

14.76 Peel, D. A, (1993): Non-Linear Risk Premia. Applied Financial Economics, 3, 201–204

14.77 Pesaran, M. H., Timmerman, A. (1992): A Simple Nonparametric Test of Predictive Performance. Journal of Business and Economic Statistics, 10, 461–465

14.78 Peters, E. E. (1994): Fractal Market Analysis: Applying Chaos Theory to Investment and Economics. John Wiley

14.79 Refenes, A. (1993): Constructive Learning and Its Application to Currency Exchange Rate Forecasting. In: Trippi, R., Turban, E. (Eds.), Neural Networks in Finance and Investing. Probus

14.80 Satchell, S., Timmermann, A. (1995): An Assessment of the Economic Value of Non-Linear Foreign Exchange Rate Forecasts. Journal of Forecasting, 14, 477–497

14.81 Sharpe, W. F. (1966): Mutual Fund Performance. Journal of Finance, 39, 119–138

14.82 Sosvilla-Rivero, S., Andrada-Félix, J., Fernández-Rodríguez, F. (1999): Further Evidence on Technical Analysis and Profitability of Foreign Exchange Intervention. FEDEA Working Paper No. 99–01 (available on line at ftp://ftp.fedea.es/pub/Papers/1999/DT99-01.pdf)

14.83 Sosvilla-Rivero, S., Fernández-Rodríguez, F., Bajo-Rubio, O. (1999): Exchange Rate Volatility in the EMS before and after the Fall. Applied Economics Letters, 6, 717–722

14.84 Sosvilla-Rivero, S., Maroto, R. (2001): Duración de los regímenes del SME. FEDEA Working Paper No. 2001-05 (available on line ftp://ftp.fedea.es/pub/Papers/2001/dt2001-05.pdf)

14.85 Stone, C. J. (1977): Consistent Nonparametric Regression. Annals of Statistics, 5, 595–620
14.86 Sugihara, G., May, R. M. (1990): Nonlinear Forecasting as a Way of Distinguishing Chaos from Measurement Error in Time Series. Nature, 344, 734–741
14.87 Sweeney, R. J. (1986): Beating the Foreign Exchange Market. Journal of Finance, 41, 163–182
14.88 Szakmary, A. C., Mathur, I. (1997): Central Bank Intervention and Trading Rule Profits in Foreign Exchange Markets. Journal of International Money and Finance, 16, 513–535
14.89 Takens, F. (1981): Detecting Strange Attractors in Turbulence. In: Rand, D. A., Young, L. S. (Eds.), Dynamical Systems and Turbulence. Springer-Verlag, 366–381
14.90 Taylor, S. J. (1992): Rewards Available to Currency Future Speculators: Compensation for Risk or Evidence of Inefficient Pricing? Economic Record, 68, 105–116
14.91 Taylor, M. P., Allen, H. (1992): The Use of Technical Analysis in the Foreign Exchange Market. Journal of International Money and Finance, 11, 304–314
14.92 Taylor, M. P., Peel, D. A. (2001): The Behaviour of Real Exchange Rates During the Post Bretton Woods Period. Journal of International Money and Finance, 19, 33–53
14.93 Trippi, R. R. (1995): Chaos & Nonlinear Dynamics in the Financial Markets. Theory, Evidence and Applications.
14.94 Wayland, R., Pickett, D., Bromley, D., Passamante, A. (1994): Measuring Spatial Spreading in Recurrent Time Series. Physica D, 79, 320–334

Part VI

Evolutionary Computation, Swarm Intelligence and Simulated Annealing

15. Discovering Hidden Patterns with Genetic Programming

Shu-Heng Chen and Tzu-Wen Kuo

AI-ECON Research Center, Department of Economics
National Chengchi University, Taipei, Taiwan 11623
email: chchen@nccu.edu.tw, kuo@aiecon.org

In this chapter, we shall review some early applications of genetic programming to financial data mining and knowledge discovery, including some analyses of its statistical behavior. These early applications are known as *symbolic regression* in GP. In this type of application, genetic programming is formally demonstrated as an engine searching for the hidden relationships among observations. Here, we find evidence of the closest step ever made toward the original motivation of John Holland's invention of genetic algorithms: *Instead of trying to write your programs to perform a task you don't quite know how to do, evolve them.*

15.1 Discovering the Hidden Law of Motion

In the first type of application, genetic programming is applied to discover the underlying law of motion, or the data-generation process. The law of motion can be a deterministic chaotic process,

$$x_t = f(x_{t-1}, ...) \tag{15.1}$$

a stochastic process,

$$x_t = f(x_{t-1}, ...) + \epsilon_t \tag{15.2}$$

$$x_t = f(x_{t-1}, y_{t-1}, z_{t-1}, x_{t-2}, y_{t-2}, z_{t-2}, ...) + \epsilon_t \tag{15.3}$$

or a regression,

$$y_i = f(x_{1,i}, x_{2,i}, ...) + \epsilon_i \tag{15.4}$$

In all these applications, GP is used to discover the underlying data-generation process of a series of observations. While this type of application is well known to econometricians, the perspective from GP is novel. As [15.14] has stated,

> An important problem in economics is finding the mathematical relationship between the empirically observed variables measuring a system. In many conventional modeling techniques, one necessarily begins by selecting the size and shape of the model. After making

> this choice, one usually then tries to find the values of certain coefficients required by the particular model so as to achieve the best fit between the observed data and the model. But, in many cases, *the most important issue is the size and shape of the model itself.* (p.57; italics added.)

Econometricians offer no general solution to the determination of *size* and *shape* (the functional form), but for Koza, finding the functional form of the model can be viewed as *searching a space of possible computer programs* for the particular computer program which produces the desired output for given inputs.

Koza employed GP to rediscover some basic physical laws from experimental data, for example, Kepler's third law and Ohm's law ([15.15]). He then also applied it to eliciting a very fundamental economic law, namely, the *quantity theory of money* or the *exchange equation* ([15.14]). Genetic programming was thus formally demonstrated as a *knowledge discovery* tool. This was probably the closest step ever made toward the original motivation of John Holland's invention: "Instead of trying to write your programs to perform a task you don't quite know how to do, *evolve them.*" Indeed, Koza did not evolve the parameters of an arbitrary chosen equation; instead, he evolved the whole equation from scratch, i.e. from the primitives of GP.

15.1.1 Deterministic Chaotic Processes

[15.7] applied GP to learn the following three chaotic equations:

$$x_{t+1} = 4x_t(1 - x_t), \quad x_t \in [0,1] \ \forall t \tag{15.5}$$

$$x_{t+1} = 4x_t^3 - 3x_t, \quad x_t \in [-1,1] \ \forall t \tag{15.6}$$

$$x_{t+1} = 8x_t^4 - 8x_t^2 + 1, \quad x_t \in [-1,1] \ \forall t \tag{15.7}$$

Four experiments were implemented for each series. For each experiment, GP was run for 1,000 generations. For series (15.5) and (15.6), GP was able to discover the underlying law of motion in all four experiments. However, the number of generations required for this discovery was different. For series (15.5), it took 7, 12, 14, and 19 generations, whereas for series (15.6), it took 29, 37, 37, and 70 generations. As for series (15.7), GP failed to discover the law of motion in three out of the four simulations, and for the only success the law of motion was discovered at the 151th generation. These experiments show the effect of the length of the LISP S-expression (the algorithmic complexity of the program) on discovery.

These three laws of motion differ in their *algorithmic size*, i.e. the *length* of their symbolic expression. To see this, we rewrite each of the equations above into the corresponding LISP S-expression:

$$(\ * \ (\ 4 \ * \ (\ x_t \ (\ - \ 1 \ x_t \ ) \ ) \ ) \ ) \tag{15.8}$$

$$(- (* 4 (* x_t (* x_t x_t))) (* 3 x_t)) \tag{15.9}$$

$$\begin{gathered}(+ (- (* 8 (* x_t (* x_t (* x_t x_t)))) \\ (* 8 (* x_t x_t))) 1)\end{gathered} \tag{15.10}$$

The *length* of a LISP S–expression is determined by counting from the left-most to the rightmost position the number of elements (atoms) in the string that makes up the S–expression. From Eqs. (15.8) to (15.10), the lengths of the LISP S-expressions are 7, 11, and 17, respectively. Therefore, in terms of algorithmic complexity, Eq. (15.5) is the simplest, while Eq. (15.7) is the most complex. [15.7] then examined how this difference might affect the performance of GP.

The *function set* originally employed by [15.7] is $\{+, -, \times, \%\}$, and the terminal set is $\{x_t, \mathcal{R}\}$, where $\mathcal{R}$ is the ephemeral random constant. If we add the function "cubic" to the function set, then the minimal description of Eq. 15.6 is simply

$$(- (* 4 (\; cubic \; x_t)) (* 3 x_t)), \tag{15.11}$$

and the program length of it is only 8. Alternatively, if we add x_t^3 to the terminal set, then the minimal description becomes

$$(- (* 4 x_t^3) (* 3 x_t)). \tag{15.12}$$

In this case, the program length of Eq. (15.6) is even shorter and is 7, which is the same as that of Eq. (15.5). It is, therefore, likely that it will discover the hidden law of series (15.6) as easily as it will discover that of series (15.5). So, depending on the user-supplied function set and terminal set, a mathematical function can have different program lengths.[1]

Formally speaking, let $\mathcal{T}$ be the terminal set and $\mathcal{F}$ be the function set, and let us denote their cardinality by $\mid \mathcal{T} \mid$ and $\mid \mathcal{F} \mid$.

$$Prob(f^* \in G_n) = P[K(f^* \mid \mathcal{F} \bigcup \mathcal{T})], \tag{15.13}$$

and

$$\frac{\Delta Prob(f^* \in G_n)}{\Delta K} \Big|_{\mathcal{F} \bigcup \mathcal{T}} \leq 0, \tag{15.14}$$

where f^* is an targeted decision rule, say, the optimal decision rule, G_n is the nth generation of population, and $K(f^* \mid \mathcal{F} \bigcup \mathcal{T})$ refers to the algorithmic complexity of the decision rule of f^* given the functional set $\mathcal{F}$ and the terminal set $\mathcal{T}$.

[1] While the assertion and the example given above are based on deterministic functions, we are conducting a study to see whether this assertion, or some modifictaion of it, can be applied to functions with noise.

Since a larger function set and a terminal set will help abbreviate the representation of a decision rule, the algorithmic complexity of a function can only be non-positively related to $\mid \mathcal{F} \mid$ and $\mid \mathcal{T} \mid$, i.e.

$$\frac{\Delta K}{\Delta \mid \mathcal{F} \mid} \leq 0, \quad and \quad \frac{\Delta K}{\Delta \mid \mathcal{T} \mid} \leq 0. \tag{15.15}$$

Going back to Eq. (15.14), it seems that a larger terminal set and a larger function set can enhance search efficiency. In fact, there is empirical evidence that suggests that this is indeed the case ([15.10]). Still, the influence of search space and population size also has to be taken into account. Let $\mathcal{S}$ be the search space, which is a collection of all potential species. The size of it, $\mid \mathcal{S} \mid$, grows exponentially with $\mid \mathcal{F} \mid$ and $\mid \mathcal{T} \mid$. Therefore, if population size does not grow exponentially with $\mid \mathcal{F} \mid$ and $\mid \mathcal{T} \mid$,

$$\lim_{\mid \Delta\mathcal{F} \mid, \mid \Delta\mathcal{T} \mid \to \infty} s = 0, \tag{15.16}$$

where

$$s = \frac{\mid G \mid}{\mid S \mid}. \tag{15.17}$$

To take into account the effect of s, let us rewrite Eq. (15.13) into a conditional density,

$$Prob(f^* \in G_n \mid s) = P[K(f^* \mid \mathcal{F} \bigcup \mathcal{T})], \tag{15.18}$$

Eq. (15.18) says that the probability of finding f^* in a finite number of generations n is *conditional upon* the population size ratio s. Since in practice the population size ratio cannot grow exponentially with the size of the function set and the terminal set, reducing the algorithmic complexity of a decision rule by enlarging the terminal and function sets may help gain a little efficiency. Therefore, constrained by the population size ratio, the sizes of $\mathcal{F}$ and $\mathcal{T}$ have to kept within certain bounds.

[15.8] conducted a series of experiments to test what has been analyzed above. They started with the function set $\mathcal{F} = \{+, -, \times, /\}$, and the terminal set $\mathcal{T} = \{X_{t-1}, 0, 1, 2, ..., 9\}$. Thus, as predicted above, in a finite number of generations, the probability of discovering (15.6) is lower than that of discovering (15.5). To show how much lower it is, [15.8] conducted two experiments. In the first experiment $\{x_t\}$ was generated from Eq. (15.5), and in the second experiment it was generated from Eq. (15.6). For each experiment, they ran 100 trials of GP. This number of runs enables us to get an estimate of $Prob(g^* \in G_{200})$ (Eq. (15.13)). As a result, they found that the chance of discovering (15.5) within 200 generations was about 83%, whereas there was only a 5% chance of discovering (15.6) with the same search intensity.

However, if we add the variable "X_{t-1}^3" to the terminal set, then the algorithmic complexity of Eq. (15.6) becomes shorter – 7, to be exact, which is the same as that of the Data Generating Mechanism (15.5). To see the effect of adding the terminal X_{t-1}^3, they conducted another two experiments. In both experiments, $\{x_t\}$ were generated from Eq. (15.5). The first experiment did not include X_{t-1}^3 in $\mathcal{T}$, but the second did. 100 trials were run for each experiment. For the experiment that did not include X_{t-1}^3, $Prob(g^* \in G_{200})$ was only 4%. However, after the inclusion of X_{t-1}^3, $Prob(g^* \in G_{200})$ increased up to 84%, which was about the same as the chance of finding (15.5) in the previous experiment.

A series of other chaotic laws of motion were also studied by George Szpiro and Mahamoud Kaboudan. [15.19] applied GP to learn and to forecast a set of chaotic time series including the *Rossler Attractor*,

$$x(t+\delta) = x(t) - [y(t) + z(t)]\delta, \tag{15.19}$$

$$y(t+\delta) = y(t) + [x(t) + ay(t)]\delta, \tag{15.20}$$

$$z(t+\delta) = z(t) + [b + x(t)z(t) - cz(t)]\delta, \tag{15.21}$$

and the *Mackey-Glass delay differential equation*,

$$\dot{x}(t) = bx(t) + \frac{ax(t-\tau)}{1 + x^c(t-\tau)}. \tag{15.22}$$

For the Rossler attractor, i.e. Eqs. (15.19)-(15.21), Szpiro simulated the time series with $a = 0.2, b = 0.2, c = 0.57$, initial values $x_0 = -1, y_0 = 0, z_0 = 0$, and $\delta = 0.02$. Genetic programming was applied to predict the time series $\{x_t\}$ and $\{z_t\}$. This produced an R^2 of 0.912 for the first series, whereas only 0.568 for the second. For Eq. (15.22), Szpiro considered the series with parameters $a = 0.2$, $b = 0.1$, $c = 10$, and $\tau = 30, 100$. [2] When $\tau = 30$, genetic programming can predict the series with an R^2 value of between 0.40 and 0.65, and when $\tau = 100$ it can produce R^2 values of between 0.45 to 0.80.

[15.21] and [15.22] also examined the GP forecasting performance over the *Henon map*,

$$x_t = 1 - 1.4x_{t-1}^2 + 0.3x_{t-2}, \tag{15.23}$$

and the *tent map*,

$$x_t = \begin{cases} cx_{t-1} & if \quad x_{t-1} < 0.5, \\ c - cx_{t-1} & if \quad x_{t-1} > 0.5. \end{cases} \tag{15.24}$$

For both, the performance of GP was surprisly good.

[2] The significance of the τ is its associated fractal dimension. For the case $\tau = 30$, it was shown that $x(t)$ is chaotic with fractal dimension 3.5, and for the case where $\tau = 100$, about 7.5. The data for the MacKey-Glass time series can be obtained by applying the Runge-Kutta method.

15.1.2 Nonlinear Stochastic Time Series

The deterministic chaotic time series can hardly satisfy the needs of economists and finance people, because it is difficult to believe that economic and financial time series are generated only in a deterministic manner. Therefore, in many applications, Eqs. (15.2) and (15.3) are assumed to be a typical environment to work with. For the former, in addition to the artificial data, GP was also applied to the real financial data, in particular the stock price and the foreign exchange rate. We shall see more of these applications in Section 15.2.

In a noisy environment, GP will not do anything miraculous. As we shall see in Section 15.2, when the signal-noise ratios become less and less, the GP performance will also get worse and worse. In the extreme case, when the data is purely obtained from pseudo-random numbers, GP will completely collapse. Therefore, like all other data mining tools, our expectations of GP should be no more than just discovering the hidden law, be it linear or nonlinear.

15.2 Statistical Behavior of Genetic Programming

15.2.1 Kaboudan's Predictability Test

[15.11] proposed a *shuffling test* that measures GP's predictive ability. The idea is quite simple. If the series $\{x_t\}$ contains a predictable pattern, then randomly shuffling the series $\{x_t\}$ will ruin predictable information in the original sequence. Therefore, the series after shuffling, say $\{z_t\}$, should be more difficult to predict than the original series $\{x_t\}$. However, if the series contains no predictable patterns, then $\{x_t\}$ and $\{z_t\}$ should be equally hard to predict. Based on this idea, a statistic is developed.

$$\eta = max\left\{0, \frac{1}{n}\sum_{i=1}^{n}\left(1 - \frac{SSE_x}{SSE_z}\right)_i\right\} \tag{15.25}$$

The η statistic is the average ratio of the (before and after shuffling) SSEs subtracted from one. If the series before shuffling contained unpredictable patterns, the ratio of the fitness measure of the *sum of squared errors* (or *SSE*) before to the SSE after running must approach *unity*.

$$\frac{SSE_x}{SSE_z} \rightarrow 1. \tag{15.26}$$

If that series contained predictable patterns, the ratio must be less than unity.

$$\frac{SSE_x}{SSE_z} < 1. \tag{15.27}$$

n refers to the sample size, which is the number of runs that GP is applied to the series $\{x_t\}$ and $\{z_t\}$. It is not difficult to see that[3]

$$0 \leq \eta \leq 1. \tag{15.28}$$

This predictability measure was applied to examine the predictability of some artificial data as well as financial data. For the artificial data, Kaboudan considered five types of laws of motion: linear, non-linear, linear-stochastic, non-linear-stochastic and pseudo-random. In addition to the Henon map, the Mackey-Glass function, and two pseudo-random numbers, there were the following six equations.

$$x_t = 1.8708 x_{t-1} - x_{t-2} \tag{15.29}$$

$$x_t = 3.9 sin(x_{t-1}) + 0.85 cos(x_{t-2}) \tag{15.30}$$

$$x_t = (1.43 - 4.5 e^{-x_{t-1}^2}) x_{t-2} \tag{15.31}$$

$$x_t = 0.6 x_{t-1} + 0.15 x_{t-2} + \epsilon_t \tag{15.32}$$

$$x_t = 0.6 x_{t-1} + 0.3 \epsilon_t x_{t-1} + \epsilon_t \tag{15.33}$$

$$x_t = \epsilon_t \sqrt{h_t}, \quad h_t = 1 + 0.25 x_{t-1}^2 + 0.7 h_{t-1} \tag{15.34}$$

$\epsilon_t \sim N(0,1)$. It was found that the linear process (15.29) was very predictable. Non-linear processes (15.30, 15.23, 15.31) free from noise were also quite predictable. The Mackey-Glass equation (15.22) generated the least predictable series among the noise-free non-linear series. Predictability decreased significantly when a stochastic component was part of the process, and the linear-stochastic (15.32) processes were more predictable than the non-linear-stochastic processes (15.33, 15.34). The random series investigated were totally unpredictable.

For financial data, Kaboudan considered three different frequencies. Prices were collected every 30 minutes, every minute, and every price change. Eight Dow Jones stocks were included in the analysis: Boeing, GE, GM, IBM, Sears, AT&T, Walmart, and Exxon. Time and quotes (TAQ) data available on CD-ROM from the NYSE, Inc. covering October 1996 through March 1997 were used. Analysis of 30-minute and one-minute data showed that none of the prices were predictable. The best which we have is one-minute AT & T with a predictability measure of 50%. Analysis of price-change returns showed significant improvements in terms of predictability.

We have a few remarks about this predictability measure. First, it is interesting to note the relationship between Theil's U statistic and Kaboudan's η statistic. Theil's U statistic would compare the SSE from GP with that

[3] Actually, it is possible that η may be less than 0. This may happen if shuffling actually introduced some pattern into an originally random data sequence. It may also happen if GP is not successful in depicting any pattern in the original data but is somehow successful in finding a pattern in the shuffled data.

from a chosen benchmark, but based on the same time series. Kaboudan's η statistic applies the same forecasting model to different series. The η statistic can be defined with other classes of models, such as neural nets. There is no guarantee that the predictability measure would be the same when GP is replaced with other models. This is probably a limitation inherent in this measure. Furthermore, as shown in [15.8], different settings of the control parameters may lead to different results. This is particularly true for the function (15.24), for which, to satisfy the closure property, one needs to include "if-then-else" in the function set. Similarly, the variable ϵ_t is required for the function (15.33). Since they were not included in the primitives in Kaboudan's application, the η statistic may not even be unique with respect to GP.

15.2.2 The Fitted Residuals of GP

The conventional diagnostics of fitted residuals were also applied to examine the statistical behavior of GP ([15.13]). Let the fitted function $\hat{f}$ derived from genetic programming, and the fitted residuals be

$$\hat{\epsilon}_t = x_t - \hat{f}(x_{t-1}, x_{t-2}, ...). \tag{15.35}$$

Then the question to be addressed here is whether the series $\hat{\epsilon}_t$ is satisfied with some standard properties from the statistical viewpoint; in particular, is the series $\hat{\epsilon}_t$ independent? If GP has successfully discovered the hidden relation, then one should expect that $\hat{\epsilon}_t$ is independent, or *white noise*, if it discovered the hidden linear relation. Therefore, the statistical behavior of the fitted residuals provides a way of evaluating the performance of GP.

The models considered in ([15.13]) are basically the same five types of models used in [15.11]. Among the noise-free equations are (15.29), (15.5), (15.23), (15.22), and the following linear threshold regression model:

$$x_t = \begin{cases} 2.5x_{t-1} + z_t & if \quad \mid x_{t-1} \mid \leq 2, \\ 0.85x_{t-1} & if \quad \mid x_{t-1} \mid > 2, \end{cases} \tag{15.36}$$

where

$$z_t = \begin{cases} 1 & if \ \ t \ \ is \ \ odd, \\ -1 & if \ \ t \ \ is \ \ even. \end{cases} \tag{15.37}$$

The linear stochastic functions involved are the AR(1) and ARMA(1,1) processes:

$$x_t = 0.9x_{t-1} + \epsilon_t, \tag{15.38}$$

and

$$x_t = 0.9x_{t-1} - 0.6\epsilon_{t-1} + \epsilon_t. \tag{15.39}$$

The non-linear stochastic functions selected for the examination are (15.33), (15.34), and the *threshold auto-regressive* (TAR) model:

$$x_t = \begin{cases} 0.9x_{t-1} + \epsilon_t & if \quad | x_{t-1} | \leq 1, \\ -0.3x_{t-1} + \epsilon_t & if \quad | x_{t-1} | > 1. \end{cases} \tag{15.40}$$

Finally, there are two pseudo-random numbers: one follows the normal distribution, and one follows the uniform distribution.

The null hypothesis that $\{\hat{\epsilon}_t\}$ is white was not rejected in (15.38), (15.39), (15.40), and (15.34). Among these four, the *normalized mean squared error* (NMSE) for Eq. (15.34) is the highest (0.70). [4] However, this result is not surprising, since, as we mentioned earlier, Eq. (15.34) is not predictable in terms of the mean. As a matter of fact, given the nature of unpredictability, we consider 0.70 low enough and it may even be low enough to indicate the possibility of *overfitting*. A similar result is also observed in the two pseudo random numbers. In both cases of normal and uniform distributions, the NMSE is 0.67, even though there is entirely nothing to learn from these two series. These empirical results suggest that overfitting in some cases can be a problem for genetic programming.

The null was also not rejected in Eq. (15.22). The NMSE in this case is 0.48, which is rather high considering that the series to be fitted is noise-free. Given this rather high NMSE, it is unlikely that the fitted residuals would be white. However, one should note that Equation (15.22) is noise free. In that case, it is questionable whether whiteness tests are still valid. This problem is also reflected in the remaining four noise-free series, in which the white noise in all cases is rejected. Actually Kaboudan warned readers not to use this to judge the adequacy of GP, because the associated normalized mean squared errors are very low in these four cases, which suggests that $\hat{f}$ actually fits f very well.

15.3 Discovering the Technical Trading Rules

In 1992 and subsequent years, partially because of the seminal work of [15.4], which showed the significant prediction potential of the simple trading strategies, applications of evolutionary computation were extended further to the study of trading strategies.[5] [15.5] and [15.1] initialized this line of research,

[4] The normalized mean squared error is the mean squared error divided by the sample variance. The NMSE is zero when a perfect prediction is generated, and is one when the resulting model is doing no better than predicting the mean value of the data set on average.

[5] [15.4] studied two of the simplest and most popular trading rules, the moving average and the trading range break, and found significant positive returns by utilizing the trading rules.

and [15.6] gave finance people the first systematic introductory book on genetic algorithms. Papers on this subject, however, did not find their way into prestigious journals until the late 1990s ([15.17], [15.2]).

15.3.1 Foreign Exchange Markets

[15.17] and [15.18] used GP techniques to find technical trading rules in the *foreign exchange market*. The former basically studied the *dollar exchange rate markets*, including the dollar/deutschemark, dollar/yen, dollar/pound, and dollar/Swiss franc markets, whereas the latter studied the EMS (European Monetary System) exchange rate markets, including the deutschemark/French franc, deutschemark/lira, deutschemark/guilder and deutschemark/pound markets. Two general issues were raised in this line of research:

1. Is it possible to identify *profitable trading rules* ex ante for these currencies using genetic programing?
2. If yes, what might be the source of the excess returns? In particular, can the observed excess returns be explained as compensation for bearing *systematic risk*?

Profitability and the Efficient Market Hypothesis. The answer is *positive* for the first issue, and *negative* for the second. These together make genetic programming particularly attractive for the investors in the foreign exchange market. [15.17] obtained rules for currencies selected in a given sample period (1975-1980), and then examined their performance over the period 1981-1995. They found that the generated rules earned *significant* out-of-sample excess returns over the period 1981-1995. [15.18] obtained rules for currencies selected in a given sample period (1979-1983), then examined their performance over the period 1983-1996. This finding is the same as that of [15.17]. In fact, in [15.18], the evidence of positive excess returns was further strengthened by careful statistical analysis involving calculating *Bayesian posterior probabilities* and the *Newey-West correction for serial correlation.* To understand whether the returns to the trading rules could be interpreted as *compensation for bearing systematic risk*, *beta* was calculated. Different benchmarks were selected for calculating beta, such as the MSCI (Morgan Stanley Capital International) world equity market index, the S&P 500, the Commerzbank of German equity, and the Nikkei. Most of the estimated betas were negative, indicating that the excess returns observed were not compensation for bearing systematic risk.

What Did GP Rules Learn?. If the excess returns were not compensation for bearing systematic risk, then what might be the source of the excess returns? To trace the source, both [15.17] and [15.18] conducted a series of analyses for the rules discovered by GP. First, *what kind of rules were found?* Were they similar to the rules commonly used by technical analysts? This question is generally not easy to answer, because the rules discovered by GP often have a rather complex nested structure. In [15.17], the mean number of nodes for the 100 best rules obtained from 100 independent runs was 45.58, and only two rules had fewer than 10 nodes. Those whose structure could be harnessed were found to be the familiar moving-average rules or double moving-average rules. To confirm whether it was these moving-average and filter rules that contributed to the excess returns observed, [15.18] compared the performance of some simple MA rules and filter rules with the GP rules. Among the ten simple rules they examined, none could match the performance of GP rules over all currencies, and, in fact, many performed rather poorly. Therefore, they concluded that GP rules perform more profitably than the simple rules in EMS currencies. *What GP discovered was certainly richer than what simple rules can tell us.*

Second, *were the rules simply exploiting known statistical properties of the data?* To answer this question, bootstrapping simulations, based on a random walk model, an ARMA model, and an ARMA-GARCH(1,1) model, were carried out. Bootstrapping permits one to determine whether the observed performance of a trading rule is likely to have been generated under a given model for the data-generating process. The results indicate that it is unlikely to replicate the observed performance from any of the three models used for bootstrapping. In other words, *what GP rules discovered were not just some well-known econometric patterns.*

Third, *were the rules discovered by GP exploiting quite general features of financial markets?* This question is mainly motivated by the claim made by technical analysts that the rules they use are not specific to any particular market. To investigate this claim, they took the successful dollar/deutschemark rules and ran them on the data for the other exchange rates. There was a marked improvement in performance over the previous rules in all cases except the dollar/yen. The authors attempted to use this result to support the claim, but the result seems too good to make such a conclusion. In fact, the result brings us a puzzle: *why did the rules trained on dataset A outperform the rules trained on dataset B when both were applied to dataset B?* We would like to draw readers' attention to [15.12] on a similar issue. It was shown in [15.12] that the evolved forecasting rules differ from day to day for the same stock and among stocks. Hence, the supporting evidence for the existence of the general rules is weak.

Fourth, from the viewpoint of knowledge discovery, *what did these GP rules actually discover*? What can we learn from these GP rules? This question was better addressed in [15.18]. [15.18] examined the structure of some of the simplest rules obtained for each currency. A feature that emerges consistently across all currencies is that the *interest differential* is the most important informational *input* to the trading rule. They confirmed the importance of the interest differential information by performing the following experiment. They ran the trading rules for all currencies over the validation period with the interest differential set to zero in the signal calculations, but included in the return calculations. The mean annual return fell to -0.24% and annual trading frequency dropped to 0.24.

Diversity of GP Rules. For each exchange rate, the authors also examined the *homogeneity* (or the hetrogeneity) of the rules, i.e. whether the rules signal the same position, long or short, at the same time. The results are quite different for different currencies. A high proportion of the 100 best dollar/deutschemark rules, obtained from 100 independent runs, identified similar patterns in the data: for most of the time, 90% or more of the rules gave the same signal. However, in the case of the dollar/yen, for a substantial amount of time, the proportion hovered around 50%, indicating the maximum differences for these rules.

Given the great diversity of the rules, it would be interesting to know whether voting would help. That is, would a majority vote over all 100 rules determine the position of the trading rule? So, a long position was taken if 50 or more of the rules signalled long, and a short position otherwise. This is also called the *median portfolio rule.* The return of a median portfolio rule was calculated by the authors, and the result was again inconsistent. For some exchange rates, this median portfolio rule produced a substantial increase in excess return. However, for certain others, it did not.

Overfitting Avoidance. [15.17] and [15.18] adopted a technique, known as *the validation procedure*, to avoid overfitting. By this procedure, one run of GP uses two time periods. The first period is called the *training period* and is used to train the genetic programs. The second period is the *selection period*, which is used to select the best performing programs and decide when to stop the training. For example, in [15.17], the termination criterion is achieved if no new best rule appears for 25 generations. However, it was found that this procedure did not eliminate the problem of overfitting, because the rule that did best in the selection period turned out to be one of the poorest in terms of performance out-of-sample. In addition to the *validation procedure*, they also experimented with different transaction costs, and found that a suitable chosen transaction cost in the training and selection periods introduced a useful friction that provides some protection against overfitting the data.

15.3.2 Stock Markets

[15.2] applied GP to the trade S&P 500 index from 3 January 1928 to 29 December 1995. This paper shares many similarities with [15.17] and [15.18]. First, they also transformed the data by dividing each day's price by a 250-day moving average. The normalized prices had an order of magnitude of around one. Second, the *validation procedure* was also adopted, but in a different style. [15.2] used ten successive training periods and reported the summarized results for each case. The five-year training and two-year selection periods started in 1929, 1934,..., 1974, with the out-of-sample test periods starting in

1936, 1941,..., 1981. With this design, the first trial used years 1929-1933 as the training period, 1934-1935 as the selection period, and 1936-1995 as the test period. For each of the ten training periods, they carried out ten trials. This procedure is known as the *rolling forward scheme*, a technical device used to tackle the issue of data-snooping.

Profitability and Transaction Costs. The major result found by [15.2] was significantly different from that found by [15.17] and [15.18]. [15.2] found that the markets for these stocks are *efficient* in the sense that it is not possible to find consistent excess returns after transaction costs over a simple buy-and-hold strategy in the holdout sample periods. In most of the periods, there are only a few rules with positive excess returns. Obviously, this result sharply contrasts with that of [15.17] and [15.18], and leaves the capabilities of GP to find successful technical trading rules *inconclusive.* It would be interesting to know what may cause this difference. Here, transaction costs seem to play an important role.

The transaction costs incorporated by [15.2] were much higher than the costs used by [15.17] and [15.18]. The former used one-way transaction costs of 0.1, 0.25, 0.5%, whereas the latter used one-way transaction costs of 0.05%. As we already saw from [15.17], transaction costs have a crucial implication for GP performance. In [15.2], some new evidence was presented. First, it was shown that the differences between "in" and "out" returns were statistically significant for the low and median transaction cost cases, but not for the high transaction cost cases. Second, trading frequency was adversely affected by the transaction costs. With low transaction costs, the trading frequency was high, with an average of 18 trades per year. However, with high transaction costs, the trading frequency dropped to an average of 1.4 trades per year.

One temptation here is to ditto [15.17] or [15.18] with higher transaction costs. However, it is not a really wise direction to move in, since we all know that keeping on increasing transaction costs will eventually discourage any possible trade. One possible solution is to report the *break-even transaction cost* proposed by [15.3], which is the level of transaction costs leading to zero trading profits.[6] By following this approach, one can test whether the break-even transaction cost in the stock market is higher than the one in the foreign exchange market. Furthermore, one can rank different markets according to their break-even transaction cost. This could be a direction to move in the future.

Diversity of Rules. Like [15.17] and [15.18], [15.2] provided a thorough analysis of the trading rules discovered by GP. Questions regarding the nature of the rules were addressed. They started with an examination of the *diversity* of the rules. In a similar way to what was found in [15.17] and [15.18], there existed a great degree of diversity in the rules discovered by

[6] [15.20] applied the break-even transaction cost to evaluate the performance of GA trading rules.

GP over different trials. The size of the rules varied from nine to 94 nodes, with a depth of between five and ten levels. While many of the rules appeared to be quite complicated at first sight, there was often a lot of redundant material. For example, some of the subtrees may never be visited when the rules are evaluated. Hence, seemingly complex tree structures may be pruned into much simpler ones. Based on that observation, they argued that *the complexity of the trading rules in general cannot be measured by the number of nodes.*[7]

Financial Knowledge Discovery with GP. At this point, they proposed an alternative measure of the complexity, i.e. to compare the behavior of the rules to a class of simple technical trading rules. Here, we are asking the same question as [15.17] did before: *did the GP trading rules tell us anything more than just the simple technical trading rules*? The answer to this question is "*it depends.*" Actually, it interestingly depends on the *transaction costs* and the *training periods.* With 0.25% transaction costs, most rules are effectively similar to a 250-day moving-average rule. With 0.1% transaction costs, half of the rules are similar to a 250-day moving-average rule, while the remaining rules resemble either 10-to-40-day moving-average rules or a trading range break rule. With 0.5% transaction costs, rules from the early training periods are similar to a 250-day moving-average rule, while the late rule from the late training periods match none of the simple rules considered.

This glossary of results leads us to the best place to see how genetic programming can automate the discovery of finance theory without too much external human tutoring or supervising. The glossary above certainly tells us much more than what simple trading rules taught us. First, general cookbooks of technical analysis would not tell us anything about the connection between transaction costs and effective trading rules. Second, while most of us believe that markets can never settle down, only genetic programming can show us how one effective trading rule was replaced by another at a different moment. These dynamics were also not taught in conventional technical analysis. Third, as we already see in many cases, there is simply no match between simple trading rules and GP rules. In these scenarios, one may learn something new from the GP-evolved rules. Fourth, even if the match exists, non-trivial differences should not be overlooked.

Actually, [15.2] analyzed a few examples of their GP rules. In one case, the rule was found to be something similar to a moving-average rule, but the window of the moving-average was not fixed. During bear markets, the rule looked for very short-term trends, with the moving-average window varying between three and four days. Then, in bull markets, the length of the time window increased to 16 days at a normalized price level of 1.1, and rose rapidly

---

[7] We believe that this is an important critique, because measuring the complexity of financial markets via measuring the complexity of evolved trading trees remains an interesting subject, and so far, the node complexity is the only measure popularly used.

thereafter to soon use all of the available price history. The interpretation of this rule given by [15.2] is that: "...this rule corresponds to a hypothesis that stock prices are in general going up but are prone to corrections after prolonged bull markets. This rule is quick to take a long position in bear markets, but is increasingly hesitant to do so when prices are high. (p. 263-264, ibid)."

Trade with A Delay. Remember that, earlier, [15.17] showed that the excess returns derived from their GP trading rules were not compensation for bearing systematic risk. Here, [15.2] also conducted a similar pursuit. They were searching for the potential explanations for the differences between "in" and "out" returns being statistically significant for the low and median transaction cost cases. To investigate this issue, the performance of the ex post trading rules was retested based on trades occurring with a delay of one day. This one-day delay was motivated by the well-known feature that there is low-order serial correlation in market indexes. This one-day delay should remove any first-order autocorrelation. It was found that, by using delayed signals, the average difference in returns during "in" and "out" days dropped from seven to two basis points. In the case of 0.1% transaction costs, a one-day delay removed all of the difference between returns during "in" and "out" days. They, therefore, concluded that the forecasting ability could be explained by low-order serial correlation in stock index returns.

Day Trading. However, this one-day delay is not applicable for the case of *day trading* based on high-frequency on-line data. [15.12] was the first to apply genetic programming to *day trading*. Using genetic programming, he established a trading system, known as a *single-day-trading-strategy* (**SDTS**). An SDTS is a strategy whereby a trader buys a stock at a price close to the *daily low* and sells at a price close to the *daily high*, regardless of the order in which trading occurs. The main task of GP is to forecast the daily low and high. The unique feature of the SDTS is to require the investor to close every position by the end of the trading day. This is mainly because trading is based on a one-day forecast. Holding no position overnight helps avoid greater gambling losses in subsequent days for which no forecast is available to base a decision on. For an empirical illustration, stocks of Chase Manhattan and General Motors from the NYSE and those of Dell Computer and MCI-Worldcom from the NASDAQ were selected to apply to test SDTS in [15.12]. For all four stocks, returns from trading based on the GP forecasts were higher than those based on a random walk.

It is worth noting here that GP was not applied to evolve trading strategy directly. Instead, it was applied to forecast the daily low and daily high. Based on the predicted daily low and daily high, the trading decision was made based on the intra-day tick-by-tick movement of the stock prices. Another difference which distinguished [15.12] from [15.2] was the trading target. What concerned [15.12] were the individual stocks of listed companies, rather than stock indexes.

15.3.3 Futures

Given the conflicting evidence on the capabilities of GP to find successful technical trading rules from the foreign exchange markets and the stock markets, [15.23] conducted another formal study of the issue: *is GP really useful in finding technical trading rules in financial markets?* It applied GP to find trading rules in S&P 500 futures, which are distinguished from their index counterpart by their sufficient marketability and liquidity. In addition, the transaction costs in futures trading are also significantly *lower* than the costs in equity trading. This consideration is important because one possible explanation for the difference between GP performance in the foreign exchange markets and the stock markets is the *transaction costs*. So far, there has been no empirical study showing whether their break-even transaction costs are the same. Therefore, by applying GP to find trading rules in futures, [15.23] is able to shed more light on whether, in addition to the foreign exchange markets ([15.17], [15.18]), GP can be useful in another liquid market with lower transaction costs. [15.23] used the daily data of the S&P 500 index, S&P 500 futures, and 3-month T-bill rates from 1983 to 1998.

Hedging. Given futures as the application domain, [15.23] further extended the application of GP from *just trading* to *hedging*, which is certainly infeasible for those earlier studies without the introduction of the futures market. This extension is empirically relevant since many investors use futures to hedge positions in the spot market, and this extension tests whether GP could be used for finding *hedging strategies*. Hence, in [15.23], GP was applied to generate trading rules for both the index and futures.

To do so, [15.23] encoded the trading rules with GP in a dramatically different way as opposed to [15.17], [15.18] and [15.2]. For the latter, the trading rules were generated by the function set, usually just the Boolean-valued functions and real-valued functions, and the terminal set of no imposed structure. However, for [15.23], one solution candidate or trading rule is represented as four hierarchical trees. Among them, the *third tree* generates a signal to be used in the *futures market*, and the *fourth tree* generates a signal to be used in the *spot market*. The root of these two sub-trees is restricted to the function *if-then-else*. Furthermore, the output of these two sub-trees is transformed into the range [0,1] by the sigmoid function. Moreover, by further dividing the range [0,1] into a few segments, one can actually decide how many units to buy or sell. This flexibility was also not shared by those earlier studies, which can only signal the time to buy and sell (one unit).

Automatically Defined Functions. The most striking feature of [15.23]'s representation of the trading rule is the employment of automatically defined functions (ADFs). The first two sub-trees are automatically defined functions. The first sub-tree is a Boolean-valued ADF, and the second sub-tree is a real-valued ADF. An ADF is one way of capturing the idea of *subroutine* and *reusable codes* in computer programming. The advantage of ADF is that the same complex function can be called from multiple nodes by a single reference, so lower algorithmic complexity can be achieved. The ADF was proposed by John Koza in [15.16], but the economic and financial applications of it were not available until very recently. [15.23] was the first to bring the attention of economists to it.

Rolling Forward Procedure. To avoid the problem of overfitting, [15.23] also adopted the *rolling forward procedure* as [15.2] did. By this procedure, we consider an investor at the beginning of 1987. The investor had 3 years of data available, and he or she defined the training period to be 1984 and 1985 and the validation period to be 1986. Ten trials were run with periods specified in this way, and 10 trading programs were generated. These 10 programs were used to trade in 1987, the out-of-sample period, and the results were reported. In this way, the editor rolled forward until 1998, which means that a total of twelve rolling forward steps were taken and 120 independent trials were implemented.

What was not made clear in [15.2] was the significance of the *rolling forward scheme*. As [15.23] mentioned, an early version of [15.2] did not use this scheme. Instead, they used one training period, one validation period, and one out-of-sample period, and they found that GP-generated rules outperformed the buy-and-hold strategy in the out-of-sample period. Once they used the rolling forward scheme to prevent data-snooping bias in the selection of time periods, the results were no longer conclusive. In an early draft of [15.23], data-snooping bias also occurred. There, the training period was defined as April 1986 to September 1987 and the validation period as 1988 and 1989. Then GP-generated rules consistently beat the buy-and-hold strategy in the *only* out-of-sample period, 1990 to 1993. However, when the rolling forward scheme was adopted, things became, once again, inconclusive.

In both [15.2] and [15.23], it was found that, under the *rolling forward scheme*, the GP-generated rules could not consistently outperform the *buy-and-hold strategy. The performance of the GP-generated trading rules varied greatly for different years.* For example, in its experimental series "Trading in Futures Only," [15.23] found that 7 out of 10 rules outperformed the buy-and-hold strategy in 1992, whereas in 1995, only 1 out of 10 rules outperformed the buy-and-hold strategy. One may wonder why, if GP really has the capability of discovering, it fails to discover the buy-and-hold strategy. The answer to this is that the training period and the out-of-sample period may not always share such similar features. If they did share the same similar features and the returns to the buy-and-hold strategy were high for the training period, then GP would eventually learn to use a buy-and-hold strategy. This was the case for 1998, when the returns in the previous 3 years were high. Five of the 10 trading rules generated from 1995 to 1997 were just buy-and-hold.

When the objective was return only and both the index and futures markets were allowed to trade, the trading rules generated by GP were more likely to lead to trade in the futures market because the transaction costs there were substantially lower. In the spot market, the low average number of position changes, from 1.6 to 6.3, indicated that the trading rules did not lead to trade much beyond the buy-and-hold strategy. In the futures market, the average number of position changes ranged from 4 to 68.5, which was much higher than the number of position changes for the buy-and-hold strategy, i.e. two. GP recognized that transaction costs were lower in the futures market and generated trading rules that more often than not resulted in trade in the futures market while the buy-and-hold strategy was adopted most of the time in the spot market.

15.4 Concluding Remarks

In this chapter, we have reviewed the two early applications of genetic programming to financial data mining and knowledge discovery. The application to time series modeling shows that GP has some power for tracking and predicting the complex dynamic behavior of financial time series, whereas the application to trading strategies indicates the potential of GP in relation to market-timing capability. For the application to trading strategies, one has generally started with a set of primitives, which may characterize state-of-the-art knowledge. From these primitives, we then let GP set in motion an evolutionary process and discover something novel for us. This style of application reveals an interesting and challenging way of doing finance in the future. As a matter of fact, recently, this style of application has just been extended to *option pricing*. For example, in their study of option pricing formulas, [15.9] included the Black-Scholes model as a primitive. The GP-evolved option pricing formulas for the equity options were hence adaptations of the Black-Scholes model. It is our belief that further progress in the theory of finance can be made in this way.

References

15.1 Allen, F., Karjalainen, R. (1993) Using Genetic Algorithms to Find Technical Trading Rules. Rodney L. White Center for Financial Research, The Wharton School, Technical Report, 20–93

15.2 Allen, F., Karjalainen, R. (1999) Using Genetic Algorithms to Find Technical Trading Rules. Journal of Financial Economics, Vol. 51, no. 2, 245–271

15.3 Bessembinder, H., Chan, K. (1995) The Profitability of Technical Trading Rules in the Asian Stock Markets. Pacific Basin Finance Journal, Vol. 3, 257–284

15.4 Brock, W.A., Lakonishok, J., LeBaron, B. (1992): Simple Technical Trading Rules and the Stochastic Properties of Stock Returns. Journal of Finance, Vol. 47, 1731–1764

15.5 Bauer, R. J. Jr., Liepins, G. E. (1992) Genetic Algorithms and Computerized Trading Strategies. In: O'Leary, D. E., Watkins, R. R. (Eds.), Expert Systems in Finance, North Holland

15.6 Bauer, R. J. Jr. (1994) Genetic Algorithms and Investment Strategies. Wiley
15.7 Chen, S.-H., Yeh, C.-H. (1997) Toward a Computable Approach to the Efficient Market Hypothesis: An Application of Genetic Programming. Journal of Economic Dynamics and Control, Vol. 21, no. 6, 1043–1063
15.8 Chen, S.-H., Kuo, T.-W. (2002) Genetic Programming: A Tutorial with the Software Simple GP. In Chen, S.-H. (Ed.), Genetic Algorithms and Genetic Programming in Computational Finance, Kluwer Academic Publishers, 55–77
15.9 Chidambaran, N., Lee, C.-W. J., Trigueros, J. (2002) Option Pricing via Genetic Programming. In: Chen, S.-H. (Ed.), Evolutionary Computation in Economics and Finance, Physica-Verlag, 383–397
15.10 Johnson, H. E., Gilbert, R. J., Winson, K., Goodacre, R., Smith, A. R., Rowland, J. J., Hall, M. A., Kell, D. B. (2000) Explanatory Analysis of the Metabolome using Genetic Programming of Simple, Interpretable Rules. Genetic Programming and Evolable Machines, Vol. 1, no 3, 243–258
15.11 Kaboudan, M. A. (1999) A Measure of Time Series's Predictability Using Genetic Programming Applied to Stock Returns. Journal of Forecasting, Vol. 18, 345–357
15.12 Kaboudan, M. A. (2002) GP Forecasts of Stock Prices for Profitable Trading. In Chen, S.-H. (Ed.), Evolutionary Computation in Economics and Finance. Physica-Verlag, 359–381
15.13 Kaboudan, M. A. (2001) Genetically Evolved Models and Normality of Their Fitted Residuals. Journal of Economic Dynamics and Control, Vol. 25, no. 11, 1719–1749
15.14 Koza, J. (1992) A Genetic Approach to Econometric Modelling. In: Bourgine, P., Walliser, B. (Eds.) Economics and Cognitive Science. Pergamon Press, 57–75
15.15 Koza, J. (1992a) Genetic Programming: On the Programming of Computers by Means of Natural Selection. The MIT Press.
15.16 Koza, J. (1994) Genetic Programming II : Automatic Discovery of Reusable Programs. The MIT Press.
15.17 Neely, C., Weller, P., Ditmar, R. (1997) Is Technical Analysis in the Foreign Exchange Market Profitable? A Genetic Programming Approach. Journal of Financial and Quantitative Analysis, Vol. 32, no. 4, 405–427
15.18 Neely, C. J., Weller P. A. (1999) Technical Trading Rules in the European Monetary System. Journal of International Money and Finance, Vol. 18, no. 3, 429–458
15.19 Szpiro, G. G. (1997a) Forecasting Chaotic Time Series with Genetic Algorithms. Physical Review E, 2557–2568
15.20 Pereira, R. (2002): Forecasting Ability But No Profitability: An Empirical Evaluation of Genetic Algorithm-Optimized Technical Trading Rules. In: Chen, S.-H. (Ed.), Evolutionary Computation in Economics and Finance. Physica-Verlag, 287–310
15.21 Szpiro, G. G. (1997b) A Search for Hidden Relationships: Data Mining with Genetic Algorithms. Computational Economics, Vol. 10, no. 3, 267–277
15.22 Szpiro, G. (2002) Tinkering with Genetic Algorithms: Forecasting and Data Mining in Finance and Economics. In: Chen S.-H. (Ed.) Evolutionary Computation in Economics and Finance. Physica Verlag.
15.23 Wang, J. (2000) Trading and Hedging in S&P 500 Spot and Futures Markets Using Genetic Programming. Journal of Futures Markets, Vol. 20, no. 10, 911–942

16. Numerical Solutions to a Stochastic Growth Model Based on the Evolution of a Radial Basis Network

Fernando Álvarez[1], Néstor Carrasquero[2], and Claudio Rocco[2]

[1] Banco Central de Venezuela, Oficina de Consultoría Económica, Venezuela
email: faapnew@hotmail.com

[2] Universidad Central de Venezuela, Facultad de Ingeniería,
Caracas 1041-A, Venezuela
email: {nestor, rocco}@neurona.ciens.ucv.ve

This chapter introduces a new heuristics for solving the optimal consumption path in one-sector growth model, a typical stochastic dynamical optimization problem in economics. The proposed method avoids the ex-ante specification of a functional form for the policy function that solves the optimization problem. This novel approach has an advantage over other approaches like the Linear Quadratic Approximation (LQA) and the Parameterized Expectation (PE) methods. Instead, the functional form arises endogenously according to the characteristic of the problem the method is seeking to deal with. The heuristics combines Radial Basis Network (RBN) as a representation of the potential solutions and an Evolutionary Strategy (ES) as a mechanism to prune the search space. Experiments were performed on different versions of a stochastic growth model and some satisfactory results were consequently obtained. In most cases the approximation obtained with the proposed method indeed outperforms the approximation reached by both the LQA and PE methods, based on not only one criterion but several different quality criteria.

16.1 Introduction

Modern economics theory relies on the use of models in order to approach economic issues. Since the problem faced by economic agents is to allocate resources under uncertainty and incomplete information through the time, it is desirable that these models are stochastic and dynamic. A family of models that properly incorporate those elements is the General Equilibrium Models (GEM) in which macroeconomic regularities emerge from the optimizing behavior of the agents and ad hoc relationships are not imposed as constraints.

The problem when dealing with this kind of model is, in general, the lack of closed analytical solutions. Therefore, numerical methods become necessary in order to find an approximation to the equilibrium associated with the model. Two of the most commonly used techniques, the Linear Quadratic Approximation (LQA) method [16.4] and the Parameterized Expectation (PE) method [16.7] predefine the functional form of the policy function that solves the problem. In this chapter, we propose a solution method in which the functional form of the policy function arises endogenously from the features of the model.

The proposed approach combines the use of the Radial Basis Network (RBN) with Evolutionary Programs [16.3, 16.8], and it is used to solve the optimum consumption path in one-sector stochastic growth models [16.13, 16.14]. The quality of the results obtained with the proposed method is compared with the quality of the approximations reached by the LQA and the PE methods.

In this chapter we deal with the version of the problem in which the leisure does not increase the family utility. In this case, the problem can be reduced to find the policy function to the control variable $c_t = f(k_t, \phi_t)$ such that:

$$\begin{aligned} &\underset{c}{Max}\ E_0 \left\{ \sum_{t=0}^{\infty} \beta^t u(c_{t,}) \right\} \\ &\text{subject to:} \qquad\qquad\qquad\qquad\qquad\qquad (16.1) \\ &\qquad c_t + k_{t+1} - (1-\delta)k_t = \phi_t F(k_t) \\ &\qquad log(\phi_{t+1}) = A(log(\phi_t)) + \varepsilon_{t+1} \end{aligned}$$

where u is the utility function, β is the discount rate, c is the consumption, k is the capital level and δ is the depreciation rate. The problem is completely defined once we give information about all the parameters of the model. In (16.1) ϕ represents a shock to the production function F and is the only source of disturbance in the system.

The rest of the chapter is organized as follow. The second section briefly introduces the proposed method. The third section shows the main results and, finally, the fourth section presents the conclusions and the final remarks.

16.2 Proposed Method

This method, that we named Evolutionary Numerical Solver (ENS), uses a specific case of RBN (a linear aggregation of Radial Basis Functions (RBF)) to represent the elements of the search space and a two-stage Evolutionary Programs (EP), based on Evolutionary Strategies (ES), as a searching technique. ES are selected among EP due to the fact that they require low-member populations, which is very suitable for the fitness evaluation of solutions. The selection of a RBN representation relies on the fact that its ability to efficiently approximate different functions has been proven [16.11].

The searching process is carried out in two dimensions: in the space of the structures and in the space of the parameters associated to them. To prune the structure space two kind of ES are used: a one member ES named

ES(1+1) and a multimember ES named ES($\mu+1$). In both cases, the parameters tune-up is carried out by an ES(1+1).

The selection of the specific kind of RBF to be used to represent a solution was done following [16.10, 16.11]. The RBF approach consists on the construction of a function F with the following form:

$$F(\boldsymbol{X}) = \sum_{i=1}^{NT} \rho_i h_i(\|\boldsymbol{X} - \boldsymbol{T}_i\|) \tag{16.2}$$

where NT is the number of processing units, ρ_i is a weight factor and h_i is a continuous function selected among:

$$F1: \quad H(\|\boldsymbol{X} - \boldsymbol{T}_i\|) = \exp(-\frac{\|\boldsymbol{X} - \boldsymbol{T}_i\|^2}{C^2}) \tag{16.3}$$

$$F2: \quad H(\|\boldsymbol{X} - \boldsymbol{T}_i\|) = \frac{1}{(C^2 + \|\boldsymbol{X} - \boldsymbol{T}_i\|^2)^p} \tag{16.4}$$

$$F3: \quad H(\|\boldsymbol{X} - \boldsymbol{T}_i\|) = (C^2 + \|\boldsymbol{X} - \boldsymbol{T}_i\|^2)^p \tag{16.5}$$

$$F4: \quad H(\|\boldsymbol{X} - \boldsymbol{T}_i\|) = \|\boldsymbol{X} - \boldsymbol{T}_i\| \tag{16.6}$$

$\boldsymbol{T}_i$ and C are the centroid and the radius of the RBF. The aggregation function (16.2) can be represented as a single-layer network as shown in Fig. 16.1.

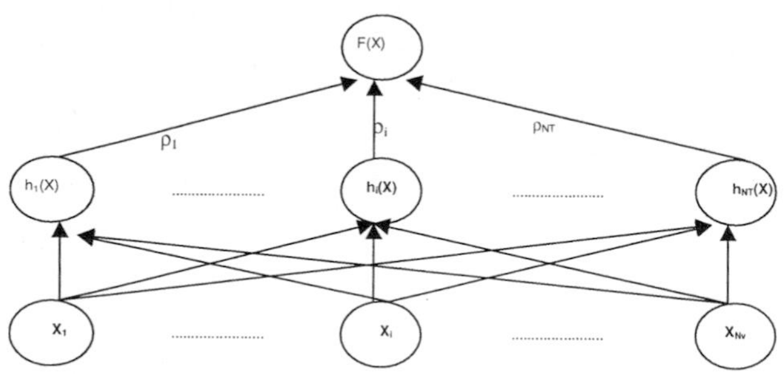

Fig. 16.1. Representation of (16.2) using a single-layered network

In the problem at hand (16.1) the input vector $\boldsymbol{X}$ has two components k_t and ϕ_t while the output is the consumption c_t. The purpose of the ENS is to evolve the proposed RBN in order to achieve the optimal consumption c_t^*. The functional form arises endogenously according to the characteristic of the problem with which the method is dealing.

ES, as EP, require the evaluation of the quality of the solutions by means of a fitness function. At the beginning of the study, we considered to use the discounted utility of the consumption as the fitness function. However, in the preliminary experiments on the deterministic version of the Mirman and Brock model [16.6, 16.13], we found that this fitness function was able to find the sequence but did not produce good policy function approximations for some values of the capital. Therefore, we decided to use the residuals of Euler's equation to measure quality, which yields values in the interval [0,1]. For this problem the Euler's equation is given by:

$$u_c(c_t) = \beta E_t \left\{u_c(c_{t+1}) \left[(1-\delta) + \phi_{t+1} F_k(k_{t+1},)\right] / \phi_t\right\} \tag{16.7}$$

It can be noticed that the fitness based on Euler's equation implies dealing with the expectation of a complex mathematical expression. In the cases in which the productivity shocks follow a continuous first order autoregressive process (AR(1)), we approximate it by a m-states Markov chain following the Tauchen approach [16.9]. By doing so, the residuals of (16.7) for each point in the state space can be easily computed.

ES also require a specific representation of the solution candidates (individuals) and the implementation of genetic operators. An individual in (16.2) can be represented using NT records, each one with the following encoding scheme:

Area 1	Area 2	Area 3	Area4	Area 5	Area 6

- Area 1 is used to define the kind of RBF used (F1 to F4)
- Area 2 contains the weight factor ρ_i
- Area 3 and 4 store the coordinates of the centroid $\boldsymbol{T}_i$
- Area 5 stores the radius C, and
- Area 6 contains the exponent p associated to RBF F2 and F3

We implemented two types of genetic operators: those that modify the structure of the RBN (NT and types of RBF used) and those that modify its parameters (centroids, weights, and so on). Among the first type, we designed three mutation operators and one crossover operator. The mutation operator can add, eliminate or transform a term from an individual chosen at random. The crossover operator randomly chooses two parents and then combines them in order to create a new individual.

The second type of genetic operator implemented is a mutation operator based on a normal perturbation of the current solution and is used to tune-up the adequate set of parameters once a structure is obtained.

The evolutionary dynamic is implemented as a two-stage process. The first stage corresponds to the RBN structure evolution using an ES($\mu + 1$) while in the second stage the parameters of each structure are tuned-up. This approach is designed in order to guarantee that the solutions will compete against each other once their parameters are good enough for their architecture.

Initially μ individuals corresponding to μ structures are randomly generated. The parameters of each structure are then tuned-up, by means of an

ES(1+1). Then, using the appropriate genetic operator, a new individual is generated and its parameters are tuned-up. Among the $\mu+1$ individuals, the best μ are then selected. The process is repeated for a specific number of generations.

During the tuning process, the mutation operator is applied with a minimum number of times and also with a maximum number times. This tuning process will also end if a number of consecutive unsuccessful mutations are reached, since this is a hint that the architecture can no longer be improved.

This dynamic can be summarized as follows:

1. Select μ ($\mu \geq 1$)
2. Create an initial population of μ individuals and tune-up their parameters
3. From 1 to GEN (a parameter denoting the number of generations)
   a) Create an offspring according to:
      - If $\mu > 1$ then select and apply the genetic operator (crossover or mutation)
      - Else, Apply one of the mutation operators
   b) Tune up the newly obtained structure.
   c) According to the fitness, update population if necessary

The tuning stage is executed following ES(1+1):

1. Given an individual
2. From 1 to ITE (a parameter denoting the number of iterations):
   a) Create a new individual using a normal perturbation on the set of parameters
   b) Choose the one with the best fitness as the current individual

16.3 Experiments and Results

To test the approximation ability of the ENS, preliminary experiments were carried out on the deterministic version of the Mirman and Brock model. In this case the analytical solution is known so it is possible to compare with the exact solution. We have named this first case *Economy I*. Then, we will focus on two variations of the stochastic case. The first of them, named *Economy II*, is a model with logarithmic utility function and partial capital depreciation rate ($\delta = 0.5$). The second stochastic case, named *Economy III*, is a model with partial depreciation rate, but the utility function is constant absolute risk aversion (CARA):

$$u(c) = -\exp(-\lambda c), \tag{16.8}$$

with the coefficient $\lambda = 0.7$. For all these two cases, the discount rate β is set to 0.95, and the capital share α is 0.33 ($F(k) = k^{0.33}$).

In terms of the nature of the random variable ϕ, two cases are studied. In the first case, named the *simple stochastic case*, the variable ϕ can take one of two possible values. The probability associated with each value is independent of the current state. In the second stochastic version, named the *Markovian approximation case*, ϕ follows an AR(1) process that is approximated by a seven-state Markov chain following [16.9].

For all cases, we compare the performance of the RBN structural evolution using an ES($\mu + 1$), with $\mu = 1$ (one-member ES) and $\mu = 6$ (multi-member ES). For multi-member cases different mutation rates were used (0.25, 0.50, 0.75 and 0.90).

As a second stage of the experimental phase, we compare the best results of the ENS method against the approximation found by LQA and PE methods. The LQA approximation was taken from Díaz [16.4], whereas the approximation for the PE method can be found in Haan and Marcet [16.6]. In the last case, the solution for the artificial *Economy III* is not available.

Table 16.1. Results for *Economy I*

	ENS	LQA	PE
Relative error for the consumption sequence	6.48E-06	7.69E-06	1.81E-04
Relative error for the policy function	1.55E-04	1.94E-01	2.18E-03
Utility	-18.88490	-18.88490	-18.88505
Fitness based on Euler's equation	0.999616	0.904808	0.988985

For comparison purposes, in the particular case of the *Economy I*, where an analytical solution is available, we use the best solution found by each method and calculate the mean deviation on the consumption sequence, the mean deviation on the policy function, the utility attained by the consumption sequence, and the fitness based on the Euler equation (16.7). Deviations are calculated based on the exact solution. In the stochastic cases (*Economy II* and *III*) the performance of the best-found solution is evaluated by three criteria: the utility attained by the consumption sequence, the expected utility of 10,000 sequences of consumption and, the deviation from the Euler's equation. To produce a sequence, 1,000 values of ϕ are randomly generated.

Results for the *Economy I*: The multi-member ES with a mutation rate of 0.9 had the best performance among all the ES trials based on the mean fitness over 10 runs. The mean fitness for this version was 0.987998. The best approximation is obtained by a RBN composed of 4 RBFs. Table 16.1 presents the numerical results and shows that the performances in terms of utility are similar for the three methods, but in terms of sequence deviations, policy function and Euler's equation error, the ES performs better.

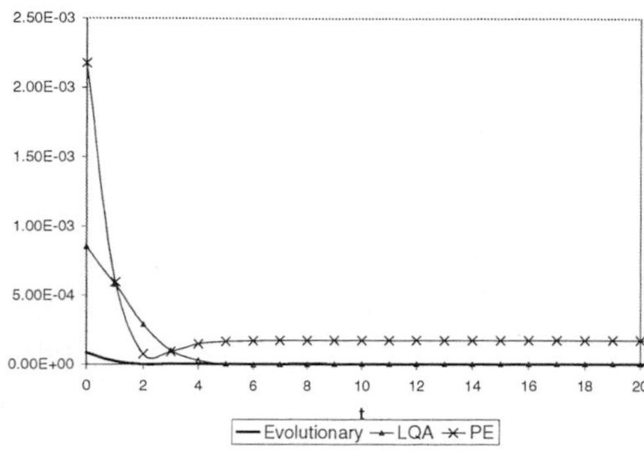

Fig. 16.2. Relative deviations on the consumption sequence in *Economy I*

The relative deviations between the exact time-sequence of consumption and the time-sequence found by the ENS and that of the LQA and PE methods are depicted in Fig. 16.2 as comparsions. It can be seen that the ENS explains the dynamic toward the steady state better than the other two methods. Fig. 16.3 shows the deviations in terms of policy function. It can be concluded that the best approximation to the exact policy function is also achieved by the ENS method, followed by the PE method, and the LQA method performs the worst. It is remarkable that in the LQA method, the further the capital is away from ist steady-state, the worse the yielded approximation is.

Results for the *Simple Stochastic* case: The one-member ES outperforms the multi-member ES for both artificial economies. In *Economy II* the average value of the fitness function is 0.97179, while in the *Economy III* the average fitness is 0.977422.

The best model attained has 8 RBF terms in the former case and 6 in the latter. Again the ENS method outperforms the LQA and the PE methods in both artificial economies by all the performance measures. Table 16.2 shows the numerical results.

Results for the *Markovian approximation* case: In this case and for both *Economy II* and *Economy III*, the performances in terms of the utility of a typical consumption sequence and of expected utility are very similar for the three methods. However, the ENS method performs better based on the error of the Euler equation.

The multi-member ES with a mutation probability value of 0.75 find the best result. The number of the RBFs in the best ENS model is 6 for the *Economy II* and 3 for the *Economy III*. Table 16.3 shows the numerical results.

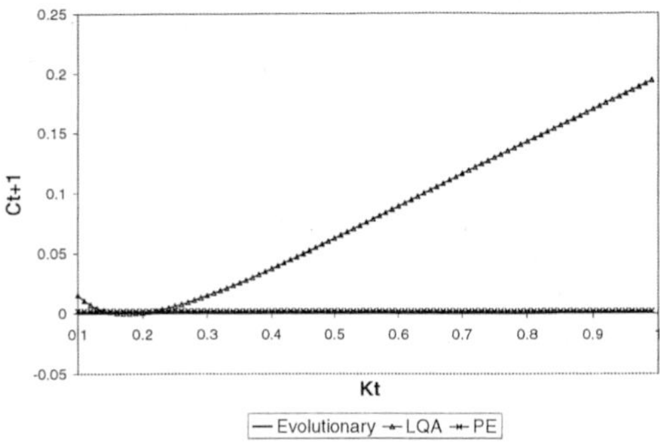

Fig. 16.3. Deviations on the policy function in *Economy I*

Table 16.2. Results for the *Simple Stochastic* case

	ENS	LQA	PE
Economy II			
Utility of one consumption sequence	-11.825	-11.825	-11.825
Expected Utility	-11.794	-11.795	-11.795
Fitness based on Euler's equation	0.9789	0.9184	0.9669
Economy III			
Utility of one consumption sequence	55.129	55.129	NA
Expected Utility	55.984	55.984	NA
Fitness based on Euler's equation	0.9417	0.9417	NA

In this case (table 16.3) the expected utility was calculated using sequences of ϕ generated by a Markov chain. In addition, the expected utility of consumption, when the productivity shocks were generated according to an AR(1) process, was also estimated. No significant differences were found, which suggests that the Markov chain is a good approximation to the continuous AR process.

16.4 Conclusions

Based upon the results obtained in this chapter, it can be concluded that the proposed ENS method can be a very useful tool to solve GEM models with a satisfactory accurate performance. The most remarkable feature of the methodology is that, unlike the LQA and PE methods, it does not assume any policy function form. This represents a great advantage due to the diversity of

Table 16.3. Results for the *Markov Approximation* case

	ENS	LQA	PE
Economy II			
Utility of one consumption sequence	-12.65327	-12.653	-12.652
Expected Utility	-12.66535	-12.665	-12.664
Fitness based on Euler's equation	0.985551	0.9187	0.964
Economy III			
Utility of one consumption sequence	55.15231	55.152	NA
Expected Utility	55.14286	55.142	NA
Fitness based on Euler's equation	0.994006	0.9420	NA

models these methods can deal with ease. This flexibility is not cost-free, since it implies a great degree of complexity from a computational perspective. However, in many cases accuracy is more important than speed.

The results also suggest that relatively small number of RBFs are good enough to obtain very accurate approximations, given the complexity of the problem we have worked with. In fact, in all of the experiments we carried out, the number of RBFs that formed the RBN is less then 10.

Another remarkable feature is that the fitness function being chosen is based on the Euler's equation which can properly guide the searching process. Unlike the one based on the utility function, it allows us to find a good approximation to the policy function those capital values which are far from the steady state.

Finally, no general conclusion can be drawn from the experiments with regard to relative performance of the two types of ES: one-member and multi-member ES. However, it seems that high mutation rates (higher than 0.50) plays a contributing role to the success in searching.

Acknowledgements

The chapter was refereed by Ya-Chi Hwang. The authors are grateful to the referee for her careful review of the chapter. The authors would also like thank the volume editors Shu-Heng Chen and Paul Wang for some of his very helpful suggestions.

References

16.1 Barro, R., Sala-I-Martin, X. (1995): Economic Growth. MCc.Graw-Hill

16.2 Cooley, T.F. (1995): Frontiers of Business Cycle Research (Ed.). Princeton University Press

16.3 Davis, L. (1991): Handbook of Genetic Algorithms (Ed.). Van Nostrand Reinhold

16.4 Díaz, J. (1999): Linear Quadratic Approximation: An Introduction. (Computational Methods for the Study of Dynamic Economies) Oxford University Press

16.5 Kidlan, F., Prescott, E. (1996): The Computational Experiment: An Econometric Tool. Journal of Economic Perspective, Vol., 10 No. 1

16.6 Marcet, A., Haan, W. D. (1990): Solving the Stochastic Growth Model by Parameterized Expectations. Journal of Business and Economics Statistics, Vol., 8 No. 1

16.7 Marcet, A., Lorenzoni, G. (1999): The Parameterized Expectation Approach: Some Practical Issues. (Computational Methods for the Study of Dynamic Economies) Oxford University Press

16.8 Michalewicz, Z. (1994): Genetic Algorithms + Data Structures = Evolution Programs. Second Edition. Springer-Verlag

16.9 Perdersen, M. (1999): Understanding Business Cycles. Institute of Economics. University of Copenhagen (Mimeo)

16.10 Poggio, T., Girosi, F. (1991): Networks for Learning: A view from theory and approximation of function. (Neural Network: Concepts, Application and Implementations). Prentice-Hall, 110–154

16.11 Rodríguez, J. S., Moreno, J. (2001): Approximating with Radial Basis Functions: An Evolutionary Approach. (Approximation, Optimization and Mathematical Economis). Physica-Verlag, 275–285

16.12 Romer, D. (1996): Advanced Macroeconomics. McGraw-Hill

16.13 Sarget, T. (1987): Dynamic Macroeconomic Theory. Harvard University Press

16.14 Stokey, N, Lucas, R. (1989): Recursive Methods in Economic Dynamic. Harvard University Press

16.15 Taylor, J, Uhlig, H. (1990): Solving Nonlinear Stochastic Growth Models: A Comparison of Alternative Solution Methods. Journal of Business and Economics Statistics, Vol., 8 No. 1

17. Evolutionary Strategies vs. Neural Networks: an Inflation Forecasting Experiment

Graham Kendall[1], Jane M. Binner[2], and Alicia M. Gazely[2]

[1] Department of Computer Science, The University of Nottingham, Nottingham, NG8 1BB, UK
email: gxk@cs.nott.ac.ukemail

[2] Department of Finance and Business Information, The Nottingham Trent University, Nottingham, NG1 4BU, UK
email: {jane.binner, alicia.gazely}@ntu.ac.uk

Previous work has used neural networks to predict the rate of inflation in Taiwan using four measures of 'money' (simple sum and three divisia measures). In this work we develop a new approach that uses an evolutionary strategy as a predictive tool. This approach is simple to implement yet produces results that are favourable with the neural network predictions. Computational results are given.

17.1 Introduction

In recent years the relationship between 'money' and the macroeconomy has assumed prominence in the academic literature and in Central Banks circles. Although some Central Bankers have stated that they have formally abandoned the notion of using monetary aggregates as indicators of the impact of their policies on the economy, research into the link between some kind of monetary aggregate and the price level is still prevalent. Attention is increasingly turning to the method of aggregation employed in the construction of monetary indices. The most sophisticated index number used thus far relies upon the formulation devised by [17.4]. The construction has it roots firmly based in microeconomic aggregation theory and statistical index number theory.

Our hypothesis is that measures of money constructed using the Divisia index number formulation are superior indicators of monetary conditions when compared to their simple sum counterparts. Our hypothesis is reinforced by a growing body of evidence from empirical studies around the world which demonstrate that weighted index number measures may be able to overcome the drawbacks of the simple sum, provided the underlying economic weak separability and linear homogeneity assumptions are satisfied. Ultimately, such evidence could reinstate monetary targeting as an acceptable method of macroeconomic control, including price regulation.

We offer an exploratory study of the relevance of the Divisia monetary aggregate for Taiwan over the period 1978 to date. In this way, we begin with a banking system that was heavily regulated by the Central Bank and the Ministry of Finance until 1989, which saw the introduction of the revised Banking Law in July. At the beginning of the 1980s, drastic economic, social and political changes took place creating a long-term macroeconomic imbalance. Rising oil prices caused consumer prices to rise by 16.3 per cent in 1981, followed by a period of near zero inflation in the mid eighties. From the nineties onwards, inflation has been fluctuating around the 5 per cent mark and hence the control of inflation has not been the mainstay of recent economic policy in Taiwan, unlike the experience of the western world. Rather, policy has focused more on achieving balanced economic and social development.

There have been major financial innovations in Taiwan as transactions technology has progressed and new financial instruments have been introduced, such as interest-bearing retail sight deposits. Although it is difficult to make a distinction between the various types of financial innovation, the effects on the productivity and liquidity of monetary assets are almost certainly different. The question we ask, in keeping with [17.8] is do the Divisia aggregates adequately capture all the financial innovations?

We adopt the principles of [17.7] by allowing both for a period of gradual learning by individuals as they adapt to the financial changes and secondly by incorporating a mechanism to accommodate the changing perceptions of individuals to the increased productivity of money. Individuals are thus assumed to adjust their holdings of financial assets until the diffusion of financial liberalisation is complete.

The novelty of this paper lies in the use of evolutionary strategies (ES) to examine Taiwan's recent experience of inflation. This is an unusual tool in this context and represents the first known application of its kind. Results are compared to those already produced for Taiwan using the Artificial Intelligence technique of neural networks [17.3] to compare the explanatory power of both Divisia and simple sum measures of broad money as indicators of inflation.

The paper concludes with a discussion of the promise of evolutionary strategies as a new tool in the macroeconomic forecasting arena.

17.2 Evolutionary Strategies

Evolutionary strategies (ES) are closely related to genetic algorithms. Originally they used only mutation, only used a population of a single individual and were used to optimise real valued variables. More recently, ES's have used a population size greater than one, they have used crossover and have also been applied to discrete variables [17.1] and [17.10]. However, their main use is still in finding values for real variables by a process of mutation, rather than crossover.

An individual in an ES is represented as a pair of real vectors, $v = (x, \sigma)$. The first vector, x, represents a point in the search space and consists of a number of real valued variables. The second vector, σ, represents a vector of standard deviations.

Mutation is performed by replacing x by

$$x^{t+1} = x^t + N(0, \sigma)$$

where $N(0, \sigma)$ is a random Gaussian number with a mean of zero and a standard deviation of σ. This mimics the evolutionary process that small changes occur more often than larger ones.

In evolutionary computation there are two variations with regard to how the new generation is formed. The first, termed $(\mu + \lambda)$, uses μ parents and creates λ offspring. Therefore, after mutation, there will be $\mu + \lambda$ members in the population. All these solutions compete for survival, with the μ best selected as parents for the next generation. An alternative scheme, termed (μ, λ), works by the μ parents producing λ offspring (where $\lambda > \mu$). Only the λ compete for survival. Thus, the parents are completely replaced at each new generation. Or, to put it another way, a single solution only has a life span of a single generation. In this work, we use a $1 + 1$ strategy and plan to develop other strategies in the future.

Good introductions to evolutionary strategies can be found in [17.2], [17.5], [17.6], [17.11] and [17.12].

17.3 Data and Model Specification

Four different monetary measures were used independently to predict future movements in the inflation rate. Monetary data consisted of three Divisia series provided by [17.7], one conventional Divisia, (DIVM2), Innovation1 and Innovation2, together with a simple sum series, constructed from component assets obtained from the Aremos-Financial Services database in Taiwan. The Divisia M2 aggregate is constructed by weighting each individual component by its own interest rate whilst Innovation1 (INN1) and Innovation2 (INN2) seek to improve upon the weighting system by capturing the true monetary services flow provided by each component asset more accurately. Thus INN1 is a development of DIVM2 inspired initially by [17.9] and used subsequently by [17.8]. A learning adjustment of the retail sight-deposit interest rate is applied to reflect the adaption of agents to the introduction of interest-bearing sight deposits in 1984. Note: the Innovation1 series does not diverge from the conventional Divisia measure until the late 1980s. The second modified Divisia series, INN2 assumes a period of gradual and continuous learning by agents as they adapt to the increased productivity of money throughout the period and corrects, at least partially, for the distortion arising from technological progress. Individuals are thus assumed to adjust their holdings of financial assets until the diffusion of financial liberalisation is complete.

Inflation was constructed for each quarter as year-on-year growth rates of prices. Quarterly data over the sample period 1970Q1 to 1995Q3 was used as illustrated in Fig. 17.1. Our preferred price series, the Consumer Price Index (CPI), was obtained from Datastream. The four monetary series were subjected to a smoothing process by taking three quarter averages to reduce noise. Finally, to avoid the swamping of mean percent error by huge values during a period of very low inflation from 1983 to 1986, the entire series was translated upwards by 5 percent and results are presented on this basis. Of the total quarterly data points available, after loss of data points due to the smoothing process and the time lag implicit in the model of up to four quarters, 96 quarters remained, of which the first 85 were used for training and the last 7 for were used as a validation set. The first 4 items were only used as a basis for the first prediction.

The aim of the evolutionary strategy is to predict the future inflation rate based on the inflation rate at the previous quarter and the previous four quarters money measure. That is

$$\Pi_t = f(M_{t-1}, M_{t-2}, M_{t-3}, M_{t-4}, \Pi_{t-1})$$

where Π is inflation, t represents time on a quarterly basis and M is the current money measure being considered (i.e. M2, DIVM2, INN1 or INN2).

The future inflation rate was predicted via a linear function with each term having a weight. The simple linear function was chosen as a useful starting point in this inflation forecasting experiment as this research is highly exploratory in nature and, to the best of our knowledge, no direct comparison between the forecasting performance of ES and neural networks has previously been undertaken. Future research will permit further experimentation with ES functions of increasing complexity. Each of the five factors used to predict the next quarter's inflation rate was given an initial, random weight and the task given to the ES was to find a set of weights that allowed future inflation to be predicted. Hence, the function given to the ES was

$$\Pi_t = f(w_1 M_{t-1} + w_2 M_{t-2} + w_3 M_{t-3} + w_4 M_{t-4} + w_5 \Pi_{t-1})$$

where w_n is the weight associated with each term that contributes to the inflation rate for the next quarter.

A given solution (set of weights) is evaluated by predicting the inflation rate for each quarter of the training set. The output from each quarter is compared against the actual value and the absolute sum of those differences is calculated. This figure represents the quality of a given solution. The aim, of course, is to minimise the evaluation function.

If a solution is found that is better than its parent, it becomes the new parent. Each weight in the parent is then mutated as described above (section 17.2). After a period of training (1,000,000 iterations in this work) the weights were used to predict the inflation rate of the validation set.

Table 17.1. ES results for full sample, $s = 85$

(a)	M2	DIVM2	INN1	INN2
RMS	0.055654	0.052813	0.052574	0.054517
MAD	0.029751	0.024886	0.024735	0.027855
MPE	25%	19%	18%	22%
(b)	M2	DIVM2	INN1	INN2
RMS	0.035367	0.016644	0.016481	0.059012
MAD	0.031768	0.013897	0.013389	0.050765
MPE	36%	16%	15%	59%

17.4 Testing and Results

The four money measures (M2, DIVM2, INN1 and INN2) were tested independently and these results compared against previous results obtained on the same data using a neural network. The neural network architecture used for all the networks was a back-propagation network with five hidden neurons in one layer; the Delta function was used for learning and Tan-H for transfer. See [17.3] for further details on the neural network methodology employed.The results reported are the arithmetic meanscalculated over ten individual runs of the ES and are divided between in-sample (the training set) and out-of-sample (the validation set). Within these two categories, three standard forecasting evaluation measures were used to compare the predicted inflation rate with the actual inflation rate, namely, Root Mean Squared Error (RMS), Mean Absolute Difference (MAD) and Mean Percent Error (MPE). The in-sample (a) and out-of-sample (b) results produced by the evolutionary strategies are presented in Table 17.1 alongside previous results [17.3] using neural networks in Table 17.2 below.

Table 17.2. NN results for full sample, $s = 85$

(a)	M2	DIVM2	INN1	INN2
RMS	0.032106	0.022578	0.018764	0.026806
MAD	0.024703	0.017493	0.013878	0.018186
MPE	30%	22%	16%	21%
(b)	M2	DIVM2	INN1	INN2
RMS	0.014801	0.016043	0.010715	0.011575
MAD	0.013540	0.014983	0.008021	0.009034
MPE	16%	17%	9%	10%

The neural network clearly has superior forecasting capabilities, producing more accurate forecasts using all forecasting evaluation methods both in and out-of sample. The four RMS errors are, on average, twice as great as those produced by the neural network. The best inflation forecast is achieved using the INN1 monetary aggregate, where the neural network RMS error is 35% lower than the ES equivalent. Figs. 17.1 and 17.2 below illustrate the best fitting (INN1) forecasts for each technique. Although the neural network is undoubtedly more powerful, the ES method clearly has potential for learning future movements in inflation. INN2 is undoubtedly the worst performing monetary aggregate, producing out-of-sample RMS errors some 3.5 times greater than INN1 on average.

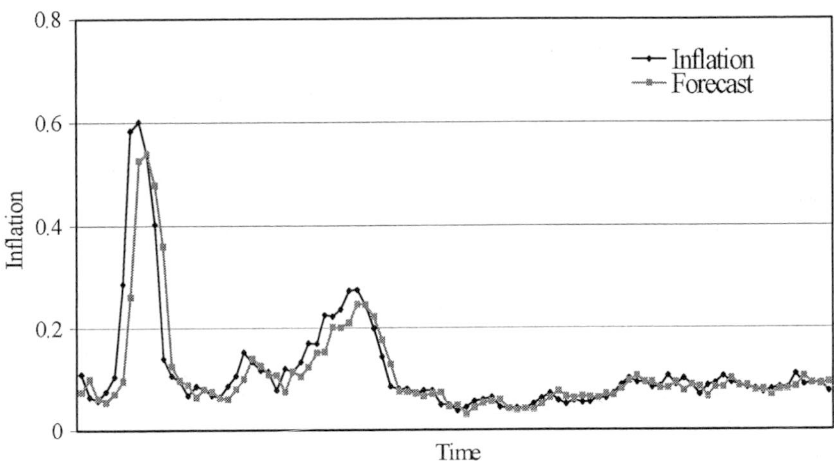

Fig. 17.1. Taiwan inflation 1972-1995: ES

To investigate and improve the fit of the algorithm further, the training data was split into smaller sample sizes. That is, the data was split into n subsets and each subset had a set of weights associated with that data. Initial experiments showed that using smaller subsets does produce results closer to the expected values. Following this insight a series of experiments were conducted where the training data was split into a number of subsets. The size of the subsets, s, took the values, $s = 10, 20, 30, 50, 70, 85$.

Tables 17.3, 17.4, 17.5 and 17.6 show the results when applying this approach to the four data sets. It is interesting to note that, in general, as the value of s increases, so does the error. This can be seen in Fig. 17.3, which plots the error of the M2 training data set. This graph is indicative of all the data sets. It is also interesting to record that $s = 1$ was tested but is not shown as this training data is learnt perfectly, but its predictive capabilities are terrible. This is to be expected as each data sample has its own set of weights, however, not one of them is able to predict accurately.

Using smaller sample sizes to allow the training data to be learnt more accurately presents a problem in that we need to decide which weights to use to predict the future inflation rate. If the sample size includes the entire training set then there is only one set of weights. If we split the training

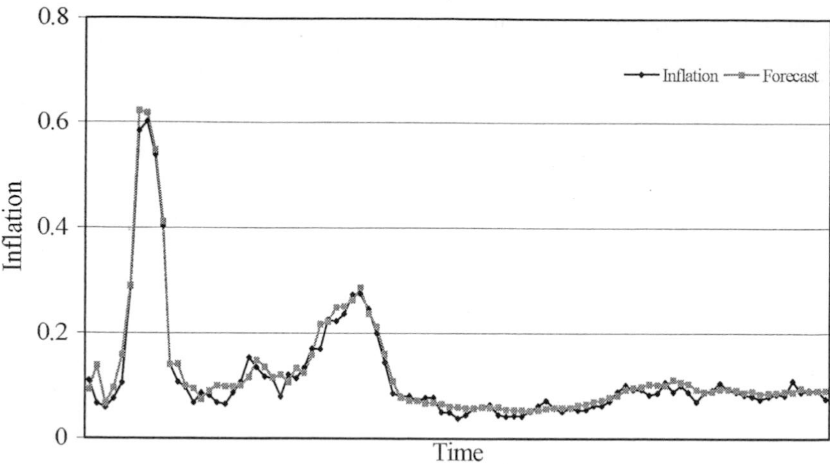

Fig. 17.2. Taiwan inflation 1972-1995: NN

Table 17.3. Different sample sizes for M2 data

(a)	$s = 10$	$s = 20$	$s = 30$	$s = 50$	$s = 70$	$s = 85$
RMS	0.029596013	0.0488479	0.0522658	0.0542072	0.054768	0.055654
MAD	0.015665693	0.0251071	0.026575	0.0276298	0.0278926	0.029751
MPE	14%	21%	22%	23%	23%	25%
(b)	$s = 10$	$s = 20$	$s = 30$	$s = 50$	$s = 70$	$s = 85$
RMS	0.03460674	0.0468069	0.0333721	0.0298592	0.0320764	0.035367
MAD	0.029368693	0.0408344	0.0282566	0.026615	0.0277392	0.031768
MPE	33%	48%	31%	30%	32%	36%

Table 17.4. Different sample sizes for DIVM2 data

(a)	$s = 10$	$s = 20$	$s = 30$	$s = 50$	$s = 70$	$s = 85$
RMS	0.029077811	0.0407076	0.0456028	0.0500104	0.0526802	0.052813
MAD	0.013509228	0.0191466	0.0214196	0.0232269	0.0241488	0.024886
MPE	13%	16%	18%	17%	18%	19%
(b)	$s = 10$	$s = 20$	$s = 30$	$s = 50$	$s = 70$	$s = 85$
RMS	0.017450212	0.013862	0.010434	0.01599	0.008025	0.016644
MAD	0.011356283	0.0095	0.0082	0.0136117	0.0061	0.013897
MPE	12%	10%	9%	16%	6%	16%

Table 17.5. Different sample sizes for INN1 data

(a)	$s = 10$	$s = 20$	$s = 30$	$s = 50$	$s = 70$	$s = 85$
RMS	0.028637794	0.0408193	0.0455279	0.0495153	0.0530051	0.052574
MAD	0.013415464	0.0191138	0.0216154	0.0233427	0.0240428	0.024735
MPE	13%	16%	18%	17%	18%	18%
(b)	$s = 10$	$s = 20$	$s = 30$	$s = 50$	$s = 70$	$s = 85$
RMS	0.019071346	0.023874	0.0158909	0.0162886	0.0160835	0.016481
MAD	0.01558306	0.0207543	0.0136517	0.0138929	0.0131833	0.013389
MPE	17%	23%	15%	16%	14%	15%

Table 17.6. Different sample sizes for INN2 data

(a)	s =10	s =20	s =30	s =50	s =70	s =85
RMS	0.029615464	0.0416456	0.0474273	0.0516671	0.0537119	0.054517
MAD	0.015299391	0.0208679	0.0241918	0.0252718	0.0262228	0.027855
MPE	15%	18%	21%	19%	20%	22%
(b)	s =10	s =20	s =30	s =50	s =70	s =85
RMS	0.104725328	0.1188458	0.04745	0.038448	0.0631411	0.059012
MAD	0.087689038	0.1061943	0.039716	0.032923	0.0539396	0.050765
MPE	103%	124%	46%	37%	63%	59%

sample into two sets then there are two sets of weights. We can continue this until the sample size is one, in which case there are n sets of weights (where n is the size of the training set).

We have tried averaging the weights from all the samples, but this yielded very poor results. Therefore, it was decided to use the final set of weights, under the premise that these weights reflect inflation rates that are temporally closer to the prediction set. These results can be seen in part b of Tables 17.3, 17.4, 17.5 and 17.6. However, using the final set of weights may not be the best approach. For example, if the sample size is 20 and the training set, n, is 85 then the last set of weights will only include 5 samples on which it has been trained. It may be better to use one of the other set of weights that is evolved using a sample size of 20. Table 17.7, summarises the predictions for all the weights for out-of-sample data for the INN1 data set with $s = 10$.

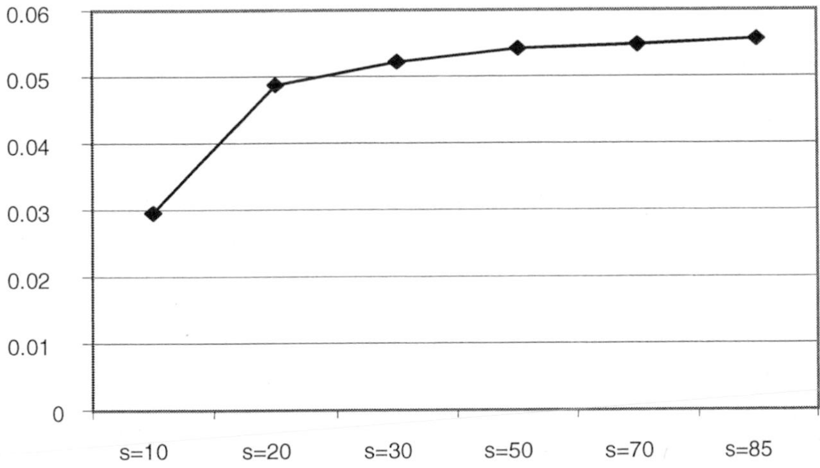

Fig. 17.3. RMS error for M2 data

This gives a set of 9 weights with the first 8 being evolved having access to 10 samples. The last set of weights has only been trained on a sample size of 5. It can be seen that the best predictions are from the final set of weights.

Furthermore, as we get temporally closer to the figures we are predicting the predictions improve. Table 17.8 shows that this observation is also true when varying s for the INN1 data set. It is interesting to note that when $s = 20$, this leads to the worst predictions overall. This is likely to be due to the fact that the fifth set of weights has only been trained on five samples, whereas when $s = 30$ the final set of weights has been trained on 25 samples, when $s = 50$ the final sample size is 35 and when $s = 70$ the sample size, although small, at 15, is still large enough to beat a larger sample size that is temporally disjoint from the data. From these results we conclude that to predict out-of-sample data the weights have to be trained on data that is temporally closer to the data being predicted but, in addition, the sample size on which those weights have been trained has to be large enough to allow the weights to learn the underlying trend. Table 17.9, shows that these observations are also true of the DIVM2 data. The other data sets also exhibit this behaviour.

Table 17.7. INN1 prediction from set of weights when s =10

	1	2	3	4	5	6	7	8	9
RMS	0.84895	0.21323	0.13490	0.14680	0.03549	0.02181	0.02038	0.01417	0.01907
MAD	0.7798	0.2071	0.1344	0.1417	0.0426	0.0283	0.0290	0.0212	0.0155
MPE	868%	232%	151%	156%	50%	33%	35%	25%	17%

Table 17.8. INN1 prediction from set of weights when s =20, 30, 50 and 70

	1	2	3	4	5
s =20					
RMS	0.141752	0.076644	0.016340	0.012454	0.023874
MAD	0.1402	0.0796	0.0237	0.0201	0.0207543
MPE	158%	91%	29%	25%	23%
s =30					
RMS	0.057670	0.015914	0.0158909		
MAD	0.0541	0.0229	0.0136517		
MPE	63%	28%	15%		
s =50					
RMS	0.032925	0.0162886			
MAD	0.0391	0.0138929			
MPE	46%	16%			
s =70					
RMS	0.017343	0.0160835			
MAD	0.0241	0.0131833			
MPE	29%	14%			

Table 17.9. DIVM2 prediction from set of weights when s =10, 20, 30, 50 and 70

	1	2	3	4	5	6	7	8	9
s =10									
RMS	0.01744	0.14721	0.10514	0.16631	0.04958	0.02645	0.01689	0.01200	0.01745
MAD	0.0147	0.1447	0.1032	0.1613	0.0480	0.0227	0.0150	0.0105	0.01136
MPE	15%	161%	114%	175%	54%	25%	16%	12%	12%
s =20									
RMS	0.20297	0.04450	0.01881	0.01392	0.01386				
MAD	0.2020	0.0408	0.0147	0.0137	0.0095				
MPE	224%	45%	16%	15%	10%				
s =30									
RMS	0.05910	0.01187	0.01043						
MAD	0.0495	0.0094	0.0082						
MPE	56%	10%	9%						
s =50									
RMS	0.02179	0.01599							
MAD	0.0170	0.01361							
MPE	19%	16%							
s =70									
RMS	0.01607	0.00803							
MAD	0.0126	0.0061							
MPE	14%	6%							

17.5 Concluding Remarks

This research provides the first evidence of its kind to our knowledge to compare the predictive performance of evolutionary strategies with neural networks. Neural networks clearly provide more accurate forecasts over larger sample sizes as they can learn the data perfectly. Evolutionary strategies are found to learn the data perfectly in extreme cases. There is clearly scope for further research into the trade off between sample size and predictive accuracy of evolutionary strategies and for the development of this technique as a new macroeconomic forecasting tool.

Evidence presented here also supports the view that Divisia indices appear to offer advantages over simple sum indices as macroeconomic indicators. It may be concluded that a money stock mismeasurement problem exists and that the technique of simply summing assets in the formation of monetary aggregates is inherently flawed. The role of monetary aggregates in the major economies today has largely been relegated to one of a leading indicator of economic activity, along with a range of other macroeconomic variables. However, further empirical work on Divisia money and, in particular, close monitoring of Divisia constructs that have been adjusted to accommodate financial innovation, may serve to restore confidence in former well established money-inflation links. Ultimately, it is hoped that money may be re-established as an effective macroeconomic policy tool in its own right. This application of evolutionary strategies vs. neural network techniques to examine the money

- inflation link is highly experimental in nature and in keeping with the pioneering work conducted by two of the current authors for the UK, USA and Italy, the overriding feature of this research is very much one of simplicity. It is virtually certain in this context that more accurate inflation forecasting models could be achieved with the inclusion of additional explanatory variables, particularly those currently used by monetary authorities around the world as leading indicator components of inflation.

References

17.1 Bäck, T., Hoffmeister, F., Schwefel, H. -P. (1991): A survey of evolution strategies. Proceedings of the Fourth Conference on Genetic Algorithms, Belew, R., Booker, L. (Eds.), Morgan Kaufmann Publishers, 2–9

17.2 Bäck, T., Fogel, D. B., Michalewicz (1997): Handbook of evolutionary computation. Oxford University Press

17.3 Binner, J. M., Gazely, A. M., Chen, S. -H. (2002): Financial innovation in Taiwan: An application of neural networks to the broad money aggregates. European Journal of Finance, Vol. 8, 238–247

17.4 Divisia, F. (1925): L'indice monetaire et al theorie de la monnaie, Rev. d'Econ. Polit., 39 980-1008

17.5 Fogel, D. B. (1998): Evolutionary computation the fossil record, IEEE Press

17.6 Fogel, D. B. (2000): Evolutionary computation: toward a new philosphy of machine intelligence, 2nd Ed., IEEE Press Marketing

17.7 Ford, J. L. (1997): Output, the price level, broad money and Divisia aggregates with and without innovation: Taiwan, 1967(1) - 1995(4). Discussion paper 97–17, Department of Economics, The University of Birmingham

17.8 Ford, J. L., Peng, W. S., Mullineux, A. W. (1992): Financial innovation and Divisia monetary aggregates. Oxford Bulletin of Economics and Statistics, 87–102

17.9 Hendry, D. F., Ericsson, N. R. (1990): Modelling the demand for narrow money in the UK and the United States. Board of Governors of the Federal Reserve System, International Finance Discussion Papers, No. 383

17.10 Herdy, M. (1991): Application of the evolution strategy to discrete optimization problems. Proceedings of the First International Conference on Parallel Problem Solving from Nature (PPSN), Lecture Notes in Computer Science, Schwefel, H. -P., Manner, R. (Eds.), Springer-Verlag, Vol. 496, 188–192

17.11 Michalewicz, Z. (1996): Genetic algorithms + data structures = evolution programs (3rd rev. and extended ed.). Springer-Verlag, Berlin

17.12 Michalewicz, Z., Fogel, D. B. (2000): How to solve it. Springer-Verlag

18. Business Failure Prediction Using Modified Ants Algorithm

Chunfeng Wang, Xin Zhao, and Li Kang

Institute of Systems Engineering, Tianjin University, Tianjin 300072, P.R. China
email: cfwang@tju.edu.cn, zhaoxin@hotmil.com

This chapter successfully introduces the ants algorithm into the business failure prediction problem domain. The original ants algorithm is also modified and improved in both transition probability and pheromone trail update mechanism. The distinct advantages of this modified ants algorithm (MAA) consist of no special demand on the problem's form, lower computer storage, and less CPU time for computation. The empirical results based on the real-world data demonstrate the effectiveness of its application to the business failure prediction problem domain and also show its advantages compared with RPA (recursive partition algorithm), DA (Discriminate Analysis) and GP (Genetic Programming).

18.1 Introduction

In recent years, much work has been done on business failure prediction in both academics and financial practice areas [18.2]. It is motivated by not only the importance of the problem, but the dramatic progress in technology also. A business failure affects a firm's entire existence and induces a high cost to the firm, the collaborators, the society, and the country's economy [18.7]. It is very important to find out an effective method to predict business failures, and therefore the failure prediction models have great importance to those managers who can take action to prevent failure and who can select firms to collaborate with or invest in. Though with a few notable efforts, a complete theory of business failure has not yet been developed due to its complexity. The common approach used by previous research in solving this problem is to formulate it as a classification problem and then employ classification techniques to solve it [18.8]. From the viewpoint of technology, the main characteristics of business failure prediction lie in three aspects: (1) the uncertainties and difficulties of variable selection in failure prediction analysis; (2) the high noisiness of the financial data; and (3) the interrelationship between financial ratios [18.9]. The characteristics of business failure prediction above have stimulated a plethora of models and techniques proposed to produce more accurate and rational classification results. Generally, these classification techniques can be divided into two categories: statistical methods and artificial intelligence methods.

Statistical models were the earliest manner employed by researchers for the prediction of business failure. According to the assumptions of the form of a discriminate function and the sample distribution of the statistical models, these models can be divided into four classes: regression analysis [18.10], multivariate discriminate analysis (MDA) [18.1, 18.3, 18.5, 18.6, 18.11], logit analysis [18.12, 18.13], and factor logistic analysis. The main disadvantage of these methods is that they need a special statistic hypothesis in the application, which often is violated in the real financial world. It is found in many literature that the financial data often have much noise and a skew distribution, which results in a statistical model failure. Furthermore, as a kind of model-driven method, statistical approaches are limited in deriving an appropriate prediction model in the absence of well-defined domain models like business failure prediction. When the assumed functional form is not correct, the statistical techniques merely confirm this, but do not predict the right functional form.

With the dramatic development of computer science, artificial intelligence (AI) has seen rapid developments and extensive applications. Because the AI methods provide a more general framework for mining relationships in data, especially to complex non-linear problems, and do not require any specification of the distribution or functional form hypothesis, they have been introduced in this domain as an effective substitution of the statistical models. AI methods, which were proposed to tackle business failure prediction, include neural networks [18.14], expert systems [18.15], recursive partitioning algorithms [18.16] and so on.

The neural network (NN) models were first employed by Odom and Sharda [18.14] to deal with the business failure prediction problem. Despite a lot of successful applications of NN, some drawbacks of NN still exist as follows. The most distinct problems of NN are their "black-box" systems and the trap of "overfitting." The resulting weights inherent in the system are not transparent and are sensitive to structural changes. The possibility of deriving an illogical network behavior in response to different variations of the input values also constitutes an important problem from a financial analysis point view [18.4, 18.17]. These problems as well as a long processing time for completing the NN training phase and the need to carry out a large number of tests to identify the NN structure can considerably limit the use of NN.

Expert system (ES), as another AI technique, was proposed to deal with the prediction of business failure by Mesier and Hansen [18.15]. The objective of ES is to take samples of a known class (failed or non-failed), which is described in terms of a fixed collection of attributes (financial ratios), and generates a production system over these attributes. By the resulting production system the new samples can be correctly classified. In the work of Mesier and Hansen [18.15], the production systems were developed based on the Quinlan's ID3 algorithm [18.18]. The main advantage of ES is that it easily explains the reason why a firm failed. The primary difficulty of it is also that of overfitting, which stems from the decision tree creation based on an ID3 algorithm. In an ID3 decision tree, the lower branches of the tree wildly capitalize on the specific characteristics of individual cases in the training sample, which are not representative of the general characteristics of the population of cases for which the tree is developed. This susceptibility to noise in the sample results in a decision tree may give poor performance when the tree is applied to new cases. Another difficulty is the computational burden when the sample set is large. In 1985, Frydman et al. proposed a recursive partition algorithm (RPA) in the decision tree modeling process. Though RPA promises to have a higher generalization ability by reducing the redundant branches of the tree, it fails when there is much more noise in the data. RPA also has a problem, which is found in most forward stepwise approaches to variable selection, that it does not review previous classifications while it introduces new classification rules and often results in a decision tree difficult to interpret and likely to be overfitting, especially to a complicate problem.

According to the characteristics of this problem domain and in order to overcome the limitations of the above methods, we design a modified ants algorithm (MAA) which is originally applied in the domain of combinatorial optimization as an effective tool to overcome "combinatorial explosion." Considering the special features of the business failure prediction problem, the original algorithm is also modified in some detail mechanism, including transition probability and pheromone trail update mechanism.

To verify the effectiveness of the proposed model, we compare the performance of the MAA with that of other AI algorithms and obtain very satisfactory results. The chapter is organized as follows: in the section 18.2, the ants algorithm is introduced; a computation experiment is given in section 18.3; the section ??, 18.5, and 18.6 present the discretisation of data and implementation of algorithm respectively; a discussion is given in the next section; the final section concerns the conclusion.

18.2 Ants Algorithm

18.2.1 Main Idea of the Ants Algorithm

The ants algorithm, simulating the foraging behavior of ants to find the optimal route linking nest and food resources, was first proposed by Dorigo [18.19, 18.20, 18.21] as a multi-agent approach to difficult combinatorial optimization problems like TSP and QAP. Ants are social insects that live in colonies and whose behavior is directed more to the colony's survival as a whole than to that of a single individual component of the colony. An important and interesting behavior of ant colonies is their foraging behavior. While walking from food sources to the nest and vice versa, ants deposit on the ground a substance called pheromone that forms a pheromone trail. Ants can smell pheromone and, when choosing the way, they tend to choose (in probability) those paths marked by strong pheromone concentrations. Furthermore, other ants can also use it to find the location of food sources.

18.2.2 The Improvement of the Ants Algorithm

Two main modifications are suggested to improve the performance of the original ant system algorithm in this chapter as follows:

(1) The Improvement of the Calculation of Transition Probability

It is known that computational efficiency is a valuable asset for any heuristic method. While this is usually achieved by careful implementation and data structure design, some higher-level considerations are possible. For the original algorithm, the transition probability can be given as follows:

$$p_{ij}^k = \begin{cases} \frac{\tau_{ij}^\alpha + \eta_{ij}^\beta}{\sum_{r \notin tabu_k} (\tau_{ir}^\alpha + \eta_{ir}^\beta)}, & if\ j \notin tabu_k \\ 0, & otherwise \end{cases} \tag{18.1}$$

where, τ is the pheromone trail; η is visibility; α and β are user-defined parameters; $tabu_k$ is the taboo list of k-th ants. As we shall become clear later (Section 18.5.2), the visibility measure considered in this chapter has noting to do with physical distance. In fact, unlike TSP, physical distance may not be meaningful in a general combinatoric optimization problem. Therefore, visibility is only used as a metaphor in this chapter, and we shall provide a visibility measure suitable for our problem in Section 18.5.2.

For the modified ants algorithm, the formula (18.1) can be converted into the following:

$$p_{ij}^k = \begin{cases} \frac{\tau_{ij}^\alpha + \eta_{ij}^{\varphi(\alpha)}}{\sum_{r \notin tabu_k} (\tau_{ir}^\alpha + \eta_{ir}^{\varphi(\alpha)})}, & if\ j \notin tabu_k \\ 0, & otherwise \end{cases} \tag{18.2}$$

where, $\varphi(\alpha)$ is the linear or exponential function on α. According to our computation experiment, $\varphi(\alpha)$ can be formulated as follows:

$$\varphi(\alpha) = 5^{(\alpha - 0.8)/0.4}. \tag{18.3}$$

As Colorni, Dorigo, and Maniezzo pointed out in 1992 [18.21], α presents an optimal range around 1 and β between 1 and 5 for formula (18.1). For formula (18.3), if the value of α changes around 1, then the value of $\varphi(\alpha)$ will change from 1 to 5. The relationship between α and $\varphi(\alpha)$ can be described in Fig. 18.1.

From Figure 18.1, we can see that if we control the value of α in its optimal range, then so is the value of $\varphi(\alpha)$. Formula (18.2) achieves the same

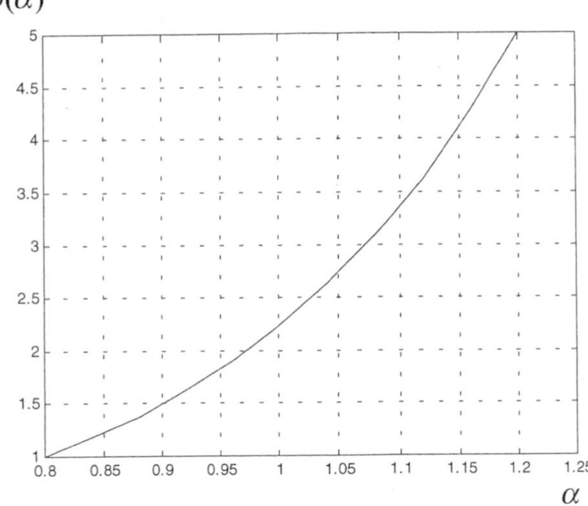

Fig. 18.1. The relationship of $\varphi(\alpha)$ to α

objective of formula (18.1) in letting the user specify a different relative importance of trail with respect to attractiveness. Using this method we can eliminate one parameter β to make the algorithm simpler and more efficient than the original algorithm.

(2) Dynamic Update Mechanism of the Pheromone Trail

The dynamic update mechanism of the pheromone trail can be formulated as:

$$\Delta\tau_{ij}(t, t+n) = \sum_{k=1}^{m} \tau_{ij}^{k}(t, t+n) + \gamma\Delta\tau_{ij}^{best}(t, t+n), k \neq best \tag{18.4}$$

Here, $\Delta\tau_{ij}^{best}(t, t+n)$ represents the contribution of the ant (who finds the best solution at the current iteration step) to construct its solution by moving from i to j. $\sum_{k=1}^{m} \tau_{ij}^{k}(t, t+n)$ is the total contribution of the other ants to construct the solutions. γ is a weight parameter of $\Delta\tau_{ij}^{best}(t, t+n)$. With the increases of iterations, γ increases, too. The formula for calculating γ is:

$$\gamma = c^{i} - 1. \tag{18.5}$$

Here, c is a constant whose value is very near to, but greater than, 1. i is the iteration times. By increasing the value of i, the changes of the value of γ can be given in Figure 18.2 (the value of c is 1.008 in this figure).

From Figure 18.2, the search process can be divided into three cases.

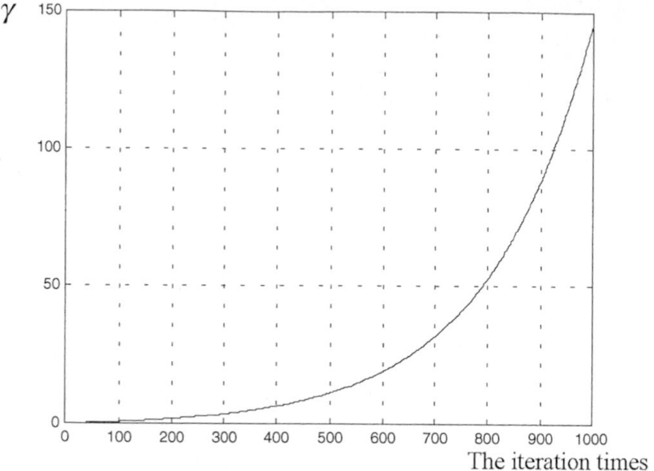

Fig. 18.2. The changes of γ by increasing the iteration times

(a) When i is less than 86, the value of γ is less than 1. From formula (18.4) we can see that $\Delta\tau_{ij}^{best}(t, t+n)$ is given a lesser weight, which can make the algorithm do a thorough search in the feasible space and enhance the diversification of the solutions so as to avoid stagnation and improve the quality of solutions.

(b) While i is greater than 87, γ is greater than 1 and increases with i, i.e., $\Delta\tau_{ij}^{best}(t, t+n)$ will be given a weight which is greater than 1, and more pheromone will be left on the optimum direction than on the others. During the process of optimization, the weight will increase, making the algorithm converge rapidly on the optimum direction to enhance the optimization efficiency of the algorithm.

(c) The value of γ is nearer to 1 ($\gamma = 1.007$) as i is equal to 87. If we keep the value of i as a constant, say 87, then formula 18.4 will be the same as that of the original algorithm towards the optimization efficiency of algorithm.

By adopting the dynamic update mechanism of the pheromone trail, the algorithm can give a glancing search on the whole feasible space in order to avoid the stagnation and trapping into the local optimum at the beginning. When increasing the iteration time, the algorithm will provide more and more searches on the optimum direction with a higher probability than on the other directions so as to converge rapidly. Obviously, the dynamic mechanism of the pheromone trail is helpful for the ants algorithm to simultaneously achieve the goal of improving the quality of solutions and search efficiency.

18.3 Computation Experiment

18.3.1 Data and Variables

The data sets used in our test come from a developed market — the U.S. The data sets selected are convenient for comparison with existing methods and the characteristics of the data samples can cover as much as possible the typical characteristics of this problem's domain.

For convenience, we adopt the same data used by Odom and Sharda [18.14]. The data is obtained from Moody's industrial Manuals and includes 129 firms, where 65 firms went bankrupt during the period 1975 through 1982. The total sample set is randomly divided into two sub-sample sets: training sub-sample set and testing sub-sample set. The training sub-sample set consists of 74 firms (36 non-bankrupt and 38 bankrupt firms). The testing sub-sample set consists of 55 firms (28 non-bankrupt and 27 bankrupt firms). Each firm is described by 5 financial ratios, the discriminate variables, which are compiled from 22 financial variables. The 5 financial ratios have been employed in previous studies such as Altman [18.1] and Odom and Sharda [18.14], according to their statistical significance.

X_1: Working capital/Total assets
X_2: Retained earnings/Total assets
X_3: EBIT/Total assets (where EBIT is earnings before interest and tax)
X_4: MVE/Total debt (where MVE is the market value of equity)
X_5: Sales/Total assets

18.3.2 The Statistic Characteristic of Data Sets

For comparison purposes, the statistic properties of the data need to be tested first. We achieve this by using the analysis of the mean and standard deviation, and the results are listed in Table 18.1 and Table 18.2. From the descriptive statistics results in the tables, the statistic characteristic or the information quality of training data can be learned. Furthermore, the more significant variables with more explanationatory abilities can be distinguished and ascertained.

Table 18.1 and Table 18.2 report descriptive statistics by bankrupt groups and non-bankrupt groups for both training and testing samples. For the training sets, a comparison of the mean value of variables in bankrupt and non-bankrupt sets indicates that working capital/total assets (X_1), retained earnings/total assets (X_2), EBIT/total assets (X_3), and MVE/total debt (X_4) are lower in the bankrupt set than those in non-bankrupt set. The p-values for the test of the mean difference between bankrupt and non-bankrupt companies are significant for each of these variables. The means for sales/total assets (X_5) are not significantly different between the bankrupt and non-bankrupt sets. For the testing sets, the means of X_1, X_2 and X_3 are lower in the bankrupt set than in the non-bankrupt set.

From the Table 18.1 and Table 18.2, we can see that the descriptive statistics of the testing sets are similar to those of the training sets except for MVE/total assets. The means for the market value variable are not significantly different between the bankrupt and non-bankrupt firms in testing sets.

The results of the Kolmogorov-Smirnov test reveal that X_1, X_2, X_3, and X_5 are the normal distributions for both non-bankrupt and bankrupt sets. According to the value of skewness and kurtosis, X_4 conforms to the distribution that have high kurtosis and a positive skewness. In Table 18.2, X_1, X_2, and X_3 are normal distribution for non-bankrupts sets and X_1, X_3, X_4, and X_5 are normal distribution for bankrupt sets according to the Kolmogorov-Smirnov test.

Table 18.1. The statistic data of training sets

	Training Set (Non-bankrupt)				Training Set (Bankrupt)				
X	Mean	Std.Dev.	Max.	Min.	Mean	Std.Dev.	Max.	Min.	p-value
X_1	0.290	0.158	0.667	0.014	0.107	0.231	0.506	−0.536	0.002
X_2	0.370	0.153	0.697	0.013	−0.077	0.344	0.282	−1.695	0
X_3	0.143	0.059	0.299	0.052	−0.050	0.145	0.203	−0.486	0
X_4	2.674	5.471	30.619	0.042	0.459	0.586	3.296	0.048	0.015
X_5	1.587	0.786	4.955	0.222	1.846	1.036	6.515	0.773	0.231

X_1: Working capital/Total assets; X_2: Retained earnings/Total assets; X_3: EBIT/Total assets; X_4: MVE/Total debt; X_5: Sales/Total assets.

Bankrupt group includes companies that experience bankruptcy, while non-bankrupt group includes non-bankrupt companies.

p-value of the test of difference in the variable means between the bankrupt and non-bankrupt groups.

Table 18.2. The statistic data of testing sets

	Testing Set (Non-bankrupt)				Testing Set (Bankrupt)				
X	Mean	Std.Dev.	Max.	Min.	Mean	Std.Dev.	Max.	Min.	p-value
X_1	0.314	0.177	0.640	0.011	0.210	0.156	0.496	−0.160	0.024
X_2	0.313	0.145	0.649	0.020	0.059	0.289	0.348	−0.981	0.001
X_3	0.121	0.072	0.300	−0.008	0.039	0.073	0.167	−0.183	0.001
X_4	2.681	6.494	34.503	0.106	0.426	0.427	2.285	0.045	0.078
X_5	1.810	1.308	7.166	0.228	1.820	0.980	5.300	0.138	0.975

18.4 Discretization of Data

In order to deal with the proposed problem, the original data sets must be in discrete or categorical form. This operates on what may be described as a decision table or information system. In order to illustrate the discrete process, a simple example is adopted and the original data of the example are listed in Table 18.3. We then can use interval ranges to cover the variables according to their statistics attribution. According to the feature of the data sets, the interval constructed for each variables is listed in Table 18.5. Using Table 18.5, the original data set is reconstructed and Table 18.4 is formed. The purpose of discretisation is to convert the proposed problem into a combination one, in order so that we can use the ants algorithm to find the partition rule that consists of interval ranges.

Table 18.3. The original data of example

Cases	WC/TA	RE/TA	EBIT/TA	MVE/TD	S/TA
1	.3922	.3778	.1316	1.0911	1.2784
2	.0574	.2783	.1166	1.3441	.2216
3	.1650	.1192	.2035	.8130	1.6702
4	.3073	.6070	.2040	14.4090	.9844
5	.4422	.1379	.0104	.2460	1.2492
6	−.0643	.1094	−.1230	.1725	1.3752
7	.2975	.3719	−.1390	.9627	2.2774

Table 18.4. The interval ranges of the example data

Cases	WC/TA	RE/TA	EBIT/TA	MVE/TD	S/TA
1	(0.2,0.4)	(0.2,0.5)	(0.1,0.2)	(1.0,1.5)	(1.0,1.5)
2	(0.0,0.2)	(0.2,0.5)	(0.1,0.2)	(1.0,1.5)	$(-\infty,0.5)$
3	(0.0,0.2)	(0.0,0.2)	$(0.2,+\infty)$	(0.5,1.0)	(1.5,2.0)
4	(0.2,0.4)	$(0.5,+\infty)$	$(0.2,+\infty)$	$(3.0,+\infty)$	(0.5,1.0)
5	(0.4,0.6)	(0.0,0.2)	(0.0,0.1)	$(-\infty,0.5)$	(1.0,1.5)
6	(−0.2,0.0)	(0.0,0.2)	(−0.2,−0.1)	$(-\infty,0.5)$	(1.0,1.5)
7	(0.2,0.4)	(0.2,0.5)	(−0.2,−0.1)	(0.5,1.0)	(2.0,2.5)

Table 18.5. The list of interval ranges and descriptors

	WC/TA	RE/TA	EBIT/TA	MVE/TD	S/TA
1	$(-\infty,-0.4)$	$(-\infty,-0.9)$	$(-\infty,-0.4)$	$(-\infty,0.5)$	$(-\infty,0.5)$
2	$(-0.4,-0.2)$	$(-0.9,-0.5)$	$(-0.4,-0.3)$	$(0.5,1.0)$	$(0.5,1.0)$
3	$(-0.2,0.0)$	$(-0.5,-0.2)$	$(-0.3,-0.2)$	$(1.0,1.5)$	$(1.0,1.5)$
4	$(0.0,0.2)$	$(-0.2,0.0)$	$(-0.2,-0.1)$	$(1.5,2.0)$	$(1.5,2.0)$
5	$(0.2,0.4)$	$(0.0,0.2)$	$(-0.1,0.0)$	$(2.0,2.5)$	$(2.0,2.5)$
6	$(0.4,0.6)$	$(0.2,0.5)$	$(0.0,0.1)$	$(2.5,3.0)$	$(2.5,3.0)$
7	$(0.6,+\infty)$	$(0.5,+\infty)$	$(0.1,0.2)$	$(3.0,+\infty)$	$(3.0,3.5)$
8			$(0.2,+\infty)$		$(3.5,+\infty)$

18.5 The Application of the Ants Algorithm in the Classification Problem

In order to apply the ants algorithm for the proposed problem in this chapter, the objective function should be constructed first. According to the feature of the classification problem, we can calculate the objective function in the following way.

18.5.1 Constructing the Objective Function

In the training samples, the non-bankrupt example set is called the positive example (PE) set. At the same time, a set made up of bankrupt examples is called the Negative Examples (NE) set. Both PE and NE are the components of the total example set $E(E = PE \cup NE)$. Suppose F is a logical classification rule. If $\forall e_j^+ \in PE$, $F(e_j^+) \rightarrow True$, then the concept described by F is the completeness for set E. If $\forall e_j^- \in NE$, $F(e_j^-) \rightarrow False$, then the concept described by F is consistent for set E. Thus, for a logical rule F of the set E, the measure of its completeness and consistency can be computed as follows:

1) Completeness: $C_{cp}(F) = |F|/|PE|$, where $|F|$ is the number of those in the positive set which are content to $F(e_j^+) \rightarrow True$ or 1, and $|PE|$ is the total number of the positive set.

2) Consistency: $C_{cs}(F) = |F|/|NE|$, where $|F|$ is the number of those in the negative set which are content to $F(e_j^-) \rightarrow False$ or 0, and $|NE|$ is the total number of the negative set.

3) Simplicity: Obviously, $C_{cp}, C_{cs} \in [0, 1]$, $C_{cp} = 1, C_{cs} = 1$ means that the concept C is content to the completeness and consistence on E. For a given concept C, the simplistic equation of C is:

$$C_{sp} = |C|/\sum_{j=1}^{n} m_j = L/\sum_{j=1}^{n} m_j \tag{18.6}$$

Here, L is the length of the partition rule; $\sum_{j=1}^{n} m_j$ is the total number of all the elements that can be selected to form the rule. Obviously, the less C_{sp} is, the simpler the rule will be. In order to obtain consistency with the above two characteristics, C_{sp} can be described as:

$$C_{sp} = (\sum_{j=1}^{n} m_j - |C|)/\sum_{j=1}^{n} m_j = (\sum_{j=1}^{n} m_j - L)/\sum_{j=1}^{n} m_j \tag{18.7}$$

The integrated estimation rule of a concept is produced by comprehensively thinking of the completeness, the consistency, and the simplicity:

$$C_{Total} = aC_{cp} + bC_{cs} + cC_{sp} \tag{18.8}$$

Here, a, b, and c represent the weight factors. These factors are used to judge the performance of rules [18.22].

18.5.2 The Description of the Application

For the proposed problem in this chapter, using the ants algorithm is done to find a classification rule, which consists of the intervals. Thus, we consider every *interval* as a *site* to which appears in the traditional ants algorithm. We can then implement the algorithm to make the intelligent agent (ants) find the combination of the intervals to form the classification rule. However, as mentioned earlier, there is no meaningful physical distance among intervals. The visibility measure defined in the TSP can no long be applied to our problem. We, therefore, propose a measure of visibility as follows:

$$\eta_{ij} = \frac{1}{\mid Tabu(j) \mid}, \tag{18.9}$$

where $\mid Tabu(j) \mid$ is the number of ants which include the site j into their tabu list in the previous tour.[1] We can then divide the implementation of the algorithm into three steps, and the detailed descriptions are given as follows:

Step 1. Initialization. Set the initial values of the trail on the sites. There are two methods to initialize the trail value. One is to set arbitrarily-chosen values (a very low value on every site). Another method is attributed to the ant's random selection. In this chapter the second method is adopted. After putting an ant on every site, they will travel to all the sites and randomly select which site should be added into the rule. One can take the k-th ant for an example; first, it randomly selects a site from all its neighbor sites as its next site and produces a random number. If the random number is greater than 0.5, then the site is put in the rule and pheromone is left on the sites. After all ants complete their travel, the initialized process is finished.

[1] Since this is not a TSP, the ant do not necessary visit all sites, as we shall see in the description of **Step 2**.

Step 2. Moving of the ants. After the process of initialization, the implementation of the algorithm begins. Still taking the k-th ant as an example, it randomly selects a site j from all the sites as its next moving site. The transition probability, $P_{ij}(t)$, is then calculated to judge whether site j is put into the rule. Next, another site is judged. When the ant has completed its travel, the new rule is produced. In one cycle of the algorithm, every ant produces its own rule. Finally, from the corresponding objective function, the change of pheromone and the visibility of every site can be worked out. We can hence select a rule whose corresponding function value is highest as a good solution in the current cycle.

Step 3. Iteration Process. Repeat Step 2 such that all the ants have the same solution. Set the rule corresponding to this solution as the classification rule (or stop the iteration at a certain generation and select a best solution from the stored good solutions as an approximate global optimal solution of the problem).

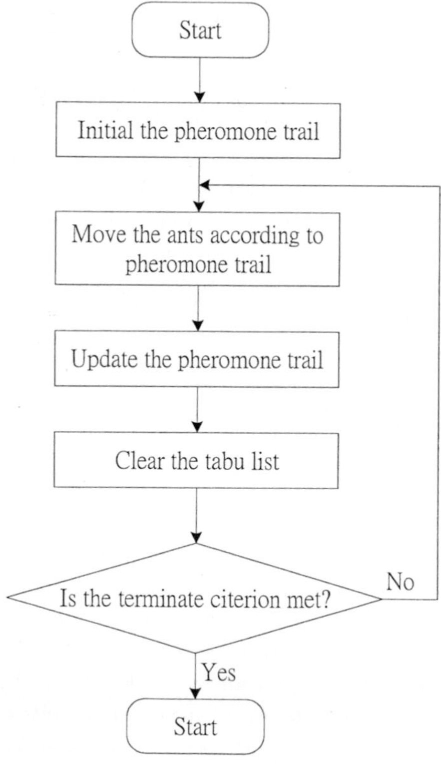

Fig. 18.3. Ants algorithm implementation

18.6 Implementation

From the previous discussion we know that the ants algorithm is in fact an intelligent search procedure that in some sense imitates the behavior of ants and applies some rules based on artificial intelligence principles. In this section we adopt MAA to the business failure prediction problem and Figure 18.3 shows the flow chart of the algorithm.

18.6.1 The Parameters of the Ants Algorithm

The main characteristics of the ant algorithm applied in this chapter have not been changed from the original algorithm, and so the related parameters applied are still its original parameters. This proves that the selection of the parameters is reasonable by several experiments. Following the paper [18.20] by Colori et al., we set the number of ants equal to that of the sites, where the value scopes of parameters α, ρ, and β are respectively $\alpha \in \{0, 0.5, 1.2\}$, $\rho \in [0.0, 1.0]$, and $\beta \in \{0.5, 1, 2, 5, 10, 20\}$.[2] According to our experiments' results, the optimal values of these two parameters are set to be $\alpha = 1$ and $\rho = 0.7$.

18.6.2 The Partition Rule and Results

By MAA we can get the classification rule which can be represented by interval ranges: $(0.0, +\infty) \wedge (0.2, +\infty) \wedge (0.0, +\infty) \wedge \big((1.0, 1.5) \vee (2.0, +\infty)\big) \wedge \big((0.5, 2.0) \vee (3.5, +\infty)\big)$. When we perform a judgment by the rule, an example will belong to good examples if four or five of its financial ratios fall into the rule.

We can testify to the obtained rules by the data given in the training sets. For the non-bankrupt group, there are three misclassifications by the rule and for the bankrupt group there are two misclassifications. For the testing sample set, there are four misclassifications of a failed firm into the non-failed group and also four, otherwise.

In this section we apply the ants algorithm to the sample sets and compare its classification ability with the GP, RPA, and DA. The classification results in the training and testing sample of the three models are shown in Table 18.6.

For the MAA, the accuracy of the classification of the training samples is not very high. However, a better result is obtained for the testing samples using MAA than with other methods. This indicates that the generalization ability of the MAA in this chapter is satisfied.

---

[2] ρ is the evaporation parameter.

Table 18.6. The results of the models

	Training Sample			Testing Sample		
Model	Type-1 Error	Type-2 Error	Classification Rate	Type-1 Error	Type-2 Error	Classification Rate
MAA	3(4.0%)	2(2.7%)	69(93.2%)	4(7.27%)	4(7.27%)	47(85.5%)
OAA	4(5.4%)	4(5.4%)	66(89.2%)	5(9.09%)	4(7.27%)	46(83.6%)
GP	1(1.4%)	2(2.7%)	71(95.9%)	5(9.09%)	5(9.09%)	45(81.0%)
RPA	1(1.4%)	1(1.4%)	73(97.3%)	7(12.7%)	5(9.09%)	43(78.2%)
DA	6(10.8%)	3(4.1%)	65(87.8%)	11(20.0%)	3(5.45%)	41(74.5%)

Type-1 error is identified as that when misclassifying a failed firm into the non-failed group. Type-2 error means a non-failed firm is classified into the failed group. The classification rate is the correct classification rate.

MAA is the modified ants algorithm; OAA is the original ants algorithm; GP is genetic programming; RPA is the recursive partition algorithm; DA is the discriminant analysis.

18.7 Discussions

18.7.1 Classification Accuracy

From the comparison results in Table 18.6, the total classification rate of MAA outperforms the GP, RPA, and DA for the testing samples. In general, the ants algorithm raised the total classification accuracy by 4.5% compared to GP, 7.3% compared to RPA, and 11% compared to DA.

The classification accuracy can also reflected in two types of classification errors made by the model. For the testing samples, the ants algorithm reduced the type-1 errors by 1.82%, 5.43%, and 12.7% from GP, RPA, and DA, respectively, and the ants algorithm reduced the type-2 errors by 1.82% compared to both RPA and GP. The comparison results indicate that the modified ants model outperform the other three models in both type-1 and type-2 errors. The importance of reducing a type-1 error in the business failure prediction domain is surely a valuable advantage of the ants algorithm in the practical application.

18.7.2 External Validity

The external validity of the ants algorithm can be discussed in these aspects: the relative small difference of the classification accuracy between the training sample and the testing sample. This can be seen from the experiment, where the RPA and GP can reach a very high classification rate (97.3% and 95.9%) in the training samples, but in the testing samples the classification rate falls by a big margin, which is about 19.1% and 14.8%. The ants algorithm also results in a high classification rate on the training samples, while in the testing sample, there is only a reduction rate of 7.7%. It is obvious that the ants algorithm has a strong generalization ability.

18.7.3 Explanatory Ability

The partition rule found by the ants algorithm in fact means an interval combination, which includes all the data characters of non-bankrupt groups. In the MAA, such an interval combination is just used to perform a classification and makes the resulting model easier to understand.

The "If ... Then" interpretation of the ants algorithm in experiment can be described as follows:

$Classifier = 0$ {$Classifier$ is a counter variable. If its value is greater than 3, then the current sample can be considered as an unfailed firm.}

If $X_{1,j} > 0.0$ then $Classifier = Classifier + 1$
  If $X_{2,j} > 0.2$ then $Classifier = Classifier + 1$
    If $X_{3,j} > 0.0$ then $Classifier = Classifier + 1$
      If $1.0 < X_{4,j} \leq 1.5$ or $X_{4,j} > 2$ then $Classifier = Classifier + 1$
        If $0.5 < X_{5,j} \leq 2.0$ or $X_{5,j} > 3.5$ then $Classifier = Classifier + 1$
        End if
      End if
    End if
  End if
End if

If $classifier > 3$ then j-th firm belongs to the non-bankrupt group. $X_{1,j}, X_{2,j}, X_{3,j}, X_{4,j}$, and $X_{5,j}$ represent values of the five financial ratios of the j-th firm.

18.7.4 Compared with the Original Ants Algorithm

In this chapter we also use the original ants algorithm to deal with the proposed problem, and we list the results in Table 18.6. The efficiency of the original ants algorithm and the modified algorithm is compared and two conclusions are obtained. Firstly, MAA is faster than the original ants algorithm in finding satisfactory solutions of the proposed problem. MAA takes 15 minutes to find its satisfactory solutions on an Intel 750 PC, while the original ants algorithm takes 22 minutes on the same computer. Thus, the MAA can save 31.8% CPU time. Secondly, better solutions can be found by MAA than by the original algorithm. From Table 18.6 we can see that MAA can reach a higher classification rate than the original ants algorithm. In the training samples, the classification rate of MAA increases by 4% to that of the original ants algorithm, while at the same time the rate rises by 1.9% in testing samples. From the above discussion, it is obvious that the MAA has its advantages over the original ants algorithm.

18.7.5 The Reduction of the Algorithm's Parameter

The elimination of the algorithm's parameter, β, makes the algorithm simpler and reduces the sensitivity of the algorithm to parameters. From computation experience we find that users can control the searching process more easily in MAA than in the original algorithm.

The ant system simply iterates a main loop where m ants construct in parallel their solutions, thereafter updating the trail levels so as to deal with the optimization problem. Although the optimal ability of the ants algorithm is not sensitive to the parameters, the proper values of parameters (such as α, relative importance to trail and attractiveness, ρ, trail persistence, $\tau_{ij}(0)$, initial trail level, and m, number of ants) will lead the algorithm to find high quality solutions at a low cost.

18.7.6 Dynamic Update Mechanism

We find that a dynamic update mechanism provides a great contribution in speeding up the search procedure and improving the quality of the solutions. Our computational experience reveals that this method performs better than the original method that adopts the static update mechanism, and the searching time for optimum solutions is reduced by at least 10%. Moreover, such a mechanism can help the algorithm avoid stagnation when finding the satisfactory solution.

18.7.7 The Termination Criterion

The termination of the ants algorithm is controlled by the uni-solution ability of the ant colony (if all optimizing agents achieve the same solution, then the algorithm should be stop). Using this termination criterion, the ants algorithm can find the global or nearly global solutions of the problem.

18.7.8 Computation Efficiency

The common feature of the ants algorithm is its emphasis on "imitating life" by subsymbolic information processing. Mimicking some rules of that "game of life" has led it to powerful optimum-seeking algorithms, and it is a heuristic and implicit parallel algorithm. Because we introduce the discretisation of the data sets, every financial ratio of MAA only has 7 to 8 values in the training procedure while such ratios from other methods have 74 values. Therefore, the seeking procedure of MAA is more efficient. At the same time, the MAA is very simple and the realization of MAA is easier than that of other methods.

18.8 Conclusion

Basing on the discussion of the characteristics of the business failure prediction problem and the main limitations of the exiting method, we have proposed a modified ants algorithm (MAA) in this chapter. As an effective tool in the domain of combinatorial optimization, the algorithm has the advantage in that it creates a potential method to tackle the difficulties of the business failure prediction. Some modifications are also done in the original algorithm to make it a better fit for this specific problem. Empirical results show that MAA is superior to the original ants algorithm and other existing business failure prediction methods. Furthermore, the computational efficiency of the algorithm is improved.

Acknowledgements

The authors are grateful to Chung-Chih Liao and Li-Cheng Sun for converting our chapter into the LaTeX format. The chapter was refereed by Chung-Chih Liao and Cheuh-Yung Tsao. The authors are grateful for their careful review of the draft, and we benefit very much for many of their constructive suggestions. Finally, the English of the chapter has been polished by the book editor, Shu-Heng Chen. His tremendous efforts to making this chapter readable is greatly acknowledged.

References

18.1 Altman, E. I. (1968): Financial Ratios, Discriminant Analysis and the Prediction of Corporate Bankruptcy. Journal of Finance, 23, 589–609
18.2 Altman, E. I. (1993): Corporate Financial Distress and Bankruptcy. John Wiley
18.3 Altman, E. I., Hadelman, R. G., Narayanan, P. (1977): Zeta Analysis, a New Model to Identify Bankruptcy Risk of Corporations. Journal of Banking and Finance, 1, 29–51
18.4 Altman, E. I., Marco, G., Varetto, F. (1994): Corporate Distress Diagnosis: Comparisons Using Linear Discriminant Analysis and Neural Networks (The Italian Experience). Journal of Banking and Finance, 18, 505–529
18.5 Altman, E. I., Avery, R., Eisenbeis, R., Stinkey, J. (1981): Application of Classification Techniques in Business, Banking and Finance, Contemporary Studies in Economic and Financial Analysis, Vol. 3. JAI Press
18.6 Deakin, E. B. (1972): A Discriminant Analysis of Predictors of Business Failure. Journal of Accounting Research, Spring, 167–179
18.7 Warner, J. B.(1977): Bankruptcy Costs: Some evidence. Journal of Finance, 32, 337–347
18.8 Dimitras, A. I., Zanakis, S. H., Zopoumidis, C. (1996): A Survey of Business Failure with Emphasis on Prediction Methods and Industrial Applications. European Journal of Operational Research, 90, 487–513

18.9 Wang, C., Kang, L. (2001): GP Decision Tree Method in Business Failure Prediction. Working Paper, Tianjin University

18.10 Moyer, R. C. (1977): Forecasting Financial Failure: A Re-examination. Financial Management, Spring, 11–17

18.11 Taffler, R. J. (1983): The Z-Score Approach to Measuring Company Solvency. Acountant's Magazine, 87, 91–96

18.12 Martin, D. (1977): Early Warning of Bank Failure: A Logit Regression Approach. Journal of Banking and Finance, 1, 249–276

18.13 Ohlson, J. A. (1980): Financial Ratios and the Probabilistic Prediction of Bankruptcy. Journal of Accounting Research, Spring, 109–131

18.14 Odom, M. D., Sharda, R. (1990): A Neural Network for Bankruptcy Prediction. International Joint Conference on Neural Networks, June 17–21, 1990, 163–168

18.15 Messier, W. F., Hansen, J. V. (1988): Inducing Rule for Expert System Development: An Example Using Default and Bankruptcy Data. Management Science, 34, 1403–1415

18.16 Frydman, H., Altman, E. I., Kao, D. -L. (1985): Introducing Recursive Partitioning for Financial Classification: The Case of Financial Distress. Journal of Finance, 51, 26–291

18.17 Dutta, S., Shekhay, S. (1992): Generalization with Neural Networks: An Application in the Financial Domain. Journal of Information Science and Technology, 1, 309–330

18.18 Quinlan, J. R., (1983): Learning Efficient Classification Procedures and Their Application to Classification Games. In: Michalski, R. S., Carbonell, J. G., Mitchell, T. M. (Eds.), Machine Learning: An Artificial Intelligence Approach. Tioga

18.19 Dorigo, M., Maniezzo, V., Colorni, A. (1991): Positive Feedback as a Search Strategy. Technical Report, Dipartimento di Elettronica, Politecnico di Milano, 91–016

18.20 Colorni, A., Dorigo, M., Maniezzo, V. (1992): Distributed Optimization by Ant Colonies. In Proceeding of the First European Conference on Artificial Life. Elsevier Publishing, 134–142

18.21 Colorni, A., Dorigo, M., Maniezzo, V. (1992): An Investigation of Some Properties of an "Ant Algorithm". PPSN 92, Brussels, Belgium, Elsevier Publishing, 509–520

18.22 Li, Minqiang (2000): Study on The Basic Theory of Genetic Algorithms and Its Applications in Knowledge Discovery. Ph.D. thesis, 153–180

19. Towards Automated Optimal Equity Portfolios Discovery in a Financial Knowledge Management System

Yi-Chuan Lu[1] and Hilary Cheng[2]

[1] Department of Information Management, Financial Data Mining Laboratory, Yuan Ze University, Chung-Li 320, Taiwan, R.O.C.
email: imylu@saturn.yzu.edu.tw

[2] Department of Business Administration, Financial Data Mining Laboratory, Yuan Ze University, Chung-Li 320, Taiwan, R.O.C.
email: hilary@saturn.yzu.edu.tw

We propose a knowledge discovery and knowledge management process for equity management institutions. We realize the process with a financial knowledge management system, FKMS, that is a system platform being able to convert various sources of data into the data warehouse, to retrieve data cubes based on different power users' commands for subsequent valuation modeling or data mining applications. We then introduced a data mining solution for equity portfolio construction using the simulated annealing algorithm. Two data sets consist of small stocks ranging from 11/86 to 10/91 and from 6/93 to 5/96 are used. The corresponding rates of return of Russell 2000 index are collected as benchmarks for evaluation based on the Sharpe ratios and the turnover ratios. The result of the simulated annealing algorithm has shown to outperform the market index as well as the gradient maximization method.

19.1 Introduction

Competition in the investment business is intense and increasing. Historically, investment managers who actively select stocks employ a team of analysts who understand various sectors of the industries and visit companies in order to learn important information in helping them to recommend which stocks to own. During the past 10 years, as computer power has increased and the cost has declined, some managers have started the employment of the quantitative techniques in evaluating stocks. Quantitative techniques involve using the mathematical and/or statistical methodologies to determine the relationships between stock characteristics and returns from historical data. These historic relationships or some models provide important information for predicting future stock returns.

Today, with the continuing surge of reduced pricing, higher-performance computer hardwares, coupled with the virtual explosion of data availability and the Internet communication, quantitative techniques are becoming more important than ever to quickly filter and translate substantial amount of data into meaningful information for investment decision making. While the Internet technology and data availability is rapidly changing, quantitative techniques also have become much more sophisticated. The data warehousing and data mining techniques are thus widely acquired across the financial services sectors and banking industry in general. Institutional investors or stock brokerages further take the advantages of comprehensive data warehousing techniques and decision methods to fine-tune their investment or trading strategies for more prudent risk management, improved efficient customer service, or keen profitability analysis.

Section 19.2 gives a motivating example to explore the need of knowledge discovery process for multi-factor equity portfolio management. In Section 19.3, we briefly describe the system architecture. A financial knowledge management system, FKMS, which was developed to provide knowledge workers and power users for easily sharing information as a basic building block for the knowledge discovery process. Section 19.4 details the methodology for data management, the characteristics of financial data, the meta data and schema management for successful data warehousing, the data conversion process, as well as the approaches of data retrieval. In Section 19.5, we propose an adaptive random search technique to select a portfolio that outperforms the market index as well as the gradient maximization method. The simulation processes and experimental results are summarized in Sections 19.6. Finally, more data mining techniques are discussed in Section 19.7 for the potential future applications.

19.2 A Case Study as Motivation

Suppose an institutional investor who wants to construct a portfolio by formulating criteria and designing the strategy. The process of constructing the portfolio involves defining a universe of stocks, dealing with data integrity issues, selecting an appropriate construction technique, determining the way of using training and testing data, setting up the construction rules and constraints, and selecting a publicly available benchmark for evaluating the test results. The whole process is known as the back-testing model in the financial investment society. An example [19.13] in managing a small-stock portfolio exploits the above process. The first is to construct a sequence of monthly universe, from January 1971 to December 1980, by ranking a database on market capitalization, then eliminating the largest 1,000 stocks and using the next 600 excluding limited partnerships, investment trusts and stocks with very low trading volumes. To meet the above requirements, a time series management system should provide the following SQL-like language:

```
select  BOTTOM 600 month (timestamp), ticker, company_name
from  {
        select  TOP 1600 the_end_of_month (market_capitalization)
```

from daily_price
where timestamp >= 1/1/1971, timestamp <= 12/31/1980,
order by the_end_of_month (market_capitalization)
}
where *order by* the_end_of_month (market_capitalization).

The resulting table, say *tmp_universe*, shall then be displayed to allow the investors to eliminate limited partnerships and investment trusts and stocks with very low trading volumes. Since the data file is the *daily_price* table, it is the responsibility of underlying system to keep data integrity. Note that the value of function *month (timestamp)* returns the latest trading date of a given month.

The second step is to include values and other proven measures to the universe. The common factors are earnings to price ratio (E/P), book value to price ratio (B/P), cash flow to price ratio (CF/P), volatility adjusted price momentum (△P/P), four quarters earnings change to price ratio (△ES/P), three-year earning change to price ratio (△EL/P). To do so, a time series management system should provide the following SQL-like language:

select timestamp, ticker, company_name, E/P, B/P, CF/P, △P/P, △ES/P, △EL/P
from daily_price, tmp_universe
where tmp_universe.ticker = daily_price.ticker,
tmp_universe.timestamp = daily_price.timestamp

The rest of work is to apply a decision model to construct monthly portfolios through time and compute the corresponding returns. The resulting returns of our portfolios are then used to compare with the selected benchmark. If the result is acceptable, the resulting knowledge and the model parameters are then stored into the system.

Fig. 19.1 shows the above back-testing process. To optimize the flow of information and "knowledge-worker-to-knowledge-worker" interaction so that companies can make better trading decisions, the specific data and modeling results should be shared and managed. This is the essence of knowledge management.

19.3 FKMS: A Knowledge Management Solution

The key in developing an efficient financial decision support system is to understand how the information flows throughout the organization, and how they are generated or integrated in the decision making process. Fig. 19.2 provides a snapshot about the different stages required to create knowledge within a research department in a security investment firm [19.2].

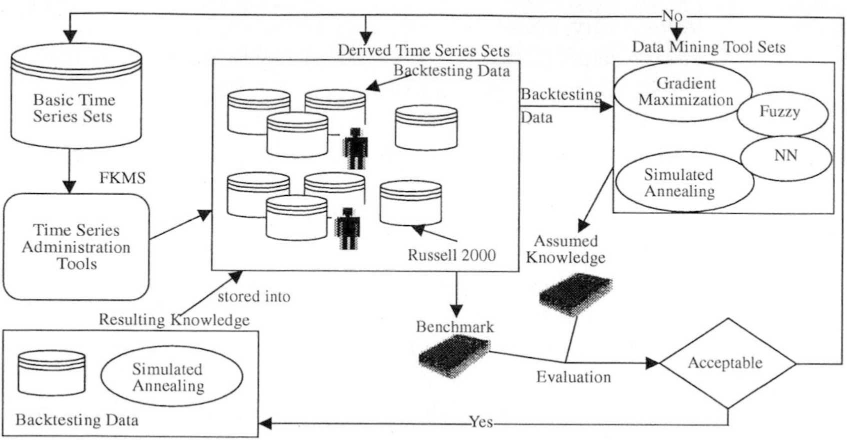

Fig. 19.1. A back-testing model for optimal equity portfolios discovery

The flow of data and shared information go as follows: securities analysts retrieve information they need for conducting top-down analysis including industry analysis and company analysis from the research database. This database integrates both structured and unstructured (either dynamic or static) data from various data sources. The data retrieved by each individual comes along with this person's domain know-how, and will automatically be saved as data cubes. Each data cube records the time this file has been created, the search criterion, the data names and types, and so on, so that colleagues can easily share and exchange their expertise with others. In the meantime, the company will keep good management of each individual worker's knowledge, and be ready for good customer relationship management. As the research results can be delivered through the middleware or via the Internet to the designated customers.

A successful knowledge management means that our data processes enhance the way people work together, enable knowledge workers and partners to share information easily so they can build on each other's ideas and work more effectively and efficiently. The goal is to draw on company proprietary knowledge to come up with the best solutions to investment problems or to quickly seize the initiative with innovative ideas.

We propose a system architecture that consists of 5 layers: the resource layer, the data conversion layer, the data storage and management layer, the knowledge/trend/pattern layer, and finally the user process layer. The system is designed to be object-oriented, COM-based, multi-tiered, parallel processing, scalable, portable and user friendly [19.10]. Since the data cubes are stored in the traditional relational database management system (RDBMS), users can easily divert the data cubes via ODBC/JDBC or XML/Schema for analytical applications at the knowledge/trend/pattern layer. Users using an-

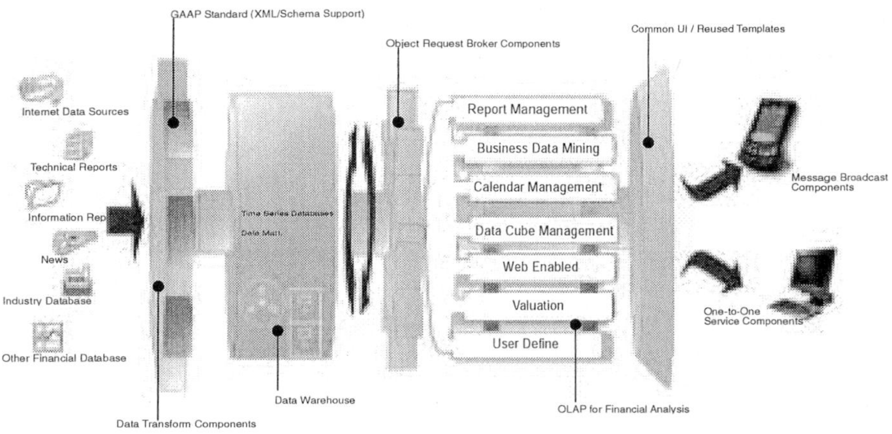

Fig. 19.2. The architecture of knowledge discovery process

alytical application software such as Matlab, SAS, SPSS, or S-Plus can easily import selected data cubes with the OLAP tools developed in the system.

19.4 Data Management

Various resources of data are used by analysts when they write research reports or run valuation models. Typical examples of these data resources are financial databases from foreign data vendors such as Bloomberg, Data Stream, First Call, or from domestic data vendors like the TEJ and the SFI, as well as other reports from competitors, and some periodically published data on the Web sites. While some data are static, meaning that they are periodically released, some are dynamic, which means that they are not periodic. On the other hand, the format of data can be classified as structured, semi-structured, and unstructured. We focus on the management of structured data, as their format is clearly defined so that data operations or manipulations can be deployed.

19.4.1 Defining Metadata

Metadata bridges the gap between data format and the physical data storage. In general, there are three levels of metadata in the data warehouse: the operational level, the core level, and the user level [19.4]. In the aggregate planning of the metadata for financial domain, we take into consideration with the following aspects for implementation [19.11]: the attributes nd the key of tables, the resources of tables, algorithms that may be used for data conversion, the update of metadata, etc.

19.4.2 Data Conversion

The next step after metadata definition is data conversion, which consists of four phases: creating the data schema, listed company management, mapping rules for conversion, and finally the conversion process. For the purpose of being expandable, scalable, and portable, the ontology of the knowledge abstraction was represented with XBRL schema [19.7]. The storage of data schema was managed by two tables: Schema_Field and Schema_NameSpace. Where there are data items with same names (such as price to book value, P/B) but from different databases (for example, Data Stream and Bloomberg, Schema_NameSpace gives a clear understanding of data sources. On the other hand, Schema_Field keeps good management of all data items in the data warehouse. Due to the increasing number of listed companies in the database, we provide a management function, which can automatically update when companies either are out of the market or change their codes. The rules mapping system not only allow the users to create relationship in between the new data source and the predefined data items, but also create mapping rules between the new data item with the same value from the ElementSourceCode in the existing schema. The data conversion process ends with the management of values of all data items, company details and the data items.

The knowledge management life cycle starts with the creation and validation of knowledge, collection and classification of knowledge, knowledge sharing, efficient informational retrieval, and ends with the knowledge improvement or the abandoning of knowledge on the extreme. The online analytical processing component for financial analysis (FOLAP) was designed to serve as an administrator to monitor different levels of access to the data sources for creation of data cubes based on power user's domain knowledge. The term, data cube, was named because of its property of three dimensions, namely, time, company, and data items.

19.4.3 Data Cube Creation

The creation of data cube takes 5 steps described as follows: Select Root Data Cube; Select Items; Screening; Add/Delete Qualified Companies; Final Review. First of all, whoever generates the new knowledge will need to define where the new knowledge comes from, and that is to define who the Root data cube(s) is (are). Once the root data cube(s) is (are) defined, the power user will then select groups of items, which will be used for analysis later. These items will be retrieved from the database once requested, and can be downloaded to either Word, Excel, or data mining tools for further

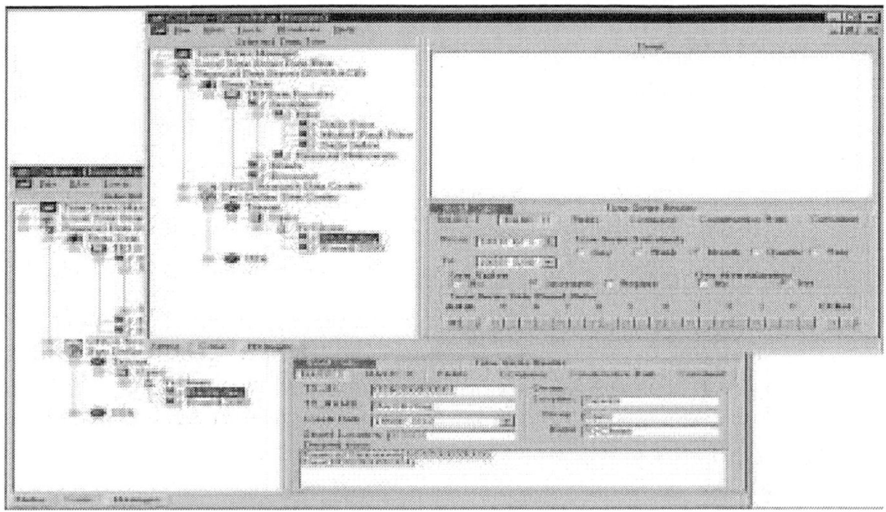

Fig. 19.3. FKMS – A TSMS for financial decision support

analysis, or simply to the Website for snapshot purpose. However, the determination of whose company data are qualified depends on the result of screening. The screening criteria can simply be a query, such as to select shares of companies with market capitalizations between 50 and 500 million, known as small-cap stocks. The screening can also be complicated, such as a set of criterion for stock portfolio selection, or for merger/acquisition purposes. Finally, the power user will review the set of qualified companies and add or delete companies based on her (his) domain expertise. FKMS allows users to save qualified securities (companies) for future uses, it also provides an option for private use or sharing with others. Here we show a few data cubes generated for specific portfolio construction purposes: companies at the turning point, and undervalued stocks.

- Companies at the turning point: To select the worst 25 percentile of companies with low operating margin last year, then query on the sales growth last quarter in descending order, output the first 25% companies.
- Undervalued stocks: To take the average of last 2 EBIT growths, discounted by 90%, then calculate the difference between the former and the projected P/E ratio, the higher the better.

Fig. 19.3 provides a snapshot of the FKMS graphic user interface (GUI).

19.5 Knowledge Discovery via Data Mining Technique

In this section, we will give an introduction about the problem domain followed by the definitions of the objective function and problem constraints.

After constructing a data set for back testing, a data mining technique is selected for constructing optimal portfolios as the process shown in Fig. 19.1. The search algorithm, namely, the simulated annealing, including the state generating function, acceptance criterion, annealing schedules are introduced accordingly.

19.5.1 Multifactor Equity Portfolio Management Issues

Building a multi-factor excess return model to select a portfolio stocks has become a widely used tool for portfolio management. The decisions of which return factors should be included in such a model vary widely from practitioners to practitioners. The common characteristic in most models is that the factor weightings are determined by linear regression. Linear regression approach is to rank stocks by expected returns and then select a portfolio from among the highest ranked stocks. The critical characteristics of regression model related to portfolio construction are that regression produces a model that is a "best" historic fit to the returns of each stock. There is no measure of portfolio return or risk to be considered in determining the regression coefficients. In addition to the portfolio risk, restrictions on factor weights, portfolio structure, transaction costs, or turnover cannot be easily incorporated in determining the regression model. In this paper, we consider the portfolio construction model as a global optimization problem as follows:

$$\max_{\theta \in \Theta} F(\theta) = E(f(\theta, \omega) | \theta \in \Theta), \tag{19.1}$$

where $f(\theta, \omega)$ is a stochastic function of unknown parameters θ (e.g. financial factors) defined on a closed subset θ on an n-dimensional space, and ω is the factor's weightings. No assumptions about the convexity of $F(\theta)$ are made, therefore, it may have several local maxima. Some information may or may not be available about the smoothness of F, values and smoothness of derivatives, *a priori* bounds on $F^* = \max(F)$, number of local optima, etc. To control risk throughout a portfolio construction process, we consider the model's objective function as to maximize the return over risk. The optimization algorithms to be examined in this paper are gradient maximization [19.1] and simulated annealing [19.8]. The model is constructed using a three-year "moving window" of stock market information. The model is then tested on the next year.

19.5.2 Objective Function and Constraints

We use the Sharpe ratio [19.14] to illustrate the concerns of portfolio managers who seek for high returns with low risks. Let p, m, r, and s stand for portfolio, market, return, and risk, respectively. The Sharpe ratio is generally defined as $(r_p - r_f)/\sigma_p$, where r_p stands for the average rate of return for the portfolio p, σ_p stands for the risk of portfolio p, and r_f is the rate of return on risk-free assets. The Sharpe excess return to the market [19.1] becomes

$$(r_p - r_m)/[\sigma_p^2 + \sigma_m^2 - 2cov(p, m)]^{1/2}, \tag{19.2}$$

where r_m and σ_m respectively stand for the average rate of return and the risk for the market m, r_m and σ_m^2 are constants, and finally, $cov(p, m)$ is the covariance between the return on the portfolio and the return on the market. In order to control risk throughout the construction process, we include the following constraints:

1. All factors weights are zero or positive real numbers.
2. Economic sector weightings within a portfolio are proportional to the selected benchmark's economic sector weightings.
3. Dividend and uninvested cash are reinvested.
4. There is a minimum or maximum weight on any one or all factors.
5. Transaction cost is considered in percent on every trade.
6. There is a fixed range of total number of stocks and amount in any portfolio.
7. A stock is bought only if it ranks above the buy threshold in any economic sector.
8. A stock is sold if it ranks below the sell threshold in any economic sector.
9. Individual holding size can not exceed an upper bound.
10. There are limitations on the maximum number of lot sizes for each economic sector.

19.5.3 Adaptive Random Search

We propose an adaptive random search technique – the simulated annealing algorithm, which can be used in conjunction with the above integrated return/risk portfolio model. The simulated annealing algorithm originated from the analogy of finding a globally minimal configuration in a combinatorial optimization problem. The key to ensuring that the system can escape from local minima lies in the use of "noise"to a probabilistic hill-climbing algorithm. As with all physical systems, the intuitive notion is that the atoms tend toward minimum energy state. At high temperatures the vigor of the atomic motion prevent this, but as the metal is gradually cooled, lower and lower energy states are assumed until finally the lowest of all possible states, a global minimum, is achieved. The cooling phase of the annealing process is that it is allowed to reach thermal equilibrium at a given temperature. For each given temperature, the thermal equilibrium can be described by a probability distribution function with respect to the occurrence of a state with energy E. This distribution is known as the Boltzmann distribution having the form similar to the exponential density function:

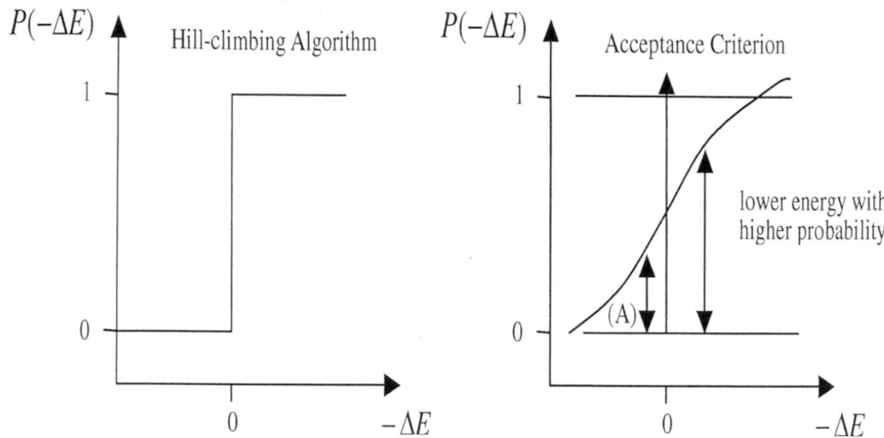

Fig. 19.4. The traditional hill-climbing algorithm only accepts E when it is less than zero, while the simulated annealing algorithm allows positive E being accepted with probability greater than 0

$$P(E) = (1/Z(T))e^{-E/(K_B T)} \tag{19.3}$$

where $Z(T)$ is a normalized constant, E is the energy of the state, $P(E)$ is the probability of finding a unit in that state, T is the temperature, and K_B is the Boltzmann's constant.

From Fig. 19.4, we see that adding the noise due to temperature effects allows jumping up the energy values, the "(A)"represents a nonzero probability for a higher energy state. In addition, the greater the energy differences between one state and a possible subsequent state, the greater the probability of transition if the second state has a lower energy. Since the Sharpe excess return could have either positive or negative returns, we write the dynamic rule in a symmetrical form

$$E = \frac{1}{1 + e^{\text{SharpeExcessReturn}}} \tag{19.4}$$

which is referred as an energy measure.

There are three steps involved in the simulated annealing process. The first step is the generation of a new state by means of a random process covering all state space. The second step is the acceptance criterion of the new state, based on the energy difference between the new and the old states. The last step will be the condition of the cooling schedule that includes the choice of the initial temperature, the equilibrium detection condition at a particular temperature, the decrease rate of the temperature, and the termination condition.

State-Generating Function. Generating a new state for the purpose of portfolio optimization is to explore the current state region around the current weighting combination. This is a process of the local search. By sequentially changing each factor's weight up and then down a bit from the original weighting combination, local search determines a combination of factor changes that identifies a best direction of improving the portfolio return. Having established a direction of local search, we next make a bigger step to weight changes in the portfolio. Every time we change the weights, we repeat the portfolio simulation using the new weighting combination and compute the new energy as well.

The energy difference between the new state and old state is of interest here because the $\triangle E$ is regarded as the magnitude of the random component of the step size which must decrease in a statistically monotonic fashion such that the simulated annealing algorithm is guaranteed to converge. These steps indeed can be regarded as a Brownian motion of a particle in physical annealing. Based on the central limiting theorem, any random variable with a bounded variance will approach the Gaussian distribution in the large sampling limit. When the second moment is taken, the Gaussian density gives the value of the temperature. Based on this observation, the annealing schedule can be conducted adaptively.

Acceptance Criterion. A new combination weighting that gives a new energy level will be used as a basis of our acceptance criterion. In general, this acceptance criterion is often referred as the Metropolis algorithm. That is, when $\triangle E$ is less than zero, we accept the new state with probability 1. If $\triangle E$ is greater than zero, then we accept the new state with the probability $e^{-\triangle E/T}$, where T represents the current temperature. To decide whether the new state is acceptable, we simply generate a random uniform deviate ranging from 0 to 1 to compare with $e^{-\triangle E/T}$. If the random number is less than $e^{-\triangle E/T}$, we accept this new state, otherwise reject it.

Annealing Schedules. The annealing schedule is critical to the performance of the algorithm. Without any knowledge of energy landscapes, one can only hope to derive an appropriate cooling schedule for a specific random process. As being a cooling schedule of the temperature $T(t)$, it must be able to decrease from a given sufficiently high temperature T_0 down to a zero degree. For a Gaussian random process, [19.6] has provided a cooling schedule of the temperature, $T(t)$, according to the inverse logarithmic formula: $T = T_0/\log(1+t)$.

There are four important criteria in the cooling process. They are the choices of initial temperature, the equilibrium detection condition at a particular temperature, the decreasing rate of the temperature and the frozen or the termination condition. We introduced an estimation method based on the known Gamma density for estimating the initial temperature.

Recall that the thermal equilibrium with a certain temperature can be detected from the probability of occurrence of a state with energy E, where the probability is given in equation (19.3). If the energy function is measured by the second moment of a random variable $U(X)$, where X is an arbitrary portfolio, $U(X)$ represents the energy of X, then this distribution can be regarded as a Gamma distribution :

$$P(E) = \lambda^r / \Gamma(r) X^{r-1} Z(T) e^{-\lambda E} \tag{19.5}$$

setting the degree r to 1, the Gamma distribution specializes to the exponential distribution with the variance $(1/\lambda^2)$:

$$P(E) = \lambda e^{-\lambda E} \tag{19.6}$$

Now, let T be $1/\lambda$, the standard deviation of exponential distribution is equivalent to the temperature T. To experiment an initial temperature, we first let the system be free running with a 100% acceptance rate for a certain number of iterations. By sampling all energy states, we can calculate the standard deviation so as to estimate an initial temperature. We implemented the cooling schedule within our proposed model using the following cooling formula [19.15]:

$$T(t) = \frac{T_0}{1+t} \tag{19.7}$$

where T_0 is the initial temperature. And, the equilibrium detection during a particular temperature will be measured by:

$$\frac{\mid T_{est} - T_c \mid}{T_c} < 0.01 \tag{19.8}$$

where T_{est} is the estimated temperature from N samples and T_c is the current temperature.

19.6 Implementation and Results

We will focus on the universe, which is designed to capture the attractive long-term return potential associated with small stocks. Typically, the universe will consist of all stocks in the Russell 2000 and S&P 500 SmallCap indices plus all domestic stocks between the high and the low market capitalization ranges of these two indices. The median market cap of the portfolio approximates $320 million. The portfolio is constructed by the proposed model to rank stocks from the most to the least attractive within each of 10 economic sectors. We only purchase a stock when it ranks above 10% in any economic sectors. A stock will be sold when it ranks below the 30% by economic sector. The round-trip trading cost is 3.6%. The total number of stocks in our portfolio is between 50 and 60.

19.6.1 Portfolio Simulation

We focus on identifying undervalue stocks – stocks with low or attractive valuations that have solid or improving businesses – by using price-to-earning ratio and price-to-book value ratio. The rationale behind value investing is that investor can benefit over time from buying stocks at discounted price. Simply buying the stocks with the lowest valuations has two important advantages. First, portfolios of stocks with low valuations have provided higher returns than the broad market over the long term. Second, prices of these stocks tend to decline less in down markets and fluctuate less than stocks with high valuations, resulting in reduced risk for the investor. However, a pure value strategy also involves some disadvantages. Low valuation can indicate deteriorating business fundamentals or failing companies. In addition, low valuation stocks may be recognized by the market for some time, holding back their good performances. Many stocks with the lowest valuations often represent a few industry groups or the same economic sector resulting in large portfolio bets, which increase risk.

In contrast, investment managers and investors have also found that growth stocks, shares of companies with above average or accelerating business growth, can also provide high returns. Since these companies have demonstrated earning growth, the stocks typically have higher valuations. Consequently, pure growth approaches involve (1) paying high valuations for stocks and the risk that a company's business may slow and disappoint the marketplace, (2) intensive monitoring of each company's progress by Wall Street, and (3) concentrated, high risk portfolio (see [19.13] for more details).

Based on the above analysis, we choose two widely used value measures and two widely used growth measures to form an integrated return/risk portfolio. The value measures are earning to price ratio and book value to price ratio, whereas the growth measures are short-term (four quarters) earning change to price ratio and long-term (three-year) earnings change to price ratio.

19.6.2 Results

Following the above methodology, we have collected two data sets. One from 11/86 to 10/91 and the other from 6/93 to 5/96. The corresponding rates of return of Russell 2000 index are collected as benchmarks. Results using the first and the second data set are presented respectively in Tables 19.1 and 19.2. The evaluation is based on the Sharpe ratio as well as the turnover ratio. The number of epochs used in Table 19.1 and Table 19.2 is 100 times. The initial temperature is estimated by Equ. (19.6). We use the cooling schedule in Equs. (19.7) and (19.8). The number of stocks in a monthly portfolio is around 50. The result shows that the simulated annealing algorithm (SA) outperforms both the gradient maximization (GM) and market index in all time periods. The results also showed the longer the time period, the lower

Table 19.1. Portfolio construction using gradient maximization and simulated annealing

Data Set (11/86 – 10/91)	Year(s)	Portfolio Return	Benchmark Return	Excess Return	Sharpe Ratio	Turnover Ratio
GM	1	-11.31%	-18.04%	6.73%	0.24	8.14%
	2	6.45%	3.21%	3.14%	0.18	8.01%
	3	11.73%	8.64%	3.09%	0.18	7.48%
	4	1.61%	-0.08%	1.69%	0.09	6.75%
	5	8.02%	6.97%	1.05%	0.05	6.16%
SA	1	-9.95%	-18.04%	8.08%	0.32	8.97%
	2	7.21%	3.21%	4.01%	0.26	9.20%
	3	13.06%	8.64%	4.42%	0.23	10.43%
	4	2.01%	-0.08%	2.09%	0.11	11.31%
	5	8.64%	6.97%	1.67%	0.09	12.06%

the Sharpe ratio, which indicates the risk is proportional to the time period. Although, the simulated annealing has a superior rate of returns, it has the highest turnover ratio.

Table 19.2. Portfolio construction using gradient maximization and simulated annealing

Data Set (11/86 – 10/91)	Year(s)	Portfolio Return	Benchmark Return	Excess Return	Sharpe Ratio	Turnover Ratio
GM	1	36.51%	4.41%	32.10%	1.06	8.67%
	2	19.31%	11.97%	7.34%	0.26	7.74%
	3	20.62%	15.81%	4.81%	0.23	6.88%
SA	1	36.11%	4.41%	31.69%	1.06	7.96%
	2	20.92%	11.97%	8.95%	0.35	8.41%
	3	24.03%	15.81%	8.22%	0.32	9.10%

19.7 Knowledge Discovery – with Other Data Mining Applications

As mentioned in Sections 19.3 and 19.4, cubes created by the power users can further be applied to statistical packages for the sake of studying causal relationships, or to artificial intelligence software such as the neural networks for clustering or mining purposes. Other Applications implemented on the FKMS include corporate credit ratings using SOFM and LVQ clustering techniques [19.2], as well as detecting potential financial crisis or projecting earnings per share via Bayesian network ([19.3],[19.9]). With the FKMS, analysts can always efficiently create and retrieve data cubes to create new knowledge, to share domain expertise with colleagues for the creation of new knowledge. Various earnings models were also implemented on the FKMS platform where investment analysts use lots of predefined cubes for corporate earnings projection. Other financial application such as corporate financial planning and forecasting models can also be implemented in the FKMS, where new schema must be defined and cubes must be generated for different department's planning and forecasting, and overall corporate aggregated planning purposes. The potential benefits to be derived from novel application indeed is quite rich.

Acknowledgements

We appreciate Prefessor Shu-Heng Chen for valuable comments and suggestions. We would also like to acknowledge the help and effort of Dr. Chia-Hsuan Yeh.

References

19.1 Brush, J. S., Schock, V. K. (1995): An Integrated Return/Risk Portfolio Construction Procedure. The Journal of Portfolio Management, summer, 89–98

19.2 Cheng, H., Lu, Y. C., Wang, W. H. (2001): Approaching a Knowledge Discovery Process for Corporate Bond Classification via Neural Clustering Technique. Proceedings of the International Conference on Artificial Intelligence, IC-AI2001, III, 1290–1296

19.3 Cheng, H., Lu, Y. C. (2001): Technical Report - Bayesian Network Applications to Stock Price Analysis, Financial Data Mining Lab, Yuan Ze University

19.4 Data Warehousing Tools Bulletin (1996): What is Metadata. ComputerWire, March 1, 41–50

19.5 Dreyer, W., Dittrich, A., Schmidt, D. (1994): An Object-oriented Data Model for A Time Series Management System. Proceedings of the 7th International Working Conference on Scientific and Statistical Database Management (SSDBMS'94), Charlottesville, Virginia, Sept., 186–195

19.6 Geman, S., Geman, D. (1984): Stochastic Relaxation, Gibbs Distributions, and the Bayesian Restoration of Images. IEEE Trans. on Pattern and Machine Intelligence, PAMI-6, Nov., 721–741

19.7 Hamscher, W., Kannon, D. V. (2000): Extensible Business Reporting Language (XBRL) Specification. XBRL Organization, July 31

19.8 Kirkpatrick, S., Gelatt, C. D., Vecchi, M. P. (1983): Optimization by Simulated Annealing. Science, 220, May, 671–680

19.9 Lu, Y. C., Cheng, H. (2001): Technical Report - Detecting Financial Failures Using Bayesian Networks. Financial Data Mining Lab, Yuan Ze University
19.10 Lu, Y. C., Cheng, H., Sheu, C., Jung, M. (1999): Financial Decision Support System: A Distributed and Parallel Approach. Workshop on Distributed System Technologies and Applications, Taiwan
19.11 Lu, Y. C., Cheng, H., Wang, W. H. (1999): Technique Report - A Time Series Management System for Financial Decision Support: Models, Techniques, and Implementations., Financial Data Mining Lab., Yuan Ze University
19.12 Schmidt, D., Marti, R. (1995): Time Series, A Neglected Issue in Temporal Database Research? Workshop on Temporal Databases, Zurich, Switzerland, Sept., 17–18
19.13 Schock, V. K., Brush, J. S. (1993): Capturing Returns and Controlling Risk in Managing a Small-Stock Portfolio. Small Cap Stocks Investment and Portfolio Strategies for the Institution Investor, Chapter 13, Editors: Robert A Klein and Jess Lederman
19.14 Sharpe, W. F. (1966): Mutual fund performance. Journal of Business, 39 #1, Part 2, January, 119–138
19.15 Szu, H., Hartley, R. (1987): Nonconvex Optimization by Fast Simulated Annealing. Proceeding of IEEE, 75(11), November

Part VII

State Space Modeling of Time Series

20. White Noise Tests and Synthesis of APT Economic Factors Using TFA

Kai Chun Chiu and Lei Xu

Department of Computer Science and Engineering, The Chinese University of Hong Kong, Shatin, N.T., Hong Kong, P.R. China
email: {kcchiu, lxu}@cse.cuhk.edu.hk

When the Temporal Factor Analysis (TFA) model is used for classical Arbitrage Pricing Theory (APT) analysis in finance, it is necessary to perform white noise tests on the residual in order to validate the model adequacy. We carry out white noise tests and obtain results that provide assurance for further statistical analysis using the TFA model. We also explore empirically the relationship between macroeconomic time series and Gaussian statistically uncorrelated hidden factors recovered by TFA. Based on the statistical hypothesis test results, we conclude that each of the four economic time series is linearly related to the uncorrelated factors. Consequently, APT economic factors can be synthesized from those statistically uncorrelated factors.

20.1 Introduction

Since its origination by Ross [20.10] in 1976, the Arbitrage Pricing Theory (APT) has drawn considerable attention of the finance community worldwide. APT relates security returns to a spectrum of several risk factors. These are often called economic factors because in literature time series of macroeconomic variables have been found probable candidates to be used as proxies for the purpose of financial modeling.

Although many arguments have been put forward for the selection of proper proxies, there remains no consensus on what should be the most appropriate proxies for those hidden economic factors over which APT seeks to model. For instance, Estep, Hanson and Johnson have considered possible factors to be changes in inflation, real growth, oil prices, defense spending and real interest rates in [20.3] in 1983. On the other hand, in 1984 Roll and Ross assert that an asset's return being directly related to unanticipated changes in four economic variables and they are inflation, industrial production, risk premiums and the slope of the term structure of interest rates respectively [20.9]. Interestingly, Chen, Roll and Ross in their paper [20.2] two years later change the last two factors to be the spread between long and short interest rates and the spread between high- and low-grade bonds. Furthermore, it has been proposed in [20.1] in 1988 that the factors can possibly be usually spread between the total monthly returns on government and corporate bonds and Treasury bills, unexpected deflation and growth rate in real final sales.

The lack of consensus and consistency over what should be the real economic factors in APT can largely be ascribed to the lack of systematic and effective methods for recovering those hidden economic factors. However, a new factor analytic technique termed Temporal Factor Analysis (TFA) proposed in [20.14] has been found to be capable of extracting statistically uncorrelated Gaussian factor scores from time series of stationary stock returns. Moreover, it provides a reasonable basis for the synthesis of stationary time series of economic variables.

The rest of this paper will be divided into five sections. Section 2 briefly reviews the arbitrage pricing theory and section 3 gives an overview of the TFA model. Statistical tests for serial correlations of the residual components for model adequacy will be presented in section 4, which is followed by the analysis and results of economic factors synthesis in section 5. Section 6 would be devoted to concluding remarks.

20.2 The Arbitrage Pricing Theory

The APT begins with the assumption that the $n \times 1$ vector of asset returns, R_t, is generated by a linear stochastic process with k factors [20.10, 20.9, 20.8]:

$$R_t = \bar{R} + Af_t + e_t \tag{20.1}$$

where f_t is the $k \times 1$ vector of realizations of k common factors, A is the $n \times k$ matrix of factor weights or loadings, and e_t is a $n \times 1$ vector of asset-specific risks. It is assumed that f_t and e_t have zero expected values so that $\bar{R}$ is the $n \times 1$ vector of mean returns. The model addresses how expected returns behave in a market with no arbitrage opportunities and predicts that an asset's expected return is linearly related to the factor loadings or

$$\bar{R} = R_f + Ap \tag{20.2}$$

where R_f is a $n \times 1$ vector of constants representing the risk-free return, and p is $k \times 1$ vector of risk premiums. Similar to the derivation of CAPM, (20.2) is based on the rationale that unsystematic risk is diversifiable and therefore should have a zero price in the market with no arbitrage opportunities.

20.3 Temporal Factor Analysis

20.3.1 An Overview of TFA

Suppose the relationship between a state $y_t \in \mathrm{R}^k$ and an observation $x_t \in \mathrm{R}^d$ are described by the first-order state-space equations as follows [20.14, 20.13]:

$$y_t = By_{t-1} + \varepsilon_t, \tag{20.3a}$$

$$x_t = Ay_t + e_t, \qquad t = 1, 2, \ldots, N. \tag{20.3b}$$

where ε_t and e_t are mutually independent zero-mean white noises with $E(\varepsilon_i \varepsilon_j) = \Sigma_\varepsilon \delta_{ij}$, $E(e_i e_j) = \Sigma_e \delta_{ij}$, $E(\varepsilon_i e_j) = 0$, Σ_ε and Σ_e are diagonal matrices, and δ_{ij} is the Kronecker delta function:

$$\delta_{ij} = \begin{cases} 1, \text{ if } i = j, \\ 0, \text{ otherwise.} \end{cases} \tag{20.4}$$

We call ε_t driving noise upon the fact that it drives the source process over time. Similarly, e_t is called measurement noise because it happens to be there during measurement. The above model is generally referred to as the TFA model.

In the context of APT analysis, (20.1) can be obtained from (20.3b) by substituting $(\tilde{R}_t - \bar{R})$ for x_t and f_t for y_t. The only difference between the APT model and the TFA model is the added (20.3a) for modeling temporal relation of each factor. The added equation represents the factor series $y = \{y_t\}_{t=1}^T$ in a multi-channel auto-regressive process, driven by an i.i.d. noise series $\{\varepsilon_t\}_{t=1}^T$ that are independent of both y_{t-1} and e_t. Specifically, it is assumed that ε_t is Gaussian distributed. Moreover, TFA is defined such that the k sources $y_t^{(1)}, y_t^{(2)}, \ldots, y_t^{(k)}$ in this state-space model are statistically independent. This constraint implies B is diagonal and ε_t is mutually independent in components. The objective of TFA [20.13, 20.14] is to estimate the sequence of y_t's with unknown model parameters $\Theta = \{A, B, \Sigma_\varepsilon, \Sigma_e\}$ through available observations. Since for Gaussian distribution statistically independent is synonymous with uncorrelated, we will use the two terms interchangeably in this paper.

In implementation, an adaptive algorithm has been suggested. At each time unit, factor loadings are estimated by cross-sectional regression and factor scores are estimated by maximum likelihood learning. Here, we adopt a simple algorithm proposed in [20.14] as shown below.

Assume $G(\varepsilon_t|0, I)$ and $G(e_t|0, \Sigma)$.

- **Step 1** Fix A, B and Σ, estimate the hidden factors y_t by

$$\hat{y}_t = [I + A^T \Sigma^{-1} A]^{-1} (A^T \Sigma^{-1} \bar{x}_t + B\hat{y}_{t-1}), \tag{20.5}$$

$$\varepsilon_t = \hat{y}_t - B\hat{y}_{t-1}, \tag{20.6}$$

$$e_t = \bar{x}_t - A\hat{y}_t, \tag{20.7}$$

- **Step 2** Fix, $\hat{y}_t$, update B, A and Σ_e by gradient descent method as follows:

$$B^{new} = B^{old} + \eta \text{diag}[\varepsilon_t \hat{y}_{t-1}^T], \tag{20.8}$$

$$A^{new} = A^{old} + \eta e_t \hat{y}_t^T, \tag{20.9}$$

$$\Sigma^{new} = (1 - \eta)\Sigma^{old} + \eta e_t e_t^T. \tag{20.10}$$

20.3.2 Model Selection vs. Appropriate Number of Factors

Central to the discussion in the paper about the number of factors in APT, TFA is superior to maximum likelihood factor analysis (MLFA) in view of its model selection ability. In the context of APT analysis, the scale or complexity of the model is equivalent to the number of hidden factors in the original factor structure. As a result, model selection refers to deciding the appropriate number of factors in APT. We can achieve the aim of model selection by enumerating the cost function $J(k)$ with k incrementally and then select an appropriate k by [20.14, 20.12, 20.15]

$$\min_k J(k) = \frac{1}{2}[k \ln(2\pi) + k + \ln |\Sigma|] \tag{20.11}$$

20.3.3 Grounds and Benefits for Using TFA in APT Analysis

Firstly, it is assumed that factors follow Gaussian distribution at each time t. There is a consensus that the noisy component in most econometric and statistical models being Gaussian distributed. The rationale comes from the central limit theorem which implies that the compounding of a large number of unknown distributions will be approximately normal. Secondly, we believe that factors recovered must be independent of each other. Although economic factors are seldom independent, it is helpful to discover statistically independent factors for the purpose of analysis because the restriction of independence will rule out many possible solutions which contain redundant elements. Furthermore, economic interpretation of factors recovered can be easily achieved by appropriate combination of those independent factors. Thirdly, we believe there exists temporal relation between factors. Equ. (20.3a) models an AR(1) time series of factors. Although formulation in this way slightly deviates from the Efficient Market Hypothesis (EMH) [20.4] on which APT is based, its rationale can be found in the literature [20.7, 20.5] which describes empirical evidence against the EMH.

Compared with MLFA, TFA has at least three distinct benefits. First, with the independence assumption in the derivation, the recovered factors are assured to be statistically independent. Second, Xu in [20.13] has shown that taking into account temporal relation effectively removes rotation indeterminacy. As a result, the solution given by TFA is unique. Theorem 3 proved by Xu in [20.13] illustrates this point. Third, it can determine the number of hidden factors via its model selection ability. Furthermore, it should be noted that MLFA is a special case of the model with $B = 0$ in (20.3a).

20.3.4 Testability of the TFA Model

The TFA model retains virtually all statistical properties of the original APT model. It is simply an extension of the APT model because it additionally includes temporal relation between factors in the APT model. Apart from that, there is no significant difference. Since the relationship between y_t and y_{t-1} described by the added equation is also linear, the entire TFA model is a linear model with both the driving noise ε_t and the measurement noise e_t assumed to be Gaussian distributed. Moreover, as both the returns and factors are stationary and the factors is assumed to be uncorrelated with idiosyncratic risks, we can safely conclude that the model is testable, just like APT [20.8, 20.11, 20.6].

20.4 Tests of White Noise Residuals

It is common in the literature of statistics that tests for model adequacy should immediately follow parameter estimation of the model under consideration. Usually the model is considered adequate if the residual component consists of white noise. In section 20.3 we have specifically emphasized the residual components of the TFA model. They are ε_t, the driving noise in (20.3a) and e_t, the measurement noise in (20.3b) respectively. For the AR(1) model as given by (20.3a) to be adequate, the estimated driving noise component should be substantially serially uncorrelated, i.e., autocorrelation of its lags should not be significantly different from zero. At the same time, for (20.3b) to be adequate, the estimated observation noise component should be largely uncorrelated among its constituents.

20.4.1 Data Considerations

We have carried out our analysis using past stock price and return data of Hong Kong. Daily closing prices of 86 actively trading stocks covering the period from January 1, 1998 to December 31, 1999 are used. The number of trading days throughout this period is 522. These stocks can be subdivided into three main categories according to different indices they constitute. Of the 86 equities, 30 of them belongs to the Hang Seng Index (HSI) constituents, 32 are Hang Seng China-Affiliated Corporations Index (HSCCI) constituents and the remaining 24 are Hang Seng China Enterprises Index (HSCEI) constituents. Before carrying out the analysis, the stock prices have been converted to stationary stock returns via $R_t = \frac{p_t - p_{t-1}}{p_{t-1}}$.

20.4.2 Test Statistics

To check if the driving noise residuals behave as a white-noise process, we will adopt the Ljung-Box modified Q-statistic shown below. The Q-statistic can be used to test whether a group of autocorrelations is significantly different from zero.

$$Q = N(N+2)\sum_{k=1}^{s}\frac{r_k^2(\hat{\varepsilon})}{N-k} \tag{20.12}$$

where N is the effective number of observations and s is the lag order. If the sample value of Q calculated above exceeds the critical value of χ^2 with $s-1$ degrees of freedom at $\alpha = 5\%$, then we can conclude that at least one value of r_k is statistically different from zero at 5% level of significance and suspect the residuals are serially correlated and not white.

On the other hand, to investigate whether each cross correlation coefficient of the observation noise residuals is not significantly different from zero, we will apply the t-test with the test statistic given by

$$t = \frac{r}{\sqrt{1-r^2}}\cdot\sqrt{n-2}, \tag{20.13}$$

where r is the correlation coefficient of a sample of n points (x_i, y_i) as given by

$$r = \frac{\sum(x_i-\overline{x})(y_i-\overline{y})}{[\sum(x_i-\overline{x})^2\sum(y_i-\overline{y})^2]^{\frac{1}{2}}}. \tag{20.14}$$

Assume that the x and y values originate from a bivariate Gaussian distribution, and that the relationship is linear, it can be shown that t follows Student's t-distribution with $n-2$ degrees of freedom. Again the predefined level of significance will be at $\alpha = 5\%$.

20.4.3 Empirical Results

30 HSI Constituents. Results of Q statistics and p-values for the driving noise residuals at lags from order 1 to 16 are shown in table 20.1. At 5% level of significance, autocorrelations of all residuals are not significantly different from zero. It implies that the underlying AR(1) model is adequate as the driving noise residuals behave as a white-noise process. On the other hand, out of 435 cross correlation coefficients, 15 of them, or 3.45% are statistically significant at $\alpha = 5\%$. As the percentage is quite small, the results are satisfactory and we accept the null hypothesis that the observation noise residuals are white. Partial results showing t-test on correlation coefficients the first two stocks with respect to the 30 HSI constituents are shown in table 20.2. Results of the other 28 stocks are omitted due to space constraint.

32 HSCCI Constituents. Results are shown in table 20.3. At 5% level of significance, autocorrelations of all residuals are not significantly different from zero at lag orders 1-16. On the other hand, out of 496 cross correlation coefficients, only 16 of them, or 3.23% are significant at $\alpha = 5\%$. Partial results for the first two stocks are shown in table 20.4. The null hypothesis is accepted based on the results.

Table 20.1. Results showing Q-statistics and the respective p-value of the residuals for 30 HSI constituents

Lag	Q-Stat Residual 1	p-value	Q-Stat Residual 2	p-value
1	0.0053	0.9418	0.0027	0.9585
2	0.5971	0.7419	0.1644	0.9211
3	2.1562	0.5406	1.9642	0.5799
4	2.6474	0.6185	2.2831	0.6838
5	3.1348	0.6792	2.6673	0.7511
6	3.1836	0.7855	3.1282	0.7926
7	4.9165	0.6701	3.1850	0.8674
8	5.2077	0.7352	3.2273	0.9193
9	5.3895	0.7991	4.4875	0.8765
10	5.5636	0.8505	6.4624	0.7750
11	5.5643	0.9008	7.0889	0.7918
12	6.7370	0.8745	7.4907	0.8235
13	8.0008	0.8435	7.8025	0.8562
14	8.0028	0.8892	7.9810	0.8903
15	8.3492	0.9090	8.0615	0.9213
16	8.3508	0.9377	8.6337	0.9277

Lag	Q-Stat Residual 3	p-value	Q-Stat Residual 4	p-value
1	0.0154	0.9012	0.0001	0.9914
2	0.7600	0.6839	0.0165	0.9918
3	0.7705	0.8565	0.2278	0.9730
4	0.8643	0.9296	0.3707	0.9848
5	1.0434	0.9590	7.4297	0.1906
6	1.0691	0.9829	8.5998	0.1974
7	1.4115	0.9852	8.9039	0.2597
8	1.5148	0.9925	8.9575	0.3459
9	2.1436	0.9890	10.4879	0.3125
10	2.1624	0.9949	10.6861	0.3825
11	2.4104	0.9965	10.9358	0.4487
12	2.4132	0.9985	13.0340	0.3666
13	4.2491	0.9882	15.1241	0.2997
14	4.3633	0.9928	19.8678	0.1344
15	5.9443	0.9807	20.5308	0.1526
16	5.9568	0.9885	20.7737	0.1875

24 HSCEI Constituents. Results are shown in table 20.5. At 5% level of significance, autocorrelations of all driving noise residuals are not significantly different from zero at lags order 1-16. On the other hand, out of 276 cross correlation coefficients, only 11 of them, or 3.99% are significant at $\alpha = 5\%$. Partial results for the first two stocks are shown in table 20.6. The null hypothesis is accepted based on the results.

Table 20.2. Partial results of t-test on the observation noise residuals for 30 HSI constituents. Only correlation coefficients the first two stocks with respect to 30 constituents are shown. Results of the other 28 stocks are omitted due to space constraint

Stock #	corr. Stock	t-stat. #1	p-value	corr. Stock	t-stat. #2	p-value
1	1.0000	–	–	-0.0041	-0.0939	0.9252
2	-0.0041	-0.0939	0.9252	1.0000	–	–
3	-0.0682	-1.5571	0.1201	-0.0215	-0.4888	0.6252
4	-0.0047	-0.1081	0.9140	0.0448	1.0212	0.3076
5	0.0275	0.6268	0.5311	-0.0041	-0.0927	0.9261
6	-0.0381	-0.8677	0.3860	-0.0112	-0.2554	0.7985
7	-0.0699	-1.5958	0.1111	0.0619	1.4133	0.1582
8	-0.0090	-0.2061	0.8368	-0.0115	-0.2609	0.7943
9	-0.0251	-0.5728	0.5670	0.0275	0.6272	0.5308
10	0.0408	0.9294	0.3531	-0.0606	-1.3840	0.1669
11	0.0307	0.7002	0.4841	0.0073	0.1674	0.8671
12	0.0001	0.0032	0.9975	-0.0138	-0.3143	0.7534
13	-0.0257	-0.5863	0.5580	-0.0217	-0.4949	0.6209
14	-0.0834	-1.9058	0.0572	-0.0270	-0.6151	0.5388
15	-0.0978	-2.2382	0.0256	-0.0303	-0.6903	0.4903
16	-0.0516	-1.1776	0.2395	0.0016	0.0373	0.9703
17	0.0821	1.8764	0.0612	0.0571	1.3030	0.1932
18	-0.0439	-1.0007	0.3175	-0.0573	-1.3075	0.1916
19	-0.0341	-0.7784	0.4367	0.0283	0.6456	0.5188
20	-0.0069	-0.1583	0.8743	-0.0484	-1.1036	0.2703
21	-0.0272	-0.6192	0.5360	0.0555	1.2662	0.2060
22	0.0115	0.2630	0.7927	0.0840	1.9205	0.0553
23	0.0567	1.2932	0.1965	-0.0156	-0.3554	0.7224
24	-0.0202	-0.4604	0.6454	0.0645	1.4719	0.1417
25	-0.0557	-1.2702	0.2046	0.1123	2.5758	0.0103
26	-0.0580	-1.3240	0.1861	0.0394	0.8982	0.3695
27	0.0139	0.3156	0.7524	-0.0030	-0.0690	0.9450
28	0.0341	0.7772	0.4374	0.0428	0.9769	0.3291
29	0.0294	0.6694	0.5035	0.0268	0.6115	0.5411
30	-0.0156	-0.3549	0.7228	-0.0394	-0.8982	0.3695

All 86 Securities. Results are shown in table 20.7. At 5% level of significance, autocorrelations of all residuals are not significantly different from zero at lag orders 1-16. It implies that the underlying AR(1) model is adequate as the driving noise residuals behave as a white-noise process. On the other hand, out of 3655 cross correlation coefficients, only 87 of them, or 2.38% are significant at $\alpha = 5\%$. The null hypothesis is accepted based on the results.

Table 20.3. Results showing Q-statistic and p-value of the driving noise residuals for 32 HSCCI constituents

Lag	Q-Stat Residual 1	p-value	Q-Stat Residual 2	p-value	Q-Stat Residual 3	p-value
1	0.0165	0.8978	0.0397	0.8421	1.2106	0.2712
2	1.6987	0.4277	0.7411	0.6903	1.2741	0.5289
3	1.7028	0.6363	0.7590	0.8592	2.5560	0.4653
4	2.4884	0.6467	3.8699	0.4239	2.5579	0.6343
5	3.1986	0.6694	4.3257	0.5036	3.4214	0.6353
6	4.5150	0.6073	6.7944	0.3403	4.0612	0.6684
7	7.2552	0.4028	6.8038	0.4496	4.2960	0.7451
8	7.5093	0.4828	6.9982	0.5368	6.9552	0.5415
9	13.5159	0.1407	15.2182	0.0852	8.9411	0.4428
10	14.2382	0.1625	18.0936	0.0535	10.8558	0.3689
11	14.4412	0.2096	18.4788	0.0712	11.3892	0.4113
12	14.4568	0.2726	18.4793	0.1020	11.8557	0.4574
13	14.6595	0.3291	21.4853	0.0639	11.9348	0.5330
14	15.2772	0.3595	21.7078	0.0849	14.4735	0.4151
15	15.2816	0.4314	21.7519	0.1146	15.0014	0.4514
16	16.4103	0.4248	22.0868	0.1405	15.0131	0.5237

20.5 Synthesis of Economic Factors

Learning via the adaptive TFA algorithm guarantees the recovered Gaussian temporal factors statistically uncorrelated. The constraint of statistical independence is important because it would eliminate the possibility of more than one solution fitting the model. Nonetheless, interpretation of the recovered uncorrelated statistical factors may be difficult as time series of common economic variables are often correlated to some degree. Moreover, as the finance community is more inclined to accept macroeconomic variables as the hidden driving force of stock returns because of both intuition and empirical evidence, attempt to build up a relationship between the recovered statistical factors and some well-known macroeconomic factors is crucial for both the theoretical analysis and practical application of the APT model. We refer to the exploration of the relationship between the economic factors and the statistically uncorrelated factors in this paper the synthesis of APT economic factors.

Admittedly, it is not easy to tell what macroeconomic time series may be treated as the most appropriate APT economic factors. Since the search for the most suitable candidates from a vast number of possible time series is always far from exhaustive, there can be no definite answer to the question of whether a specific economic time series should be the optimum candidate. Moreover, it is possible that the so-called economic factors such as inflation rate, interest rate and unemployment rate are also being affected by some hidden driving force. This argument is similar to that stock returns can be

Table 20.4. Partial results of t-test on the observation noise residuals for 32 HSCCI constituents. Only correlation coefficients the first two stocks with respect to 32 constituents are shown. Results of the other 30 stocks are omitted due to space constraint

Stock #	corr. Stock	t-stat. #1	p-value	corr. Stock	t-stat. #2	p-value
1	1.0000	–	–	-0.0693	-1.5827	0.1141
2	-0.0693	-1.5827	0.1141	1.0000	–	–
3	-0.0085	-0.1944	0.8460	-0.0160	-0.3650	0.7153
4	-0.0096	-0.2192	0.8265	-0.0120	-0.2727	0.7852
5	-0.0296	-0.6756	0.4996	-0.0372	-0.8482	0.3967
6	-0.0418	-0.9541	0.3405	-0.0634	-1.4465	0.1486
7	-0.0038	-0.0874	0.9304	0.0075	0.1717	0.8637
8	0.0162	0.3698	0.7117	-0.0040	-0.0906	0.9278
9	0.0054	0.1221	0.9029	0.0387	0.8829	0.3777
10	-0.0386	-0.8792	0.3797	0.0434	0.9899	0.3227
11	0.0046	0.1050	0.9164	-0.0026	-0.0592	0.9528
12	0.0092	0.2106	0.8333	0.0153	0.3497	0.7267
13	-0.0363	-0.8281	0.4080	0.0410	0.9338	0.3508
14	-0.0045	-0.1018	0.9189	-0.0162	-0.3689	0.7123
15	-0.0358	-0.8162	0.4148	-0.0080	-0.1821	0.8556
16	0.0148	0.3373	0.7360	0.0067	0.1522	0.8791
17	0.0716	1.6357	0.1025	-0.0359	-0.8195	0.4129
18	-0.0350	-0.7985	0.4249	0.0169	0.3845	0.7007
19	-0.0135	-0.3075	0.7586	-0.0260	-0.5922	0.5540
20	0.0261	0.5939	0.5529	0.0315	0.7180	0.4731
21	-0.0045	-0.1018	0.9190	-0.0777	-1.7750	0.0765
22	-0.0565	-1.2881	0.1983	0.0056	0.1272	0.8988
23	0.0151	0.3440	0.7310	0.0359	0.8176	0.4140
24	0.0534	1.2187	0.2235	0.0087	0.1992	0.8422
25	0.0320	0.7285	0.4666	-0.0104	-0.2373	0.8125
26	-0.0264	-0.6016	0.5477	0.0305	0.6943	0.4878
27	0.0222	0.5069	0.6125	-0.0241	-0.5500	0.5825
28	-0.0638	-1.4555	0.1461	0.0279	0.6368	0.5245
29	0.0476	1.0862	0.2779	-0.0202	-0.4612	0.6449
30	0.0103	0.2336	0.8154	0.0489	1.1153	0.2652
31	0.0508	1.1587	0.2471	-0.0296	-0.6735	0.5009
32	-0.0266	-0.6054	0.5452	-0.0188	-0.4283	0.6686

conceived as being affected by some unknown hidden factors. Broadly speaking, security returns can also be considered as sorts of macroeconomic time series. Thus, if given sufficient macroeconomic time series, it may be possible that TFA can be used to recover those hidden common independent factors for those macroeconomic time series and comparison can be made on their correlations with those statistically independent factors affecting stock returns.

Table 20.5. Results showing Q-statistic and p-value of the driving noise residuals for 24 HSCEI constituents

Lag	Q-Stat Residual 1	p-value	Q-Stat Residual 2	p-value
1	2.0511	0.1521	0.0639	0.8004
2	4.4732	0.1068	0.0650	0.9680
3	5.9503	0.1141	0.1283	0.9882
4	5.9518	0.2028	0.7236	0.9484
5	5.9692	0.3093	1.7719	0.8797
6	6.9439	0.3261	1.8497	0.9330
7	7.4245	0.3861	4.3465	0.7391
8	7.5826	0.4753	5.0000	0.7576
9	9.4875	0.3936	6.1129	0.7286
10	10.1511	0.4274	6.5875	0.7637
11	10.8101	0.4593	9.9673	0.5333
12	12.4066	0.4136	12.1570	0.4332
13	14.4841	0.3407	14.8576	0.3164
14	15.5100	0.3443	16.1930	0.3018
15	15.9087	0.3882	16.3040	0.3622
16	17.9654	0.3260	20.3909	0.2032

Lag	Q-Stat Residual 3	p-value	Q-Stat Residual 4	p-value
1	0.0048	0.9447	0.0110	0.9166
2	0.4872	0.7838	0.4207	0.8103
3	0.5800	0.9010	0.4266	0.9347
4	8.1902	0.0849	5.4541	0.2438
5	10.2321	0.0689	5.5303	0.3547
6	10.3302	0.1115	8.3598	0.2129
7	10.7253	0.1511	8.5396	0.2875
8	10.7254	0.2178	9.4638	0.3047
9	10.8287	0.2877	10.2432	0.3312
10	10.9531	0.3612	11.7718	0.3007
11	11.2581	0.4219	15.6198	0.1559
12	11.4262	0.4928	17.6307	0.1274
13	12.1782	0.5131	17.6722	0.1704
14	12.6480	0.5544	18.4682	0.1864
15	12.7657	0.6204	19.0036	0.2137
16	14.2880	0.5773	19.3154	0.2527

20.5.1 Methodology and Test Statistics

Since we are only able to collect few economic time series, the proposition of factor correlations comparison seems not viable. An alternative to explore the relationship between economic factors and the statistically uncorrelated factors is by means of statistical test on the significance of the coefficients of a regression between an economic time series and the statistically independent factors extracted from stock returns series via TFA. As usual, the time series

Table 20.6. Partial results of t-test on the observation noise residuals for 24 HSCEI constituents. Only correlation coefficients the first two stocks with respect to 24 constituents are shown. Results of the other 22 stocks are omitted due to space constraint

Stock #	corr. Stock	t-stat. #1	p-value	corr. Stock	t-stat. #2	p-value
1	1.0000	–	–	0.0182	0.4150	0.6783
2	0.0182	0.4150	0.6783	1.0000	–	–
3	0.0076	0.1735	0.8623	-0.0108	-0.2463	0.8055
4	0.0094	0.2134	0.8311	-0.0251	-0.5716	0.5678
5	-0.0317	-0.7222	0.4705	0.0245	0.5577	0.5773
6	-0.0111	-0.2519	0.8012	-0.0487	-1.1100	0.2675
7	-0.0173	-0.3933	0.6943	0.0055	0.1260	0.8998
8	0.0089	0.2019	0.8400	-0.0133	-0.3040	0.7613
9	-0.0512	-1.1670	0.2437	0.0609	1.3895	0.1653
10	0.0738	1.6852	0.0925	0.0282	0.6425	0.5209
11	0.0271	0.6172	0.5374	-0.0156	-0.3554	0.7224
12	-0.0090	-0.2056	0.8372	0.0369	0.8415	0.4004
13	-0.0114	-0.2602	0.7948	0.0201	0.4571	0.6478
14	0.0611	1.3954	0.1635	-0.0030	-0.0673	0.9464
15	-0.0022	-0.0504	0.9598	0.0199	0.4525	0.6511
16	-0.0047	-0.1063	0.9154	0.0555	1.2670	0.2057
17	0.0337	0.7671	0.4434	-0.0054	-0.1232	0.9020
18	-0.0112	-0.2553	0.7986	0.0082	0.1873	0.8515
19	0.0146	0.3321	0.7399	-0.0506	-1.1542	0.2490
20	0.0220	0.5002	0.6171	-0.0688	-1.5702	0.1170
21	0.0088	0.2008	0.8409	0.0180	0.4091	0.6826
22	0.0378	0.8624	0.3889	0.0713	1.6281	0.1041
23	0.1204	2.7629	0.0059	0.0233	0.5300	0.5963
24	0.0372	0.8476	0.3971	-0.0140	-0.3193	0.7496

have been preprocessed so that stationarity is guaranteed. The t-statistic would be used to test for the individual significance while the F-statistic could be used to test for the joint significance of the coefficients. We will examine the results of both tests at levels of significance of 5%. The null hypothesis is that all factor loadings are simultaneously zero. Thus the alternate hypothesis is H_1 : There exist nonzero constants $a_1, a_2, \cdots, a_k$ such that

$$TS_t^{(j)} = a_1^{(j)} y_{1t} + a_2^{(j)} y_{2t} + \cdots + a_k^{(j)} y_{kt} \tag{20.15}$$

where $TS_t^{(j)}$ is the value of the j-th transformed time series at time t.

20.5.2 Empirical Results

We have used four economic time series during the same period from January 1, 1998 to December 31, 1999 for empirical test. They are the 1 month Hong Kong Inter-Bank Middle Rate (Series A), 1 year Hong Kong Inter-Bank

Table 20.7. Results showing Q-statistics and the respective p-value of the residuals for all 86 securities

Lag	Q-Stat Residual 1	p-value	Q-Stat Residual 2	p-value
1	0.0743	0.7852	0.2992	0.5844
2	1.8990	0.3869	1.2295	0.5408
3	3.7295	0.2922	1.4767	0.6877
4	4.5944	0.3315	1.5594	0.8161
5	5.5482	0.3527	4.8766	0.4311
6	8.0917	0.2315	4.9022	0.5564
7	8.1760	0.3174	8.9754	0.2545
8	8.3028	0.4045	9.0244	0.3403
9	8.6708	0.4682	9.3282	0.4076
10	10.8504	0.3693	11.5515	0.3162
11	13.7944	0.2446	13.9240	0.2373
12	14.7830	0.2536	14.3835	0.2770
13	14.7841	0.3211	18.3171	0.1459
14	16.1057	0.3070	20.0656	0.1282
15	16.1238	0.3739	20.7264	0.1459
16	16.2176	0.4379	22.9507	0.1151

Lag	Q-Stat Residual 3	p-value	Q-Stat Residual 4	p-value	Q-Stat Residual 5	p-value
1	0.0005	0.9818	0.0795	0.7780	3.5885	0.0582
2	0.0666	0.9672	3.9236	0.1406	3.6315	0.1627
3	0.0733	0.9948	3.9439	0.2676	3.9939	0.2621
4	8.9685	0.0619	4.2149	0.3777	4.1009	0.3925
5	9.0195	0.1083	6.8929	0.2288	4.1699	0.5252
6	9.0344	0.1717	7.7789	0.2548	4.1714	0.6535
7	10.6932	0.1526	9.1086	0.2450	7.4628	0.3824
8	11.5813	0.1709	11.4475	0.1777	10.6244	0.2240
9	11.5813	0.2380	11.5153	0.2421	11.5409	0.2405
10	12.0479	0.2819	12.4976	0.2532	11.6855	0.3067
11	12.0507	0.3599	16.5561	0.1218	11.7112	0.3858
12	12.6850	0.3924	17.0175	0.1490	12.3739	0.4162
13	12.9104	0.4548	19.5611	0.1068	12.5800	0.4808
14	12.9815	0.5280	19.7313	0.1389	12.6173	0.5569
15	13.1612	0.5899	20.2641	0.1621	12.6222	0.6314
16	13.2102	0.6573	20.7273	0.1893	13.1958	0.6584

Middle Rate (Series B), the Hang Seng Index (Series C) and the Dow Jones Industrial Average (Series D) respectively. We use the five independent factors recovered from all 86 securities for regression. The results showing the p-values of each coefficient are shown in table 20.8. Since at $\alpha = 5\%$ the coefficients are both individually and jointly significant, the null hypothesis is rejected and we can reasonably conclude that there is linear relationship between the time series under test and the statistically uncorrelated factors recovered by TFA.

Table 20.8. Results of t-test and F-test of regression coefficients for real economic time series

	p-value	of	t-test			p value of
	a_1	a_2	a_3	a_4	a_5	F-test
A	0.0000	0.0011	0.0000	0.0000	0.0006	0.0000
B	0.0001	0.0001	0.0000	0.0005	0.0000	0.0000
C	0.0003	0.0000	0.0071	0.0000	0.0032	0.0000
D	0.0005	0.0023	0.0001	0.0009	0.0103	0.0000

20.6 Conclusion

We have carried out the white noise tests on the residual of the TFA model for model adequacy. The results provide assurance for further statistical analysis using the TFA model. Based on the statistical test results, the null hypothesis is rejected and we henceforth accept the alternative hypothesis that each of the four economic time series is linearly related to the statistically uncorrelated factors as determined via the TFA model. Therefore, it is prudent to say that the APT economic factors can be synthesized from uncorrelated Gaussian temporal factors as determined via the TFA model.

References

20.1 Berry, M., Burmeister, E., McElroy, M. (1988): Sorting out risks using known APT factors. Financial Analysts Journal, 44, 29–42
20.2 Chen, N. F., Roll, R., Ross, S. (1986): Economic forces and the stock market. Journal of Business, 59, 383–403
20.3 Estep, T., Hansen, N., Johnson, C. (1983): Sources of value and risk in common stocks. Journal of Portfolio Management, 9, 5–13
20.4 Fama, E. (1970): Efficient capital markets: a review of theory and empirical work. Journal of Finance, 25, 383–417
20.5 La Porta, R., Lakonishok, J., Shliefer, A., Vishny, R. (1997): Good news for value stocks: further evidence on market efficiency. Journal of Finance, 52, 859–874
20.6 Lehmann, B. N., Modest, D. M. (1988): The empirical foundations of the arbitrage pricing theory. Journal of Financial Economics, 21, 213–254
20.7 Roll, R. (1988): R^2. Journal of Finance, 43, 541–566
20.8 Roll, R., Ross, S. (1980): An empirical investigation of the arbitrage pricing theory. Journal of Finance, 35, 1073–1103

20.9 Roll, R., Ross, S. (1984): The arbitrage pricing theory approach to strategic portfolio planning. Financial Analysts Journal, 40, 14–26
20.10 Ross, S. (1976): The arbitrage theory of capital asset pricing. Journal of Economic Theory, 13, 341–360
20.11 Shanken, J. (1982): The arbitrage pricing theory: is it testable. Journal of Finance, 37, 1129–1140
20.12 Xu, L. (1998): Bayesian Ying-Yang learning theory for data dimension reduction and determination. Journal of Computational Intelligence in Finance, 6(5), 6–18
20.13 Xu, L. (2000): Temporal BYY learning for state space approach, hidden markov model and blind source separation. IEEE Transactions on Signal Processing, 48, 2132–2144
20.14 Xu, L. (2001): BYY harmony learning, independent state space and generalized APT financial Analysis. IEEE Transactions on Neural Networks, 12(4), 822–849
20.15 Xu, L. (2001): Best harmony, unified RPCL and automated model selection for unsupervised and supervised learning on Gaussian mixtures, three-layer nets and ME-RBF-SVM models. International Journal of Neural Systems, 11, 43–69

21. Learning and Monetary Policy in a Spectral Analysis Representation

Andrew Hughes Hallett[1] and Christian R. Richter[2*]

[1] Department of Economics, Vanderbilt University, Nashville, TN, USA, and CEPR, London
email: A.HughesHallett@Vanderbilt.edu

[2] Department of Economics, University of Strathclyde, Glasgow, UK
email: Christian.Richter@Uni-Klu.ac.at

In this paper we expand the well-known methodology of deriving cross spectra from autoregressive lag models to the class of time-varying parameter models. That enables us to analyse all frequency properties at all points in time (in contrast to the wavelet methodology, where one can only analyse certain frequencies at all points in time). This allows us to separate the evolution of an equilibrium from dynamic adjustments to it and the process of learning about it. Using these results, we analyse the behaviour of short term interest rates in response to monetary policy changes in Britain during and following the European Exchange Rate Mechanism (ERM) crisis of 1992/3. We find that the British monetary transmission mechanism is very stable even in an event like the ERM crisis. This is possible due to the adjustment of the risk premium.

21.1 Introduction

A standard paradigm in economics is efficient markets. However, in practice many markets turn out to be inefficient. This can happen in financial markets because of "informational inefficiencies" due to the cost of gathering information; the costs of adjusting to a new equilibrium; uncertainty and the process of learning; or simply to the asymmetries in the possession of information ([21.2], [21.3]). Other markets - typically labour or certain goods markets - may also be inefficient because such informational problems cause rigidities in behaviour ([21.1]).

One might therefore expect market behaviour, particularly in financial markets, to show complex dynamics as agents adjust towards their equilibrium behaviour, given the information available to them, and as they acquire new information, or decide to acquire new information, to help them determine what those equilibrium relationships should be. On this "behavioural finance" view, systematic and significant deviations from efficiency can be expected and may well persist ([21.21], [21.20], [21.23]). Indeed movements to equilibrium will only arise when the profit opportunities are large enough

* Financial Support from the Austrian Science Foundation (project no P12745-OEK) is gratefully acknowledged.

to compensate investors (traders) for the costs of gathering information and then trading to adjust their stocks of assets (skills, goods). But, as they do so, they are likely to affect what other agents then perceive to be their equilibrium state. Indeed if information deficiencies underlie these disequilibrium movements, and changes in the perceived equilibrium path, it would hardly be sensible to model these markets as if they had strictly rational expectations and efficiency ([21.4]), and as if prices reflect all the available information.

The dynamics of any financial market should reflect adjustments to an equilibrium path which is itself adjusting to learning, the resolution of uncertainty and the costs of information gathering. That would imply different behaviour and different adjustment characteristics at different frequencies. For example, the short run behaviour of the agents may differ from the longer run convergence pattern of the market as a whole, because agents adjust to eliminate disequilibria in their current positions, and then adjust their understanding of what their long run equilibrium positions should be.

The only way to analyse the input of variations in these different layers of dynamic behaviour, and the relations between them, is via a spectral analysis in the frequency domain. It is not possible to disentangle the different elements in a time domain representation. Normally however, that route is closed in economics because of the long samples of data needed to make direct estimates of economic spectra. As a rule of thumb one would require a sample period seven to eight times larger than the longest cycle present in the data - and that is when the underlying economic/market structure is constant, which it clearly is not if learning and information gathering lies at the heart of the convergence process.

The purpose of this paper, therefore, is to examine a technique for calculating the necessary spectra indirectly; that is from the kind of dynamic regression or VAR models that have proved so popular in the literature. This can be done in the case where the parameters of the underlying market structure are time-varying for the reasons we have mentioned: that is when we get different parameter values at different points in time in the conventional economic relationships. [21.13], [21.15], and [21.26] have each shown that a dynamic regression equation can be used to derive the spectra and cross-spectra needed for a frequency domain analysis of the data. That results in a time-varying spectral analysis designed to capture the within and between period adjustments of our time series model.

Finally, we use these tools to analyse monetary policy in Britain in the 1990s. Not only do we learn something useful about the market's learning process in each country; we are also able to compare the learning and adjustment that has taken place in the financial markets in each country. What we find is that the British markets show evidence of learning and hence of an ability to adapt in the short and medium term.

This paper is structured as follows: the second section briefly summarises the relationship between the time domain and the frequency domain. The third section puts these results in a time-varying framework. In the fourth section, we apply these methods to certain relationships from the British financial market, and draw some monetary policy implications from the contrast between the behaviour at the short-end of the British term structure and how that has changed since the ERM crisis of 1992.

21.2 The Relationship Between the Time Domain and the Frequency Domain

21.2.1 Spectra and Cross-spectra in Our Analysis

In this paper, we are interested in the relationship between different variables, Y_t and X_t, which are not necessarily assumed to be stationary and related in the following way:[1]

$$Y_t = A(L)\, X_t + u_t. \tag{21.1}$$

where $A(L)$ is a filter, and L is the lag operator such that $LY_t = Y_{t-1}$. In this case, we have to consider the sample *cross-covariance* function (see for example [21.7]):

$$\gamma_{YX}(\tau) = E\left(Y_t X_{t-\tau}\right). \tag{21.2}$$

But if Y_t follows 21.1, we can substitute 21.1 in 21.2 to get:

$$\gamma_{YX}(\tau) = E\left(A(L)\, X_t X_{t-\tau}\right) + E\left(u_t X_{t-\tau}\right). \tag{21.3}$$

Since u_t is i.i.d. $(0, \sigma^2)$, $E\left(u_t X_{t-\tau}\right) = 0$. Hence 21.3 simplifies to

$$\gamma_{YX}(\tau) = \sum_{j=0}^{\infty} a_j \gamma_{XX}(\tau - j). \tag{21.4}$$

In Eq. (21.4), a_j is a sequence of coefficients from $A(L)$ and $\gamma_{XX}\ (\tau - j) = E\left(X_t X_{t-(\tau-j)}\right)$. Hence the cross-covariance is depending on the coefficients a_j from Eq. (21.1). Thus, if we interpret Eq. (21.1) as an estimated equation, then each cross-covariance depends on the estimated coefficients of the distributed lag model Eq. (21.1). Consequently, if the estimates obtained for coefficients in Eq. (21.1) are efficient, then the estimated cross-covariance will also be efficient too (given knowledge of γ_{XX}).

The technique we use is that, using estimated coefficients from Eq. (21.1) vastly simplifies estimation of the spectra. Indeed one of the biggest disadvantages of a direct estimation approach is the large number of observations that

[1] On theoretical grounds, interest rates cannot be integrated of order one, since they do not have infinite variance and they are bounded to zero. On the other hand, economic time series often exhibit lots of structural breaks, so that the series appears to be non-stationary ([21.16]). Furthermore, [21.22] showed that even in case of integrated time series, the OLS estimator is consistent (under certain circumstances).

would otherwise be necessary to carry out the necessary frequency analysis. We now show how we can get round that problem by starting from regression based estimates. We exploit a technique which we analysed in greater detail in [21.9]. It depends on the fact that we can write the *cross spectral density* of the exogenous variable as being proportional to the Fourier transform of the lag coefficients (see also [21.15]): i.e.

$$g_{YX}(z) = \sum_{\tau=-\infty}^{\infty} \sum_{j=0}^{\infty} a_j \gamma_{XX}(\tau - j) z^{\tau} = A(z) g_{XX}(z), \tag{21.5}$$

$z = e^{-i\omega}$, with $i^2 = -1$ and ω is the frequency in radians per unit of time, where $A(z)$ is called the *frequency response function*, and $\mid A(z) \mid^2$ is called the *transfer function* of the filter $A(L)$. But Eq. (21.5) can also be written in terms of the spectra involved. Using the fact that

$$f_{YX}(\omega) = \frac{1}{2\pi} g_{YX}(z) \tag{21.6}$$

and that

$$f_{XX}(\omega) = \frac{1}{2\pi} g_{XX}(z), \tag{21.7}$$

we get the cross-spectrum $f_{YX}(\omega)$ defined as

$$f_{YX}(\omega) = A(z) f_{XX}(\omega) \tag{21.8}$$

where $A(z)$ is the Fourier transform of the weights $a_j; j = -\infty..\infty$. From Eq. (21.8) we get

$$A(z) = \frac{f_{YX}(\omega)}{f_{XX}(\omega)}. \tag{21.9}$$

Eq. (21.9) implies that

$$\mid A(z) \mid = \frac{\mid f_{YX}(\omega) \mid}{f_{XX}(\omega)}. \tag{21.10}$$

The function $\mid A(\omega) \mid$ is sometimes called the *gain*. This gain is equivalent to the regression coefficient for each frequency ω. It measures the amplification of the frequency components of the X-process to obtain the corresponding components of the Y-process. However, it is often more convenient to express the gain in terms of the complex scalar z, i.e. $\mid A(\omega) \mid = \mid A(z) \mid$.

However, a direct calculation of the gain can create problems. But, we can always rewrite Eq. (21.10) as ([21.13], [21.15], [21.26])

$$\mid A(z) \mid = \sqrt{\mid A(z) \mid^2} = \sqrt{A(z) A(\bar{z})} \tag{21.11}$$

where $\bar{z}$ is the conjugate complex of z, i.e. $\bar{z} = e^{i\omega}$. Thus, in order to calculate the gain, all we have to know is the sequence of the coefficients a_j from Eq. (21.1). The question is how these coefficients can be estimated. We show that in section 21.2.3 below.

Meanwhile, $\phi(\omega)$ is called the *phase angle*. This phase angle reveals the lead and lag relationship between two variables at different frequencies. The phase angle $\phi(\omega)$ can be expressed in terms of the cospectrum $(C_{YX}(\omega))$ and the quadrature spectrum $(Q_{YX}(\omega))$:

$$\phi(\omega) = tan^{-1} \frac{-Q_{YX}(\omega)}{C_{YX}(\omega)}. \tag{21.12}$$

The cospectrum $(C_{YX}(\omega))$ is defined as ([21.26]):

$$C_{YX}(\omega) = f_{XX}(\omega) \sum_{j=0}^{\infty} a_j \cos \omega j; \tag{21.13}$$

whereas the quadrature spectrum $(Q_{YX}(\omega))$ is defined as

$$Q_{YX}(\omega) = -f_{XX}(\omega) \sum_{j=0}^{\infty} a_j \sin \omega j. \tag{21.14}$$

The phase angle is therefore

$$\phi(\omega) = \tan^{-1} \left(\frac{\sum_{j=0}^{\infty} a_j \sin \omega j}{\sum_{j=0}^{\infty} a_j \cos \omega j} \right). \tag{21.15}$$

In this paper, however, we analyse the "standardised" phase angle, or *phase shift*:

$$\tau(\omega) = \frac{\phi(\omega)}{\omega}. \tag{21.16}$$

The question is of how to derive the phase shift from the time domain. We show that in section 21.2.3.

21.2.2 Confidence Intervals

In order to calculate confidence intervals for the gain and the phase shift, we have to know the coherence and the spectra of the individual time series. Assuming that the spectra are known (see below), the coherence is:[2]

$$K_{12}^2(\omega) = \frac{1}{1 + (f_{YY}(\omega) / \mid A(z) \mid f_{XX}(\omega))}. \tag{21.17}$$

The $100(1-\alpha)$ percent confidence intervals for the gain are therefore

$$\mid A(z) \mid \pm \mid A(z) \mid \sqrt{\frac{2}{\nu - 2} f_{2,\nu-2}(1-\alpha) \left(\frac{1 - K_{12}^2(\omega)}{K_{12}^2(\omega)} \right)}, \tag{21.18}$$

where ν is the number of degrees of freedom associated with the smoothing of the output spectrum and $f_{2,\nu-2}(1-\alpha)$ is the upper $100(1-\alpha)$ percentage

[2] see for example [21.12]

point of the $F_{2,\nu-2}$ distribution.[3] The associated $100\,(1-\alpha)$ percent confidence interval for the phase shift is

$$\tau(\omega) = \pm \arcsin \sqrt{\frac{2}{\nu-2} f_{2,\nu-2}(1-\alpha) \left(\frac{1-K_{12}^2(\omega)}{K_{12}^2(\omega)} \right)}. \tag{21.19}$$

21.2.3 The Indirect Estimation Technique

In order to calculate the sequence of coefficients a_j from Eq. (21.1) we have to analyse the lag structure of the underlying economic model. In order to do this, we start from a general linear model of distributed lags:

$$V(L)\,Y_t = U(L)\,X_t + \varepsilon_t, \tag{21.20}$$

where $V(L) = \sum_{s=0}^{p} v_s L^s, v_0 = 1$, and $U(L) = \sum_{r=0}^{q} u_r L^r$. Thus as long as all eigenvalues of the characteristic equation of $V(L)$ are less than one, we can write

$$Y_t = \frac{U(L)}{V(L)} X_t + \frac{1}{V(L)} \varepsilon_t \tag{21.21}$$

We are particularly interested in the first ratio of Eq. (21.21) in order to derive the gain of the variable X_t:

$$\frac{U(L)}{V(L)} = W(L) \tag{21.22}$$

where $W(L) = \sum_{j=0}^{k} w_j L^j$ is the weight function.[4] The sequence $w_j; j = 1, ..., k$ defines the model's lag structure. The coefficients w_j show the impact which results from a change of the explanatory variable j periods ago. In particular, w_0 is the instantaneous reaction coefficient. In order to achieve a sensible economic interpretation, it is required that if $X_t = X_{t-1} = ... = X$, i.e. in an (deterministic) equilibrium, then the dependant variable Y should also be constant (and finite).

In order to calculate w_j in terms of u_r and v_s, we make use of the following relationship [21.8], [21.13]:

$$\left(\sum_{j=0}^{k} w_j L^j \right) \left(\sum_{s=0}^{p} v_s L^s \right) = \sum_{r=0}^{q} u_r L^r. \tag{21.23}$$

Equating powers of L on the two sides of Eq. (21.23), and noting that $v_0 = 1$, we get the following recursive equations:

$w_0 = u_0;$

$w_1 + w_0 v_1 = u_1;$

$w_2 + w_1 v_1 + w_0 v_2 = u_2$

...

[3] ν is 34 for Britain

[4] k may tend to infinity, but does not have to.

Solving for the unknown coefficients w_j, we have
$w_0 = u_0$;
$w_1 = u_1 - u_0 v_1$;
$w_2 = u_2 - (u_1 - u_0 v_1) v_1 - u_0 v_2$;
and so on.

Although, we can calculate the w_j coefficients, their statistical and distributional properties are not clear. Indeed the above equations show that it will actually be impossible to calculate a closed form expression for their distributions. On the other hand, [21.8] performed a Monte Carlo simulation in order to determine the size of the coefficient bias. It turned out that coefficients are only slightly biased given sample sizes as small as 20. However, they converged almost completely to an unbiased coefficient estimate for a sample size of 80. The largest bias he reported was less than 0.02 (for a true coefficient of 0.9) at a sample size of 80. That compares to a simulation standard error of less than 0.006.

Given the lag structure in Eq. (21.23), we are now able to generate the gain according to Eq. (21.11):

$$r(z) = \sqrt{\sum_{j=0}^{k} w_j z^j \sum_{j=0}^{k} w_j z^{-j}} \tag{21.24}$$

where $z = e^{-i\omega}$, and the phase angle is according to Eq. (21.13)

$$p(\omega) = \tan^{-1}\left(\frac{\sum_{j=0}^{k} w_j \sin \omega j}{\sum_{j=0}^{k} w_j \cos \omega j}\right). \tag{21.25}$$

21.3 Econometric Implementation: a Time-Varying Approach to the Term Structure of Interest Rates

So far we have described the case of time-invariant parameters only. We now look at the case where the parameters are time-varying, i.e. Eq. (21.20) changes to

$$V(L)_t Y_t = U(L)_t X_t + \varepsilon_t. \tag{21.26}$$

In what follows, we have estimated Eq. (21.26) using Kalman filter techniques. We have used the Kalman filter because we need to test certain hypotheses about the time-varying nature of underlying parameters. It is entirely possible that the agents in the financial markets will change their behaviour, and hence the way in which interest rates are determined in those markets, depending on which policy regime or information regime they face. The Lucas critique in other words: if we change the way in which monetary policy is set, the way in which agents determine the relevant short and long term interest rates (and their risk premia) will also be changed.

In what follows we apply these estimation techniques to analyse the behaviour of UK short term interest rates over the period 1982 - 1998. That period includes the soft (but adjustable) European Monetary System (EMS), the "hard" EMS regime, the collapse of the Exchange Rate Mechanism (ERM) wide bands, and the introduction of inflation targeting in the UK. It is important to be able to take the consequences of all these regime changes into account in our model of interest rate determination. The advantage of using the Kalman filter algorithm to do this is that it assumes that agents form one-period ahead forecasts. These forecasts are then compared with the corresponding (new) observation for the same variable. According to the Kalman gain, the coefficients can then be systematically updated in order to minimise the one period ahead forecast error. That property makes the Kalman filter convenient for modelling learning and the acquisition of new information:[5] it incorporates rational learning behaviour by market participants, in terms of minimising short-run forecasting errors. Hence we estimated the following state space model:

$$i_t = D_t X_t + \varepsilon_{1,t}, \tag{21.27}$$

where 21.27 is the measurement equation, and

$$D_t = D_{t-1} + \varepsilon_{2,t}, \tag{21.28}$$

with $\varepsilon_{a,t} \sim i.i.d.\left(0, \sigma^2_{\varepsilon_a}\right)$ for $a = 1, 2$
is the state equation. In Eq. (21.27) it is the British two year interest rate; and X_t is a set of determining variables such as the British base rate (the monetary instrument in the British case), the German two year interest rate (to represent ERM influences), the British ten year interest rate, or the US two year interest rate (to represent world markets). D_t is a matrix of estimated parameters, including a time-varying constant term, which represents a time-varying country-specific risk premium. In either case, the rationale of 21.28 is that agents only update the parameters of the model once an unforeseen shock occurs ([21.14]).

The question now is how the parameters may be updated to reflect learning. [21.24] shows that, in the case of an exogenous shock, the parameters are optimally updated as

$$d_{t|t} = d_{t|t-1} + K_t\left(i_t - X_t d_{t|t-1}\right) \tag{21.29}$$

where $d_{t|s}$ denotes the estimate of the state d at time t conditional on the information available at time s. The interesting part of Eq. (21.29) is the term in brackets. It shows the forecast error. Hence, the current parameters are updated according to the forecast error resulting from an estimated parameter which did not contain the additional information revealed in the current

[5] Indeed the Kalman filter is widely used in finance and macroeconomics as a learning algorithm, see for example [21.14], [21.5], [21.6], [21.25]. [21.19] shows that the Kalman filter is a special neural network. Hence the Kalman filter can be regarded as a procedural learning algorithm.

period. This forecast error in turn affects the Kalman gain. Thus the Kalman gain may be calculated according to

$$K_t = P_{t|t-1} X_t' \left(X_t P_{t|t-1} X_t' + \Xi \right)^{-1} \tag{21.30}$$

where $P_{t|s}$ is the variance of the forecast error at time t conditioned on the system at time s and Ξ is the covariance matrix of $\varepsilon_{2,t}$. In other words, the updating process depends on the one period forecast error and its distribution in the past.

As it stands, the measurement Eq. (21.27) would allow us to test the expectations hypothesis, as well as test for uncovered interest parity, time varying risk premia, and monetary efficiency. In this paper we focus on monetary efficiency, in terms of how the short end of the term structure is affected by the monetary instrument.[6]

To estimate our term structure model, Eqs. 21.27 and 21.28, we used monthly data from the Bank of England, Federal Reserve, and the Bundesbank. The sample runs from 1982:1 to 1998:10. In the following section, we analyse the effects of one (significant) shock, namely the collapse of the ERM in 1992. We investigate the parameter changes before the shock, during the shock, after the shock in 1992:10; and compare them with the parameters at the end of the sample in 1998:10. These different regression results then allow us to infer the changes in the gain and in the phase shift over different periods of time.

21.4 Empirical Results

21.4.1 Results in the Time Domain

The measurement Eq. (21.27) for Britain has the following form for all periods:

$$\begin{aligned} i_{t,2}^{brit} &= \alpha_{1,t} + \alpha_{2,t} BaseRate_t + \alpha_{3,t} i_{t,2}^{ger} + \alpha_{4,t} i_{t,10}^{brit} + \alpha_{5,t} i_{t,2}^{US} \\ &+ \alpha_{6,t} BaseRate_{t-1} + \alpha_{7,t} i_{t-1,2}^{brit} + \alpha_{8,t} i_{t-1,2}^{US} + \alpha_{9,t} i_{t-1,10}^{brit} + \varepsilon_t \end{aligned} \tag{21.31}$$

where $i_{t,m}^{a}$ is the interest rate of country a (Britain, Germany, USA) at time t with a term to maturity of m years. The derivation of the structural form and all tests can be found in [21.18]. In this paper, we are particularly interested in the parameter values which held during the ERM crisis in 1992. The following table gives the estimated parameter values at different points in time.

[6] see also [21.9], who examine some of the other characteristics.

Table 21.1. Parameter values for the British regression for different points in time

Time	α_1	α_2	α_3	α_4	α_5	α_6	α_7	α_8	α_9
1992 : 08	0.585	0.433	0.038	0.449	0.572	−0.271	0.673	−0.341	−0.261
1992 : 10	0.576	0.435	0.036	0.405	0.590	−0.274	0.666	−0.321	−0.257
1993 : 01	0.570	0.441	0.036	0.392	0.596	−0.268	0.669	−0.327	−0.261
1998 : 10	0.481	0.422	0.030	0.408	0.599	−0.290	0.689	−0.332	−0.280

21.4.2 The Gain and the Phase Shifts in the Implementation of British Monetary Policy 1992–98

[21.17] points out that interpreting a time-dependent spectrum is not easy. However, even when an equilibrium has not been reached, we can still interpret different spectral gains as being valid at a particular point in time. For example, looking at 1998:10 in Fig. 21.1, we see that variations in the base rate have caused various cycles in the interest rate payable on bonds with two years maturity. Although these cycles may not have died down before the next shock or base rate change hits the system, it is still interesting to know what cycles were caused in the short-term what the consequences of dynamic behaviour were. However, if the parameters do eventually converge, then the end of a cycle will be reached. In that case, the gain (and the phase angle) can be interpreted as their steady state or permanent values as usual.

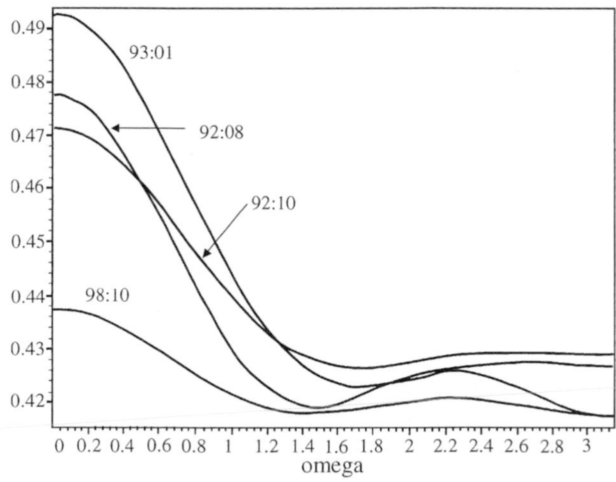

Fig. 21.1. Gains at different points in time for the UK base rate

In practice, the calculated gains for the impact of the UK base rate on interest rates with a two year maturity value, reported in Fig. 21.1, show that these cycles were quite stable (the gain is always less than unity). In fact, changes in the UK base rate set up both long and short cycles in short term interest rates, which have amplitudes of less than half the size of the original movements in the base rate. However, the short cycles (less than 6 months in length) are clearly weaker than the long cycles (i.e. those with a periodicity of 1 year and upwards). Hence UK monetary policy has had both long term and short term consequences for monetary conditions in the economy.

Of more interest perhaps are the results which show how these long and short term cyclical characteristics changed around the ERM crisis in September 1992. Fig. 21.1 demonstrates that the immediate effect of the ERM crisis was a reduction of the long-term value of the gain. However, the slope was also affected, i.e. shorter cycles gained weight, relative to longer cycles. That would reflect greater short term uncertainty or confusion. The question though is whether this change is significant. The confidence intervals provide a convenient method to analyse possible changes of the gain at each frequency. This is the reason why we look for structural breaks in the frequency domain. It is possible that structural breaks appear only for certain frequencies but not for all.

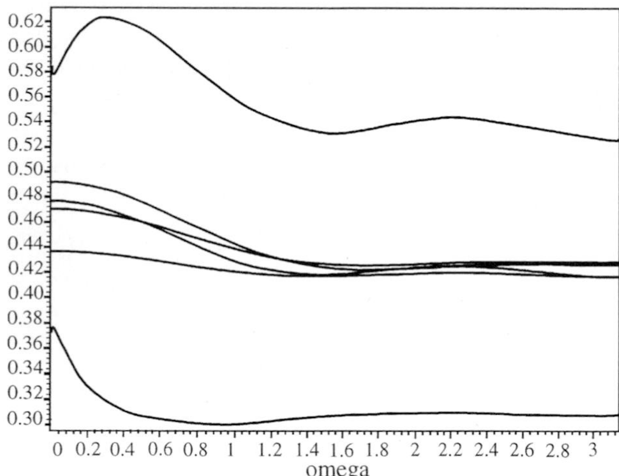

Fig. 21.2. The gains and the confidence intervals

So, when we take the confidence intervals into account (Fig. 21.2), we see that all these changes are actually well within the limits of their 95% confidence intervals.[7] So while there has no doubt been some changes in behaviour, the ability of agents to learn and adapt has not produced statistically significant changes in the underlying behaviour of interest rates. That confirms our conclusion that, contrary to popular wisdom, British monetary policy and market behaviour has actually remained remarkably stable throughout the tensions and uncertainties of the 1990s. This result is remarkable since

[7] We only report the confidence interval for the 92:08 results here, since a possible structural change would cause the gain to leave that confidence interval.

the Kalman filter is explicitly designed to detect structural breaks when they occur. In this case, however, we do observe small changes in behaviour even if they turn out not to be significant. Without a time-varying approach we would simply have had to assume that the estimated coefficients were all constant.[8] In other words, the Kalman filter has actually shown that they are constant, although there was no restriction to force that result.

We now turn our attention to the phase shift.

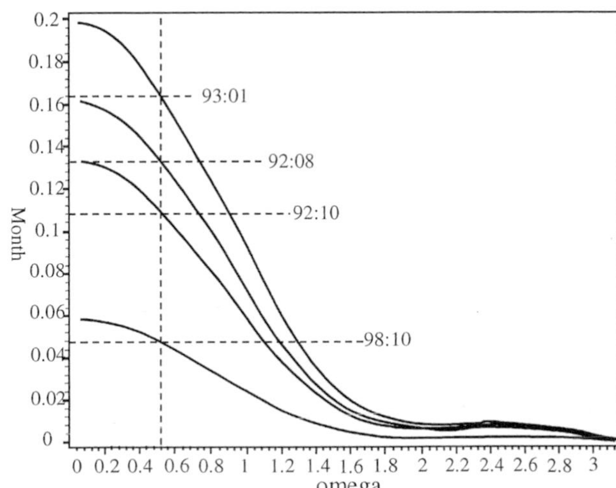

Fig. 21.3. Phase shifts for different points in time of the base rates

Fig. 21.3 shows the phase shift, or lead-lag relationship, between the base rate and the two year interest rate. Before the ERM crisis (1992:08), one year cycles of the base rate were 4.5 days (i.e. 0.14 of a 30 day month) ahead of the two year interest rate. But the immediate effect of the crisis was to shorten this lead to 3.3 days (1992:10). However, the base rate's lead then increased to 5.1 days after the crisis (1993:01). And, five years later, learning had shortened that lead again: it was down to 1.5 days, for one year cycles, in 1998.

The main result therefore remains that learning leads to shorter lead times. That shows that agents were able to anticipate quite quickly what was going on in the market. However, as in the changes to the gain, the changes of the phase shift are well within the limits of the confidence interval (Fig. 21.4). Hence, from a statistical point of view, these changes cannot be distinguished from one another; and the underlying relationships between the monetary variables remains effectively unchanged, before and after the crisis.

[8] That also implies that no learning could probably take place. We explicitly allow for learning, which means we may well depart from rational expectations' behaviour. Rational expectations are just a special case in our analysis.

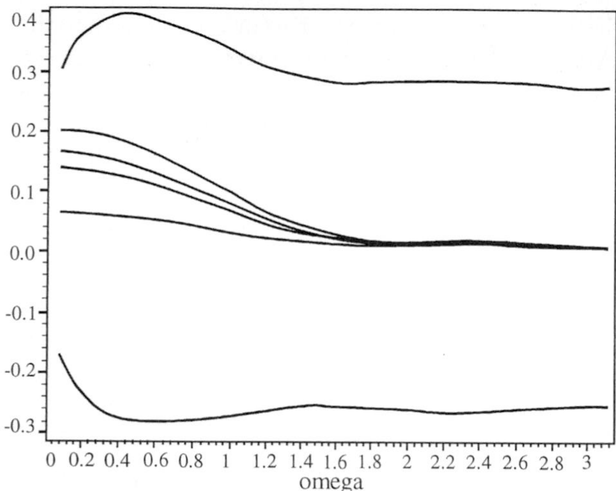

Fig. 21.4. Phase shifts and the confidence intervals

21.4.3 The Risk Premium on the UK Interest Rate

Finally, the risk premium can be modelled as the time-varying constant in Eq. (21.31). In this formulation, the risk premium contains liquidity preferences as well as uncertainty about changes of the expected exchange rate. Hence, the risk premium may be time varying due to changing preferences, or changing expectations, or both.

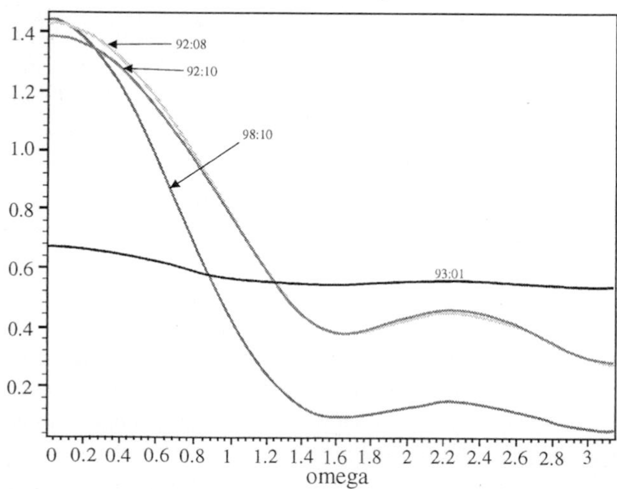

Fig. 21.5. Impact of the risk premium on British rates at different points in time

From Fig. 21.5 we can see that, at the beginning of the sample, the effect of the risk premium is mainly stabilising, but has (as one might expect) a stronger impact on the longer cycles than on the shorter cycles. This is still the case during the ERM crisis. However, an adjustment at that point led to a more evenly spread impact over the entire band by 93:01 - especially after the pound had left the ERM system altogether. Later on, the gain gradually returned to its previous shape. Although, the impact on longer cycles is now substantially less than before, and decreasing, the risk premium remains more stabilising in the short term since the impact on shorter cycles has been reduced.

Finally, Fig. 21.6 shows that by 1998 the risk premium term had moved outside its lower confidence band for values of ω bigger or equal 0.6. That implies a statistically significant and permanent reduction in short and medium term risk in the British financial markets. Before that, i.e. by 1993 but after the onset of the ERM crisis, there had been a significant, but temporary, reduction in the long term risk (i.e. for 1993 and $\omega < 0.9$) - the crisis having made the financial markets focus on the short term implications of any changes in monetary policy, to the neglect of the long term consequences. Later, as learning begins to be effective, we move to the reductions in short term risk revealed in the 1998 results, with no increase in long term risk.

The interesting finding here is that the risk premium is the only variable whose changes were big enough for them to leave their confidence interval entirely. This was particularly obvious for 1993:01. And although the risk premium moved back into the interval in 98:10, it is still outside the confidence interval for frequencies higher than 0.6 (i.e. for cycles of 10 months or less). From this we can conclude that, in Britain, adjustments to the ERM crisis took place through changes in the risk premium. The risk premium/liquidity preference variable is the only variable which changed significantly during the ERM crisis. It acted to absorb the volatility or uncertainty effects of that crisis, and allowed the underlying interest rates to remain constant.

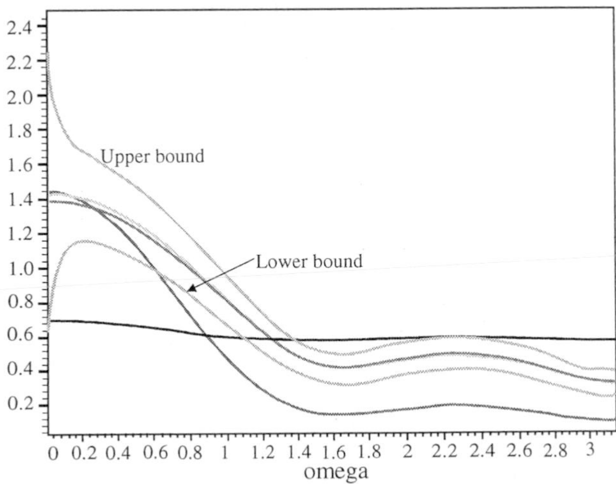

Fig. 21.6. Confidence intervals for the British risk premium

In the context of our model, that means agents in the British markets learnt rather rapidly how the new monetary regime was going to work, and how to adapt their behaviour to fit.

21.5 Conclusion

In this paper we analysed British monetary policy by assuming that agents are learning. We modelled learning by using the Kalman filter. We find that British monetary policy has mainly long term effects. That in turn implies that short term volatility is relatively small. Furthermore, British monetary policy appears to be very stable in terms of changes in cyclical movements. The ERM crisis did not lead to significant changes of the monetary transmission mechanism. However, the ERM crisis did lead to a change of the risk premium. We can therefore conclude that shocks are absorbed by the risk premium. The change of the risk premium enables the monetary transmission mechanism to remain constant. We showed how to derive these results starting with a time domain regression and analysed its dynamic properties in the frequency domain. The advantage of this method is that possible structural changes could be detected with a relatively small sample. Structural changes could even be identified towards the end of the sample where many conventional time domain tests would fail. In this paper, we only analysed the British monetary transmission mechanism. The question is whether there are significant differences of the transmission mechanism across countries (see for example [21.11]). This is from particular importance for the Euro countries. However, if Britain wants to join the Euro then the transmission mechanisms should be similar. The importance of these results for academic research is that they could not have been uncovered without a technique which allows a frequency domain analysis to be applied to the problem. We have adopted an indirect estimation approach based on autoregressive distributed lag regression estimates for that purpose. The importance for the policy makers is that if monetary transmission differences exist - perhaps contrary to popular wisdom, then they can tackle the problem of how to design institutional arrangements that can accommodate these differences. Again that cannot be done without the appropriate frequency domain analysis.

References

21.1 Akerlof, G. (1970): The Market for Lemons: Qualitative Unvertainty and the Market Mechanism, Quarterly Journal of Economics, 89, 488–500
21.2 Black, F. (1989): Noise, Journal of Finance, 48, 540–555
21.3 Easley, D., O'Hara, M. (1992): Adverse Seletion and Large Trade Volume: The Implications for Market Efficiency, Journal of Financial and Quantitative Analysis, 27, 185–208

21.4 Frydman, R. (1983): A Distinction between the Unconditional Expectational Equilibrium and the Rational Expectations Equilibrium, In: Frydman, R., Phelps, E. S. (Eds.): Individual Forecasting and Aggregate Outcomes, 139–146, Cambridge University Press

21.5 Garratt, A., Hall, S. G. (1997): E-equilibria and Adaptive Expectations: Output and Inflation in the LBS Model, Journal of Economic Dynamics and Control, 21, 1149–1171

21.6 Garratt, A., Hall, S. G. (1997): The Stability of Expectational Equilibria in the LBS Model, In: Allen, C., Hall, S. (Eds.): Macroeconomic Modelling in a Changing World, 217–246, Wiley.

21.7 Hamilton, J. (1994): Time Series Analysis, Princeton University Press

21.8 Hendry, D. F. (1995): Dynamic Econometrics, Oxford University Press

21.9 Hughes Hallett, A., Richter, C. (2001): Spectral Analysis as a Tool for Financial Policy: an Analysis of the Short End of the British Term Structure, Paper given the 7th Conference of the Society for Computational Economics, Yale University, and under submission to Computational Economics

21.10 Hughes Hallett, A., Richter, C. (2001): Are Capital Markets Efficient? Evidence from the Term Structure of Interest Rates in Europe, Ludwig Boltzmann Institute Working Paper no 2001.22, Vienna, December 2001

21.11 Hughes Hallett, A., Richter, C. (2002): A Comparative Dynamics Analysis of British and German Monetary Policy in the 1990: A Spectral Analysis Approach, mimeo

21.12 Jenkins, G. M., Watts, D. G. (1968): Spectral Analysis and its Applications, Holden-Day

21.13 Laven, G., Shi, G. (1993): Zur Interpretation von Lagverteilungen, University of Mainz discussion paper, No. 41

21.14 Lucas, R. E. (1976): Econometric Policy Evaluation: A Critique, In: Brunner, K., Meltzer, A. (Eds.): The Phillips Curve and Labor Markets, 19–46, North-Holland

21.15 Nerlove, M., Grether, D. M., Carvalho, J. L. (1995): Analysis of Economic Time Series Academic Press

21.16 Perron, P. (1989): The Great Crash, the Oil Price Shock and the Unit Root Hypothesis, Econometrica, 57, 1361–1401

21.17 Priestley, M. B. (1996): Wavelets and Time Dependent Spectral Analysis, Journal of Time Series Analysis, 17, 1, 85–103

21.18 Richter, C. (2001): Learning and the Term Structure of Interest Rates in Britain and Germany, PhD thesis, University of Strathclyde,

21.19 Salmon, M. (1995): Bounded Rationality and Learning: Procedural Learning, In: Kirman, A, Salmon, M. (Eds.): Learning and Rationality in Economics, 236–275

21.20 Shefrin, H. (1999): Beyond Greed and Fear: Understanding Bahvioral Finance and the Psychology of Investing, Harvard Business School Press

21.21 Shleifer, A. (2000): Inefficient Markets, Oxford University Press

21.22 Stock, J. H., Watson, M. W. (1991): Variable Trends in Economic Time Series, In: Engle, R. F.,Granger, C. W. J. (Eds.): Long-Run Economic Relationships, 17–49, Oxford University Press

21.23 Thaler, R. H. (1994): Quasi Rational Economics, Russell Sage Foundation

21.24 Wells, K. (1996): The Kalman Filter in Finance, Kluwer Academic Publishers

21.25 Whitley, J. D. (1994): A Course in Macroeconomic Modelling and Forecasting, Harvester Wheatsheaf

21.26 Wolters, J. (1980): Stochastic Dynamic Properties of Linear Econometric Models, Springer- Verlag

22. International Transmission of Business Cycles: a Self-organizing Markov-Switching State-Space Model

Morikazu Hakamata[1] and Genshirou Kitagawa[2]

[1] The Graduate University for Advanced Studies, 4-6-7 Minami-Azabu, Minato-ku, Tokyo 106-8569, Japan
email: hakamata@ism.ac.jp

[2] The Institute of Statistical Mathematics, 4-6-7 Minami-Azabu, Minato-ku, Tokyo 106-8569, Japan

In this paper, we incorporate a Self-organizing state-space model into the Markov-switching model, and propose the Self-organizing Markov-switching state-space (SOMS) model. The approximate Monte Carlo filter is applied for state estimation, including the latent Markov chain, of this model. The SOMS model allows us to evaluate complex systems. We apply it to an analysis of international transmission of business cycles between the U.S. and Germany.

22.1 Introduction

International transmission of the business cycle is an important issue in macroeconomic analysis. Resolving the puzzle of the relationships and linkages between business cycles among some industrial countries is important for policy development and for business managers. In this paper, we propose the Self-organizing Markov-switching state-space (SOMS) model to analyze the international transmission of the business cycle. This approach captures the behavior of bivariate time series including their time-varying correlation. Our SOMS model is derived by incorporating a Self-organizing state-space model into the Markov-switching model.

Although many important problems in time series analysis can be solved by using the ordinary state-space model, more complex systems such as a Markov-switching model are sometimes required in general nonlinear or non-Gaussian situations [22.6]. Recent improvements in computing power and the development of Monte Carlo-based algorithms [22.7] have made the use of nonlinear and non-Gaussian time series models feasible. The Self-organizing state-space model was proposed by Kitagawa [22.8] to deal with the model identification problem associated with the use of the recursive Monte Carlo method. In this approach, the state vector is augmented by the unknown parameters of the model and the state and the parameters are estimated simultaneously by the recursive filter and smoother. In the method used in this paper, accurate approximations of the marginal posterior densities of the state and the parameters are obtained by the Monte Carlo filter [22.7].

The Markov-switching model is a useful tool for capturing the dynamic structures of time series that change under the unknown discrete Markov chain regimes. Hamilton [22.3] first applied a Markov-switching autoregressive model to the analysis of the US business cycle by using quarterly data. Hamilton's [22.3] model assumes that the GNP data shifts between the low and high growth states or regimes according to a first-order Markov process. Variants of Hamilton's [22.3] model have been used in single country analyses of the business cycle. Kim and Yoo [22.5] and Kontolemis [22.9] extended the univariate Markov-switching model to the multivariate version to capture the co-movement of some economic indicators.

The paper is organized as follows. In section 22.2, we present the sequential Monte Carlo filter and smoother. The Self-organizing state-space model, including the method of estimating the unknown state vector and the structural parameters, is described in section 22.3. The SOMS model is presented in section 22.4. Section 22.5 provides an application. Finally, section 22.6 concludes the paper.

22.2 Monte Carlo Filter/Smoother

22.2.1 General State-Space Model

Consider a nonlinear non-Gaussian state-space model [22.6] for the time series y_n as follows

$$x_n = F(x_{n-1}, v_n), \tag{22.1a}$$
$$y_n = H(x_n, w_n), \tag{22.1b}$$

where x_n is an unknown state vector, v_n and w_n are the system noise and the observation noise with densities $q(v)$ and $r(w)$, respectively. (22.1a) and (22.1b) are known as the system model and the observation model, respectively. The initial state x_0 is assumed to be distributed according to the density $p_0(x)$. $F(x, v)$ and $H(x, w)$ are possibly nonlinear functions of the state and the noise inputs.

The above nonlinear non-Gaussian state-space model specifies the conditional density of the state given the previous state $Q(x_n|x_{n-1})$ and that of the observation given the state $R(y_n|x_n)$. This is the essential feature of the state-space model, and it is sometimes convenient to express the model in this general form using conditional distributions

$$x_n \sim Q(x_n|x_{n-1}), \tag{22.2a}$$
$$y_n \sim R(y_n|x_n). \tag{22.2b}$$

With this model, it is also possible to treat the discrete process [22.6].

22.2.2 Sequential Monte Carlo Filter/Smoother

The filtering formula for general state-space models can easily be implemented by using a sequential Monte Carlo method [22.1] [22.2] [22.7]. This method approximates each distribution by using many "particles," which can be considered as realizations from that distribution. Specifically, assume that each distribution is expressed by using m "particles" as follows

$$\begin{array}{lll} \{v_n^{(1)}, \ldots, v_n^{(m)}\} & \sim p(v_n) & \text{System noise} \\ \{p_n^{(1)}, \ldots, p_n^{(m)}\} & \sim p(x_n|Y_{n-1}) & \text{Predictor} \\ \{f_n^{(1)}, \ldots, f_n^{(m)}\} & \sim p(x_n|Y_n) & \text{Filter} \\ \{s_{n|N}^{(1)}, \ldots, s_{n|N}^{(m)}\} & \sim p(x_n|Y_N) & \text{Smoother} \end{array}$$

That is, we approximate the distributions with the empirical distributions determined by the m particles. Then it can be shown that a set of realizations expressing the one step ahead predictor $p(x_n|Y_{n-1})$ and the filter $p(x_n|Y_n)$ can be obtained recursively as follows

[Monte Carlo Filter]

1. Approximate the initial distribution by $f_0^{(j)} \sim p_0(x)$ for $j = 1, \ldots, m$.
2. Repeat the following steps for $n = 1, \ldots, N$.
   a. Generate system noise $v_{0,n}^{(j)} \sim p(v_0)$ and $v_{1,n}^{(j)} \sim p(v_1)$ for $j = 1, \ldots, m$.
   b. Approximate the predictive distribution by $p_n^{(j)} = Q(x_n|f_{n-1}^{(j)})$ for $j = 1, \ldots, m$.
   c. Compute the Bayes importance weight by $\alpha_n^{(j)} = R(y_n|p_n^{(j)})$ for $j = 1, \ldots, m$.
   d. Approximate the filter distribution, generating $f_n^{(j)}$, $j = 1, \ldots, m$ by the re-sampling of $p_n^{(1)}, \ldots, p_n^{(m)}$ with weight proportional to $\alpha_n^{(j)}$, $j = 1, \ldots, m$.

Fig. 22.1 illustrates one cycle of the Monte Carlo filtering with $m = 9$ for simplicity. The particle approximating the predictor is generated from a pair of particles approximating the previous filtered state and the system noise. Then the predictive distribution is approximated by the empirical distribution $m^{-1} \sum_{j=1}^{m} I(x, p_n^{(j)})$. Note that $I(x, a)$ is the indicator function defined by $I(x, a) = 0$ if $x < a$ and $I(x, a) = 1$ otherwise. In the filtering step, the filter distribution is obtained by changing these equal weights to ones proportional to $\alpha_n^{(j)}$. Finally, by re-sampling (sampling with replacement), the filter distribution is re-approximated by using the particles $f_n^{(j)}$ with equal weights $1/m$.

The state-space model, (22.2a) and (22.2b), contains several unknown parameters. The vector of such unknown parameters is hereafter denoted by θ. The likelihood of time series model parameterized by θ is given by

$$L(\theta) = p(y_1, \ldots, y_N|\theta) = \prod_{n=1}^{N} p(y_n|Y_{n-1}, \theta), \tag{22.3}$$

where $p(y_n|Y_{n-1}, \theta)$ is the conditional density of y_n given Y_{n-1} and is given by

$$\begin{aligned} p(y_n|Y_{n-1}, \theta) &= \int p(y_n|x_n, \theta) p(x_n|Y_{n-1}, \theta) dx_n \\ &\approx \frac{1}{m} \sum_{j=1}^{m} \alpha_n^{(j)}, \end{aligned} \tag{22.4}$$

where $\alpha_n^{(j)}$ is the importance weight of the j-th particle obtained in the Monte Carlo filter. In this case, because the likelihood contains the sampling error due to the approximation (22.4), it is difficult to obtain precise maximum likelihood estimates of the parameters. This problem will be considered later in the next section.

[Monte Carlo Smoother]
An algorithm for smoothing [22.7] is obtained by replacing Step 2-d. of the filtering algorithm by

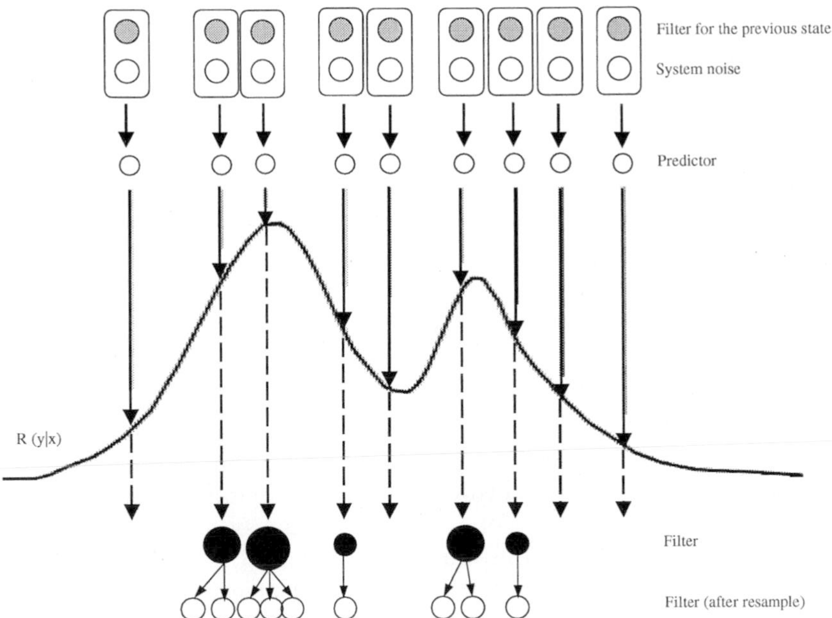

Fig. 22.1. One cycle of Monte Carlo filter ($m = 9$)

2-d*. For fixed L, generate $\{(s_{n-L|n}^{(j)}, \cdots, s_{n-1|n}^{(j)}, s_{n|n}^{(j)})^T, j = 1, \ldots, m\}$ by the re-sampling of $\{(s_{n-L|n-1}^{(j)}, \cdots, s_{n-1|n-1}^{(j)}, p_n^{(j)})^T, j = 1, \ldots, m\}$ with $f_n^{(j)} = s_{n|n}^{(j)}$.

This is equivalent to applying the L-lag fixed lag smoother. Increasing the lag, L, will improve the accuracy of the $p(x_n|Y_{n+L})$ as an approximation to $p(x_n|Y_N)$, while it is very likely to decrease the accuracy of $\{s_{n|N}^{(1)}, \cdots, s_{n|N}^{(m)}\}$ as representatives of $p(x_n|Y_{n+L})$. Because $p(x_n|Y_{n+L})$ usually converges rather quickly to $p(x_n|Y_N)$ as L increases, it is recommended to use a not-so-large L.

22.3 Self-organizing State-Space Model and Self-Tuning of Parameters

Rather than estimating θ by maximum likelihood, we consider Bayesian estimation by augmenting the state vector as

$$z_n = \begin{bmatrix} x_n \\ \theta \end{bmatrix}. \tag{22.5}$$

The state-space model for this augmented state vector z_n is given by

$$z_n = F^*(z_{n-1}, v_n), \tag{22.6a}$$

$$y_n = H^*(z_n, w_n), \tag{22.6b}$$

where the nonlinear functions $F^*(z, v)$ and $H^*(z, w)$ are defined by $F^*(z, v) = [F(x, v), \theta]^T$ and $H^*(z, w) = H(x, w)$, respectively.

Assume that we obtain the posterior distribution $p(z_n|Y_N)$. Since the original state vector x_n and the parameter vector θ are included in the augmented state vector z_n, the marginal posterior densities of the parameter and of the original state are given by

$$p(x_n|Y_{n-1}) = \int p(x_n, \theta|Y_{n-1})d\theta, \tag{22.7a}$$

$$p(\theta|Y_{n-1}) = \int p(x_n, \theta|Y_{n-1})dx_n. \tag{22.7b}$$

This Bayesian approach to the simultaneous estimation of the parameters of the state-space model can be easily extended to a time-varying parameter situation; that is, where $\theta = \theta_n$ evolves over time n. In this case, it is necessary to specify an appropriate model for the evolution over time of the time-varying parameter, such as the random walk model, $\theta_n = \theta_{n-1} + u_n$. In addition, if the Monte Carlo filter is used to implement the Self-organizing state-space model, the original formulation of the Self-organizing state-space model, which lacks system noise for the parameter, does not work well in practice. Hence, inclusion of the system noise enables application of the Monte Carlo filter and smoother.

22.4 Self-organizing Markov-Switching State-Space Model

We first consider a bivariate Markov switching model for the time series $y_{1,n}$ and $y_{2,n}$, which are for different countries' economic indicators, as follows

$$y_{k,n} = t_{k,n} + w_{k,n}, \qquad (k = 1, 2) \tag{22.8}$$

where $t_{k,n}$ is an unknown individual trend component, $w_{k,n}$ is Gaussian observation noise with zero mean and variance σ_k^2 for each time series. For simplicity, we assume that the trend component $t_{k,n}$ follows the first-order stochastic trend model with individual drift as follows

$$t_{k,n} = t_{k,n-1} + v_{k,n}, \tag{22.9}$$

where $v_{k,n}$ is Gaussian system noise depending on a $\{0,1\}$-valued Markov chain $S_{k,n}$; that is,

$$p(v_{k,n}|S_{k,n} = j_k) \sim \begin{cases} N(\mu_{k,0}, \tau_k^2), & \text{if } j_k = 0, \\ N(\mu_{k,1}, \tau_k^2), & \text{if } j_k = 1. \end{cases} \tag{22.10}$$

The Markov chain $S_{k,n}$ indicates the "regime" of economic activity. If $j_k = 0$, the economy of the k-th country is expanding. If $j_k = 1$, it is contracting. The transition probability of $S_{k,n}$ is given by

$$\Pr(S_{1,n} = j_1, S_{2,n} = j_2 | S_{1,n-1} = i_1, S_{2,n-1} = i_2) = p_{i_1 i_2, j_1 j_2} \tag{22.11}$$

with

$$\sum_{j_1=0}^{1} \sum_{j_2=0}^{1} p_{i_1 i_2, j_1 j_2} = 1. \tag{22.12}$$

This can be rewritten in transition matrix form as follows

$$P = \begin{pmatrix} p_{00,00} & p_{11,00} & p_{01,00} & p_{10,00} \\ p_{00,11} & p_{11,11} & p_{01,11} & p_{10,11} \\ p_{00,01} & p_{11,01} & p_{01,01} & p_{10,01} \\ p_{00,10} & p_{11,10} & p_{01,10} & p_{10,10} \end{pmatrix}. \tag{22.13}$$

The Markov switching model allows for several types of transmission of the business cycle between two countries; for example, precedent, positive, and negative simultaneous regime shifts. Given the assumption of this model specification, the joint conditional distribution of $t_{k,n}$ and $S_{k,n}$ given the previous values $t_{k,n-1}$, $S_{1,n-1}$ and $S_{2,n-1}$ is given by

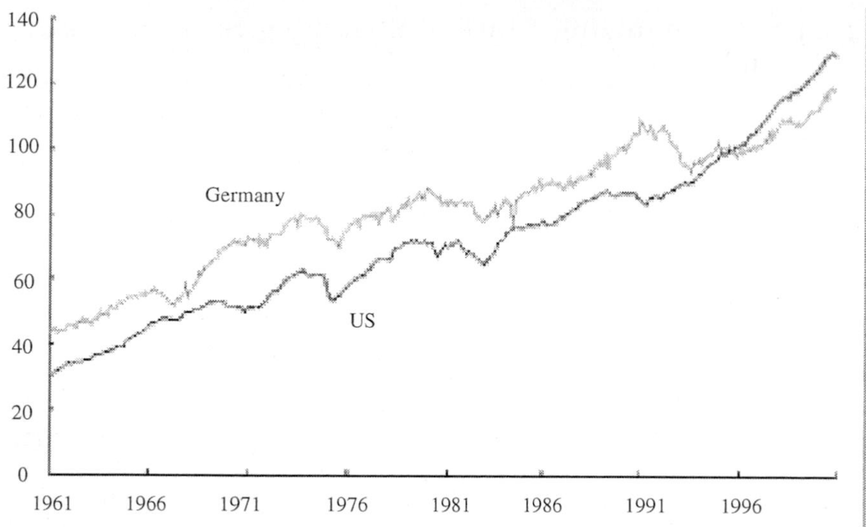

Fig. 22.2. U.S. and German industrial production indices

$$\begin{aligned} p(t_{1,n}, t_{2,n}, S_{1,n}, S_{2,n} | t_{1,n-1}, t_{2,n-1}, S_{1,n-1}, S_{2,n-1}) \\ = \Pr(S_{1,n}, S_{2,n} | S_{1,n-1}, S_{2,n-1}) \\ \times p(t_{1,n} | t_{1,n-1}, S_{1,n}) p(t_{2,n} | t_{2,n-1}, S_{2,n}), \end{aligned} \quad (22.14)$$

The terms on the right-hand-side of the above equation are specified by (22.10) and (22.11), respectively.

Then, by defining the state vector x_n as

$$x_n = (t_{1,n}, t_{2,n}, S_{1,n}, S_{2,n})^T, \quad (22.15)$$

our bivariate Markov-switching model can be expressed in terms of the general state-space model in (22.2a) and (22.2b). In addition, we can use the Self-organizing state-space model for simultaneous estimation of the state and the parameters. Here, we obtain the Self-organizing Markov switching state-space (SOMS) model by incorporating the Self-organizing state-space form into the Markov switching model. For that purpose, we define an augmented state vector by $z_n = (x_n^T, \theta_n^T)^T$ where x_n is the original state vector defined in (22.15) and θ is the parameter vector defined in the previous section. We assume a (vector) random walk model

$$\theta_n = \theta_{n-1} + u_n, \quad (22.16)$$

where u_n is Gaussian white noise with zero mean and covariance diag$\{\xi_1, \ldots, \xi_{20}\}$. All of the structural parameters, σ_k^2, τ_k^2, $\mu_{k,0}$, $\mu_{k,1}$ and $p_{i_1 i_2, j_1 j_2}$ ($i_1, i_2, j_1 \in \{0, 1\}, j_2 = 0$) are assumed to be time-varying, and the parameter vector is defined by

$$\theta_n = (\sigma_{k,n}^2, \tau_{k,n}^2, \mu_{k,0,n}, \mu_{k,1,n}, p_{i_1 i_2, j_1 j_2, n})^T. \quad (22.17)$$

[Monte Carlo Filter for the SOMS Model]

We can easily implement the Monte Carlo filter for the SOMS model by replacing step 2-b of the general version described in the previous section with following steps

b[1]. Generate a Markov chain $S_{1,n}^{(j)}$ and $S_{2,n}^{(j)} \in \{0, 1\}$ given $S_{1,n-1}^{(j)}$, $S_{2,n-1}^{(j)}$ and the time-varying transition probabilities $p_{i_1 i_2, j_1 j_2, n-1}$, for $j = 1, \ldots, m$.

b[2]. Approximate the predictive distribution by $p_n^{(j)} \sim N(\mu_{k,j_k}, \tau_k^2; j_k = 0, 1)$ given $S_{1,n}^{(j)}$ and $S_{2,n}^{(j)}$, for $j = 1, \ldots, m$.

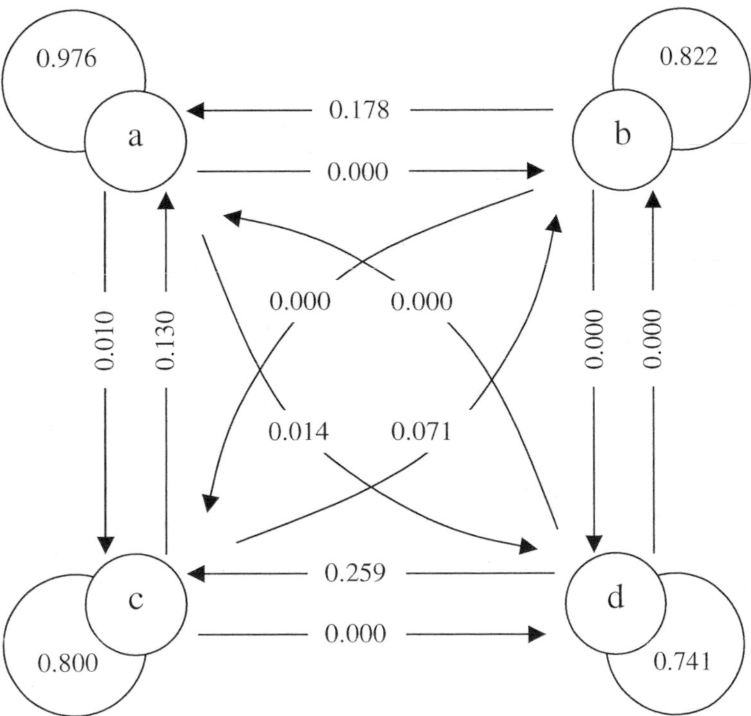

Fig. 22.3. Transition probabilities of the approximate maximum likelihood estimate. Note circle a is for $S_{1,n} = 0$ and $S_{2,n} = 0$, circle b is for $S_{1,n} = 1$ and $S_{2,n} = 1$, circle c is for $S_{1,n} = 0$ and $S_{2,n} = 1$, circle d is for $S_{1,n} = 1$ and $S_{2,n} = 0$

22.5 Application

22.5.1 Data

In this section, we apply the SOMS model to analyze transmission of the business cycle between the U.S. and Germany. Because each country's business cycle is not observed, we must use the observable economic indicators as proxy variables. We use monthly industrial production indices (seasonally adjusted, and equal to 100 in the base year, 1995) for the U.S. and Germany from January 1961 to December 2000. The indices are provided by the OECD and Fig. 22.2 shows the original time series.

22.5.2 Results

Before moving on to the results of the SOMS model, it is desirable to describe the maximum likelihood estimate (MLE) of the Markov switching model in (22.8) and (22.9) with fixed-value parameters, which is not the SOMS model. Fig. 22.3 illustrates the estimated transition probabilities between four regimes. These approximate MLE are obtained by the Hamilton's filter [22.3] and the maximum likelihood method.

We then turn to an application of the SOMS model. The higher the dimension of the model, the greater the computational burden; therefore, we attempt to self-turn the parameters with the SOMS model. Fig. 22.4 shows the smoothed joint probabilities $\Pr(S_{1,n} = j_1, S_{2,n} = j_2)$ for $j_1, j_2 = 0, 1$ from the SOMS model with $\xi_1 = \cdots = \xi_{20} = 0.01$. The shadowed bars (in Fig. 22.4) denote those from Kim's smoothing [22.4] algorithm given the approximate MLE above. Note that the number of particles is $m = 20,000$, and the initial mean values of the parameter vector are $\sigma_{1,0}^2 = 0.03, \sigma_{2,0}^2 = 0.75, \tau_{1,0}^2 = 0.45^2, \tau_{2,0}^2 = 0.60^2, \mu_{1,0,0} = 0.30, \mu_{2,0,0} = 0.30, \mu_{1,1,0} = -0.50, \mu_{2,1,0} = -0.50$ and

$$P_0 = \begin{pmatrix} 0.97 & 0.01 & 0.01 & 0.01 \\ 0.01 & 0.97 & 0.01 & 0.01 \\ 0.01 & 0.01 & 0.97 & 0.01 \\ - & - & - & - \end{pmatrix}. \tag{22.18}$$

Although we set the initial distributions of the parameter vector in (22.17) approximately, the smoothed joint probabilities from the SOMS model are quite similar to the results from the approximate MLE. We obtained the results from a single run of the Monte Carlo filter and smoother. From Fig. 22.4, it is apparent that all regimes are highly persistent, and that the SOMS model succeeds in classifying the whole period into four regimes.

Therefore, it can be said that the SOMS model allows the estimate of the unobserved regimes of the business cycles between two countries without a MLE, and the capture of the international transmission and relationships of business cycles.

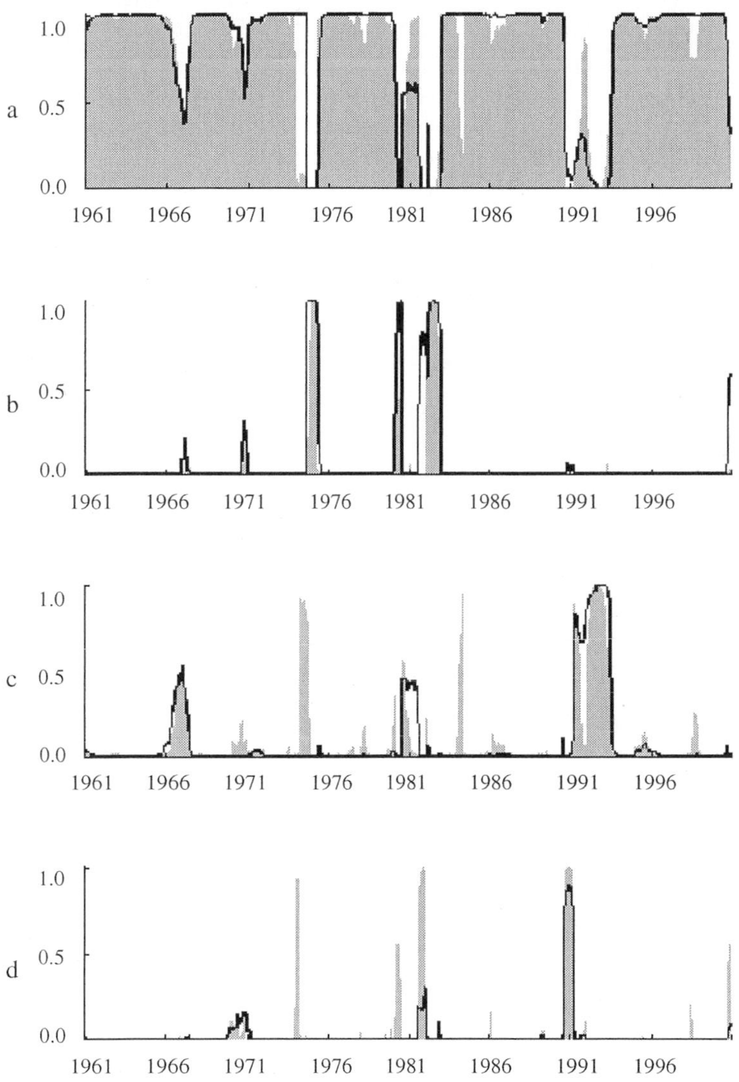

Fig. 22.4. Smoothed joint probabilities of the SOMS model. Note that a is for $\Pr(S_{1,n} = 0, S_{2,n} = 0)$, b is for $\Pr(S_{1,n} = 1, S_{2,n} = 1)$, c is for $\Pr(S_{1,n} = 0, S_{2,n} = 1)$, d is for $\Pr(S_{1,n} = 1, S_{2,n} = 0)$. The shadowed bars denote the smoothed join probabilities of the approximate maximum likelihood estimate

22.6 Conclusions

In this paper, we proposed the Self-organizing Markov-switching state-space (SOMS) model to evaluate international transmission of the business cycle. Our SOMS model is an effective analytical tool in understanding complex systems. Although, application of the Monte Carlo filter to the Markov-switching model may make parameter estimation difficult, we dealt with this by incorporating a Self-organizing state-space model into the Markov-switching model. In the SOMS model, the state and unknown parameters are estimated simultaneously. We applied our model to an analysis of US and German business cycles, and successfully captured characteristics of the transmission of, and relationship between, business cycles of different countries.

References

22.1 Doucet, A., de Freitas, N., Gordon, N. (2001): Sequential Monte Carlo Methods in Practice, Springer-Verlag

22.2 Gordon, N. J., Salmond, D. J., Smith, A. F. M. (1993): Novel approach to nonlinear/non-Gaussian Bayesian state estimation, IEEE Proceedings–F, 140(2), 107–113

22.3 Hamilton, J. D. (1989): A New Approach to the Economic Analysis of Nonstationary Time Series and the Business Cycle. Econometrica, 57(2), 357–384

22.4 Kim, C. J. (1994): Dynamic linear models with Markov-switching. Journal of Econometrics. 60, 1–22

22.5 Kim, M. J., Yoo, J. S. (1995): New index of coincident indicators: A multivariate Markov-switching factor model approach. Journal of Monetary Economics. 36, 607–630

22.6 Kitagawa, G. (1987): Non-Gaussian state-space modeling of nonstationary time series (with discussion). Journal of the American Statistical Association. 82(400), 1032–1063

22.7 Kitagawa, G. (1996): Monte Carlo filter and smoother for non-Gaussian nonlinear state-space model. Journal of Computational and Graphical Statistics. 5, 1–25

22.8 Kitagawa, G. (1998): Self-Organizing State-space Model. Journal of the American Statistical Association. 93(443), 1203–1215

22.9 Kontolemis, Z. G. (2001): Analysis of the US Business Cycle with a Vector-Markov-switching Model. Journal of Forecasting. 20, 47–61

Part VIII

Agent-Based Models

23. How Information Technology Creates Business Value in the Past and in the Current Electronic Commerce (EC) Era

Yasuo Kadono[1,2] and Takao Terano[1]

[1] University of Tsukuba, Otsuka 3-29-1, Bunkyo-ku, Tokyo 113-0012, Japan
[2] Management Science Institute Inc. (MSI)

The purpose of this paper is to discuss how IT, information technology, creates additional business value. Firstly, in order to articulate the IT added value creation mechanism, we introduce a new framework, called the "3C-DRIVE," composed of two axes: (1) 3C: Company, Competitor, and Customer, and (2) Direction, Readiness, Information Technology, and Value Evaluation. The 3C-DRIVE framework has bird's-eye-view, dynamic, and emerging characteristics. Secondly, we apply this framework to the existing research and perspectives regarding the relationship between business and IT. Thirdly, we design a mathematical model for computer simulation task and apply it to several conceivable cases, taking into consideration of the factors such as organization, readiness and business environment in addition to the IT investment. As a result we may conclude that complementary factors other than the IT investment greatly enhance the added business value from IT especially in the light of the perspective of the current EC era.

23.1 Introduction

What would be a compelling driving force for a corporate senior manager to positively address the benefits of the employment of so called IT to business development. This is a long standing worldwide issue. However, frameworks and perspectives to solve IT-related issues from corporate senior managers' viewpoints are still insufficient for the current Electronic Commerce (EC) era. We are still lacking a viewpoint of linkage between management and IT. As a result, most of IT-related decisions, including management decision, have been excessively dependent on the proposals from providers of IT products and services. This situation could be a cause of ineffective and excessive social costs. It is easy to imagine that the today's IT management is getting more and more complicated for top management because of the aspects such as personalization of IT (e.g., personal computers and the Internet) and multi-vendor computing. However, many companies react to EC and IT revolutions simply by increasing IT investments and blindly following the advanced cases, and most of them do not even evaluate what business values are created from the IT investments.

To study the added business value creation from IT, the following two points should be and must be discussed adequately:

- Definition of the business value created from IT and methods to measure it, and
- Factors related to the business value creation from IT and the creation mechanism.

In this paper, we focus on the latter point and discuss the following three issues:

- A hypothesis about factors and frameworks as related to business value creation from IT,
- A study of the outcomes and future tasks of existing research and the perspectives in comparison with our hypothesis, and
- The results and the interpretations of simulations with a simple model of our framework.

In the next section, we propose a hypothesis for a new framework: "3C-DRIVE." The 3C-DRIVE is a hypothesis about overall factors to discuss the business value creation from IT. This aims at an universal framework applicable for seedtime through the EC era of IT. In section 3, we survey representative existing research on IT and its value creation in comparison with 3C-DRIVE. In particular, we focus our attention on factors considered in existing researches and relationship among these factors. We discuss the applicability and limitations of existing frameworks in the EC era. In section 4, we develop a simple model based on 3C-DRIVE and simulate conceivable cases of companies. In this model, we consider complementary factors that affect the effectiveness of IT investments, such characteristics in 3C-DRIVE as IT investment stance of top management of a company and the competitiveness of the industry of which a company belongs to. Then, we interpret the simulation results along with typical cases of companies. Finally, we conclude our discussion and suggest the future work in the last section.

At the timing when we are getting into the EC era, we would think it is important to consider how an individual company ought to bring its structures and environments inside and outside the company to confront its need to this new situation, especially the IT-related issues, and hence to create new business values effectively. However, we have not done enough.

23.2 Framework of 3C-DRIVE

In this section, we first set the requirements for the framework of the business value creation from IT. Based on the extracted requirements, we propose 3C-DRIVE, a hypothesis about the framework to grasp overall picture of business value creation from IT [23.7]. We set the following requirements to discuss value creation mechanism, which are universally applicable to seedtime through the EC era.

- Bird's-eye view: To discuss the business value creation from IT in the EC era, the framework needs to have overlooking view on both corporate management and IT. This is because the management of a company is relativized through the context of customers and competitors, and because business and IT are two sides of the same coin. These tendencies are accelerated in the EC era, since one of key factors for success for a company is to pursue the unique value creation through networks which connect the company with the outside of the company.
- Dynamics and emergency: The relations between all factors in above stated overlooking view become easy to change in a dynamic and emergent fashion in the EC era. From the viewpoint of business management, the relations between companies, competitors/cooperators, and customers are more changeable in the EC era than it used to be. It may be one of the causes of these phenomena that switching costs in business activities tend to go down in the EC era. Besides, the sequence of factors from business through IT-related factors could be easy to change. In fact, IT could be a driver of business in an EC era, while IT used to be usually an enabler of business.

We propose a new framework, 3C-DRIVE, which satisfies the above stated requirements (Table 23.1). To maintain its bird's eye view, 3C-DRIVE explains overall picture of the IT value creation using the independent two axes: Management and IT. The factors in these axes are mutually exclusive and collectively exhaustive. We also indicate that the relations between these factors are dynamic and emergent.

23.2.1 Vertical Axis

The vertical axis of 3C-DRIVE is a corporate management axis, which represents basic structure of competitiveness. The context starts from a company's products/services and customers' needs, and the supply-demand relationship is built up. Then, competitive relationships arise over the customers' needs, and companies pursue sustainable competitive advantages by deciding the strategic direction and the relationship with other companies (competition or cooperation). The vertical axis is composed of the following three organizational levels [23.10].

Company. Company represents an organizational level that has individual business goals. In this paper, we mainly examine what mechanism creates business value from IT in individual Company level. This means that the following two levels are positioned as the external structures for this level.

Competitor. Competitor represents groups of companies that are in competitive relationships over customers' needs. In other words, this is a context of competitive environment and IT is a part of it. Since competitive and/or cooperative relationships change dynamically because of relatively

Table 23.1. Framework of 3C-DRIVE

	Direction	Readiness	Information Technology	Value Evaluation
Company	· Corporate vision · Business strategy · Business portfolio	· Management leadership · Organizational strategy · Business Process Redesign	· IT investment · IT strategy/ development/ management · IT organization	· Cost, speed, productivity · Sales, profit · Share of Market · Intellectual capital
Competitor	· Industrial forces · Industry value chain	· Industry-wide standardization · Industrial custom	· EDI, CALS, EC · Industrial IT solution · Package software	· Market growth · Industrial global competitiveness
Customer	· Demand-Supply relationship · Customer behavior · Social change	· Regulation · Standardization · Security · Social custom	· Platform, infrastructure · Broadband, Mobile · Information capital · IT literacy IT literacy	· GDP growth · Global competitiveness

low switching costs in the EC era, it is possible that the IT environment of a competitor suddenly becomes the one of a cooperator. Therefore, this level implies the industry structure and its IT environment that affects the company interactively or in an emergent fashion.

Customer. Customer means customers for the Company and it represents a community or an entire society level that the individual customers belong to. In the case of Business to Consumers (B2C) EC model, it could be a country, a region, a society, or IT environment that the customers belong to. In the case of Business to Business (B2B) EC model, it could be the social structure or its IT environment that the employees of the company belong to. It is essential for us to bring social structure and IT environment of Customers into our scope when individual users could be participants of business value creation as customers of the Internet transactions in the EC era.

In conventional businesses models, the providers (Companies and Competitors) supply products or services to the Customers. However, in the EC era, Customers could decide conditions of competition by transmitting information to providers by themselves (e.g., Internet auctions). Also, startup companies might accelerate the rapid changes of a competitive position because mega-capital is not necessarily one of key factors for success in networking society. Therefore, 3C-DRIVE supposes dynamic and emergent relations that above stated three levels affect interactively.

23.2.2 Horizontal Axis

The horizontal axis of 3C-DRIVE is composed of the four factors that represent business value creation from IT (Table1). Conventionally, system deployment procedures are used to be understood as waterfall-type procedures sequentially from Direction to Value Evaluation, using IT as an enabler. Nowadays, business model creation and business process reengineering use IT as a driver. For example, IT sometimes could be a good chance to formulate a new business strategy. In other words, above stated four factors are in dynamic and emergent relations in the EC era.

Direction. Direction indicates strategic intentions in a Company level (such as corporate vision, business strategies, and business portfolio), or structural changes in Competitor or Customer levels (such as changes of customers' purchase behavior) in the vertical axis of 3C-DRIVE.

Readiness. Readiness indicates prerequisites that organizations and operations must prepare in order to implement Direction and lead IT to success. In Company level, IT achieves its effects only when IT Readiness is satisfied, such as leadership of top management, Business Process Reengineering (BPR), and organization structure. For example, in Competitor or Customer level, governmental policies and regulations, standardization of operation procedures, and IT literacy level of the society are the important factors.

Generally, Readiness has not been explicitly treated as a factor to determine the success or failure of IT deployment. This paper recognizes it as an important factor and extracts it into the horizontal axis. Readiness is a part that is difficult for other companies to imitate, in comparison with IT, and will be the source of differentiation based on business value creation from IT.

Information Technology(IT). In a Company level, IT indicates the company's overall commitment to IT on human, material, and capital resources: such as investments, deployments, strategies, developments, administrations, and organizations for IT. In Competitor and Customer level, this indicates overall IT environment that the company can utilize: such as penetration of EDI, CALS, and EC, status of platform implementation, human resources on IT, and their skill levels in the industry or the society.

Value Evalution. In a Company level, the elements of Value Evaluation include all of business values: short/middle/long term, tangible/intangible. For example, annual performance of a company such as turnover, recurring profit, ROA, and ROE; the Key Performance Indices of operation side such as productivity, speed, and lead-time; and long-term asset value of organization side such as intellectual property or knowledge. For instance, the elements of Value Evaluation include market growth rate in Competitor level and GDP growth rate, international competitiveness in Customer level. The measurement methods of these elements are outside the scope of this paper.

Applying 3C-DRIVE to an individual company enables us to grasp the strengths and the weaknesses of the company to acquire business value from IT. Then, this enables us to clarify the tasks to be addressed. When we suppose networking society in the EC era, the bottleneck of IT deployment for the company could exist in Competitor (or industry) or Customer (or society) levels than Company level. Therefore, 3C-DRIVE enables us to address the issues of the Company with an impartial view from Competitor and Customer levels. Furthermore, 3C-DRIVE can be also used as a framework to discuss the order of policy to solve multiple issues on 3C-DRIVE.

We put emphasis on that the result of applying 3C-DRIVE is unique to each company. The relationships among Company, Competitor and Customer are always unique and relative to each other. Especially, it is one of the key factors for success for a Company to acquire unique value from the network externality. However, we often observe that senior managers of different companies, with different organizational cultures and existing IT infrastructures, expect very similar blue prints of IT deployment. A Company must always keep it in mind that the high Value creation from IT is acquired only when the company makes optimal selections of Direction, Readiness and IT of the Company, keeping eyes on its Competitors and Customers.

In this section, we have described 3C-DRIVE, which is a hypothesis for a framework to organize management and IT related issues. This is a universal concept consists of two axes of management and IT. 3C-DRIVE is equipped with bird's-eye view and dynamic and emergent characteristics. Applying this framework leads each company to the unique results.

23.3 Survey of Existing Research and Perspectives

In this section, we conduct a survey of representative research so far and give perspectives covering business management and IT by using the framework 3C-DRIVE introduced in the preceding section. We intend to devote special attention to the constituent elements in the IT value creation mechanism and their inter-relations, which are examined in such prior research in performing our evaluation of the applicability and limits of the existing framework in an EC era. The following five perspectives will be employed in our survey:

1. Perspective of direct quantitative comparison of IT investment and business value,
2. Perspective of information economics,
3. Perspective of alignment approach,
4. Perspective of IT-organization correlation, and
5. Other perspectives,

23.3.1 Perspective of Direct Quantitative Comparison of IT Investment and Value

First, we conduct a survey of the prior research that discusses the relationship between IT and Value Evaluation of 3C-DRIVE. An age of "productivity paradox" persisted in the period from the 1980s to the first half of the 1990s. Advocates of this paradox argued that there was no correlation between IT investment and productivity since the vast amounts of IT investment that the American industry was making in this period did not help improve its productivity. Strassmann [23.17] proved the absence of any correlation between IT investment and corporate profits by using data obtained by questionnaire surveys of businesses, thus providing support to the "productivity paradox" in IT and Value relations of Companies as well as Competitor levels. He established quantitatively that the IT cost and revenue ratio (%), an index of the level of corporate commitment to IT investment, did not possess any correlation with the return on assets (ROA, %) or return on shareholders' equity (ROE, %). His study served to warn corporate senior management who tended to make IT investment without much forethought. Strassman's study also came up with some valuable insight; For example, it pointed out that a well-performing firm had a more pronounced tendency to be aggressive about IT investment for 'operational' purposes and not for 'administrative' applications than low-performing companies.

Subsequently, however, Hitt and Brynjolfsson verified a correlation between IT investment and productivity (by studying 370 American companies; 1988–1992) and a series of studies with similar findings followed [23.6]. At present, affirmative assessments of IT investment are prevalent in the United States as a result of the favorable economic periods, and the productivity paradox seldom surfaces in academic discourse [23.18], [23.19]. Hitt and Brynjolfsson suggest that while IT indeed does create value, it can negatively impact corporate profits, and hint at the need for a closer analysis of successes or failures and the mechanism involved therein rather than talking about the effectiveness of IT investment.

Similar case studies on newly industrializing economies (NIES) in the Asia discuss differences in national policies for technological development. It is reported, for example, that familiar indexes such as ROE, ROA and ROS show significant variances between market-driven Hong Kong and Malaysia on one hand and planning-reliant Singapore and Taiwan on the other [23.21].

These studies tend to show that the effectiveness of IT investment can no longer be fully grasped only by focusing on such investment in the EC era. In other words, the financial value of IT cannot be assessed solely by the direct relationship between IT and Value Evaluation for Company or Competitor level; it would also be necessary to include in our analysis Direction and Readiness, as well as Customer.

23.3.2 Perspective of Information Economics

Parker and Benson introduce the concept of "information economics" in their attempt at correlating IT and corporate performance by designing decision-making processes for IT investment and its evaluation [23.12]. Their approach would place its primary emphasis on quantification of Value Evaluation at Company in 3C-DRIVE. This methodology usually deals with the business strategy in Direction and the elements contained in IT, but does not take Readiness into account. The size of organization for their analysis could be on the level of a division or an infrastructure-system unit in a Company.

Information Economics determines the value of a system as an aggregate of 1) simple rate of investment return, 2) assessment of its business, and 3) assessment of technology. The simple rate of investment return 1) is composed of assessment of conventional cost-benefit analysis (e.g., the return on investment (ROI), net present value (NPV)), value consolidation effect (an ancillary effect in other divisions), value acceleration effect (an effect due to causal relationship), value reorganization effect (a value created by reorganization of jobs and functions), and innovation effect (a value created by introduction of innovative systems). The assessment of the business 2) is composed of strategic support value, competitive advantage value, administration information value, competitive action value, and organization-related risks. The assessment of the technology 3) takes into account the strategic systems infrastructure value, technological uncertainties, and risks associated with IT infrastructure.

However, when we try to apply these assessment methods in the EC era fraught with increasing levels of uncertainty, we cannot but admit the limitation of the process of quantifying some unexpected effects by the simple rate of investment return 1). Similarly, it remains uncertain how much quantification can be performed on value consolidation effect, value acceleration effect, value reorganization effect, and innovation effect in specific cases. We must also be wary of the possibility of repetitive inclusions of operating profits in each term of the equation of 1) the simple rate of investment return + 2) assessment of business + 3) assessment of technology.

In summary, it has to be pointed out that any methodologies that rely on reduction of certain pre-conceived, static elements suffer from limitations in studying IT value in the EC era, which presupposes occurrence of dynamic and emergent phenomena.

23.3.3 Perspective of Alignment Approach

This perspective considers the relations and order of priority among Direction, Readiness, and IT at Company in 3C-DRIVE. Qualitative discussions and proofs would predominate on Value Evaluation and no quantitative case-study approach would be adopted. The notable framework applicable in this field is the Strategic Alignment Model (SAM) proposed by Henderson and

Venkatraman as part of management study programs at MIT in the 1990s [23.5]. This scheme deals with the alignment among the following four domains: i.e., 1) business strategy, 2) organizational foundation and process, 3) IT strategy, and 4) IS foundation and process. These domains contain what is stated below, respectively. Incidentally, the domains 1) and 3) are denoted as external and the domains 2) and 4) as internal.

1) Business strategy: selected business fields in product market strategy, competitive capability to differentiate self from others, and operational controls such as collaboration with others.
2) Organizational foundation and process: organizational structure, organizational foundation including control techniques, and control process and skills.
3) IT strategy: applicable scope of IT, advantages over others, and IT control capability including partnership with others.
4) Information Systems (IS) foundation and process: system foundation including hardware, software, and networks, their architectural processes, and supporting skills.

Henderson and Venkatraman discuss the manner of alignment by using several perspectives which are combinations of the domains 1) through 4) above.

Earl and others in the United Kingdom started to issue warnings about IT strategies in the second half of the 1980s. They pointed out the fact that IT strategies tended to over-emphasize technical issues and terms while neglecting to pay enough attention to application needs and business mentality. In other words, the IS strategy which deals with WHAT, or business needs and raison-d'etre of applications, must be set aside from the IT strategy which deals with HOW, or technology and infrastructure. They also proposed to add the IM strategy, a discipline to examine separately organizational setups and control regimes for IT and IS. As the 1990s dawned, Earl reformulated his views presented above into his Organizational Fit Framework (OFF) concept [4]. The OFF consists of the four domains of IT strategy (How), IS strategy (What), IM strategy (Who), and organizational strategy (Wherefore).

Studies in the alignment-approach perspective, including SAM and OFF, have brought forward some frameworks that encompass the two domains of strategy and IT and take into account the order of priority among the elements involved. Such proposed frameworks have given us much food for thought since they seem to be applicable in an EC era. While the determinants to be used for finding the match between companies and one of the perspectives of the frameworks may be largely affected by the conditions of Readiness in 3C-DRIVE, explicit discussion was not necessarily presented on this point in the studies.

23.3.4 Perspective of IT-Organization Correlation

This perspective deals primarily with readiness and IT at Company in 3C-DRIVE. Close attention is being paid to the relationship between organization and change management in Readiness, and care is taken to penetrate into the level of unexpected benefits and spontaneous serendipitous ideas and discoveries.

Sproull and Kiesler has proposed a theory of two-staged technological development by focusing on the effects of improving the network infrastructure [23.16]. The effects on the first stage are predictable. During the second stage, there are unpredictable effects caused by changes in people's behavior induced in turn by new technology under the circumstances of changed social contacts and interactions.

To cope with these unpredictable effects, Orlikowski and Hofman proposed an Improvisational Model for change management by using cases of groupware adoption for IT [23.11]. Their model, taking a cue from improvisations in Jazz sessions, is unique in that it incorporates emergent changes in the metamorphosis and value associated with IT application. They argue that the conventional models dealing with shifts in technological bases are all plan-centric and assume that pre-defined events and steps would take place sequentially, and that when one postulates an open-loop system with customizable technologies, complicated and unprecedented changes would materialize. Their Improvisational Model categorizes potential changes in organizations into the following three scenarios:

- The anticipated change: what is being planned for the expansion of the mainstay system, start of e-mail, etc.
- The emergent change: spontaneous - localized innovations, pro-active collaboration, grapevines for rumors and information, etc.
- The opportunity-based change: what is impossible to foresee but intentionally adopted - training, World Wide Web, etc.

These three types of changes are normally observed continuously over some period of time, and the course of such change is importantly affected by the relationship between change model, organization, and technology at Company level. Determination of whether or not the improvisational model would fit a given organization requires a closer study of the specifics of Readiness (e.g., organizational architecture and culture) at an individual Company in 3C-DRIVE.

This proposed framework adopts the ideas of unexpected effects and improvisations in assessing changes and values brought about by IT deployment and, is, for that reason, expected to increase its value as analytical tool in an uncertain situation of an EC era.

23.3.5 Other Perspectives

Bensaou and Earl compared Western management with Japanese management regarding attitude of IT adoption from the perspective of international comparison [23.2]. The result shows the difference of country level background such as management theme, organization structure/operation which top management faces, utilization of new technologies and PC diffusion significantly influences management's attitude toward IT adoption. This also suggests the importance of watching Customer (or society) level influence of 3C-DRIVE regularly, which changes rapidly in accordance with the development of an EC era. In order to solve these overlooking issues of social and public nature, it is essential to adopt basic concepts of economics, business administration, organization, such as network externality, architecture, platform and organization of information and knowledge [23.3], [23.8], [23.9], [23.13], [23.14], [23.15], and [23.20].

In this section, we have surveyed main research and perspectives regarding management and IT, and evaluated the possibility and limit of adopting them for the EC era. As a result, we have found that it is difficult to explain the value of IT just by studying IT investment, and confirmed necessity of complementary factors other than the IT investment. Namely, it is indicated that such approach as stepping into mechanism of value creation by adding the bird's eye view, emergent and dynamic perspectives, just like 3C-DRIVE, on top of existing partial, fixed and static perspective has become more important. We can say it is demanded to tackle proceeding to the EC era, taking ever-changing environment around business and IT inside and outside of a company into consideration.

23.4 Simulation Study and Results

In this section, we will describe an agent-based simulation model based on 3C-DRIVE, and carry out simulation experiments about conceivable company cases. In this model, we consider both IT investment stance of top management in a company and the characteristics in 3C-DRIVE, such as competitiveness of the industry a company belongs to. We simulate what kind of influence will be brought to business value of IT by difference of these factors. Finally, we discuss the results applying possible situations of companies, which are facing the EC era [23.7].

23.4.1 Simulation Objectives and Experimental Setup

The objectives here are to observe how competitiveness (or survivor) derived from IT value is determined both by IT investment stance of a company and by other factors in 3C-DRIVE, such as competitive situations of the real industrial world.

We use an agent-based simulation tool to experiment the following cases. We hereby give definitions of three conceivable IT investment stances of companies in different competitive situations of industries. Three IT investment stances of top management in a company are summarized as follows:

- IT-driven: pursuing high IT value opportunities taking high risk
- Balanced: pursuing middle IT value opportunities taking middle risk
- Non-IT-driven (Strategic/process-driven): pursuing low IT value opportunities taking low risk Several competitive situations of industries can be defined continuously in the simulation tool.

23.4.2 Descriptions of Competitive Environments

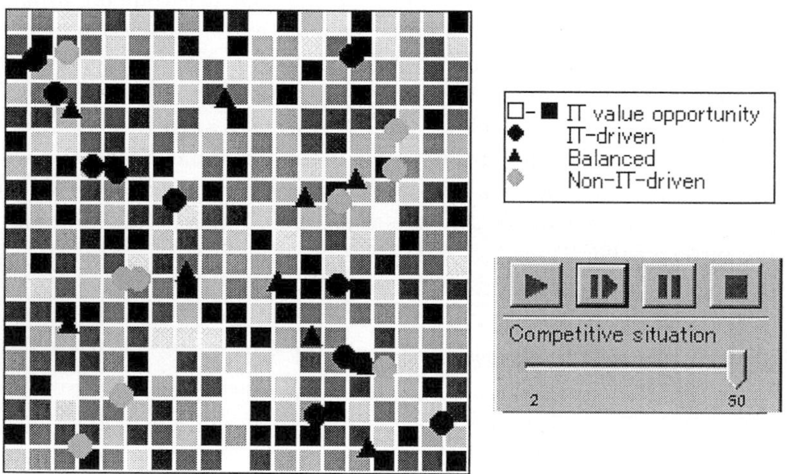

Fig. 23.1. Initial display of the simulation model

The implementation of the simulation model is summarized as follows:

- Initial IT value opportunity is assigned randomly on each cell in the 20 by 20 matrix. This forms the virtual world. Fig. 23.1 shows the screen of the initial state.
- IT value opportunity of a cell increases one unit per time step, up to the upper bound (e.g., 10) depending on the competitive situation, which can

be defined in the tool. The setup enables us to tune the growth rate of the industrial environments to uncover the behaviors of each kind of companies.
- IT value opportunity of a cell decreases by the amount of the IT value opportunity an agent captures.

23.4.3 Characteristics of the Agents

- There are three types of agents depending on IT investment stances: IT-driven, Balanced and Non-IT-driven (Strategy/process-driven).
- Initial number of agents is 10 for each IT investment stance.
- Initial energy level of any agent is 30 units
- Energy level of an agent decreases one unit per time step to be alive. When energy level of an agent is zero, the agent retires.
- To survive in the competition, an agent needs to capture energy level by utilizing IT value opportunity.
- Visible length of an agent is defined 5 cell-lengths as IT-driven, 2 cell-lengths as Balanced, and 1 cell-length as Non-IT-driven.
- Efficiency of capturing IT value opportunity is defined as 3 for IT-driven, 5 for Balanced, and 7 for Non-IT-driven, respectively. The efficiency value is used to compare the behaviors of the companies with efficient/no-efficient management strategies to acquire the IT values.
- Therefore, an IT-driven (resp., Non-IT-driven) agent is relatively the stronger (resp., weaker) sighted and relatively captures the less (resp., more) efficiently IT value opportunities.

23.4.4 Action Rules of the Agents

At each time step, an agent moves around and captures opportunity of IT value in the matrix based on the following rules.

Step 1: Find the most resourceful cell within the visible length depending on the type of IT investment stance of each agent.
Step 2: Move into the cell the agent found at Step 1.
Step 3: Increase energy level by eating the resource of the cell at Step 2.
Step 4: Decrease energy level as mobile cost.
Step 5: If energy level of an agent is more than zero, the agent survives (go to Step 1). Otherwise, the agent dies.

23.4.5 Simulation Results

We have carried out the following cases in accordance with the procedures stated above and compare the results from the viewpoint of survivors and average energy levels over the time. The performance will be interpreted in connection with conceivable cases.

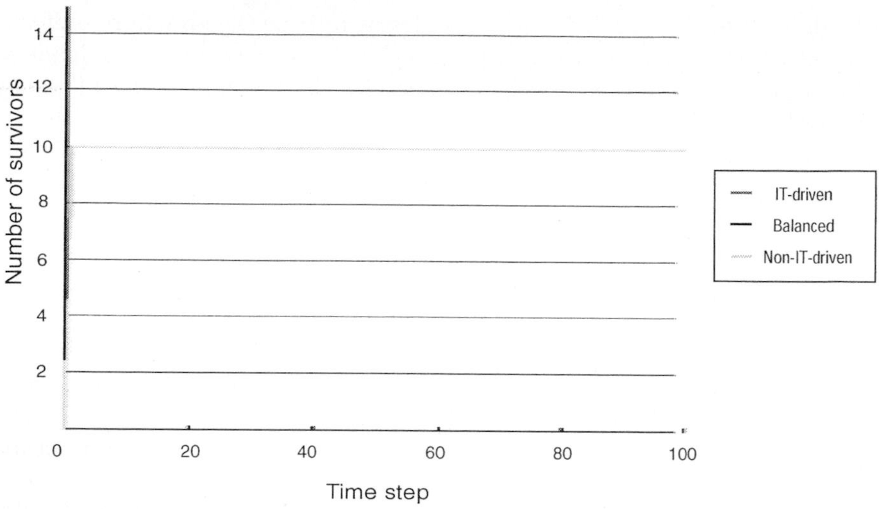

Fig. 23.2. Number of survisers (case 1)

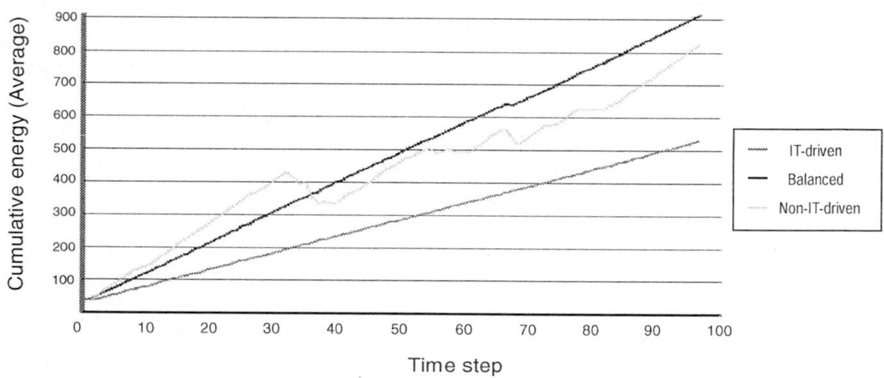

Fig. 23.3. Average of cumulative energy of survivors (case 1)

Case 1. Figs. 23.2 and 23.3 are the results of the case with competitiveness = 2 and time step = 100. The results state that Balanced > Non-IT-driven (Strategy/Process-driven) > IT-driven.

Figs. 23.2 and 23.3 suggest that for a company in a growth industry, the Balanced stance would capture the most efficient business strategies from IT in three IT investment stances. And the most remarkable point in this case is that IT-driven approach could create the least business value from IT of all IT investment stances. As a matter of fact, many senior managers in the industries under strong regulations in Japan do not seem to convince themselves to justify the IT investment. At the same time, we often observe the cases that such established companies, who tried to invest on huge IT projects in IT-driven way, resulted in disastrous situations

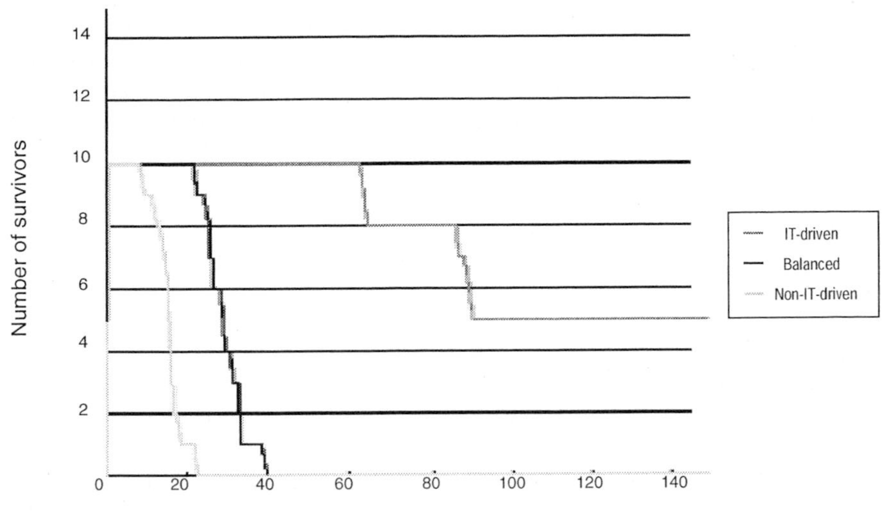

Fig. 23.4. Number of survisers (case 2)

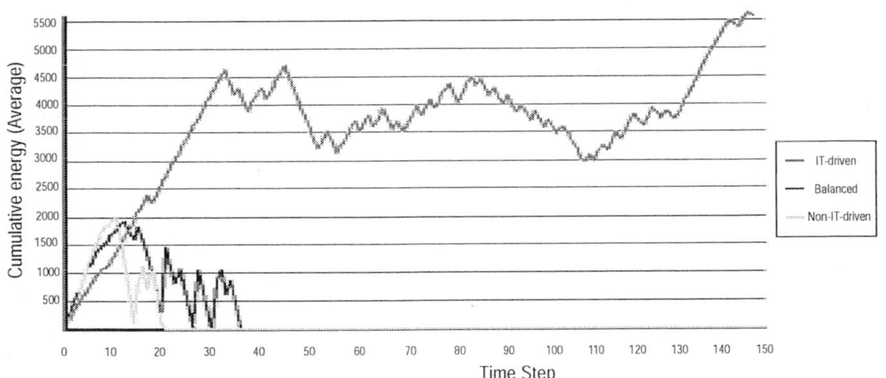

Fig. 23.5. Average of cumulative energy of survivors (case 2)

Case 2. Figs. 23.4 and 23.5 are results of the case with competitiveness = 50 and time step = 150. The results state that IT-driven > Balanced > Non-IT-driven (Strategy/Process-driven).

Figs. 23.4 and 23.5 indicate that a company in an emerging industry would take the more advantage of IT investment, the more the company focuses on the IT-driven approach. We can imagine that some companies in the competitive industry in the U.S. (e.g., IT industries) increase their competitiveness by adopting the leading edge IT-driven business model taking a risk.

Moreover, from the viewpoint of global competition, a company with slow management style under domestically regulated environment needs to compete with another company with agile management style in the global marketplace. In this situation, significant gaps of IT values between two companies could be made, just like differences between Fig. 23.2 and Fig. 23.4.

In this section, we have conducted simple modeling of 3C-DRIVE and simulations of conceivable cases of companies. The simulation results suggested that complementary factors such as competitiveness of the industry a company belongs to, other than the IT investment, could greatly influence the business value from IT. Therefore, a company should take an appropriate IT investment stance depending on the characteristics of factors in 3C-DRIVE especially in the Electronic commerce (EC) era.

23.5 Conclusion

This paper has discussed the perspective of a mechanism as to how information technology may create the added business value. We have proposed the strategy of 3C-DRIVE, which is a proposed framework in order to organize much improved management and its associated IT related issues. This is a generalized concept which consists of two axes of management and IT. 3C-DRIVE is equipped with emerging focused overview as well as some visible emergent characteristics. Unlike the conventional frameworks to focus only on some specific parts of factors as relating to IT value creation. This new strategy is really qualified to be identified as an integrated whole. Then, we have surveyed representative existing research and perspectives regarding management and IT, and evaluated the possibility and the limit of adopting them for the EC era. As a result, we have found it difficult to explain the value of IT in an EC era just by studying IT investment, and confirmed necessity of complementary factors other than IT investment. Next, we have described an agent-based simulation model based on 3C-DRIVE, and have carried out simulation experiments about conceivable company cases. Through our simulations, we have suggested that complementary factors in 3C-DRIVE other than the IT investment could greatly influence the business value from IT especially in the EC era. Therefore we strongly recommend a company should take an appropriate IT investment stance depending on the 3C-DRIVE situations.

Our future work includes (1) to conduct empirical studies to better understand the constituent elements in the IT value creation mechanism and their inter-relations in 3C-DRIVE, and (2) to enhance the agent-based models and simulations of 3C-DRIVE to enable us to grasp multi-agent and multi-level problems in more detail [23.1].

Finally, we would like to continue to undertake the challenge of bridging the gap between business and IT playing field such that we can better address the strategical, operational, organizational, and IT-related issues on 3C-DRIVE as an integral whole rather than isolated parts. It is conceivable that this methodology can help senior leaders in a convincing manner in responding to unpredictable future in the EC era.

Acknowledgements

The authors gratefully acknowledge the insightful comments and suggestion of the late Professor Yui Kimura, the skillful programming of an agent-based tool of Ms. Reiko Hishiyama, the stimulating comments on a simulation modeling of Dr. Setsuya Kurahashi at Tsukuba University. They are also grateful for valuable suggestions of the editors and reviewers during the review process.

References

23.1 Axelrod, R. (1997): The Complexity of Cooperation: Agent-Based Models of Competition and Cooperation, Princeton University Press
23.2 Bensaou, M., Earl, M. (1978): The Right Mind-set for Managing information Technology, Harvad Business Review
23.3 Nicholas, E. (1996): The Economics of Networks, International Journal of Industrial Organization, Vol. 14
23.4 Earl, M. J. (1996): IntegratingIS and the Organization:A Framework of Organizational Fit, In: Earl, M. J. (Ed.), Information Management, Oxford University Press
23.5 Hendeson, J. C., Venkatraman, N., Oldach, S. (1993): Continuous Strategic Alignment, European Management Journal, Vol. 11, No. 2
23.6 Hitt, L. and Brynjolfsson, E. (1995): Productivity without Profit. Three Measures of information Technology's Value, MIS Quarterly
23.7 Kadono, Y. and Terano, T. (2000): How Information Technology Creates Values in Business Management, Proc First International Workshop on Computational Intelligence in Economic and Finace
23.8 Katz, M., Shapiro, C. (1985): Network Extemalities, Competition and Compatibility, The American Economic Review, Vol. 75, No. 3
23.9 Nelson, R. R., Winter, S. G. (1982): An Evolutionary Theory of Economic Change, Cambridge, Belknap Press
23.10 Ohmae, K. (1982): The Mind of the Strategist, McGraw-Hill
23.11 Orlikowski, W., Hofman, J. D. (1997): An Improvisational Model for Change Management, Sloan Management Review, winter
23.12 Parker, M. M., Benson, R. J. (1988): Information Economics: Linking Business Performance to Information Technology, Prentice Hall
23.13 Porter, M. E. (1980): Competitive Strategy, New York, Free Press
23.14 Shapiro, C., Varian, H. (1998): Information Rules, Harvard Business School Press
23.15 Simon, H. A. (1976): Administrative Behavior, 3rd Edirion, Free Press
23.16 Sproull, L., Kiesler, S. (1993): New Ways of Working in the Networked Organization, MIT Press

23.17 Strassman, P. (1990): Business Value of Computers, Information Economic Press
23.18 U. S. Department of Commerce, The Emerging Digital Economy, 1999
23.19 U. S. Department of Commerce, The Emerging Digital Economy, 2000
23.20 Umesao, T. (1983): Information Civilization, (in Japanese) Chyuoh-Kohronsya
23.21 Kar Yan Tam (1998): The Impact of Information Technology Investments on Firm Performance and Evaluation: Evidence from Newly Industrialized Economies, Information Systems Research, Vol. 9, No. 1, March

Author Index

Subject Index